RAS and Cancer in the 21st Century

A subject collection from *Cold Spring Harbor Perspectives in Medicine*

OTHER SUBJECT COLLECTIONS FROM *COLD SPRING HARBOR PERSPECTIVES IN MEDICINE*

Enteric Hepatitis Viruses
Bone: A Regulator of Physiology
Multiple Sclerosis
Cancer Evolution
The Biology of Exercise
Prion Diseases
Tissue Engineering and Regenerative Medicine
Chromatin Deregulation in Cancer
Malaria: Biology in the Era of Eradication
Antibiotics and Antibiotic Resistance
The p53 Protein: From Cell Regulation to Cancer
Aging: The Longevity Dividend
Epilepsy: The Biology of a Spectrum Disorder
Molecular Approaches to Reproductive and Newborn Medicine
The Hepatitis B and Delta Viruses
Intellectual Property in Molecular Medicine
Retinal Disorders: Genetic Approaches to Diagnosis and Treatment
The Biology of Heart Disease

SUBJECT COLLECTIONS FROM *COLD SPRING HARBOR PERSPECTIVES IN BIOLOGY*

Cytokines
Immune Memory and Vaccines: Great Debates
Circadian Rhythms
Cell–Cell Junctions, Second Edition
Prion Biology
The Biology of the TGF-β Family
Synthetic Biology: Tools for Engineering Biological Systems
Cell Polarity
Cilia
Microbial Evolution
Learning and Memory
DNA Recombination
Neurogenesis
Size Control in Biology: From Organelles to Organisms
Mitosis
Glia
Innate Immunity and Inflammation
The Genetics and Biology of Sexual Conflict

RAS and Cancer in the 21st Century

A subject collection from *Cold Spring Harbor Perspectives in Medicine*

EDITED BY

Linda Van Aelst
Cold Spring Harbor Laboratory

Julian Downward
The Francis Crick Institute

Frank McCormick
University of California, San Francisco

COLD SPRING HARBOR LABORATORY PRESS
Cold Spring Harbor, New York • www.cshlpress.org

RAS and Cancer in the 21st Century

A subject collection from *Cold Spring Harbor Perspectives in Medicine*
Articles online at www.perspectivesinmedicine.org

Printed in the United States of America

Executive Editor	Richard Sever
Managing Editor	Maria Smit
Senior Project Manager	Barbara Acosta
Permissions Administrator	Carol Brown
Production Editor	Diane Schubach
Production Manager/Cover Designer	Denise Weiss
Publisher	John Inglis

Front cover artwork: Cover image depicts a ribbon model of RAS. (Artwork [original size: 27.5″ × 39.4″] created by Gabi Haeusser, SAP-SE, Walldorf, Germany; photographed and adapted by Annette Buschlinger, Heidelberg, Germany.) See also chapter by Scheffzek and Shivalingaiah for details.

Library of Congress Cataloging-in-Publication Data

Names: Van Aelst, Linda, editor.
Title: RAS and cancer in the 21st century / edited by Linda Van Aelst, Cold Spring Harbor Laboratory, Julian Downward, The Francis Crick Institute and Frank McCormick, University of California, San Francisco.
Description: Cold Spring Harbor, New York : Cold Spring Harbor Laboratory Press, [2018] | Series: A subject collection from Cold Spring Harbor perspectives in medicine | Includes bibliographical references and index.
Identifiers: LCCN 2017056421 | ISBN 9781621822219 (hardcover : alk. paper)
Subjects: LCSH: Ras oncogenes.
Classification: LCC RC268.44.R37 R36 2018 | DDC 616.99/4--dc23
LC record available at https://lccn.loc.gov/2017056421

All World Wide Web addresses are accurate to the best of our knowledge at the time of printing.

For a complete catalog of all Cold Spring Harbor Laboratory Press publications, visit our website at www.cshlpress.org.

Contents

Preface

THE IMPORTANCE OF *RAS* GENES AND THE PROTEINS they encode in human cancer was firmly established in the 20th century, but more than 30 years later, we are still working to understand precisely how RAS proteins cause cancer and how we can intervene therapeutically. The contributions from the authors included in this volume highlight many of the complexities that have to be faced when we consider RAS proteins as drug targets and the progress that is being made toward effective therapies.

RAS proteins are essential for signal transduction in all cells, and oncogenic mutants are structurally very similar to their normal counterparts. This is one of many challenges. RAS proteins are also members of a broader superfamily of closely related proteins that play essential roles in many aspects of cell biology. Therapies aimed at oncogenic RAS proteins therefore need to avoid targeting these relatives as well as RAS proteins in normal cells. The RAS proteins themselves do not offer obvious opportunities for drug targeting, as they do not have active sites which can be exploited by traditional medicinal chemistry. Early efforts that instead targeted RAS processing failed, but new insights into the enzymes that regulate their critical posttranslational modification as well as a better understanding of how RAS proteins interact with lipid components in the membrane may yet lead to targeting opportunities.

Targeting pathways downstream is a strategy that may dodge the problems of targeting RAS proteins directly. Fortunately, the two major pathways downstream of RAS, the Raf/MAPK pathway and the PI3-kinase pathway, present many tractable drug targets, many of which have already been tested in the clinic. However, these pathways have proven unexpectedly complicated, with layers of redundancy and feedback that have, so far, defeated attempts to exploit them. Furthermore, tonic activation of these pathways by mutant RAS does not present a clear therapeutic window relative to transient activation in normal cells, so that drugs blocking enzymes in these pathways have, so far, been unacceptably toxic. Unbiased approaches to find vulnerabilities in RAS-driven cancers have also been disappointing in terms of identifying druggable enzymes that RAS tumors depend on uniquely, but again, new technologies and approaches are being applied that may identify opportunities in some RAS-driven cancers at least. Any new targets obviously have to be validated, preferably in vivo. Here again, new technologies offer far superior systems for testing the importance of potential targets and will certainly be critical to the development of the next generation of therapeutic candidates. Finally, the dream of harnessing the immune system to eradicate RAS-driven cancers is beginning to become reality and is also discussed in this volume.

It is our sincere hope that effective strategies for killing RAS-driven cancers will be in clinical practice before the 21st century advances much further. The editors, along with all of the excellent contributors to this volume and the entire RAS research community will continue to try and make this happen. We are very grateful to Barbara Acosta and her colleagues at Cold Spring Harbor Laboratory Press for all their patient efforts in assembling this volume, and we hope that some of the ideas and challenges described here will inspire others to join the fight to eliminate RAS-driven cancers.

LINDA VAN AELST
JULIAN DOWNWARD
FRANK MCCORMICK

From Ras to Rap and Back, a Journey of 35 Years

Johannes L. Bos

Molecular Cancer Research, University Medical Center Utrecht, Universiteitsweg 100, 3584 CG Utrecht, The Netherlands

Correspondence: j.l.bos@umcutrecht.nl

Our laboratory has studied Ras and Ras-like proteins since the discovery of the Ras oncogene 35 years ago. In this review, I will give an account of what we have done in these 35 years and indicate the main papers that have guided our research. Our efforts started with the early analysis of mutant Ras in human tumors followed by deciphering of the role of Ras in signal transduction pathways. In an attempt to interfere in Ras signaling we turned to Rap proteins. These proteins are the closest relatives of Ras and were initially identified as Ras antagonists. However, our studies revealed that the Rap signaling network primarily is involved in spatiotemporal control of cell adhesion, in part through regulation of the actin cytoskeleton. More recently we returned to Ras, trying to interfere in Ras signaling by combinatorial drug testing using the organoid technology.

THE BEGINNING

In 1982, the stunning result that a single-point mutation converted a normal H-Ras gene into an oncogene was published by the Weinberg laboratory (Parada et al. 1982). The experimental setup was based on the assumption that cancer is caused by mutations in genes and that these "oncogenes," when transfected into normal cells, should be able to convert these cells in tumor cells. For this assay, Weinberg and colleagues transferred genomic DNA from a human kidney tumor cell line into NIH-3T3 mouse fibroblasts. This resulted in morphological transformation of NIH-3T3 cells, suggesting the presence of an oncogene in the genomic DNA of these tumor cells. However, the identification of such an oncogene was a painstaking process that requested multiple rounds of selection, to reduce the amount of human DNA present in the transformed mouse cell. The next step was the separation of the human DNA from the mouse DNA based on human-specific repetitive sequences, followed by cloning and sequence analysis. Many groups followed a similar path, resulting in the discovery of the H-ras-related genes K-ras and N-ras (Hall et al. 1983; Shimizu et al. 1983). Many tumor cell lines and tumor tissues were subsequently screened for oncogenes. But the assay remained extremely labor-intensive and the success rate was low. At that time, I was working in the laboratory of Alex van der Eb on cell transformation by the adenovirus E1 region. To switch to human oncogenes would have been an obvious choice, but jumping on a very labor-intensive bandwagon was not very attractive. I was, however, struck by a paper of Bruce Wallace showing a method to identify specific mutations in sickle cell anemia using synthetic oligonucleotides

Cite this article as *Cold Spring Harb Perspect Med* doi: 10.1101/cshperspect.a031468

(Conner et al. 1983). The basic principle is that the thermal stability of a 20-mer oligonucleotide annealed to genomic DNA is reduced by a couple of degrees when one base pair is mismatched. As mutations were only found in codon 12 and 61 in H-, K-, and N-Ras, we needed a limited number of oligonucleotides representing all possible single base pair changes. At that time oligonucleotide synthesis was possible in only a few laboratories, one of which was the laboratory of Jacques van Boom next door. With some help from my thesis supervisor Piet Borst, I convinced him to make the required 72 different oligonucleotides, covering all possible codon 12 and 61 mutations. Detecting mutant Ras with this approach turned out to work quite well for cell lines and for really pure tumor samples, but it also had its limits of detection (Bos et al. 1984). The discovery of the polymerase chain reaction (PCR) method (Saiki et al. 1985) resolved this limitation and we were able to identify Ras mutations in a large variety of tumors by simple dot-blot assays (Verlaan-de Vries et al. 1986). As a PCR machine, we used a robot arm moving between water baths of different temperatures as our first PCR machine, with homemade Klenow polymerase to be added after each cycle. One of the highlights was the colon study with Bert Vogelstein showing K-ras mutations to occur in 50% of the colon tumors (Bos et al. 1987), an observation independently made by Manual Perucho and coworkers (Forrester et al. 1987). Interestingly, Ras mutations were also present in 50% of benign adenomas. Together with other genetic alterations detected by Vogelstein, this led to the postulation of the sequential activation of oncogenes in colorectal cancer (Vogelstein et al. 1988). Another highlight was the study of lung adenocarcinoma together with Sjoerd Rodenhuis, revealing 30% K-ras mutations (Rodenhuis et al. 1987). However, the clinical relevance/diagnostic value of Ras mutations became only apparent with the development of epidermal growth factor receptor (EGFR) inhibitors like cetuximab and erlotinib. The presence of a Ras mutation is currently an exclusion criterion for these drugs (Eberhard et al. 2005; Karapetis et al. 2008). Ras mutations were most frequently found in pancreatic cancer, as first described by Perucho and coworkers (Almoguera et al. 1988).

In these early studies, we already observed genetic heterogeneity in tumors. For instance, in melanoma we detected two different Ras mutations in one primary tumor that separated in the metastatic lesions (van 't Veer et al. 1989). In seminomas, a morphologically very homogeneous tumor, Ras mutations were found in some parts of the primary tumor, but not in other parts (Mulder et al. 1989). In acute myeloid leukemias (AMLs), 30% of which carry N-Ras mutations, we observed that in some cases the Ras mutation remained detectable after complete remission. Interestingly, in these cases, Ras mutations were also present in the apparently unaffected lymphoid lineage, suggesting that the Ras mutation is present in an early hematopoietic stem cell not affected by the treatment. In other cases of AML, the Ras mutation disappeared after treatment, and did not recur after relapse. We concluded from these studies, first, that Ras mutations can contribute to the oncogenic process at different stages, second, that mutant Ras apparently has no effect on the lymphoid lineage, and third, that treatment kills all the aggressive blast cells of the leukemia, but not the underlying premalignant cell clone (Yunis et al. 1989). These and other early studies have been summarized in Bos (1989).

Ras REGULATION

After identification of mutated Ras as oncogene, much effort was put in understanding the function of Ras, a small GTPase that normally cycles between an active GTP-bound and an inactive GDP-bound state, but when oncogenically mutated, is stalled in the active GTP-bound state. The obvious rationale being that understanding the normal function of Ras would provide insight into how to treat tumors harboring oncogenic Ras. The kinetics of Ras cycling between a GTP- and a GDP-bound state in vitro indicated the existence of accessory proteins to regulate Ras loading. The first regulatory protein identified was the CDC25 protein, a guanine nucleotide exchange protein for Ras in yeast (Broek et al. 1987; Robinson et al. 1987). Soon after, Frank McCormick discovered a p120 GTPase-activating protein (p120GAP) guided by the observation that the in vivo GTPase activity of Ras was much higher than the in vitro activity (Trahey and McCormick 1987; Wong et al. 1992).

At that time, GDP/GTP cycling of Ras was studied in vitro and a method measuring Ras activation in vivo was initially lacking. Julian Downward developed a method based on labeling of cells with radioactive orthophosphate followed by immunoprecipitation of Ras proteins. GDP and GTP were eluted and separated by thin layer chromatography. Using this procedure, it was first found that T-cell receptor engagement was a strong inducer of Ras activity (Downward et al. 1990). We followed by showing that insulin could activate Ras, connecting Ras to receptor tyrosine kinases (Burgering et al. 1991). Interestingly, many other extracellular stimuli were shown to activate Ras, including stimuli that use G-protein-coupled receptors (van Corven et al. 1993). This showed that Ras activation was not part of a signaling pathway induced by one specific ligand, but contributes to signaling induced by many factors controlling cellular responses. Subsequent genetic studies by others revealed several mammalian counterparts to the yeast CDC25 protein that mediate Ras activation, such as the Sos proteins, all having a characteristic CDC25 homology domain as catalytic domain. The various Ras guanine nucleotide exchange factors (GEFs) and GTPase activating proteins (GAPs) and their mode of action have been reviewed extensively (Bos et al. 2007).

Ras EFFECTORS

In the early 1990s, we learned from Chris Marshall that extracellular regulated kinase (ERK), also known as microtubule-associated protein kinase (MAPK), was activated in cells transfected with mutant Ras (Leevers and Marshall 1992). At that time, we were developing a system pioneered by Marino Zerial to use vaccinia virus to introduce dominant negative Ras (RasN17) into cells. Indeed, in cells expressing RasN17 insulin-induced ERK activation was completely blocked: This made ERK the first "proven" downstream biochemical effect of Ras in mammalian cells (de Vries-Smits et al. 1992). However, Ras and ERK did not directly interact and subsequently Raf was found to be the effector of Ras that, through MAPK/ERK kinase (MEK), activates ERK (Moodie et al. 1993).

Soon hereafter, multiple Ras effectors were identified, including phosphatidylinositol 3-kinase (PI3K) (Kodaki et al. 1994) and Ral guanine nucleotide dissociation stimulator (RalGDS) (Hofer et al. 1994). The discovery by Boudewijn Burgering and Paul Coffer in our laboratory that the product of PI3K, phosphatidylinositol 3,4,5, triphosphate, activates protein kinase B (Burgering and Coffer 1995) opened a completely new line of research, resulting in the identification of FoxO transcription factors as main targets for protein kinase B–mediated phosphorylation (Kops et al. 1999).

The importance of multiple effectors for Ras function was beautifully shown by Michael White using effector domain mutants that selectively bind to the various effectors: when transfected separately they had hardly any transforming activity, but in combination they did (White et al. 1995).

In the years thereafter, many aspects of the signaling networks in which Ras plays a central role and the role of Ras in cancer were elucidated (Cox et al. 2014; Papke and Der 2017). The most updated version of the Ras signaling network can be found at the National Cancer Institute (www.cancer.gov/research/key-initiatives/ras/ras-central/blog/2017/mccormick-ras-pathway-v3).

HOW TO INTERFERE IN MUTANT Ras SIGNALING

Knowledge on the regulation of Ras rapidly accumulated but did not yet result in treatment options for cancer patients. Initially, it was thought that Ras action could be inhibited by displacing the GTP moiety by a small molecule, but this idea was rapidly abandoned when it became clear that the affinity of GTP (and GDP) for Ras is within the picomolar range. Instead, most attention was focused on interfering in the lipid modification of Ras, particularly the carboxy-terminal farnesylation and later geranyl-geranylation. When it became clear that Ras belongs to a superfamily of small GTPases with over a hundred members almost all similarly modified, the enthusiasm for this approach diminished. These studies have recently been discussed by Cox and coworkers (Cox et al. 2014). Nowadays most attention is focused on interfering in signaling downstream of Ras (see below).

Rap PROTEINS

Rap proteins were first identified by Veronique Pizon and Pierre Chardin and, based on their similarity to Ras, the authors suggested that they may interact with the same effectors as Ras (Pizon et al. 1988). Subsequently, Noda and coworkers identified K-ras revertant 1 (K-rev1), which was identical to Rap1, in a screen for genes that can revert morphological transformation of mutant K-ras cells (Kitayama et al. 1989). This led to the hypothesis that Rap1 might be a decoy that traps Ras effectors in an inactive complex and could be a possible way to interfere in mutant Ras tumors? We became interested in Rap after our finding that in certain cell types cyclic adenosine monophosphate (cAMP) can inhibit growth-factor-induced (and Ras-mediated) activation of ERK (Burgering et al. 1991). Combined with the observation of Daniel Altschuler that cAMP could activate Rap1 (Altschuler et al. 1995), this suggested a role for Rap1 in inhibiting Ras following cAMP stimulation of cells. However, despite the fact that in vitro Rap1 can bind to the Ras-binding domain of Raf1, we were unable to show a direct involvement of Ras effector signaling (Zwartkruis et al. 1998).

A function of Rap proteins distinct from Ras is further supported by the notion that Ras and Rap were already present as separate proteins in the hypothetical last eukaryotic common ancestor (van Dam et al. 2011). Later in evolution at the opisthokonta stage the Rap branch separated in Rap1 and Rap2 and at the vertebrata stage in Rap1A and B and Rap2A, B, and C (Fig. 1). Research on Rap orthologs in many species has provided pivotal knowledge about human Rap proteins (Frische and Zwartkruis 2010), but especially the analysis of the single Rap protein in budding yeast, Rsr1, has led to a framework for the regulation and function of these small GTPases. Budding always occurs adjacent to a scar of a previous bud. This bud site selection is however abolished in cells mutated for Rsr1 (Bud1), for a GEF (Bud5) or for a GAP (Bud2) for Rsr1. It was found that these proteins form a complex that recognizes special landmark proteins in the rim of the bud scar. The signal is further propagated to the CDC42 complex, which facilitates the local recruitment of the actin cytoskeleton (Bi and Park 2012). In addition, in *Drosophila*, Rap proteins play a role in cell migration and the proper localization of adherence junctions (Asha et al. 1999; Knox and Brown 2002). This led to the hypothesis that also mammalian Rap proteins are involved in spatial control of cell adhesion, most likely through actin-driven processes.

REGULATION OF Rap

Our initial studies were devoted to find stimuli that can activate Rap1 and we used platelets as they abundantly express Rap1. To measure Rap activity, we developed a nonradioactive pull-down assay, in which active GTP-bound Rap1 is precipitated with a GST-coupled Ras-binding domain (Franke et al. 1997). Such pull-down assays have subsequently been developed for Ras and many other GTPases and are now the most widely used assay to measure activity of a variety of small GTPases, including those from the Rho family. We found that all stimuli that resulted in a rapid activation of platelets also activated Rap1. As one of the hallmarks of platelet activation is increased integrin-mediated adhesion and spreading, we postulated that Rap1 may control integrin activation (Franke et al. 1997). Subsequent studies revealed that a large variety of stimuli are able to activate Rap proteins in all cell types tested.

The observation that, similar to Ras, Rap1 is activated by many stimuli opened the quest for

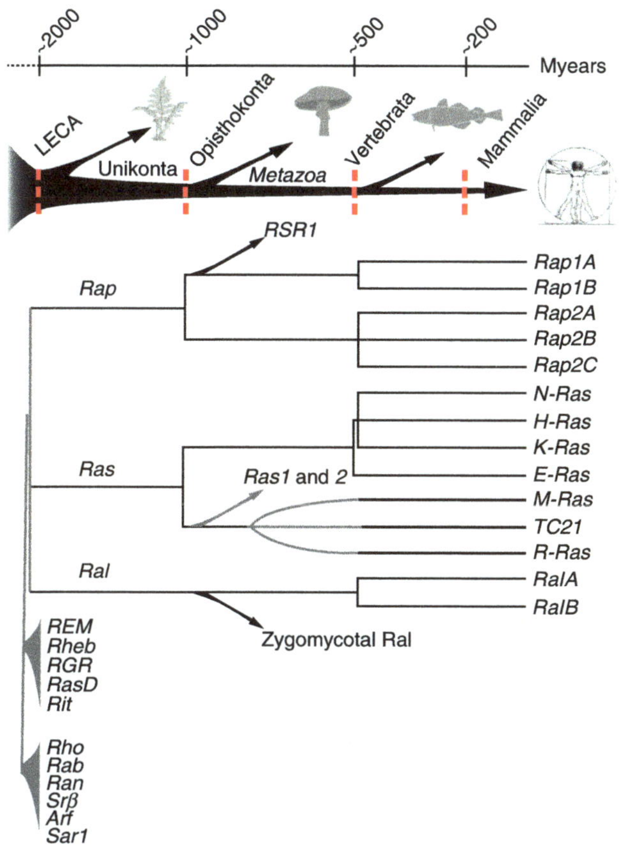

Figure 1. Evolutionary timeline of the Ras family, Ras, Rap, and Ral proteins. The last eukaryotic common ancestor (LECA) likely contained already representatives of these proteins. Later duplications are responsible for the different members in mammalian cells. (From van Dam et al. 2011; adapted, with permission, from the authors.)

guanine nucleotide exchange factors (GEFs) and GAPs that could regulate this activity. The first Rap-specific GEF, C3G, was identified by Michiyuki Matsuda and coworkers (Gotoh et al. 1995). Among others, it mediates signals from receptor tyrosine kinases through the adaptor protein Crk. Currently at least nine Rap-selective GEFs have been identified (Fig. 2), which have, similar to Ras GEFs, a CDC25 homology domain. However, both in vivo and in vitro there is a strong specificity for Rap, with CalDAGGEF3 (RasGrp3) as an exception as it acts both on Ras and Rap. In addition, almost all these proteins have protein–protein or protein–lipid interaction domains, suggesting multiple interaction partners. RapGAPs have a different structure and a different mechanism of catalysis compared to RasGAPs; whereas RasGAPs provide a critical arginine (arginine finger) for hydrolysis of GTP, RapGAPs provide an aspartate (aspartate thumb). Again, despite similarities between Ras and Rap, the GAPs are specific for their cognate GTPase (Bos et al. 2007). Exceptions are the GAP1(IP4BP) family members a Rap-like GAP, which upon membrane binding adopts activity toward Ras (Sot et al. 2013).

EXCHANGE PROTEIN DIRECTLY ACTIVATED BY cAMP (Epac)

At one time, it was a dogma that all mammalian cAMP effects were mediated by protein kinase A. However, we noted that inhibition of protein

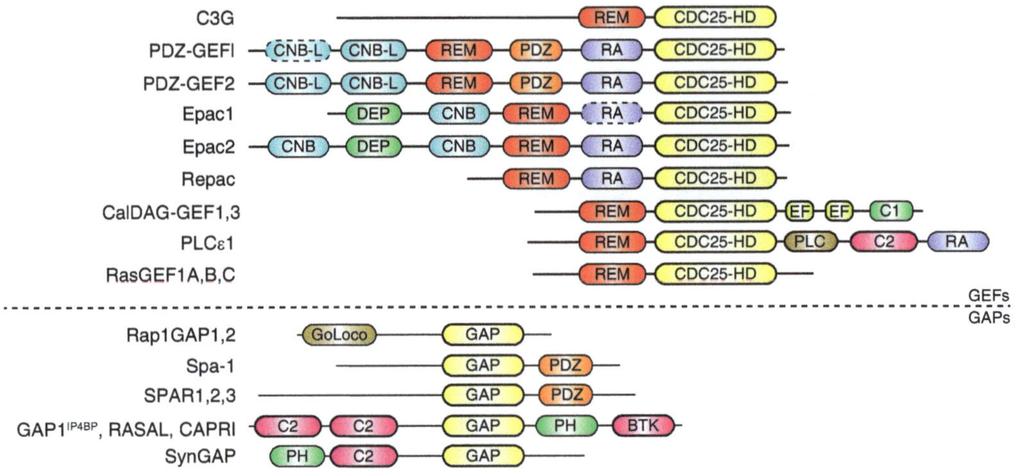

Figure 2. Rap guanine nucleotide exchange factors (GEFs) and GTPase activating proteins (GAPs). Indicated are the domain structures of Rap-specific GEFs and GAPs. (From Bos et al. 2007; adapted, with permission, from the authors.)

kinase A did not affect cAMP-induced Rap1 activity. In a subsequent search in publicly available databases of the human genome project, we found cAMP-binding domains in close vicinity of CDC25 homology domains. Using purified protein encoded by the corresponding complementary DNA (cDNA), we could show a Rap-specific GEF with activity that was completely dependent on the presence of cAMP. We called this protein exchange protein directly activated by cAMP (Epac) (de Rooij et al. 1998). Two human Epac proteins were identified: one with a single cAMP-binding domain (Epac1) and one with two cAMP-binding domains (Epac2). Subsequent structural analysis, in collaboration with Fred Wittinghofer, revealed a detailed mechanism on how Epac is regulated (Fig. 3A). In the absence of cAMP, Epac is in a closed conformation with the cAMP-binding domain occluding the Rap-binding site. When cAMP binds, Epac adopts an open conformation, where Rap can bind and to become activated (Rehmann et al. 2006, 2008). In addition, our studies for the first time revealed how ingenuously cAMP acts: the cyclic phosphate group of cAMP releases a brake that allows a major conformational change that is further stabilized by interactions with the adenosine group (Rehmann et al. 2003). Simultaneously, cAMP also induces the translocation of Epac1 (but not Epac2) from the cytosol to the plasma membrane (Ponsioen et al. 2009). This is facilitated by a cAMP-induced conformational change in the Dishevelled, EGL-10, and pleckstrin (DEP) domain (Li et al. 2011), resulting in an increased affinity for membrane-bound phosphatidic acid (Consonni et al. 2012). Several additional membrane anchors have been identified for Epac1, including activated Ezrin, showing that, dependent on the conditions, Epac1 can localize differently (Gloerich et al. 2010).

DEVELOPMENT OF 007

Further studies on the function of Epac1 were hampered by the fact that cAMP activates both protein kinase A (PKA) and Epac. However, Stein Døskeland pointed out to us a peculiarity in the cAMP-binding domain, where a highly conserved glutamate that forms hydrogen bounds with the 2′-hydroxyl of the ribose of cAMP was absent in the cAMP-binding domain of Epac1 and the high-affinity, second binding site of Epac2. This suggested that these hydrogen bonds, and thus the 2′OH of the ribose may be required specifically for cAMP to interact with PKA, but not with Epac. Together

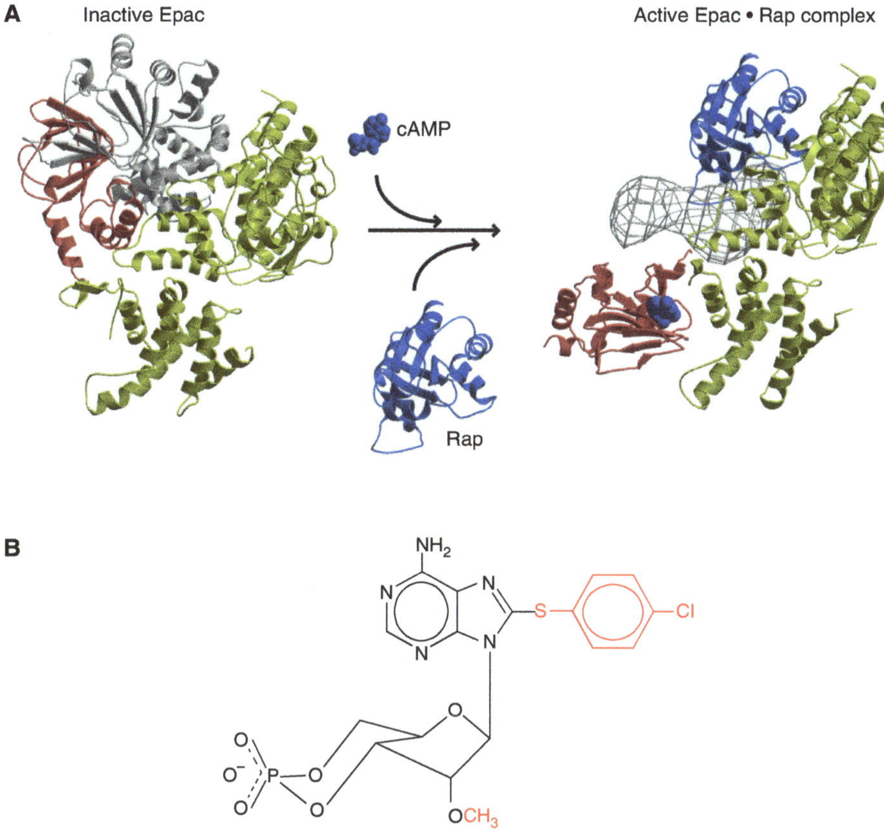

Figure 3. (*A*) Crystal structure of Epac 2 in the closed (*left*) and open (*right*) conformation. The gray area (*right*) represents the first cyclic adenosine monophosphate (cAMP) domain and Dishevelled, EGL-10, and pleckstrin (DEP) domain visualized by cryoelectron microscopy (Rehmann et al. 2006, 2008). (*B*) Structure of 8CPT-2′OMe-cAMP (007).

with Frank Schwede and Hans-Gottfried Genieser, we started to modify cAMP, resulting in a highly Epac-selective compound 8-CPT-2′OMe-cAMP (Fig. 3B), which we called 007 (Enserink et al. 2002). Using this compound, we demonstrated that cAMP through Epac is able to modify integrin-mediated cell-substratum adhesion and E-cadherin-mediated cell–cell adhesion (Rangarajan et al. 2003; Price et al. 2004). Subsequently, many studies have been published using this compound thereby showing the versatile role Epac proteins play in cAMP signaling, including insulin secretion, endothelial barrier function, memory, and heart rhythm (for further details, see Gloerich and Bos 2010).

DOWNSTREAM TARGETS OF Rap1 PROTEINS

007 did not induce activation of ERK, indicating that B-raf is not a downstream target of Rap1 (Enserink et al. 2002). In this respect, a claim that B-raf is a downstream target of Rap1 is incorrect. The first genuine effector for a Rap protein identified was the *Drosophila* Canoe protein (Boettner et al. 2003). This protein is similar to the mammalian junctional protein AF6 with ubiquitin-like Ras association (RA) domains to bind active Rap1. Subsequently, AF6 was found to be one of the effectors of Rap1 that mediate some, but not all, responses of Rap1 activation. Indeed, of the many RA-

domain-containing proteins multiple have been implicated to transduce Rap1 signaling. These effectors, together with the existence of multiple Rap1-GEFs and GAPs, prompted us to extend the hypothesis based on the yeast studies. We proposed that a cell can have multiple independent Rap modules consisting of a Rap protein, a GEF, and probably a GAP, each recognizing different landmarks (spatial cues) to regulate localized recruitment of the actin cytoskeleton in time and in space using a variety of effector systems. The presence of additional regulatory and interaction domains, particularly in the various GEFs, provide ample opportunities for specific landmark recognition and regulation. The consequence of this hypothesis was that we had to search for a specific Rap signaling module in a defined model system. Although we explored Rap1-induced integrin-mediated cell adhesion and Rap1-induced cell spreading in epithelial cells, we particularly focused our attention on endothelial barrier function. Endothelial cells are highly responsive to 007 as measured by, for instance, an increased electrical resistance of the monolayer. This increased barrier function is accompanied by a shift that is observed from radial stress fibers to circumferential actin, which could explain the increased tightness of endothelial cell–cell junctions. Using small-interfering RNA (siRNA) screens, we found that Rap1 interacted with a complex of two related RA-domain-containing proteins, Rasip1 and Radil. Interestingly, both Radil and Rasip1 are homologous to Canoe, the first effector found for Rap1 in *Drosophila*. Importantly, Radil directly interacts with ArhGAP29, a Rho-specific GAP, and we could show that ArhGAP29 mediates 007-induced decrease in radial stress fibers and increased barrier function (Fig. 4, left) (Post et al. 2013, 2015). Intriguingly, multiple Rap1 modules may control barrier function simultaneously. For instance, Mochizuki and coworkers discovered that circumferential actin formation is mediated by the CDC42-GEF FGD5 (Ando et al. 2013). How FGD5 is activated is still unclear, but, in overexpression studies, we observe that FGD5 also forms a complex with Radil, suggesting that the Radil/Rasip complex regulates both pathways. Also, the Rap1 effector AF6 is involved in this process (Birukova et al. 2013), but how these different effectors cooperate is still unclear. Surprisingly, Krit1, previously identified as a Rap1 effector in a similar assay in Huvec cells (Glading et al. 2007), did not affect barrier function in our experiments (Pannekoek et al. 2011).

Several other effectors for Rap1 have been identified, such as the RA-domain-containing

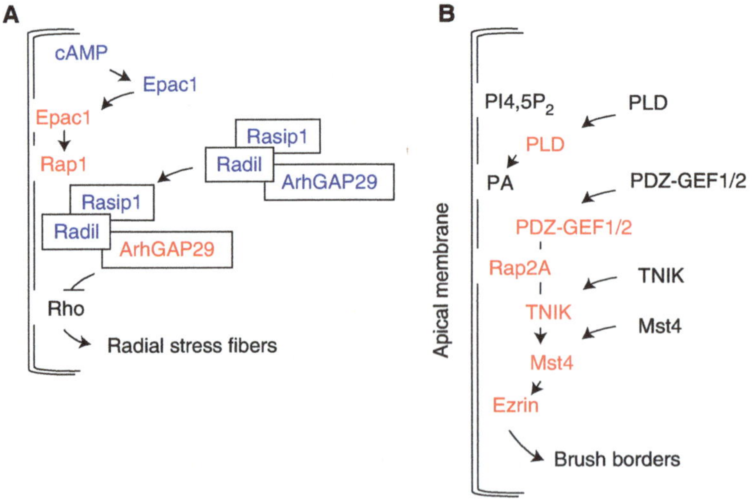

Figure 4. Graphical summary of (*A*) Epac1-mediated regulation of radial stress fibers through the Radil-Rasip1-ArgGAP29 complex, and of (*B*) Rap2A-mediated intestinal brush borders formation (for explanation see text).

protein Riam, which plays an important role in integrin-mediated cell adhesion (Lafuente et al. 2004).

Rap2

The early deviation of Rap1 and Rap2 in evolution suggests that they may have different functions and indeed, although Rap2 is regulated by the same GEFs and GAPs as Rap1, it is clearly different as Rap2(-GTP) binds to a different set of effector proteins lacking an RA domain: the Traf2 and Nck-interacting kinases (TNIKs) (Taira et al. 2004). We stumbled on Rap2A in the analysis of brush-border formation in W4 intestinal epithelial cells. In these cells, brush-border formation is induced by doxycycline-induced activation of the LKB–Strd1α complex (Baas et al. 2004). When we tested the involvement of Rap proteins in this process, we found that siRNA mediated depletion of Rap2A, but the other Rap proteins abolished this process. Subsequent studies revealed a pathway, which is induced by PIP2 enriched apical membrane formation. PIP2 serves as anchor for phospholipase D, which produces phosphatidic acid. Phosphatidic acid is then the anchor for the RapGEF PDZ-GEF. As a consequence, PDZ-GEF is recruited to the apical membrane, activates Rap2A, which in turn recruits and activates the serine-threonine kinase TNIK. TNIK activates the serine-threonine kinase Mst4, which induces the phosphorylation and activation of Ezrin. This activation of Ezrin is a critical step as active Ezrin can rescue Rap2 depletion. Ezrin recruits the actin cytoskeleton for brush-border formation (Fig. 4, right) (Gloerich et al. 2012). The similarity with the yeast model is evident: A landmark is recognized in two steps by a GEF for Rap followed by effector binding, which in several steps leads to the recruitment of actin.

ARE WE FINISHED WITH Ras?

Although our analysis and those of many other research groups have provided much insight into the role Rap proteins play in cellular responses, the initial idea that Rap1 counteracts mutant Ras signaling and could be used as a route to interfere in mutant Ras signaling in cancers turned out to be too naïve. However, despite huge efforts by many groups and many pharmaceutical industries, the alternative approach to interfere in the Ras pathway by selective inhibitors has not worked either as we are still unable to treat Ras-mutant cancers with targeted drugs (Stephen et al. 2014). Ras itself is still largely "undruggable," and the alternative to inhibit downstream targets of mutant Ras, most notably Raf, MEK, ERK, and PI3K, have thus far failed in the clinic, especially when single drugs are used. For such an approach to be successful, it is essential that mutant Ras-containing tumors cells do have a specific sensitivity for these pathway drugs. Indeed, genetic and drug screens in mutant cell lines did identify several possible targets for mutant Ras tumor cell lines, but these have not been translated into clinical targets (Downward 2015). Perhaps the commonly used mutant Ras-containing cell lines identify sensitivities that do not occur in tumors. We therefore shifted to the tumor organoid model system developed by Hans Clevers (Sato et al. 2009; van de Wetering et al. 2015). The advantage of this model system is that tumor organoids can be rapidly established from almost all individual (epithelial) tumors and thus can capture the genetic diversity of these tumors. Moreover, drug responses can be compared with organoids from normal tissue to determine the therapeutic window. Initial studies using a number of colon tumor organoids already indicated the power of this technology (van de Wetering et al. 2015). As a first proof-of-concept, we tested various targeted drugs that are currently being used in clinical trials for inhibiting mutant Ras-containing colorectal cancer, most notably pan-EGFR inhibitors in combination with MEK inhibitors (Sun et al. 2014). We found that these combinations gave a clear cell-cycle arrest of mutant K-ras-containing organoids, supporting the observation that this combination has an effect on mutant Ras tumor cells. However, organoids with nonmutated K-Ras remain far more sensitive for this drug combination. Furthermore, we noted that organoids from normal tissue were similarly sensitive for this combination than tumor organoids with nonmutated K-ras, and thus much more sensitive than mutant K-ras organoids, indicating that for this combination the therapeutic window is limited (Verissimo et al. 2016).

It is now more than 35 years ago that Ras was identified as an oncogene in human cancers and we still have not solved the problem of how to target mutant Ras in cancer. As a consequence, patients carrying mutant Ras in their tumors are excluded from therapies targeting the EGFR signaling cascade. Moreover, when patients carrying normal Ras in their tumors are treated, the appearance of mutant Ras is a frequent resistance mechanism. To make personalized medicine a success, we have to come up with a solution. New model systems like tumor organoids may be helpful in the identification of new drug combinations, but also in finding those patients that may uniquely respond to such combinations.

ACKNOWLEDGMENTS

I thank all members of my laboratory in past years for the tremendous contributions they have made and are still making to understand Ras-like small GTPases and hopefully to find a drug combination that works for mutant Ras-containing tumors. I also thank my colleagues Fried Zwartkruis, Willem Jan Pannekoek, and Boudewijn Burgering for critically reading this manuscript.

REFERENCES

Almoguera C, Shibata D, Forrester K, Martin J, Arnheim N, Perucho M. 1988. Most human carcinomas of the exocrine pancreas contain mutant c-K-ras genes. *Cell* **53:** 549–554.

Altschuler DL, Peterson SN, Ostrowski MC, Lapetina EG. 1995. Cyclic AMP-dependent activation of Rap1b. *J Biol Chem* **270:** 10373–10376.

Ando K, Fukuhara S, Moriya T, Obara Y, Nakahata N, Mochizuki N. 2013. Rap1 potentiates endothelial cell junctions by spatially controlling myosin II activity and actin organization. *J Cell Biol* **202:** 901–916.

Asha H, de Ruiter ND, Wang MG, Hariharan IK. 1999. The Rap1 GTPase functions as a regulator of morphogenesis in vivo. *EMBO J* **18:** 605–615.

Baas AF, Kuipers J, van der Wel NN, Batlle E, Koerten HK, Peters PJ, Clevers HC. 2004. Complete polarization of single intestinal epithelial cells upon activation of LKB1 by STRAD. *Cell* **116:** 457–466.

Bi E, Park HO. 2012. Cell polarization and cytokinesis in budding yeast. *Genetics* **191:** 347–387.

Birukova AA, Tian X, Tian Y, Higginbotham K, Birukov KG. 2013. Rap-afadin axis in control of ρ signaling and endothelial barrier recovery. *Mol Biol Cell* **24:** 2678–2688.

Boettner B, Harjes P, Ishimaru S, Heke M, Fan HQ, Qin Y, Van Aelst L, Gaul U. 2003. The AF-6 homolog canoe acts as a Rap1 effector during dorsal closure of the *Drosophila* embryo. *Genetics* **165:** 159–169.

Bos JL. 1989. *ras* oncogenes in human cancer: A review. *Cancer Res* **49:** 4682–4689.

Bos JL, Verlaan-de Vries M, Jansen AM, Veeneman GH, van Boom JH, van der Eb AJ. 1984. Three different mutations in codon 61 of the human N-ras gene detected by synthetic oligonucleotide hybridization. *Nucleic Acids Res* **12:** 9155–9163.

Bos JL, Fearon ER, Hamilton SR, Verlaan-de Vries M, van Boom JH, van der Eb AJ, Vogelstein B. 1987. Prevalence of ras gene mutations in human colorectal cancers. *Nature* **327:** 293–297.

Bos JL, Rehmann H, Wittinghofer A. 2007. GEFs and GAPs: critical elements in the control of small G proteins. *Cell* **129:** 865–877.

Broek D, Toda T, Michaeli T, Levin L, Birchmeier C, Zoller M, Powers S, Wigler M. 1987. The *S. cerevisiae* CDC25 gene product regulates the RAS/adenylate cyclase pathway. *Cell* **48:** 789–799.

Burgering BM, Coffer PJ. 1995. Protein kinase B (c-Akt) in phosphatidylinositol-3-OH kinase signal transduction. *Nature* **376:** 599–602.

Burgering BM, Medema RH, Maassen JA, van de Wetering ML, van der Eb AJ, McCormick F, Bos JL. 1991. Insulin stimulation of gene expression mediated by p21ras activation. *EMBO J* **10:** 1103–1109.

Conner BJ, Reyes AA, Morin C, Itakura K, Teplitz RL, Wallace RB. 1983. Detection of sickle cell β S-globin allele by hybridization with synthetic oligonucleotides. *Proc Natl Acad Sci* **80:** 278–282.

Consonni SV, Gloerich M, Spanjaard E, Bos JL. 2012. cAMP regulates DEP domain-mediated binding of the guanine nucleotide exchange factor Epac1 to phosphatidic acid at the plasma membrane. *Proc Natl Acad Sci* **109:** 3814–3819.

Cox AD, Fesik SW, Kimmelman AC, Luo J, Der CJ. 2014. Drugging the undruggable RAS: Mission possible? *Nat Rev Drug Discov* **13:** 828–851.

de Rooij J, Zwartkruis FJ, Verheijen MH, Cool RH, Nijman SM, Wittinghofer A, Bos JL. 1998. Epac is a Rap1 guanine-nucleotide-exchange factor directly activated by cyclic AMP. *Nature* **396:** 474–477.

de Vries-Smits AM, Burgering BM, Leevers SJ, Marshall CJ, Bos JL. 1992. Involvement of p21ras in activation of extracellular signal-regulated kinase 2. *Nature* **357:** 602–604.

Downward J. 2015. RAS synthetic lethal screens revisited: Still seeking the elusive prize? *Clin Cancer Res* **21:** 1802–1809.

Downward J, Graves JD, Warne PH, Rayter S, Cantrell DA. 1990. Stimulation of p21ras upon T-cell activation. *Nature* **346:** 719–723.

Eberhard DA, Johnson BE, Amler LC, Goddard AD, Heldens SL, Herbst RS, Ince WL, Janne PA, Januario T, Johnson DH, et al. 2005. Mutations in the epidermal growth factor receptor and in KRAS are predictive and prognostic indicators in patients with non-small-cell lung cancer treated with chemotherapy alone and in combination with erlotinib. *J Clin Oncol* **23:** 5900–5909.

Enserink JM, Christensen AE, de Rooij J, van Triest M, Schwede F, Genieser HG, Doskeland SO, Blank JL, Bos JL. 2002. A novel Epac-specific cAMP analogue demonstrates independent regulation of Rap1 and ERK. *Nat Cell Biol* **4:** 901–906.

Forrester K, Almoguera C, Han K, Grizzle WE, Perucho M. 1987. Detection of high incidence of K-ras oncogenes during human colon tumorigenesis. *Nature* **327:** 298–303.

Franke B, Akkerman JW, Bos JL. 1997. Rapid Ca^{2+}-mediated activation of Rap1 in human platelets. *EMBO J* **16:** 252–259.

Frische EW, Zwartkruis FJ. 2010. Rap1, a mercenary among the Ras-like GTPases. *Dev Biol* **340:** 1–9.

Glading A, Han J, Stockton RA, Ginsberg MH. 2007. KRIT-1/CCM1 is a Rap1 effector that regulates endothelial cell–cell junctions. *J Cell Biol* **179:** 247–254.

Gloerich M, Bos JL. 2010. Epac: Defining a new mechanism for cAMP action. *Annu Rev Pharmacol Toxicol* **50:** 355–375.

Gloerich M, Ponsioen B, Vliem MJ, Zhang Z, Zhao J, Kooistra MR, Price LS, Ritsma L, Zwartkruis FJ, Rehmann H, et al. 2010. Spatial regulation of cyclic AMP-Epac1 signaling in cell adhesion by ERM proteins. *Mol Cell Biol* **30:** 5421–5431.

Gloerich M, ten Klooster JP, Vliem MJ, Koorman T, Zwartkruis FJ, Clevers H, Bos JL. 2012. Rap2A links intestinal cell polarity to brush border formation. *Nat Cell Biol* **14:** 793–801.

Gotoh T, Hattori S, Nakamura S, Kitayama H, Noda M, Takai Y, Kaibuchi K, Matsui H, Hatase O, Takahashi H, et al. 1995. Identification of Rap1 as a target for the Crk SH3 domain-binding guanine nucleotide-releasing factor C3G. *Mol Cell Biol* **15:** 6746–6753.

Hall A, Marshall CJ, Spurr NK, Weiss RA. 1983. Identification of transforming gene in two human sarcoma cell lines as a new member of the ras gene family located on chromosome 1. *Nature* **303:** 396–400.

Hofer F, Fields S, Schneider C, Martin GS. 1994. Activated Ras interacts with the Ral guanine nucleotide dissociation stimulator. *Proc Natl Acad Sci* **91:** 11089–11093.

Karapetis CS, Khambata-Ford S, Jonker DJ, O'Callaghan CJ, Tu D, Tebbutt NC, Simes RJ, Chalchal H, Shapiro JD, Robitaille S, et al. 2008. K-ras mutations and benefit from cetuximab in advanced colorectal cancer. *N Engl J Med* **359:** 1757–1765.

Kitayama H, Sugimoto Y, Matsuzaki T, Ikawa Y, Noda M. 1989. A *ras*-related gene with transformation suppressor activity. *Cell* **56:** 77–84.

Knox AL, Brown NH. 2002. Rap1 GTPase regulation of adherens junction positioning and cell adhesion. *Science* **295:** 1285–1288.

Kodaki T, Woscholski R, Hallberg B, Rodriguez-Viciana P, Downward J, Parker PJ. 1994. The activation of phosphatidylinositol 3-kinase by Ras. *Curr Biol* **4:** 798–806.

Kops GJ, de Ruiter ND, De Vries-Smits AM, Powell DR, Bos JL, Burgering BM. 1999. Direct control of the Forkhead transcription factor AFX by protein kinase B. *Nature* **398:** 630–634.

Lafuente EM, van Puijenbroek AA, Krause M, Carman CV, Freeman GJ, Berezovskaya A, Constantine E, Springer TA, Gertler FB, Boussiotis VA. 2004. RIAM, an Ena/VASP and Profilin ligand, interacts with Rap1-GTP and mediates Rap1-induced adhesion. *Dev Cell* **7:** 585–595.

Leevers SJ, Marshall CJ. 1992. Activation of extracellular signal-regulated kinase, ERK2, by p21ras oncoprotein. *EMBO J* **11:** 569–574.

Li S, Tsalkova T, White MA, Mei FC, Liu T, Wang D, Woods VL Jr, Cheng X. 2011. Mechanism of intracellular cAMP sensor Epac2 activation: cAMP-induced conformational changes identified by amide hydrogen/deuterium exchange mass spectrometry (DXMS). *J Biol Chem* **286:** 17889–17897.

Moodie SA, Willumsen BM, Weber MJ, Wolfman A. 1993. Complexes of Ras.GTP with Raf-1 and mitogen-activated protein kinase kinase. *Science* **260:** 1658–1661.

Mulder MP, Keijzer W, Verkerk A, Boot AJ, Prins ME, Splinter TA, Bos JL. 1989. Activated ras genes in human seminoma: Evidence for tumor heterogeneity. *Oncogene* **4:** 1345–1351.

Pannekoek WJ, van Dijk JJ, Chan OY, Huveneers S, Linnemann JR, Spanjaard E, Brouwer PM, van der Meer AJ, Zwartkruis FJ, Rehmann H, et al. 2011. Epac1 and PDZ-GEF cooperate in Rap1 mediated endothelial junction control. *Cell Signal* **23:** 2056–2064.

Papke B, Der CJ. 2017. Drugging RAS: Know the enemy. *Science* **355:** 1158–1163.

Parada LF, Tabin CJ, Shih C, Weinberg RA. 1982. Human EJ bladder carcinoma oncogene is homologue of Harvey sarcoma virus ras gene. *Nature* **297:** 474–478.

Pizon V, Lerosey I, Chardin P, Tavitian A. 1988. Nucleotide sequence of a human cDNA encoding a *ras*-related protein (*rap*1B). *Nucleic Acids Res* **16:** 7719.

Ponsioen B, Gloerich M, Ritsma L, Rehmann H, Bos JL, Jalink K. 2009. Direct spatial control of Epac1 by cyclic AMP. *Mol Cell Biol* **29:** 2521–2531.

Post A, Pannekoek WJ, Ross SH, Verlaan I, Brouwer PM, Bos JL. 2013. Rasip1 mediates Rap1 regulation of Rho in endothelial barrier function through ArhGAP29. *Proc Natl Acad Sci* **110:** 11427–11432.

Post A, Pannekoek WJ, Ponsioen B, Vliem MJ, Bos JL. 2015. Rap1 spatially controls ArhGAP29 to inhibit Rho signaling during endothelial barrier regulation. *Mol Cell Biol* **35:** 2495–2502.

Price LS, Hajdo-Milasinovic A, Zhao J, Zwartkruis FJ, Collard JG, Bos JL. 2004. Rap1 regulates E-cadherin-mediated cell–cell adhesion. *J Biol Chem* **279:** 35127–35132.

Rangarajan S, Enserink JM, Kuiperij HB, de Rooij J, Price LS, Schwede F, Bos JL. 2003. Cyclic AMP induces integrin-mediated cell adhesion through Epac and Rap1 upon stimulation of the β2-adrenergic receptor. *J Cell Biol* **160:** 487–493.

Rehmann H, Prakash B, Wolf E, Rueppel A, de Rooij J, Bos JL, Wittinghofer A. 2003. Structure and regulation of the cAMP-binding domains of Epac2. *Nat Struct Biol* **10:** 26–32.

Rehmann H, Das J, Knipscheer P, Wittinghofer A, Bos JL. 2006. Structure of the cyclic-AMP-responsive exchange factor Epac2 in its auto-inhibited state. *Nature* **439:** 625–628.

Rehmann H, Arias-Palomo E, Hadders MA, Schwede F, Llorca O, Bos JL. 2008. Structure of Epac2 in complex with a cyclic AMP analogue and RAP1B. *Nature* **455:** 124–127.

Robinson LC, Gibbs JB, Marshall MS, Sigal IS, Tatchell K. 1987. CDC25: A component of the RAS-adenylate cyclase pathway in Saccharomyces cerevisiae. *Science* **235:** 1218–1221.

Rodenhuis S, van de Wetering ML, Mooi WJ, Evers SG, van Zandwijk N, Bos JL. 1987. Mutational activation of the K-ras oncogene. A possible pathogenetic factor in adenocarcinoma of the lung. *N Engl J Med* **317:** 929–935.

Saiki RK, Scharf S, Faloona F, Mullis KB, Horn GT, Erlich HA, Arnheim N. 1985. Enzymatic amplification of β-globin genomic sequences and restriction site analysis for diagnosis of sickle cell anemia. *Science* **230:** 1350–1354.

Sato T, Vries RG, Snippert HJ, van de Wetering M, Barker N, Stange DE, van Es JH, Abo A, Kujala P, Peters PJ, et al. 2009. Single Lgr5 stem cells build crypt-villus structures in vitro without a mesenchymal niche. *Nature* **459:** 262–265.

Shimizu K, Birnbaum D, Ruley MA, Fasano O, Suard Y, Edlund L, Taparowsky E, Goldfarb M, Wigler M. 1983. Structure of the Ki-ras gene of the human lung carcinoma cell line Calu-1. *Nature* **304:** 497–500.

Sot B, Behrmann E, Raunser S, Wittinghofer A. 2013. Ras GTPase activating (RasGAP) activity of the dual specificity GAP protein Rasal requires colocalization and C2 domain binding to lipid membranes. *Proc Natl Acad Sci* **110:** 111–116.

Stephen AG, Esposito D, Bagni RK, McCormick F. 2014. Dragging ras back in the ring. *Cancer Cell* **25:** 272–281.

Sun C, Hobor S, Bertotti A, Zecchin D, Huang S, Galimi F, Cottino F, Prahallad A, Grernrum W, Tzani A, et al. 2014. Intrinsic resistance to MEK inhibition in KRAS mutant lung and colon cancer through transcriptional induction of ERBB3. *Cell Rep* **7:** 86–93.

Taira K, Umikawa M, Takei K, Myagmar BE, Shinzato M, Machida N, Uezato H, Nonaka S, Kariya K. 2004. The Traf2- and Nck-interacting kinase as a putative effector of Rap2 to regulate actin cytoskeleton. *J Biol Chem* **279:** 49488–49496.

Trahey M, McCormick F. 1987. A cytoplasmic protein stimulates normal N-ras p21 GTPase, but does not affect oncogenic mutants. *Science* **238:** 542–545.

van Corven EJ, Hordijk PL, Medema RH, Bos JL, Moolenaar WH. 1993. Pertussis toxin-sensitive activation of p21ras by G protein-coupled receptor agonists in fibroblasts. *Proc Natl Acad Sci* **90:** 1257–1261.

van Dam TJ, Bos JL, Snel B. 2011. Evolution of the Ras-like small GTPases and their regulators. *Small GTPases* **2:** 4–16.

van de Wetering M, Francies HE, Francis JM, Bounova G, Iorio F, Pronk A, van Houdt W, van Gorp J, Taylor-Weiner A, Kester L, et al. 2015. Prospective derivation of a living organoid biobank of colorectal cancer patients. *Cell* **161:** 933–945.

van 't Veer LJ, Burgering BM, Versteeg R, Boot AJ, Ruiter DJ, Osanto S, Schrier PI, Bos JL. 1989. N-ras mutations in human cutaneous melanoma from sun-exposed body sites. *Mol Cell Biol* **9:** 3114–3116.

Verissimo CS, Overmeer RM, Ponsioen B, Drost J, Mertens S, Verlaan-Klink I, Gerwen BV, van der Ven M, Wetering MV, Egan DA, et al. 2016. Targeting mutant RAS in patient-derived colorectal cancer organoids by combinatorial drug screening. *eLife* **5:** e18489.

Verlaan-de Vries M, Bogaard ME, van den Elst H, van Boom JH, van der Eb AJ, Bos JL. 1986. A dot-blot screening procedure for mutated ras oncogenes using synthetic oligodeoxynucleotides. *Gene* **50:** 313–320.

Vogelstein B, Fearon ER, Hamilton SR, Kern SE, Preisinger AC, Leppert M, Nakamura Y, White R, Smits AM, Bos JL. 1988. Genetic alterations during colorectal-tumor development. *N Engl J Med* **319:** 525–532.

White MA, Nicolette C, Minden A, Polverino A, Van Aelst L, Karin M, Wigler MH. 1995. Multiple Ras functions can contribute to mammalian cell transformation. *Cell* **80:** 533–541.

Wong G, Muller O, Clark R, Conroy L, Moran MF, Polakis P, McCormick F. 1992. Molecular cloning and nucleic acid binding properties of the GAP-associated tyrosine phosphoprotein p62. *Cell* **69:** 551–558.

Yunis JJ, Boot AJ, Mayer MG, Bos JL. 1989. Mechanisms of ras mutation in myelodysplastic syndrome. *Oncogene* **4:** 609–614.

Zwartkruis FJ, Wolthuis RM, Nabben NM, Franke B, Bos JL. 1998. Extracellular signal-regulated activation of Rap1 fails to interfere in Ras effector signalling. *EMBO J* **17:** 5905–5912.

Ras-Mediated Activation of the Raf Family Kinases

Elizabeth M. Terrell and Deborah K. Morrison

Laboratory of Cell and Developmental Signaling, NCI-Frederick, Frederick, Maryland 21702

Correspondence: morrisod@mail.nih.gov

The extracellular signal-regulated kinase (ERK) cascade comprised of the Raf, MEK, and ERK protein kinases constitutes a key effector cascade used by the Ras GTPases to relay signals regulating cell growth, survival, proliferation, and differentiation. Of the ERK cascade components, the regulation of the Raf kinases is by far the most complex, involving changes in subcellular localization, protein and lipid interactions, as well as alterations in the Raf phosphorylation state. The Raf kinases interact directly with active, membrane-localized Ras, and this interaction is often the first step in the Raf activation process, which ultimately results in ERK activation and the downstream phosphorylation of cellular targets that will specify a particular biological response. Here, we will examine our current understanding of how Ras promotes Raf activation, focusing on the molecular mechanisms that contribute to the Raf activation/inactivation cycle.

Members of the Raf serine/threonine kinase family are the initiating enzymes in the three-tiered extracellular signal-regulated kinase (ERK) cascade, and in mammalian cells, there are three Raf proteins, A-Raf, B-Raf, and C-Raf (also known as Raf-1) (Wellbrock et al. 2004). The Raf kinases were first discovered when the *raf-1* gene was identified as the cellular counterpart of the murine retroviral oncogene, *v-raf* (Rapp et al. 1983). Later, this kinase family rose to prominence when they were found to be direct effectors of activated Ras, mediating signal transmission from Ras to the downstream kinases MEK and ERK (Marshall 1996). Not surprisingly, given their central position in such a major signaling pathway, mutations in the Raf kinases also contribute to human disease states. In particular, somatic mutations in B-Raf are important drivers of human cancer (Dhomen and Marais 2007), and germline mutations in either B-Raf or C-Raf can be causative for certain developmental disorders in the Rasopathies spectrum (Tartaglia et al. 2011).

Despite their critical function in cell signaling, determining how Raf activity is regulated has been a challenging task—resulting from the complexity of the Raf activation process. In this article, we will review the molecular mechanisms involved in Raf kinase regulation. Our discussion will include an examination of the presignaling inactive state of the Rafs as well as the events that contribute to Raf kinase activation, with a focus on the crucial role that the Ras GTPases play in promoting Raf activation under normal growth conditions as well as in human disease states. Whenever possible, we will also comment on the regulatory differences that have been identified for the individual Raf members,

Cite this article as *Cold Spring Harb Perspect Med* doi: 10.1101/cshperspect.a033746

and we will conclude with a description of the mechanisms that attenuate Raf activity following signal transmission.

THE RAF PRESIGNALING INACTIVE STATE

All members of the Raf kinase family contain three conserved regions, CR1, CR2, and CR3, and each Raf protein can be divided into two functional domains—an amino-terminal regulatory domain and a carboxy-terminal catalytic domain (Fig. 1A) (Daum et al. 1994). Both the CR1, which consists of a Ras-binding domain (RBD) and a cysteine-rich domain (CRD), and the CR2, a region rich in serine/threonine residues, are found within the amino-terminal regulatory domain, whereas the CR3 comprises the carboxy-terminal kinase domain. The Raf proteins do not contain any intrinsic subcellular localization motifs, and in quiescent cells, they localize to the cytosol as monomers (Nan et al. 2013). Before pathway activation, the Raf monomers are maintained in a stable, presignaling inactive state through multiple regulatory mechanisms, including autoinhibition, phosphorylation of negative regulatory sites, and binding of inhibitory proteins (Figs. 1B and 2).

Raf Autoinhibition

The first indication that the Raf amino-terminal domain might possess autoinhibitory activity came from the observation that this domain is absent in the oncogenic v-Raf protein (Rapp et al. 1983). Subsequently, it was shown that deletion of the amino-terminal domain could convert any of the mammalian Raf proteins into constitutively active kinases capable of inducing cellular transformation (Stanton et al. 1989; Heidecker et al. 1990). Further studies in

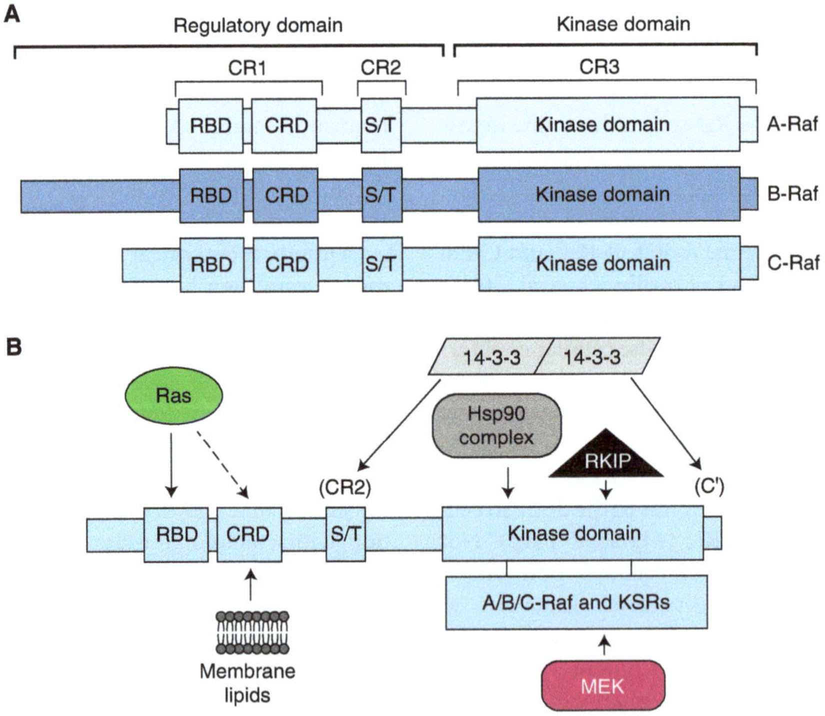

Figure 1. Raf domain structure and protein interactions. (*A*) The Raf kinases can be divided into two functional domains: an amino-terminal regulatory domain and a carboxy-terminal kinase domain and they contain three conserved regions (CRs): CR1, which contains a Ras-binding domain (RBD) and a cysteine-rich domain (CRD); CR2, an area rich in serine/threonine residues (S/T); and CR3, which contains the protein kinase domain. (*B*) Numerous Raf-binding partners have been identified.

 Cite this article as *Cold Spring Harb Perspect Med* doi: 10.1101/cshperspect.a033746

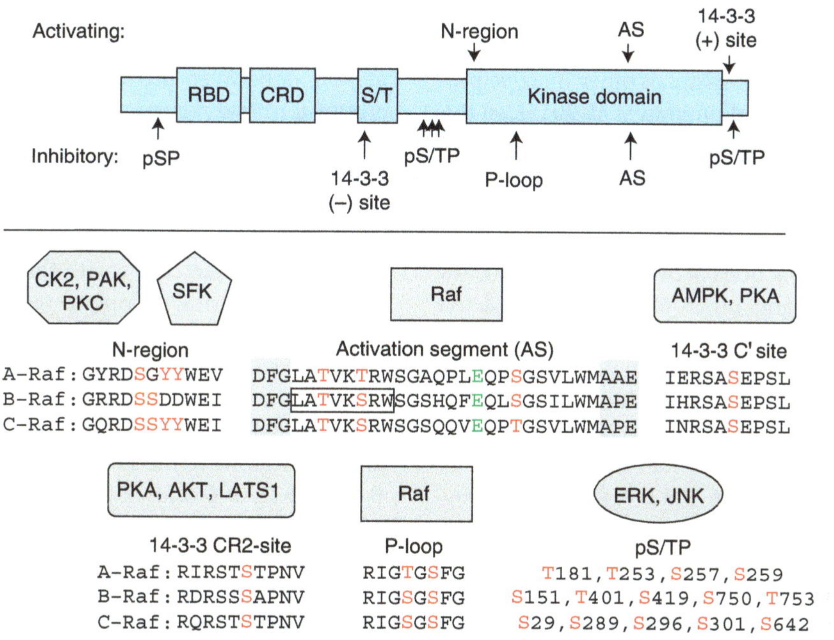

Figure 2. Regulation of the Raf kinases by phosphorylation. (*Top*) Raf is both positively and negatively regulated by phosphorylation on numerous sites. (*Bottom*) Shown are the amino acid sequence alignments of the sites of Raf phosphorylation, with the residues phosphorylated shown in red, and the relevant kinases indicated above the sequences. A glutamic acid residue that can act as a phosphomimetic to promote the active conformation of the activation segment (AS) is also shown in green.

which the amino- and carboxy-terminal domains of B-Raf or C-Raf were expressed as individual proteins revealed that the isolated amino-terminal domains of the Rafs could bind their respective kinase domains and block signal transmission to MEK and ERK (Cutler et al. 1998; Chong and Guan 2003; Tran and Frost 2003; Tran et al. 2005). Regions in the CR1, including the CRD and parts of the RBD, are critical for the amino-terminal autoinhibitory activity, and phosphorylation of sites in the carboxy-terminal kinase domain contribute prominently to the relief of autoinhibition, with phosphorylation of sites in the negative charge regulatory region (N-region) being particularly important (Cutler et al. 1998; Tran and Frost 2003; Tran et al. 2005).

In addition to the autoinhibition mediated by the amino-terminal domain, structural data indicate that the Raf kinase domain itself is held in an inactive conformation. Like all other protein kinases, the Raf kinase domain consists of an amino-terminal lobe (N-lobe) and a carboxy-terminal lobe (C-lobe) that are connected to one another through a flexible hinge region (Shaw et al. 2014). The N-lobe contains five antiparallel β-strands and the regulatory αC-helix, whereas the C-lobe is comprised primarily of α-helices and contains the activation segment (AS). For all kinases to assume the active/ON state conformation, spatially conserved hydrophobic residues spanning both lobes of the kinase domain must align to form two structural entities called the regulatory and catalytic spines (McClendon et al. 2014). The regulatory spine (R-spine) consists of four residues, with each residue derived from a key catalytic element in the kinase domain—a residue from the β4-strand, the αC-helix, the DFG motif, and the HRD sequence. In many inactive kinase structures, the R-spine is misconfigured and for the spine residues to align, both the αC-helix and the AS must change positions. Although the structure of an authentic OFF state Raf kinase has not been solved, the structure of a monomeric, inhibitor-bound B-Raf kinase domain has been determined that has features characteristic of other inactive protein kinases (Thevakumaran et al. 2015). In this monomeric B-Raf structure, the αC-helix is shifted to an outward position and residues adjacent to the DFG motif in the AS form a small helix called the AS-H1-helix (Fig. 3). Moreover, extensive interactions between the αC-helix, the AS-H1-helix and other residues in the active site, including residues in the conserved DFG and HRD motifs, stabilize the αC-helix in the inactive OUT position and inhibit the formation of an essential salt bridge that defines the active state (E501-K483 for B-Raf, E393-K375 for C-Raf, E354-K336 for A-Raf). A similarly configured AS-H1-helix has been observed in the inactive conformation of the epidermal growth factor receptor (EGFR) kinase (Zhang et al. 2006), and it is important to note that most activating B-Raf mutations identified in human cancer occur in residues comprising the AS-H1-helix (Lavoie and Therrien 2015). Thus, it is tempting to speculate that in the context of full-length Raf proteins, Raf monomers are held in an inactive state through intramolecular interactions occurring within the kinase domain as well as between the kinase domain and the amino-terminal regulatory domain.

14-3-3 Binding to the CR2 Site

The autoinhibited Raf conformation is further stabilized by interactions with the 14-3-3 family of phosphoserine/phosphothreonine-binding proteins (Muslin et al. 1996). All Raf kinases contain two high-affinity, phosphorylation-dependent 14-3-3 binding sites, one in the CR2 (S365 of B-Raf, S259 of C-Raf, S214 of A-Raf) and one following the kinase domain (S729 of B-Raf, S621 of C-Raf, S582 of A-Raf), both of which are phosphorylated in the presignaling inactive state (Morrison et al. 1993; Dougherty et al. 2004; Ritt et al. 2010). 14-3-3 proteins form constitutive dimers with each protomer of the dimer containing an independent binding channel, and it has been proposed that a 14-3-3 dimer can interact with both the CR2 and carboxy-terminal sites simultaneously to stabilize the autoinhibited conformation (Tzivion et al.

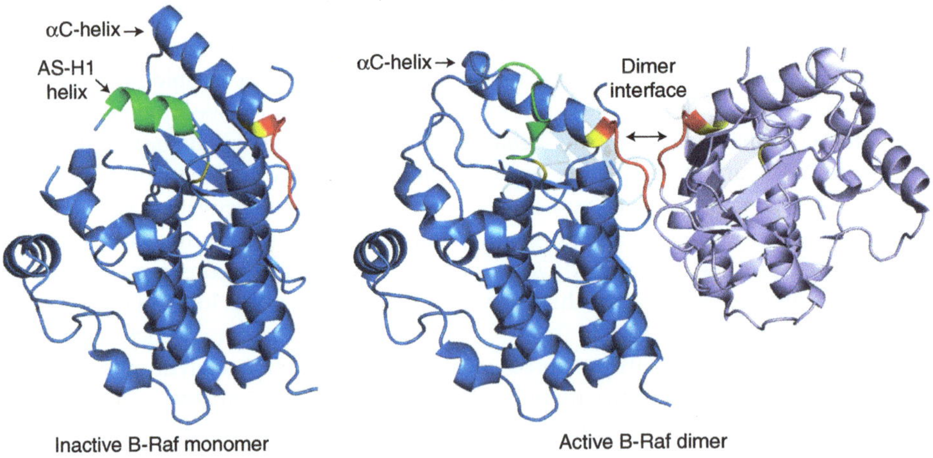

Figure 3. Regulation of the Raf kinase domain by dimerization. Shown are the crystal structures of a monomeric, B-Raf kinase domain in an inactive conformation (PDB, 4WO5) and of dimerized B-Raf kinase domains in an active conformation (taken from the B-Raf–MEK complex PDB, 4MNE). In the monomeric, inactive state, the αC-helix is shifted outward and residues adjacent to the DFG motif in the activation segment (AS) form a small helix known as the AS-H1-helix. When Raf kinase domains dimerize in a side-to-side manner, the αC-helix shifts inward to the active position and the AS-H1-helix is disrupted, allowing residues in the R-spine to align. The conserved Raf dimer interface motif RKTR is shown in red, the R-spine residues in the αC-helix (L505) and in the DFG motif (D595) are indicated in yellow, and the sequences comprising the AS-H1-helix are shown in green.

Cite this article as *Cold Spring Harb Perspect Med* doi: 10.1101/cshperspect.a033746

1998). Moreover, binding of 14-3-3 to the CR2 site has a known negative regulatory function and appears to mask regions within the amino terminus, such as the CRD, which are required for Raf activation (Michaud et al. 1995; Clark et al. 1997; McPherson et al. 1999; Light et al. 2002). Mutation of the CR2 serine site or residues that comprise the 14-3-3 binding motif have been identified in tumors and Rasopathies as gain-of-function mutations (Lavoie and Therrien 2015), resulting in increased Ras/Raf binding, membrane recruitment, and Raf activation (Michaud et al. 1995; Dhillon et al. 2002a; Light et al. 2002; Pandit et al. 2007; Razzaque et al. 2007). Protein kinase A (PKA) and AKT have been identified as the primary cellular kinases that phosphorylate the CR2 site; however, the CR2 site can also be phosphorylated by LATS1 through signaling cross talk with the MST/Hippo pathway (Zimmermann and Moelling 1999; Dhillon et al. 2002b; Dumaz and Marais 2003; Romano et al. 2014).

Hsp90 Chaperone Complex

In their presignaling inactive state as well as in their active signaling state, all members of the Raf kinase family interact with Hsp90 complexes, which include Hsp90 proteins, the Cdc37 targeting subunit, and the peptidyl-prolyl *cis-trans* isomerase FKBP5 (Stancato et al. 1994; Eisenhardt et al. 2016; Diedrich et al. 2017). The Hsp90 complex binds to the Raf kinase domain and this interaction is required for Raf protein stability (Schulte et al. 1995).

B-Raf–MEK1 Complex

Another protein that has been found to associate with inactive B-Raf complexes is MEK1 (Haling et al. 2014; Eisenhardt et al. 2016). In experiments investigating whether preformed Raf/MEK complexes exist before Ras pathway activation, MEK1 was found to preferentially interact with B-Raf proteins found in the cytosol of quiescent cells or in cancer cell lines that contain low levels of active MEK (Haling et al. 2014). When cells were treated with growth factors to activate the pathway, the B-Raf–MEK complex was disrupted in a manner that correlated with the appearance of active phosphoMEK, but could reform when phosphoMEK levels returned to baseline, suggesting that phosphorylation and activation of MEK disrupts the interaction. Structural analysis of the B-Raf–MEK1 complex indicates that the two kinase domains interact in a face-to-face manner (Haling et al. 2014), with primary points of contact being the AS of both proteins and their respective αG-helices, a structural element implicated in kinase–substrate interactions. Although it is not clear why MEK1 preferentially binds inactive B-Raf, it is interesting to note that the RKIP inhibitory protein has been found to interact with the C-Raf kinase domain in a manner that is mutually exclusive with MEK binding (Yeung et al. 2000). Phosphorylation of RKIP on S153 by protein kinase C (PKC) disrupts the RKIP–C-Raf interaction and appears to account for the release of RKIP from C-Raf under signaling conditions (Corbit et al. 2003).

From the above findings, the following model has emerged: In quiescent cells, members of the Raf kinase family exist as autoinhibited monomers, with the inactive conformation stabilized by the phosphorylation-dependent binding of 14-3-3 dimers as well as by the interaction with Hsp90 complexes. A pool of B-Raf and MEK1 exists in preformed complexes, whereas MEK may be precluded from C-Raf complexes as a result of RKIP binding. The presignaling, inactive Raf complexes localize to the cytosol, poised ready for recruitment in response to upstream pathway activation (Fig. 4).

THE RAF ACTIVATION PROCESS

For the Rafs to become active enzymes, autoinhibition mediated by the amino-terminal domain must be relieved and the kinase domain must assume an active/ON state conformation. After more than 30 years of research, we now know that the cell accomplishes this task through an intricate series of events that include a change in Raf subcellular localization, protein and lipid interactions (Fig. 1B), as well as regulatory phosphorylation/dephosphorylation events (Fig. 2).

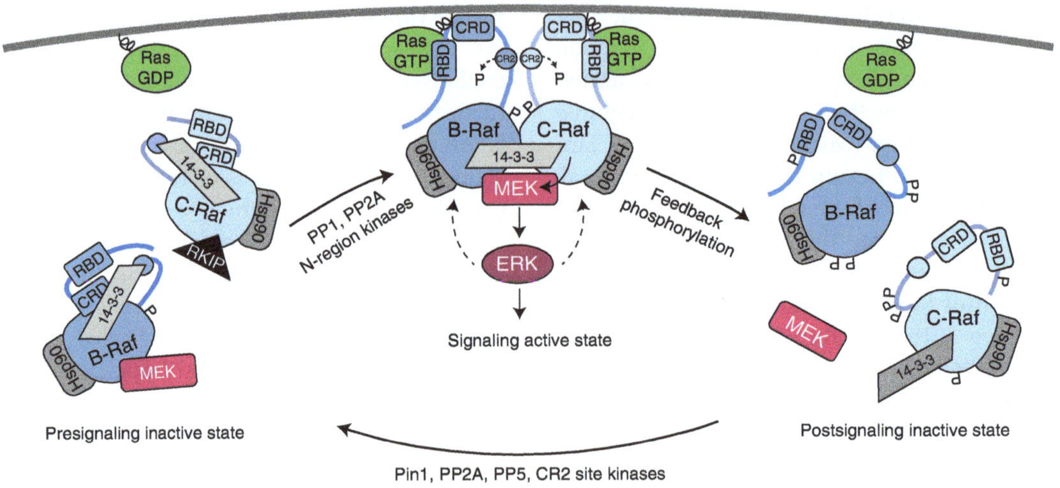

Figure 4. The Raf activation–inactivation cycle. A model depicting the Raf kinases in a presignaling inactive state, a signaling active state, and a postsignaling inactive state is shown (see text for details). GDP, Guanosine diphosphate; GTP, guanosine triphosphate; RBD, Ras-binding domain; CRD, cysteine-rich domain; ERK, extracellular signal-regulated kinase.

Ras Binding

A major breakthrough in understanding Raf activation came with the discovery that all Raf kinases selectively interact with active guanosine triphosphate (GTP)-bound Ras (Van Aelst et al. 1993; Vojtek et al. 1993; Zhang et al. 1993). Binding to active Ras does not stimulate Raf enzymatic activity directly, but instead localizes the normally cytosolic Raf proteins to the plasma membrane. This change in localization is a critical step in the Raf activation process as artificially targeting Raf to the plasma membrane results in a kinase that is constitutively active in a Ras-independent manner (Leevers et al. 1994; Stokoe et al. 1994). The initial contact with Ras is made by the Raf-RBD, which adopts a conserved ubiquitin-like structure that binds with high affinity (18 nM) to the Ras effector domain in its GTP-bound state (Herrmann et al. 1995; Nassar et al. 1995). In addition to the RBD, the Raf CRD makes distinct contacts with lipid modified Ras. The Raf CRD is a zinc-coordinated structure of the atypical type (Mott et al. 1996), and as an isolated domain, the CRD shows constitutive low-affinity binding to Ras with no preference for its nucleotide state (Williams et al. 2000). However, in the context of the RBD, the CRD can further facilitate the membrane recruitment of the Rafs by selectively binding the farnesyl groups of processed Ras (Luo et al. 1997; Williams et al. 2000). Ras binding also allows the CRD zinc finger motifs to directly contact phosphatidyl serine in the plasma membrane (Ghosh et al. 1996; Hekman et al. 2002). Full activation of Raf requires that both the RBD and the CRD are engaged, and CRD interactions with Ras and phosphatidyl serine are critical for relief of autoinhibition as well as stably linking Raf to the plasma membrane (Roy et al. 1997; Cutler et al. 1998; Bondeva et al. 2002). Notably, more than 40% of activating B-Raf mutations identified in the Rasopathy cardio-facio-cutaneous (CFC) syndrome occur in the CRD, further demonstrating the biological importance of this domain (Sarkozy et al. 2009).

Displacement of 14-3-3 and Dephosphorylation of the CR2 Site

In addition to changing the subcellular localization of the Rafs, experimental evidence suggests that the interaction with Ras and membrane phospholipids displaces 14-3-3 from the CR2 site, thereby exposing the site to phosphatase

 Cite this article as *Cold Spring Harb Perspect Med* doi: 10.1101/cshperspect.a033746

activity (Rommel et al. 1996; McPherson et al. 1999). In response to epidermal growth factor signaling, a complex consisting of active M-Ras, Shoc2, and the catalytic subunit of protein phosphatase 1 (PP1) plays a prominent role in mediating the dephosphorylation of the CR2 site (Rodriguez-Viciana et al. 2006). However, the CR2 site is also a substrate for the heterotrimeric protein phosphatase PP2A (Abraham et al. 2000; Jaumot and Hancock 2001; Ory et al. 2003a). PP2A has been identified in active Raf complexes and may contribute to the dephosphorylation of the CR2 site in specific cell types or under certain signaling conditions (Ory et al. 2003a; Eisenhardt et al. 2016).

Raf Dimerization and Activating Phosphorylation

The interaction with Ras also serves a critical function in promoting the phosphorylation of the Raf kinase domain on activating sites and is often a prerequisite for Raf dimerization. Phosphorylation sites that are particularly important for Raf activation are found in the negative charge regulatory region (N-region). The N-region of all Raf proteins contains either negatively charged amino acids or residues that become negatively charged as a result of phosphorylation (Mason et al. 1999). Negative charges in this region contribute to the relief of autoinhibition (Cutler et al. 1998; Tran and Frost 2003; Tran et al. 2005) and are needed to form intermolecular contacts that stabilize the Raf dimer structure (Baljuls et al. 2011; Hu et al. 2013). The N-regions of both C-Raf and A-Raf are comprised of phosphoacceptor residues that only become phosphorylated at the cell surface in response to signaling stimuli (Mason et al. 1999). In contrast, the N-region of B-Raf maintains a constant charge because of the presence of two aspartic acid residues and a serine site (S445) that is constitutively phosphorylated by the CK2 kinase (Mason et al. 1999; Ritt et al. 2007). The tyrosine residues in the N-region of C-Raf and A-Raf are phosphorylated primarily by members of the SRC family kinases, whereas the N-region serine site can be phosphorylated by members of the PAK and PKC families, or by CK2 if the tyrosine sites are phosphorylated (Fabian et al. 1993; Marais et al. 1995; Diaz et al. 1997; King et al. 1998; Chaudhary et al. 2000; Chiloeches et al. 2001; Hamilton et al. 2001; Ritt et al. 2007).

A second major breakthrough in understanding Raf activation came with the discovery that under most conditions, dimerization of the Raf kinases is required. Structural studies indicate that the Raf kinase domains form side-to-side dimers with interactions mediated by both the N- and C-lobes of the kinase domain (Rajakulendran et al. 2009). Although it is unclear whether any contacts are made between the Raf amino-terminal domains, binding to Ras, concurrent with the membrane clustering of active Ras proteins, is thought to increase the local concentration of Raf at the cell surface and induce conformational changes that promote Raf dimerization (Plowman et al. 2005; Terai and Matsuda 2005; Tian et al. 2007; Hibino et al. 2009). In addition to the interaction with Ras, the retained binding of 14-3-3 to the Raf carboxy-terminal site is also required for Raf dimerization in cells (Rushworth et al. 2006). Whether the bound 14-3-3 dimers facilitate contact between the Raf kinase domains or act to stabilize the dimerized structure is currently unknown; however, Raf proteins that contain mutations in the carboxy-terminal serine site fail to dimerize and cannot be activated (Garnett et al. 2005; Rajakulendran et al. 2009; Ritt et al. 2010). The Raf kinases can homodimerize or heterodimerize with any other Raf family member, and they can form side-to-side dimers with the closely related kinase suppressor of Ras (KSR) proteins (Hu et al. 2013). Although the factors that determine the dimerization preferences are poorly understood, it has been shown that B-Raf/C-Raf heterodimers predominate in Ras-mediated signaling and appear to have the highest catalytic activity (Weber et al. 2001; Rushworth et al. 2006; Freeman et al. 2013).

Structural studies have provided much insight regarding how dimerization promotes Raf activation. As stated above, for protein kinase domains to adopt an active/ON conformation, the hydrophobic residues comprising the regulatory R-spine must align, and for many kinases, this requires that the αC-helix in the N-lobe and the AS in the C-lobe shift to an active position. For the Raf kinases, the αC-helix and the dimer interface are allosterically linked in that the hydrophobic R-spine residue in the αC-helix (L505 for B-Raf, L397 for C-Raf, L358 for A-Raf) lies adjacent to the conserved RKTR motif in the dimer interface (Rajakulendran et al. 2009), and when dimerization occurs, contacts made at the dimer interface are thought to promote key conformational changes (Fig. 3). In particular, dimer interface contact is thought to shift the αC-helix into the active IN position and disrupt its interactions with the AS-H1-helix, which in turn may help destabilize the helix (Thevakumaran et al. 2015). Interestingly, a protomer does not need to possess catalytic activity to transactivate the second protomer as oncogenic mutations that abolish the kinase activity of B-Raf can promote tumorigenesis through the allosteric activation of C-Raf (Wan et al. 2004). B-Raf proteins that are inactivated by inhibitor treatment can also transactivate C-Raf in a Ras-dependent manner to promote paradoxical ERK cascade signaling (Hatzivassiliou et al. 2010; Heidorn et al. 2010; Poulikakos et al. 2010). Notably, the key determinant for whether a protomer in a dimer complex can function as an activator for the second protomer is the presence of charged and/or phosphorylated residues in the N-region (Hu et al. 2013).

Similar to many other kinases, it is also thought that phosphorylation of the AS is required for full Raf activity, functioning to expose the catalytic cleft and to promote the alignment of the R-spine residue in the DFG motif. Two conserved AS residues have been implicated, T598/S601 for B-Raf, T491/S494 for C-Raf, and T452/T455 for A-Raf, with substitution of phosphomimetic residues at these sites resulting in increased kinase activity (Zhang and Guan 2000; Chong et al. 2001). Given that these residues localize to the region of the AS that forms the AS-H1-helix, it has been proposed that phosphorylation of these sites may be required to disrupt the AS-H1-helix, which binds and stabilizes the αC-helix in an inactive OUT position (Thevakumaran et al. 2015). Through in vitro kinase assays, one study has provided evidence that the phosphorylation of these sites occurs in *cis* after Raf dimerization (Hu et al. 2013). Strikingly, however, detecting these phosphorylation events in cells has been particularly difficult. For example, these phosphorylation sites have not been identified in any large-scale phosphoproteomic study despite the frequent detection of other phosphorylation sites, including the Raf N-region sites (Lavoie and Therrien 2015). Thus, it is possible that phosphorylation of the AS sites occurs at a very low level or is highly transient, or based on the B-Raf/MEK1 structure may not always be required. In most Raf crystals the AS is largely unstructured; however, in the B-Raf–MEK1 complex, residue E611 in the B-Raf AS acts as a phosphomimetic to form an intramolecular salt bridge with R575 in the conserved HRD motif, allowing the AS to assume an active conformation (Haling et al. 2014). The E611 residue aligns with AS phosphorylation sites known to be required for the activation of other kinases, including PKCα, and given that residues analogous to E611 and R575 are conserved in all Raf members, it is possible that the AS of the Raf kinases can adopt an "active" conformation in the absence of phosphorylation. Thus, although phosphorylation of the AS sites would promote the active conformation, their phosphorylation may not be essential under all signaling conditions or for all Raf proteins.

Nonetheless, it is clear that the conformation of the AS is a critical determinant in Raf kinase activation, as many oncogenic B-Raf mutations occur in this segment, with V600E being the most prevalent. In the case of V600E, not only would this substitution act as a phosphomimetic to disrupt the AS-H1-helix, structural studies indicate that the negatively charged side chain of the glutamic acid can form an intramolecular salt bridge with K507, a component of the RKTR dimer interface motif that lies at the tip of the αC-helix. As a result of this interaction, both the αC-helix and the AS stably adopt an active conformation in the absence of dimerization (Haling et al. 2014; Thevakumaran et al. 2015). Moreover, recent studies indicate that mutations to the B-Raf V600 residue are among the few disease-associated Raf mutations that promote Raf activation in a dimerization-independent manner (Yao et al. 2015).

POSTSIGNALING INACTIVATION OF RAF

Once Raf has been activated and the signal transmitted to the downstream kinases MEK and ERK, it is critical that the Rafs return to an inactive state, as constitutive activation can result in tumorigenesis. Several mechanisms that contribute to Raf inactivation have been identified. As expected, activating phosphorylation sites must be dephosphorylated, and protein phosphatase 5 (PP5) has been reported to dephosphorylate the critical S338 N-region site of C-Raf (von Kriegsheim et al. 2006). Raf autophosphorylation has also been implicated as a mechanism for inactivating the Rafs. In particular, a report examining inhibitor-bound B-Raf dimers has identified autoinhibitory sites in the ATP-binding P-loop (Holderfield et al. 2013), and a phosphoproteomic study characterizing the V600E-B-Raf mutant has identified an autoinhibitory site in the AS (S614-B-Raf) (Dernayka et al. 2016). Further studies will be needed to determine the prevalence and contribution of these autophosphorylation events in the context of Ras-mediated signaling.

Another important mechanism that attenuates the activity of all Raf family members involves a negative feedback loop in which active ERK phosphorylates the Rafs on multiple S/TP sites (Brummer et al. 2003; Dougherty et al. 2004; Ritt et al. 2010). Phosphorylation of the S/TP sites disrupts the interaction with Ras as well as Raf dimerization, and results in Raf proteins that are signaling incompetent (Dougherty et al. 2004; Ritt et al. 2010). Of note, some of these S/TP sites are also phosphorylated by the c-Jun amino-terminal kinases (JNKs) to prevent Raf signaling in times of cellular stress, and this stress-mediated desensitization of the Rafs can be induced by certain cancer therapeutics, such as rigosertib and paclitaxel (Ritt et al. 2016).

Ultimately, the postsignaling hyperphosphorylated Raf proteins must be recycled to a presignaling inactive state that is competent to respond to a new round of activating signals. Dephosphorylation of the ERK target sites is mediated by PP2A in a manner that requires the Pin1 prolyl-isomerase (Dougherty et al. 2004). More specifically, the WW-domain of Pin1 binds to the pS/TP sites, allowing Pin1 to isomerize the pS/T–proline bond to a *trans* conformation that can then be dephosphorylated by PP2A. Whether PP2A also serves to dephosphorylate the sites of autophosphorylation is currently unknown; however, once dephosphorylated, the Raf proteins can then reestablish the intramolecular contacts that mediate the autoinhibited state. Finally, with the rephosphorylation of the CR2 site and the rebinding of 14-3-3 to this site, the Raf kinases are returned to their presignaling, inactive conformation.

OTHER DIMER PARTNERS: KSR

Members of the KSR family are close relatives of the Raf kinases, and like the mechanisms mediating Raf activation, the function of KSR in Ras signaling is complex. KSR was first discovered as a positive modulator of the Ras pathway through genetic studies performed in *Drosophila* and *Caenorhabditis elegans* (Kornfeld et al. 1995; Sundaram and Han 1995; Therrien et al. 1995) and two KSR proteins (KSR1 and KSR2) are present in mammalian cells. The KSR family shares significant sequence homology with the Rafs, and like the Rafs, have a carboxy-terminal kinase domain and an amino-terminal domain that contains a CRD followed by a region rich in serine/threonine residues (Therrien et al. 1995). However, the KSRs do not have an RBD, but instead contain a coiled-coil fused to a sterile-α motif (CC-SAM) domain and a region rich in proline residues (Therrien et al. 1995; Koveal et al. 2012). Although it was initially thought that the KSRs would function as typical protein kinases, the mammalian KSR proteins were found to lack a lysine residue that is normally required for the phosphotransfer reaction (Therrien et al. 1995). Moreover, when residues known to be critical for enzyme catalysis were mutated in *C. elegans* or *Drosophila* KSR, little to no effect on the biological function of these KSR family members was observed (Stewart et al. 1999; Roy et al. 2002), raising the question of whether the KSRs indeed possess intrinsic kinase activity—a question that is still debated.

Subsequently, reports emerged indicating that the KSR proteins may have scaffolding activities, interacting with the three core components of the ERK cascade (Morrison 2001). In quiescent cells, KSR is found in a multiprotein complex containing MEK, 14-3-3, and components of the Hsp90 complex (Xing et al. 1997; Denouel-Galy et al. 1998; Yu et al. 1998; Cacace et al. 1999; Stewart et al. 1999). Similar to the Rafs, binding of 14-3-3 to sites in the KSR amino-terminal domain sequesters the inactive KSR complex in the cytosol (Müller et al. 2001; Ory et al. 2003b), and as with B-Raf, the kinase domains of KSR and MEK interact in a face-to-face manner (Brennan et al. 2011). On signal activation, KSR rapidly translocates to the plasma membrane coincident with the dephosphorylation of one of the 14-3-3 binding sites (Ory et al. 2003b). The CRD and the CC-SAM domain are critical for membrane recruitment of KSR (Zhou et al. 2002; Koveal et al. 2012), and at the membrane, KSR serves to increase the local pool of MEK, thus facilitating MEK activation. Moreover, KSR proteins can also form side-to-side dimers with the Rafs (Rajakulendran et al. 2009; Brennan et al. 2011), and if KSR is phosphorylated on residues analogous to the Raf N-region sites, it can function as an allosteric activator of Raf (Hu et al. 2013). Importantly, the KSR proteins also contain FxFP docking sites for activated ERK (Jacobs et al. 1999), and binding of ERK to the membrane-localized KSR complexes facilitates the phosphorylation of KSR and the Rafs on S/TP sites (McKay et al. 2009). Phosphorylation of these feedback sites disrupts the signaling complexes and promotes the release of KSR and the Rafs from the cell surface (McKay et al. 2009). Thus, KSR can modulate the dynamics of Ras pathway signaling by both potentiating and attenuating Raf activity and signal transmission to MEK and ERK.

CONCLUDING COMMENTS

The past decades have witnessed tremendous advances in our understanding of the Raf activation/inactivation process (Fig. 4). The early years of research were focused on elucidating the biochemical properties of the Rafs, determining their subcellular localization, phosphorylation state, and binding partners. Subsequently, the pace of discovery accelerated with the characterization of disease-associated mutations and the analysis of Raf crystallographic structures. Although much has been learned, outstanding questions still remain. In particular, it is unclear what factors determine Raf dimerization preferences and whether these preferences change depending on the signaling environment. Moreover, the structural features of a full-length Raf protein are unknown as are the structural changes induced by Ras binding and membrane interactions. The ultimate goal is that with further insight regarding the complexities of Raf regulation, new strategies for therapeutic intervention will emerge for targeting Ras-mediated Raf activation in human disease states.

REFERENCES

Abraham D, Podar K, Pacher M, Kubicek M, Welzel N, Hemmings BA, Dilworth SM, Mischak H, Kolch W, Baccarini M. 2000. Raf-1-associated protein phosphatase 2A as a positive regulator of kinase activation. *J Biol Chem* **275:** 22300–22304.

Baljuls A, Mahr R, Schwarzenau I, Muller T, Polzien L, Hekman M, Rapp UR. 2011. Single substitution within the RKTR motif impairs kinase activity but promotes dimerization of RAF kinase. *J Biol Chem* **286:** 16491–16503.

Bondeva T, Balla A, Várnai P, Balla T. 2002. Structural determinants of Ras–Raf interaction analyzed in live cells. *Mol Biol Cell* **13:** 2323–2333.

Brennan DF, Dar AC, Hertz NT, Chao WC, Burlingame AL, Shokat KM, Barford D. 2011. A Raf-induced allosteric transition of KSR stimulates phosphorylation of MEK. *Nature* **472:** 366–369.

Brummer T, Naegele H, Reth M, Misawa Y. 2003. Identification of novel ERK-mediated feedback phosphorylation sites at the C-terminus of B-Raf. *Oncogene* **22:** 8823–8834.

Cacace AM, Michaud NR, Therrien M, Mathes K, Copeland T, Rubin GM, Morrison DK. 1999. Identification of constitutive and Ras-inducible phosphorylation sites of KSR: Implications for 14-3-3 binding, mitogen-activated protein kinase binding, and KSR overexpression. *Mol Cell Biol* **19:** 229–240.

Chaudhary A, King WG, Mattaliano MD, Frost JA, Diaz B, Morrison DK, Cobb MH, Marshall MS, Brugge JS. 2000. Phosphatidylinositol 3-kinase regulates Raf1 through Pak phosphorylation of serine 338. *Curr Biol* **10:** 551–554.

Chiloeches A, Mason CS, Marais R. 2001. S338 phosphorylation of Raf-1 is independent of phosphatidylinositol 3-kinase and Pak3. *Mol Cell Biol* **21:** 2423–2434.

Chong H, Guan K-L. 2003. Regulation of Raf through phosphorylation and N-terminus–C terminus interaction. *J Biol Chem* **278:** 36269–36276.

Chong H, Lee J, Guan KL. 2001. Positive and negative regulation of Raf kinase activity and function by phosphorylation. *EMBO J* **20:** 3716–3727.

Clark GJ, Drugan JK, Rossman KL, Carpenter JW, Rogers-Graham K, Fu H, Der CJ, Campbell SL. 1997. 14-3-3 ζ negatively regulates Raf-1 activity by interactions with the Raf-1 cysteine-rich domain. *J Biol Chem* **272:** 20990–20993.

Corbit KC, Trakul N, Eves EM, Diaz B, Marshall M, Rosner MR. 2003. Activation of Raf-1 signaling by protein kinase C through a mechanism involving Raf kinase inhibitory protein. *J Biol Chem* **278:** 13061–13068.

Cutler REJ, Stephens RM, Saracino MR, Morrison DK. 1998. Autoregulation of the Raf-1 serine/threonine kinase. *Proc Natl Acad Sci* **95:** 9214–9219.

Daum G, Eisenmann-Tappe I, Fries HW, Troppmair J, Rapp UR. 1994. The ins and outs of Raf kinases. *Trends Biochem Sci* **19:** 474–480.

Denouel-Galy A, Douville EM, Warne PH, Papin C, Laugier D, Calothy G, Downward J, Eychene A. 1998. Murine Ksr interacts with MEK and inhibits Ras-induced transformation. *Curr Biol* **8:** 46–55.

Dernayka L, Rauch N, Jarboui MA, Zebisch A, Texier Y, Horn N, Romano D, Gloeckner CJ, Kriegsheim A, Ueffing M, et al. 2016. Autophosphorylation on S614 inhibits the activity and the transforming potential of BRAF. *Cell Signal* **28:** 1432–1439.

Dhillon AS, Meikle S, Yazici Z, Eulitz M, Kolch W. 2002a. Regulation of Raf-1 activation and signalling by dephosphorylation. *EMBO J* **21:** 64–71.

Dhillon AS, Pollock C, Steen H, Shaw PE, Mischak H, Kolch W. 2002b. Cyclic AMP-dependent kinase regulates Raf-1 kinase mainly by phosphorylation of serine 259. *Mol Cell Biol* **22:** 3237–3246.

Dhomen N, Marais R. 2007. New insight into *BRAF* mutations in cancer. *Curr Opin Genet Dev* **17:** 31–39.

Diaz B, Barnard D, Filson A, MacDonald S, King A, Marshall M. 1997. Phosphorylation of Raf-1 serine 338-serine 339 is an essential regulatory event for Ras-dependent activation and biological signaling. *Mol Cell Biol* **17:** 4509–4516.

Diedrich B, Rigbolt KT, Roring M, Herr R, Kaeser-Pebernard S, Gretzmeier C, Murphy RF, Brummer T, Dengjel J. 2017. Discrete cytosolic macromolecular BRAF complexes exhibit distinct activities and composition. *EMBO J* **36:** 646–663.

Dougherty MK, Müller J, Ritt DA, Zhou M, Zhou XZ, Copeland TD, Conrads TP, Veenstra TD, Lu KP, Morrison DK. 2004. Regulation of Raf-1 by direct feedback phosphorylation. *Mol Cell* **17:** 215–224.

Dumaz N, Marais R. 2003. Protein kinase A blocks Raf-1 activity by stimulating 14-3-3 binding and blocking Raf-1 interaction with Ras. *J Biol Chem* **278:** 29819–29823.

Eisenhardt AE, Sprenger A, Roring M, Herr R, Weinberg F, Kohler M, Braun S, Orth J, Diedrich B, Lanner U, et al. 2016. Phospho-proteomic analyses of B-Raf protein complexes reveal new regulatory principles. *Oncotarget* **7:** 26628–26652.

Fabian JR, Daar IO, Morrison DK. 1993. Critical tyrosine residues regulate the enzymatic and biological activity of Raf-1 kinase. *Mol Cell Biol* **13:** 7170–7179.

Freeman AK, Ritt DA, Morrison DK. 2013. The importance of Raf dimerization in cell signaling. *Small GTPases* **4:** 180–185.

Garnett MJ, Rana S, Paterson H, Barford D, Marais R. 2005. Wild-type and mutant B-RAF activate C-RAF through distinct mechanisms involving heterodimerization. *Mol Cell* **20:** 963–969.

Ghosh S, Strum JC, Sciorra VA, Daniel L, Bell RM. 1996. Raf-1 kinase possesses distinct binding domains for phosphatidylserine and phosphatidic acid. *J Biol Chem* **271:** 8472–8480.

Haling JR, Sudhamsu J, Yen I, Sideris S, Sandoval W, Phung W, Bravo BJ, Giannetti AM, Peck A, Masselot A, et al. 2014. Structure of the BRAF–MEK complex reveals a kinase activity independent role for BRAF in MAPK signaling. *Cancer Cell* **26:** 402–413.

Hamilton M, Liao J, Cathcart MK, Wolfman A. 2001. Constitutive association of c-N-Ras with c-Raf-1 and protein kinase Cε in latent signaling modules. *J Biol Chem* **276:** 29079–29090.

Hatzivassiliou G, Song K, Yen I, Brandhuber BJ, Anderson DJ, Alvarado R, Ludlam MJ, Stokoe D, Gloor SL, Vigers G, et al. 2010. RAF inhibitors prime wild-type RAF to activate the MAPK pathway and enhance growth. *Nature* **464:** 431–435.

Heidecker G, Huleihel M, Cleveland JL, Kolch W, Beck TW, Lloyd P, Pawson T, Rapp UR. 1990. Mutational activation of c-*raf*-1 and definition of the minimal transforming sequence. *Mol Cell Biol* **10:** 2503–2512.

Heidorn SJ, Milagre C, Whittaker S, Nourry A, Niculescu-Duvas I, Dhomen N, Hussain J, Reis-Filho JS, Springer CJ, Pritchard C, et al. 2010. Kinase-dead BRAF and oncogenic RAS cooperate to drive tumor progression through CRAF. *Cell* **140:** 209–221.

Hekman M, Hamm H, Villar AV, Bader B, Kuhlmann J, Nickel J, Rapp UR. 2002. Associations of B- and C-Raf with cholesterol, phosphatidylserine, and lipid second messengers. *J Biol Chem* **277:** 24090–24102.

Herrmann C, Martin GA, Wittinghofer A. 1995. Quantitative analysis of the complex between $p21^{ras}$ and the Ras-binding domain of the human Raf-1 protein kinase. *J Biol Chem* **270:** 2901–2905.

Hibino K, Shibata T, Yanagida T, Sako Y. 2009. A RasGTP-induced conformational change in C-RAF is essential for accurate molecular recognition. *Biophys J* **97:** 1277–1287.

Holderfield M, Merritt H, Chan J, Wallroth M, Tandeske L, Zhai H, Tellew J, Hardy S, Hekmat-Nejad M, Stuart DD, et al. 2013. RAF inhibitors activate the MAPK pathway by relieving inhibitory autophosphorylation. *Cancer Cell* **23:** 594–602.

Hu J, Stites EC, Yu H, Germino EA, Meharena HS, Stork PJ, Kornev AP, Taylor SS, Shaw AS. 2013. Allosteric activation of functionally asymmetric RAF kinase dimers. *Cell* **154:** 1036–1046.

Jacobs D, Glossip D, Xing H, Muslin AJ, Kornfeld K. 1999. Multiple docking sites on substrate proteins form a modular system that mediates recognition by ERK MAP kinase. *Genes Dev* **13:** 163–175.

Jaumot M, Hancock JF. 2001. Protein phosphatases 1 and 2A promote Raf-1 activation by regulating 14-3-3 interactions. *Oncogene* **20:** 3949–3958.

King AJ, Sun H, Diaz B, Barnard D, Miao W, Bagrodia S, Marshall MS. 1998. The protein kinase Pak3 positively regulates Raf-1 activity through phosphorylation of serine 338. *Nature* **396:** 180–183.

Kornfeld K, Hom DB, Horvitz HR. 1995. The *ksr-1* gene encodes a novel protein kinase involved in Ras-mediated signaling in *C. elegans*. *Cell* **83:** 903–913.

Koveal D, Schuh-Nuhfer N, Ritt D, Page R, Morrison DK, Peti W. 2012. A CC-SAM, for coiled coil-sterile α motif, domain targets the scaffold KSR-1 to specific sites in the plasma membrane. *Sci Signal* **5:** ra94.

Lavoie H, Therrien M. 2015. Regulation of RAF protein kinases in ERK signalling. *Nature Rev Mol Cell Biol* **16:** 281–298.

Leevers SJ, Paterson HF, Marshall CJ. 1994. Requirement for Ras in Raf activation is overcome by targeting Raf to the plasma membrane. *Nature* **369:** 411–414.

Light Y, Paterson H, Marais R. 2002. 14-3-3 Antagonizes Ras-mediated Raf-1 recruitment to the plasma membrane to maintain signaling fidelity. *Mol Cell Biol* **22:** 4984–4996.

Luo Z, Diaz B, Marshall MS, Avruch J. 1997. An intact Raf zinc finger is required for optimal binding to processed Ras and for Ras-dependent Raf activation in situ. *Mol Cell Biol* **17:** 46–53.

Marais R, Light Y, Paterson HF, Marshall CJ. 1995. Ras recruits Raf-1 to the plasma membrane for activation by tyrosine phosphorylation. *EMBO J* **14:** 3136–3145.

Marshall CJ. 1996. Ras effectors. *Curr Opin Cell Biol* **8:** 197–204.

Mason CS, Springer CJ, Cooper RG, Superti-Furga G, Marshall CJ, Marais R. 1999. Serine and tyrosine phosphorylations cooperate in Raf-1, but not B-Raf activation. *EMBO J* **18:** 2137–2148.

McClendon CL, Kornev AP, Gilson MK, Taylor SS. 2014. Dynamic architecture of a protein kinase. *Proc Natl Acad Sci* **111:** E4623–E4631.

McKay MM, Ritt DA, Morrison DK. 2009. Signaling dynamics of the KSR1 scaffold complex. *Proc Natl Acad Sci* **106:** 11022–11027.

McPherson RA, Harding A, Roy S, Lane A, Hancock JF. 1999. Interactions of c-Raf-1 with phosphatidylserine and 14-3-3. *Oncogene* **18:** 3862–3869.

Michaud NR, Fabian JR, Mathes KD, Morrison DK. 1995. 14-3-3 is not essential for Raf-1 function: Identification of Raf-1 proteins that are biologically activated in a 14-3-3 and Ras-independent manner. *Mol Cell Biol* **15:** 3390–3397.

Morrison DK. 2001. KSR: A MAPK scaffold of the Ras pathway? *J Cell Sci* **114:** 1609–1612.

Morrison DK, Heidecker G, Rapp UR, Copeland TD. 1993. Identification of the major phosphorylation sites of the Raf-1 kinase. *J Biol Chem* **268:** 17309–17316.

Mott HR, Carpenter JW, Zhong S, Ghosh S, Bell RM, Campbell SL. 1996. The solution structure of the Raf-1 cysteine-rich domain: A novel ras and phospholipid binding site. *Proc Natl Acad Sci* **93:** 8312–8317.

Müller J, Ory S, Copeland T, Piwnica-Worms H, Morrison DK. 2001. C-TAK1 regulates Ras signaling by phosphorylating the MAPK scaffold, KSR1. *Mol Cell* **8:** 983–993.

Muslin AJ, Tanner JW, Allen PM, Shaw AS. 1996. Interaction of 14-3-3 with signaling proteins is mediated by the recognition of phosphoserine. *Cell* **84:** 889–897.

Nan X, Collisson EA, Lewis S, Huang J, Tamguney TM, Liphardt JT, McCormick F, Gray JW, Chu S. 2013. Single-molecule superresolution imaging allows quantitative analysis of RAF multimer formation and signaling. *Proc Natl Acad Sci* **110:** 18519–18524.

Nassar N, Horn G, Herrmann C, Scherer A, McCormick F, Wittinghofer A. 1995. The 2.2 Å crystal structure of the Ras-binding domain of the serine/threonine kinase c-Raf1 in complex with Rap1A and a GTP analogue. *Nature* **375:** 554–560.

Ory S, Zhou M, Conrads TP, Veenstra TD, Morrison DK. 2003a. Protein phosphatase 2A positively regulates Ras signaling by dephosphorylating KSR1 and Raf-1 on critical 14-3-3 binding sites. *Curr Biol* **13:** 1356–1364.

Ory S, Zhou M, Conrads TP, Veenstra TD, Morrison DK. 2003b. Protein phosphatase 2A positively regulates Ras signaling by dephosphorylating KSR1 and Raf-1 on critical 14-3-3 binding sites. *Curr Biol* **13:** 1356–1364.

Pandit B, Sarkozy A, Pennacchio LA, Carta C, Oishi K, Martinelli S, Pogna EA, Schackwitz W, Ustaszewska A, Landstrom A, et al. 2007. Gain-of-function RAF1 mutations cause Noonan and LEOPARD syndromes with hypertrophic cardiomyopathy. *Nat Genet* **39:** 1007–1012.

Plowman SJ, Muncke C, Parton RG, Hancock JF. 2005. H-Ras, K-Ras, and inner plasma membrane raft proteins operate in nanoclusters with differential dependence on the actin cytoskeleton. *Proc Natl Acad Sci* **102:** 15500–15505.

Poulikakos PI, Zhang C, Bollag G, Shokat KM, Rosen N. 2010. RAF inhibitors transactivate RAF dimers and ERK signalling in cells with wild-type BRAF. *Nature* **464:** 427–430.

Rajakulendran T, Sahmi M, Lefrancois M, Sicheri F, Therrien M. 2009. A dimerization-dependent mechanism drives RAF catalytic activation. *Nature* **461:** 542–545.

Rapp UR, Goldsborough MD, Mark GE, Bonner TI, Groffen J, Reynolds J, Stephenson FH Jr. 1983. Structure and biological activity of v-*raf*, a unique oncogene transduced by a retrovirus. *Proc Natl Acad Sci* **80:** 4218–4222.

Razzaque MA, Nishizawa T, Komoike Y, Yagi H, Furutani M, Amo R, Kamisago M, Momma K, Katayama H, Nakagawa M, et al. 2007. Germline gain-of-function mutations in *RAF1* cause Noonan syndrome. *Nat Genet* **39:** 1013–1017.

Ritt DA, Zhou M, Conrads TP, Veenstra TD, Copeland TD, Morrison DK. 2007. CK2 Is a component of the KSR1 scaffold complex that contributes to Raf kinase activation. *Curr Biol* **17:** 179–184.

Ritt DA, Monson DM, Specht SI, Morrison DK. 2010. Impact of feedback phosphorylation and Raf heterodimerization on normal and mutant B-Raf signaling. *Mol Cell Biol* **30:** 806–819.

Ritt DA, Abreu-Blanco MT, Bindu L, Durrant DE, Zhou M, Specht SI, Stephen AG, Holderfield M, Morrison DK. 2016. Inhibition of Ras/Raf/MEK/ERK pathway signaling by a stress-induced phospho-regulatory circuit. *Mol Cell* **64:** 875–887.

Rodriguez-Viciana P, Oses-Prieto J, Burlingame A, Fried M, McCormick F. 2006. A phosphatase holoenzyme comprised of Shoc2/Sur8 and the catalytic subunit of PP1 functions as an M-Ras effector to modulate Raf activity. *Mol Cell* **22:** 217–230.

Romano D, Nguyen LK, Matallanas D, Halasz M, Doherty C, Kholodenko BN, Kolch W. 2014. Protein interaction switches coordinate Raf-1 and MST2/Hippo signalling. *Nat Cell Biol* **16:** 673–684.

Rommel C, Radziwill G, Lovric J, Noeldeke J, Heinicke T, Jones D, Aitken A, Moelling K. 1996. Activated Ras displaces 14-3-3 protein from the amino terminus of c-Raf-1. *Oncogene* **12:** 609–619.

Roy S, Lane A, Yan J, McPherson R, Hancock JF. 1997. Activity of plasma membrane-recruited Raf-1 is regulated by Ras via the Raf zinc finger. *J Biol Chem* **272:** 20139–20145.

Roy F, Laberge G, Douziech M, Ferland-McCollough D, Therrien M. 2002. KSR is a scaffold required for activation of the ERK/MAPK module. *Genes Dev* **16:** 427–438.

Rushworth LK, Hindley AD, O'Neill E, Kolch W. 2006. Regulation and role of Raf-1/B-Raf heterodimerization. *Mol Cell Biol* **26:** 2262–2272.

Sarkozy A, Carta C, Moretti S, Zampino G, Digilio MC, Pantaleoni F, Scioletti AP, Esposito G, Cordeddu V, Lepri F, et al. 2009. Germline *BRAF* mutations in Noonan, LEOPARD, and cardiofaciocutaneous syndromes: Molecular diversity and associated phenotypic spectrum. *Hum Mutat* **30:** 695–702.

Schulte TW, Blagosklonny MV, Ingui C, Neckers L. 1995. Disruption of the Raf-1-Hsp90 molecular complex results in destabilization of Raf-1 and loss of Raf-1-Ras association. *J Biol Chem* **270:** 24585–24588.

Shaw AS, Kornev AP, Hu J, Ahuja LG, Taylor SS. 2014. Kinases and pseudokinases: Lessons from RAF. *Mol Cell Biol* **34:** 1538–1546.

Stancato LF, Chow YH, Owens-Grillo JK, Yem AW, Deibel MR Jr, Jove R, Pratt WB. 1994. The native v-Raf.hsp90.p50 heterocomplex contains a novel immunophilin of the FK506 binding class. *J Biol Chem* **269:** 22157–22161.

Stanton VP, Nichols DW, Laudano AP, Cooper GM. 1989. Definition of the human *raf* amino-terminal regulatory region by deletion mutagenesis. *Mol Cell Biol* **9:** 639–647.

Stewart S, Sundaram M, Zhang Y, Lee J, Han M, Guan KL. 1999. Kinase suppressor of Ras forms a multiprotein signaling complex and modulates MEK localization. *Mol Cell Biol* **19:** 5523–5534.

Stokoe D, Macdonald SG, Cadwallader K, Symons M, Hancock JF. 1994. Activation of Raf as a result of recruitment to the plasma membrane. *Science* **264:** 1463–1467.

Sundaram M, Han M. 1995. The *C. elegans ksr-1* gene encodes a novel Raf-related kinase involved in Ras-mediated signal transduction. *Cell* **83:** 889–901.

Tartaglia M, Gelb BD, Zenker M. 2011. Noonan syndrome and clinically related disorders. *Best Pract Res Clin Endocrinol Metab* **25:** 161–179.

Terai K, Matsuda M. 2005. Ras binding opens c-Raf to expose the docking site for mitogen-activated protein kinase kinase. *EMBO Rep* **6:** 251–255.

Therrien M, Chang HC, Solomon NM, Karim FD, Wassarman DA, Rubin GM. 1995. KSR, a novel protein kinase required for RAS signal transduction. *Cell* **83:** 879–888.

Thevakumaran N, Lavoie H, Critton DA, Tebben A, Marinier A, Sicheri F, Therrien M. 2015. Crystal structure of a BRAF kinase domain monomer explains basis for allosteric regulation. *Nat Struct Mol Biol* **22:** 37–43.

Tian T, Harding A, Inder K, Plowman S, Parton RG, Hancock JF. 2007. Plasma membrane nanoswitches generate high-fidelity Ras signal transduction. *Nat Cell Biol* **9:** 905–914.

Tran NH, Frost JA. 2003. Phosphorylation of Raf-1 by p21-activated kinase 1 and Src regulates Raf-1 autoinhibition. *J Biol Chem* **278:** 11221–11226.

Tran NH, Wu X, Frost JA. 2005. B-Raf and Raf-1 are regulated by distinct autoregulatory mechanisms. *J Biol Chem* **280:** 16244–16253.

Tzivion G, Luo Z, Avruch J. 1998. A dimeric 14-3-3 protein is an essential cofactor for Raf kinase activity. *Nature* **394:** 88–92.

Van Aelst L, Barr M, Marcus S, Polverino A, Wigler M. 1993. Complex formation between Ras and Raf and other protein kinases. *Proc Natl Acad Sci* **90:** 6213–6217.

Vojtek AB, Hollenberg SM, Cooper JA. 1993. Mammalian Ras interacts directly with the serine/threonine kinase Raf. *Cell* **74:** 205–214.

von Kriegsheim A, Pitt A, Grindlay GJ, Kolch W, Dhillon AS. 2006. Regulation of the Raf-MEK-ERK pathway by protein phosphatase 5. *Nat Cell Biol* **8:** 1011–1016.

Wan PT, Garnett MJ, Roe SM, Lee S, Niculescu-Duvaz D, Good VM, Jones CM, Marshall CJ, Springer CJ, Barford D, et al. 2004. Mechanism of activation of the RAF-ERK signaling pathway by oncogenic mutations of B-RAF. *Cell* **116:** 855–867.

Weber CK, Slupsky JR, Kalmes HA, Rapp UR. 2001. Active Ras induces heterodimerization of cRaf and BRaf. *Cancer Res* **61:** 3595–3598.

Wellbrock C, Karasarides M, Marais R. 2004. The RAF proteins take centre stage. *Nat Rev Mol Cell Biol* **5:** 875–885.

Williams JG, Drugan JK, Yi GS, Clark GJ, Der CJ, Campbell SL. 2000. Elucidation of binding determinants and functional consequences of Ras/Raf-cysteine-rich domain interactions. *J Biol Chem* **275:** 22172–22179.

Xing H, Kornfeld K, Muslin AJ. 1997. The protein kinase KSR interacts with 14-3-3 protein and Raf. *Curr Biol* **7:** 294–300.

Yao Z, Torres NM, Tao A, Gao Y, Luo L, Li Q, de Stanchina E, Abdel-Wahab O, Solit DB, Poulikakos PI, et al. 2015. BRAF mutants evade ERK-dependent feedback by different mechanisms that determine their sensitivity to pharmacologic inhibition. *Cancer Cell* **28:** 370–383.

Yeung K, Janosch P, McFerran B, Rose DW, Mischak H, Sedivy JM, Kolch W. 2000. Mechanism of suppression of the Raf/MEK/extracellular signal-regulated kinase pathway by the Raf kinase inhibitor protein. *Mol Cell Biol* **20:** 3079–3085.

Yu W, Fantl WJ, Harrowe G, Williams LT. 1998. Regulation of the MAP kinase pathway by mammalian KSR through direct interaction with MEK and ERK. *Curr Biol* **8:** 56–64.

Zhang BH, Guan KL. 2000. Activation of B-Raf kinase requires phosphorylation of the conserved residues Thr598 and Ser601. *EMBO J* **19:** 5429–5439.

Zhang XF, Settleman J, Kyriakis JM, Takeuchi-Suzuki E, Elledge SJ, Marshall MS, Bruder JT, Rapp UR, Avruch J. 1993. Normal and oncogenic $p21^{ras}$ proteins bind to the amino-terminal regulatory domain of c-Raf-1. *Nature* **364:** 308–313.

Zhang X, Gureasko J, Shen K, Cole PA, Kuriyan J. 2006. An allosteric mechanism for activation of the kinase domain of epidermal growth factor receptor. *Cell* **125:** 1137–1149.

Zhou M, Horita DA, Waugh DS, Byrd RA, Morrison DK. 2002. Solution structure and functional analysis of the cysteine-rich C1 domain of kinase suppressor of Ras (KSR). *J Mol Biol* **315:** 435–446.

Zimmermann S, Moelling K. 1999. Phosphorylation and regulation of Raf by Akt (protein kinase B). *Science* **286:** 1741–1744.

PI3K: A Crucial Piece in the RAS Signaling Puzzle

Agata Adelajda Krygowska and Esther Castellano

Centre for Cancer and Inflammation, Barts Cancer Institute, Queen Mary University of London, London EC1M 6BQ, United Kingdom

Correspondence: e.castellano-sanchez@qmul.ac.uk

RAS proteins are key signaling switches essential for control of proliferation, differentiation, and survival of eukaryotic cells. RAS proteins are mutated in 30% of human cancers. In addition, mutations in upstream or downstream signaling components also contribute to oncogenic activation of the pathway. RAS proteins exert their functions through activation of several signaling pathways and dissecting the contributions of these effectors in normal cells and in cancer is an ongoing challenge. In this review, we summarize our current knowledge about how RAS regulates type I phosphatidylinositol 3-kinase (PI3K), one of the main RAS effectors. RAS signaling through PI3K is necessary for normal lymphatic vasculature development and for RAS-induced transformation in vitro and in vivo, especially in lung cancer, where it is essential for tumor initiation and necessary for tumor maintenance.

The superfamily of small guanosine triphosphate (GTP)-binding proteins plays a pivotal role in cellular signal transduction. Within this superfamily, RAS proteins (HRAS, NRAS, and Kirsten rat sarcoma [KRAS]) are the key regulators of a large number of processes such as cell-cycle progression, cell division, and apoptosis. RAS cycles between an active, GTP-bound state (RAS-GTP), and an inactive, guanosine diphosphate (GDP)-bound state (RAS-GDP) (Buday and Downward 2008; Vigil et al. 2010; Rojas et al. 2011). Guanine nucleotide exchange factors (GEFs) catalyze the exchange of GDP for GTP, whereas GTPase-activating proteins (GAPs) increase the rate of GTP hydrolysis to GDP (Fig. 1) (Quilliam et al. 1995).

RAS proteins have >80% homology sequence and differ mostly at the carboxyl terminus hypervariable region (HVR), a stretch of 25 amino acids known as the CAAX box (C, cysteine; A, aliphatic amino-acid; X, any amino acid) (Parker and Mattos 2015). To become fully active, newly synthetized RAS proteins undergo a number of posttranslational modifications at the CAAX box, including the addition of a farnesyl group to the cysteine residue of the CAAX box, the cleavage of the three terminal amino residues, and, finally, the carboxymethylation of the new carboxy-terminal cysteine residue (Tamanoi et al. 1988). HRAS and NRAS undergo an additional palmitoylation modification (Dudler and Gelb 1996). All these modifications are required for binding to the lipid membrane and dictate membrane localization of RAS proteins to specific microdomains. Each RAS isoform localizes to a distinct, nonoverlapping plasma membrane microdomain, where it contacts a unique pool of effector and regulatory proteins (Parker and Mattos 2015). The hypothesis that each isoform recruits effector proteins to the membrane via lipid and structural reorganization (Li and Gorfe 2012; Zhou et al.

Cite this article as *Cold Spring Harb Perspect Med* doi: 10.1101/cshperspect.a031450

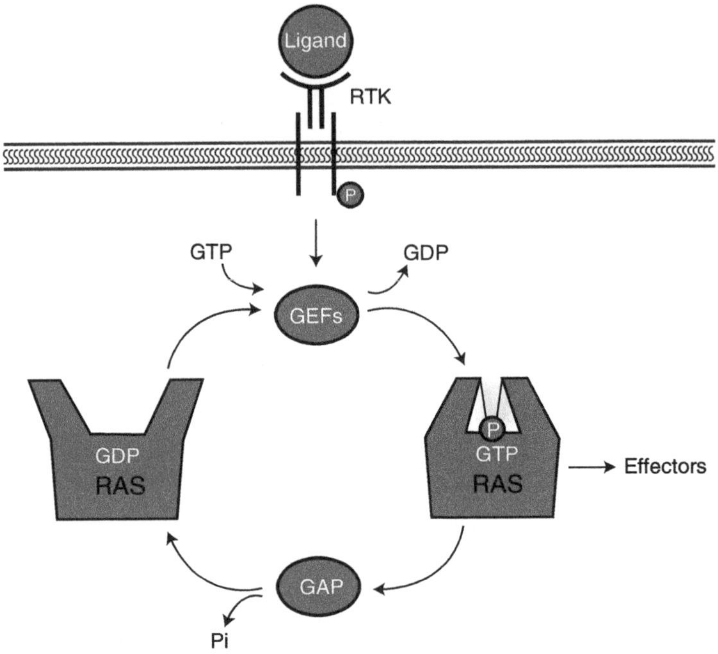

Figure 1. RAS activation/deactivation cycle by guanine nucleotide exchange factors (GEFs) and guanosine triphosphate (GTP)ase-activating proteins (GAPs). RAS proteins cycle between an inactive (guanosine diphosphate [GDP]-bound) and an active (GTP-bound) state. After receptor tyrosine kinase (RTK) engagement, RAS are activated by GEFs, which stimulate the exchange of GDP for GTP, and inactivated by GAPs that bind to RAS and stimulate its intrinsic GTPase activity.

2014) has emerged as an organizational scheme to produce overlapping but distinct signaling outputs from RAS nanoclusters. However, isoform-specific protein–lipid interaction and its connection to RAS-mediated signaling cascades have not yet been explored.

Wild-type RAS proteins are crucial for several physiological processes, but it is their involvement in cancer that has attracted the most attention (Land et al. 1983a,b; Ruley 1983; Trahey and McCormick 1987; Hirakawa and Ruley 1988). RAS proteins are mutated in ~30% of human cancers. Point mutations in certain amino acid residues, most commonly at positions G12, G13, and Q61, lead to the expression of constitutively active proteins (Bos 1989; Prior et al. 2012) that signal to downstream effectors even in the absence of extracellular stimuli (Barbacid 1987; Riely et al. 2009). Remarkably, these mutations predominantly affect the KRAS locus, with oncogenic KRAS mutations being detected in 25%–30% of all screened tumor samples (Forbes et al. 2011). RAS oncoproteins have also been implicated in the development of cancer by signaling downstream with increased intensity or duration (Downward 2003). The importance of RAS signaling in cancer is accentuated by the incidence of mutations in many of its regulators or effector pathways.

Aberrant RAS signaling is also implicated in several developmental disorders known as cardio-facio-cutaneous diseases (i.e., neurofibromatosis-type 1, Costello syndrome, and Noonan syndrome) (Cox and Der 2010; Fernandez-Medarde and Santos 2011).

RAS EFFECTOR PATHWAYS

RAS proteins play a key part in multiple signaling networks, linking upstream signals to a broad set of downstream pathways that control cell-cycle progression (Coleman et al. 2004), growth, migration, apoptosis, and more (Fig. 2) (Fernandez-Medarde and Santos 2011).

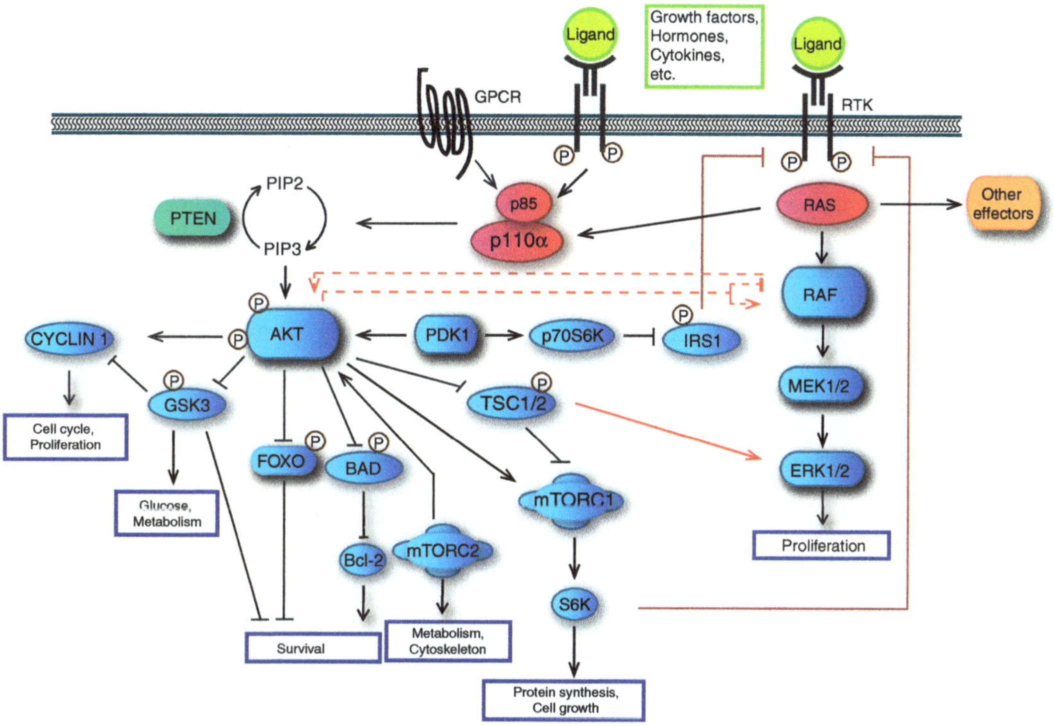

Figure 2. RAS effector pathways. Once activated, RAS proteins signal through multiple effector pathways, thus activating many different signal transduction pathways, such as mitogen-activated protein kinase (MAPK) and phosphatidylinositol 3-kinase (PI3K). The RAS signaling network gets more complex because PI3K is also directly activated by receptor tyrosine kinases (RTKs) or G protein–coupled receptors (GPCRs). Besides, there are several points in which the MAPK and PI3K pathways cross talk, leading to either activation or inhibition at different levels. Raf can be inhibited by PI3K in different situations. PI3K pathway can also be activated or inhibited by Raf under certain conditions. Another level of connection of these two pathways involves target of rapamycin complex 1 (mTORC1) and insulin receptor substrate 1 (IRS1). When these proteins get inhibited, a feedback signal through RTKs is initiated that activates MAPK in a RAS-dependent manner, thus provoking extracellular signal-regulated kinase (ERK) and AKT phosphorylation and activation of both pathways. PTEN, Phosphatase and tensin homolog; PDK1, PI3K-dependent kinase 1; MEK1/2, mitogen-activated protein kinases 1 and 2.

The cross talk and relative balance between these pathways determines cellular responses (Fig. 2) (Rajalingam et al. 2007; Stites and Ravichandran 2009).

RAS effectors share little sequence homology but all contain a RAS-binding domain (RBD) that shows a ubiquitin fold (Herrmann 2003). Recent nuclear magnetic resonance (NMR) studies provide evidence that RAS prefers certain effectors more than others in the presence of multiple binding partners. This preference changes with oncogenic mutations such as G12V, explaining how small changes in RAS affinity could substantially modify the outcome of integrated RAS signaling networks (Marshall et al. 2012a; Smith and Ikura 2014). The enhancement of specific effector pathways plays a critical role in maintaining an appropriate biological response (Morrison and Davis 2003). The specificity in RAS-induced signaling is primarily determined by the balance between RAS affinity for each of its effectors and the local concentrations of those effectors (Rodriguez-Viciana et al. 2004). Scaffold proteins also guide activation of specific effector pathway(s) (Li et al. 2000; Elion 2001).

In humans, there are ~100 genes containing putative RAS-association (RA) or RBDs, although some of these proteins might not actually bind to RAS with physiologically relevant affinities. Actual interactions of RAS with about 20 effectors have been confirmed by direct affinity measurements (Kiel et al. 2005; Wohlgemuth et al. 2005).

One of the first identified mammalian effectors of RAS was Raf (Fig. 2) (Moodie et al. 1993; Warne et al. 1993). There are three closely related RAF proteins, C-Raf1, B-Raf, and A-Raf, which are known to be activated by RAS-GTP (Roskoski 2010). Activated Raf phosphorylates and activates downstream targets such as mitogen-activated protein kinases 1 and 2 (MEK1 and MEK2). MEK1 and MEK2 then phosphorylate and activate other mitogen-activated protein kinases (MAPKs), such as extracellular signal-regulated kinases 1 and 2 (ERK1 and ERK2) (Alessi et al. 1994). The substrates for ERK1/2 include both nuclear and cytosolic proteins, of which transcription factors, such as ELK1 and c-Jun, have been widely studied (Murphy and Blenis 2006; Yoon and Seger 2006). Activation of all of these transcription factors promotes cell-cycle progression and proliferation of the cells (Pruitt and Der 2001).

In addition to the RAF/MAPK effector pathway, RAS can also interact with phosphatidyl inositol 3-kinases (PI3Ks) (Rodriguez-Viciana et al. 1994; Pacold et al. 2000). PI3K activity is a necessity for the activation of Akt family Ser/Thr kinases, which, in turn, are necessary in the inhibition of apoptosis and the promotion of cell survival (Khwaja et al. 1997; Datta et al. 1999). Other PI3K targets include the Akt-activating kinase PDK1 (PI3K-dependent kinase 1) and GEFs that target members of the Rho family of RAS superfamily GTPases (Rodriguez-Viciana et al. 1996a; Downward 1997, 2004; Hamad et al. 2002; Karnoub and Weinberg 2008; Yuan and Cantley 2008; Zhao and Vogt 2008). The functional importance of the RAS/PI3K signaling pathway will be more extensively discussed in the next sections.

The third class of well-known effector of RAS includes three exchange factors for RAS-related RAL proteins: RAL guanine nucleotide dissociation stimulator (RALGDS), RGL2/RLF, and RALGDS-like protein (RGL/RSB2) (Hofer et al. 1994; Feig et al. 1996; Takaya et al. 2007). RAL proteins regulate vesicular trafficking within the cell and nuclear factor (NF)-κB activation.

Other proteins known to directly interact with RAS-GTP are PLC, AF6, TIAM1, RIN, NORE1, among others (Castellano and Downward 2010; Stephen et al. 2014). The functional significance of many of these interactions has not been properly determined yet.

PI3K SIGNALING PATHWAY

PI3Ks are members of a conserved family of lipid kinases that are split in three classes according to their substrate preference and sequence homology (Vanhaesebroeck et al. 2010b). The class I PI3Ks (the most studied in cancer) comprises p110α (encoded by *PIK3CA*), p110β (*PIK3CB*), p110γ (*PIK3CG*), and p110δ (*PIK3CD*). p110 subunits were originally divided into a class IA group (p110α, p110β, and p110δ), which bind the p85 type of regulatory subunit, and a class IB group (p110γ), which binds to either p101 or p87. Class II PI3Ks includes PI3K-C2α, PI3K-C2β, and PI3K-C2γ. "Signals feeding into vacuolar protein sorting 34 (VPS34), the only class III PI3K, are becoming apparent, but its physiological importance is unclear" (Vanhaesebroeck et al. 2010a).

Dissecting the role of RAS proteins in the activation of PI3K has been particularly complicated because these kinases can be activated by multiple upstream signals, both dependent and independent of RAS proteins: G protein–coupled receptors (GPCRs) and receptor tyrosine kinases (RTKs) can activate PI3K independently of RAS by its binding to p85. However, RTKs can also bind to RAS, which in turn bind to p110α subunit and activate PI3K signaling (Fig. 2) (Hemmings and Restuccia 2015). In addition, although H-, N-, and KRAS activate p110α and p110γ subunits of PI3K (Rodriguez-Viciana et al. 2004), the closely related R-RAS proteins activate the p110α, p110γ, and p110δ subunits (Marte et al. 1997), and p110β is regulated by the related proteins Rac1 and Cdc42 (Yang et al. 2012; Fritsch et al. 2013).

RAS-dependent activation of PI3K requires a cascade of events that start with the dimerization of RTKs in response to growth factor stimulation. This leads to autophosphorylation of the RTKs at tyrosine residues and activates Src homology 2 (SH2) domain-containing molecules, like GRB2. The signaling cascade then proceeds through Son of Sevenless (SOS) onto RAS, which then interacts directly with the RBD in the p110 subunit of PI3K and activates it. This process does not depend on the p85 regulatory subunit of PI3K, although GRB2 can be found in a complex with a scaffolding protein (e.g., GAB), which interacts with p85, enhancing the chance of p110 activation (Ong et al. 2001; Castellano and Downward 2010).

When active, PI3Ks phosphorylate phosphatidylinositol (4,5)-bisphosphate (PI-4,5-P_2) and phosphatidylinositol (4)-phosphate (PI(4)P) on the inositol 3′-hydroxyl group, thereby generating phosphatidylinositol (3,4,5)-trisphosphate (PIP3) and phosphatidylinositol (3,4)-bisphosphate (PI(3,4)P2), respectively (Fig. 2). An important second messenger is PIP3, which binds to pleckstrin homology (PH) domains of effector proteins, causing their activation and/or translocation (Zhao et al. 2006; Fritsch and Downward 2013; Cescon et al. 2015). These lipids act on different pathways often through activation of either the PDK1, the Ser/Thr kinase Akt, or the mammalian target of rapamycin complex 1 (mTORC1) (Martini et al. 2014; Gyori et al. 2017). Akt regulates prosurvival signaling pathways by mediating various cellular processes such as epithelial to mesenchymal transition, glycogen metabolism, autophagy, and glucose uptake, whereas mTORC1 stimulates protein synthesis by phosphorylating the eukaryotic initiation factor 4E and the ribosomal S6 protein, among others. Although mTORC1 relays signals following PI3K/Akt activation, a second mTOR complex, mTORC2, contributes to complete Akt activation by phosphorylating Akt on serine 473. Activation of the mTORC1 target S6 kinase negatively feeds back to decrease PI3K activation (Polivka and Janku 2014; Mayer and Arteaga 2016).

Negative regulation of PI3K is accomplished primarily via action of the phosphatase and tensin homolog (PTEN) deleted on chromosome ten. PTEN encodes a lipid and protein phosphatase whose primary lipid substrate is PIP3. Mutations within the phosphatase domain of PTEN are frequent in many human cancers and impair its function resulting in elevated Akt activity and abnormal cell proliferation (Vanhaesebroeck et al. 2012; Martini et al. 2014).

A common feature of cellular signaling systems is feedback control, and MEK and PI3K signaling networks provides many examples. PI3K/Akt and RAS/MAPK pathways influence each other, both negatively and positively, at different stages of signal propagation resulting in dynamic and complex cross talk whose coordinated action determines the cell fate (Fig. 2). Such cross talk is often revealed when one pathway is chemically blocked, thereby releasing the cross-inhibition and effectively activating the other pathway (Castellano and Downward 2010; Mendoza et al. 2011; Aksamitiene et al. 2012; Fruman and Rommel 2014).

Oncogenic compensation can severely limit the anticancer efficacy of PI3K/Akt/mTOR inhibitors. Conversely, active PI3K signaling is a central mechanism of resistance to various targeted therapies (Tan and Yu 2013; Wood 2015). An important consequence of this feedback is that inhibitors of different components of the PI3K signaling pathway can cause elevated expression and activity of growth factor receptors, which leads to increased PI3K activity and RAS signaling, in addition to alternative survival pathways in cancer cells (Chandarlapaty et al. 2011; Rodrik-Outmezguine et al. 2011). There are multiple potential strategies to overcome the "rebound" signaling that occurs in response to PI3K/Akt/mTOR inhibitors. These include vertical inhibition of several signaling nodes and combination approaches.

RAS–PI3K INTERACTION: STRUCTURE AND BINDING INTERFACES

RAS-GTP binds to class I PI3K through its RBD, present in the amino terminal; p110α, p110δ, and p110γ are each stimulus-dependent RAS effectors (Vanhaesebroeck et al. 2010a). In contrast, the p110β RBD interacts with Rab5 GTPase (Christoforidis et al. 1999) and Rho GTPase family members (Fritsch et al. 2013) and, specifically, Rac potentiates p110β GPCR responses (Guillermet-Guibert et al. 2008; Dbouk et al. 2012; Fritsch et al. 2013). Class II PI3Ks also harbor an RBD, although these signaling inputs are not well characterized; the RBD domain is not present in class III PI3K (Vanhaesebroeck et al. 2010a; Thorpe et al. 2015).

RAS isoforms HRAS, NRAS, and KRAS share 100% sequence identity in the effector lobe and interact with effectors and regulators via conserved switch I and switch II regions. The switch I and II regions change conformation on GTP binding and serve as binding sites for a diverse array of downstream effectors.

PI3K's RBD is a small globular domain of ~100 residues in length that consists of a five-stranded mixed β sheet (Rb1-Rb5) flanked by two α helices (Rα1 and Rα2) (Walker et al. 1999; Pacold et al. 2000). RAS complexes with RBDs that come from different effectors form a similar model in which β sheets from RAS and the RBD align to form a single β-sheet connecting both proteins. Contacts between the switch I region of RAS and the RBD stabilize the interaction and ensure its dependence on RAS-GTP. Despite the commonalities of different RAS–RBD interactions, RAS discriminates among its effectors by rotating with respect to the RBD of a given effector. In the case of PI3Kγ, the length of the Rα1 helix and the size of the subsequent loop cause a significant rotation of RAS relative to the PI3Kγ RBD and results in switch II interactions for which there are no equivalents in the other RAS effector RBDs (Pacold et al. 2000; Djordjevic and Driscoll 2002). In addition to the allosteric effect, RAS orients the PI3K molecule in a way that allows it to more efficiently access its lipid substrate inserted in the membrane (Herrmann 2003).

Until recently, little was known about the considerably weaker interactions between RAS-GDP and RAS effectors. A double mutant of c-Raf N71R/A85K designed to bind RAS-GDP with a 100-fold increased affinity made it possible to solve for the first time the structure of a RAS-GDP/effector complex (Filchtinski et al. 2010). Surprisingly, this revealed that when bound to the N71R/A85K Raf mutant the switch I region of RAS-GDP is found in a conformation similar to that of RAS-GTP, and not RAS-GDP. Moreover, the structure indicates an increased mobility of the switch I region as shown by increased B factors. This increased flexibility compared with that of the same loop in RAS-GTP likely explains the low affinity of RAS effectors toward RAS-GDP.

RAS/PI3K SIGNALING IN HEALTH

After nearly 20 years of research, we are now getting a clearer picture of how and when RAS proteins mediate PI3K activation in normal and cancer cells. The use of RAS mutants that are defective in PI3K binding, but retain other effector functions, has significantly added to our knowledge of the specific functional role of RAS/PI3K signaling. Suire et al. (2006) generated a mouse model with a mutated p110γ RBD in which neutrophils presented decreased production of PI(3,4)P2 and PIP3, decreased activation of Akt, and diminished chemotaxis compared with wild-type cells. Additionally, the production of reactive oxygen species in response to agonist stimulation was decreased in p110γ RBD mutant mice. This revealed a significant role for the interaction in the normal function of neutrophils (Suire et al. 2006).

Further evidence of the importance of the RAS–PI3K interaction in vivo has been provided by genetic mouse models in which the RBD of p110α has two point mutations (T208D and K227A) that prevent its interaction with RAS without affecting its enzymatic activity (referred to herein as RBD-mutant model) (Gupta et al. 2007). Homozygous RBD-mutant animals often died shortly after birth owing to defects in the branching and development of the lymphatic vasculature system; however, surviving pups reached adulthood with no further obvious defects (Gupta et al. 2007). Similar observations were seen in mice expressing a kinase-dead version of p110α, a modification that is embryonic lethal because of impaired angiogenic remodeling defects (Graupera et al. 2008). Furthermore, deletion of *Pik3ca* alleles exclusively in endothelial cells resulted in a similar outcome. Intact RAS/PI3K signaling was also found to be essential for proper angiogenesis in zebrafish (Liu et al. 2008).

The role of RAS/PI3K/Akt signaling in cell survival and prevention of apoptosis is well known and was long established (Cox and Der 2003; Downward 2004; Zhang et al. 2011). Activation of Akt leads to both inhibition of proapoptotic proteins and activation or antiapoptotic proteins. Thus, by increasing the ability for growth and decreasing the capacity for apoptosis, RAS/PI3K signaling supports tumorigenesis.

Autophagy is a process by which cytoplasmic components (macromolecules and organelles) are degraded by the lysosome. Autophagy is essential in physiological processes that range from adaptation to starvation, cell differentiation and development, tumor suppression, innate and adaptive immunity, life span extension, and cell death (Mizushima 2007; Mizushima and Levine 2010; Wang et al. 2011; Boya et al. 2013; Nezis et al. 2014). The cross talk between RAS and autophagy is highly complex. RAS has been shown to be both a positive and negative regulator of autophagy, depending on the cell type, cellular context, and effector activation (Schmukler et al. 2014). However, it is clear that RAS inhibition of autophagy is controlled through PI3K activation and subsequent inhibition of the ULK1/Atg13/FIP200 complex (Furuta et al. 2004; Berry and Baehrecke 2007; Shao et al. 2016).

RAS signaling through PI3K is essential for hematopoiesis. The first evidence that RAS regulates eritropoiesis came from studies using primary fetal liver erythroid progenitors, in which knockout of KRAS impaired Akt activation and delayed erythroid differentiation (Zhang and Lodish 2005). Similar observations in zebrafish confirmed that KRAS knockout resulted in hematopoietic and angiogenic defects, including impaired expression of the erythroid-specific genes *gata1* and *βe3-hemoglobin*, reduced blood circulation and disorganized blood vessels (Liu et al. 2008). Recently, the role of RAS–PI3K interaction in hematopoiesis has been shown in a mouse model of myeloid leukemia in which RAS/PI3K signaling is necessary for maturation and maintenance of erythroblasts (Gritsman et al. 2014).

RAS/PI3K SIGNALING IN CANCER

Activating point mutations in the genes encoding RAS proteins contribute to the formation of a large proportion of human tumors. The repertoire of their downstream effectors, upstream activators, and molecular players regulating their signaling pathway are genetically modified in an even larger proportion of human tumors.

The most frequently mutated kinase in human cancer is P110α with particularly high incidences in cancers of the breast, the colon, or the endometrium; alterations are also found in tumors of the brain, liver, stomach, lung, and ovary (Ligresti et al. 2009). There are three hotspot mutations (E542K, E545K, and H1047R) that comprise 80% of all defects found in the *PIk3ca* gene. The E542 and E545K mutations map to the helical domain of p110α, whereas H1047R is located in the kinase domain. These mutations have been shown to increase proliferation and invasion of cells in vitro and in vivo by hyperactivating the downstream target Akt (Kang et al. 2005; Samuels et al. 2005; Zhao and Vogt 2010). Helical and kinase domain mutations induce gains-of-function by different molecular mechanisms. Activity of kinase domain mutants is RAS-independent, but requires p85 binding, whereas the helical domain mutants are p85-independent but their full activation requires RAS (Zhao and Vogt 2010). Pang et al. (2009) showed that expression of p110α E545K in a breast cancer cell line leads to a more severe metastatic phenotype than that induced by expressing p110α H1047R. On the other hand, H1047R, but not E545K, enhances HER2-mediated transformation of immortalized mammary epithelial cells (Chakrabarty et al. 2010). Recently, it was shown that p110α helical domain mutants bind to insulin receptor substrate 1 (IRS1) in the membrane, but the mutant proteins also require RAS-binding to fully exert their oncogenic function (Hao et al. 2013).

RTKs and RAS have been shown to cooperate in activating PI3K in KRAS mutant human colorectal (Ebi et al. 2011) and lung cancer cell lines (Molina-Arcas et al. 2013). Insulin-like growth factor 1 receptor (IGF-1R) was predominantly responsible for PI3K activation through binding to IRS adaptor proteins, with KRAS making a relatively minor contribution. It is likely that both RAS and receptor-mediated p110 activation is required to achieve PI3K activation levels high enough for tumor growth, which can be further amplified by loss of PTEN. The fact that alternative means of PI3K activation persist in the context of oncogenic KRAS may be caused by cancer cells losing their addiction to KRAS.

A useful insight about the importance mutant RAS interacting with endogenous p110α during tumor development has come from the use of KRAS mutants that do not bind PI3K but retain other effector functions. Shannon and colleagues showed that these mutants can still drive T-lineage acute lymphoblastic leukemias, with loss of PTEN compensating for the lack of RAS-p110α binding (Shieh et al. 2013).

Functional studies show that PI3K activation is vital for tumor initiation. Expression of a dominant-negative p85α lacking the p110-binding domain inhibited RAS-mediated transformation (Rodriguez-Viciana et al. 1997; Katso et al. 2001). Additionally, RBD-mutant mice are highly resistant to formation of lung and skin tumors induced by mutant KRAS (Gupta et al. 2007). Similarly, deletion of *Pik3r1* and *Pik3r2* abrogated $KRAS^{G12D}$-induced lung tumorigenesis (Engelman et al. 2008). Although PI3K activation may be necessary for KRAS-induced tumorigenesis, different studies have shown that inhibition of PI3K by itself is not sufficient to shrink established tumors in vivo or effectively treat *KRAS* mutated cancer cells in vitro (Engelman et al. 2008; Ihle et al. 2009). Experiments that use an inducible version of the RBD-mutant mice showed that disruption of RAS/PI3K in established lung tumors causes partial regression of tumors, followed by long-term stasis, but it is insufficient to cause complete regression of established tumors in the autochthonous setting, underscoring the differences between blocking tumorigenesis and eliminating established tumors.

Tumors harboring mutations in p110α also garner mutations in components of the PI3K pathway, like RAS or loss of PTEN (Cescon et al. 2015). Mutations of p110α and RAS are mutually exclusive in breast cancer, but often coexist in colorectal, endometrial, and lung cancers (Parsons et al. 2005; Oda et al. 2008; Janku et al. 2011; Chaft et al. 2012; Wang et al. 2014; Zhang et al. 2015). It is plausible that additional mutations enhance oncogenic transformation by increasing PI3K pathway activity levels beyond those caused by oncogenic RAS, thus activating additional downstream pathways. Recently, it was described that inhibition of p110α in colorectal cancer cell (CRC) lines harboring KRAS mutations or concomitant *KRAS/PIK3CA* mutations does not have the same effect. Functional responses to PI3K p110α silencing were distinct depending on the genetic background for *KRAS* and *PIK3CA*. *PIK3CA* silencing in CRC cells harboring KRAS/PIK3CA mutations induced apoptosis; in contrast, cell-cycle arrest at G_2/M and a concomitant reduction in the number of cells at G_1 phase was observed in CRC cells harboring KRAS mutations (Fernandes et al. 2016).

RAS/PI3K Signaling in the Tumor Stroma

RAS has been the prime example of a potent cell-autonomous oncogene for decades, but it is now evident that its effects stretch further to include non-cell-autonomous changes in the cellular microenvironment that have essential roles in both tumor initiation and progression.

The mechanisms by which RAS activation initiates and sustains proangiogenic processes are complex and rely on the modulation of endothelial growth factors and also on the increase of local inflammation and stromal remodeling (Pylayeva-Gupta et al. 2011). Blocking RAS signaling through PI3K in the tumor stroma impairs growth of tumors in which RAS/PI3K signaling is intact, as a consequence of inefficient angiogenesis (Murillo et al. 2014). Similar tumor growth impairment was observed in a p110α kinase-dead mouse model (Soler et al.

2013). Accordingly, PI3K inhibitors block tumor growth not only by targeting tumor cells but also by impairing tumor vasculature formation (Garlich et al. 2008; Schnell et al. 2008; Yuan et al. 2008; Lane et al. 2009; Soler et al. 2015).

Data obtained both in human cancer and transgenic mouse models indicate that RAS-driven tumors are able to overcome host-protecting adaptive immune responses (Clark et al. 2009). One mechanism by which oncogenic RAS may impair antitumor immune response is through recruitment of immunosuppressive regulatory T cells (Tregs) and myeloid-derived suppressor cells (MDSCs) (Clark et al. 2007; Tran Thang et al. 2010). Recent evidence suggests that PI3Ks play an important role in regulating the immune-suppressive, tumor-associated myeloid cell and Tregs, respectively (Gyori et al. 2017). Interestingly, RAS/PI3K signaling, either in the tumor (Castellano et al. 2013) or stromal compartment (Murillo et al. 2014), is required for macrophage recruitment, which may explain at least in part the decrease in tumor growth and angiogenesis observed on blocking the interaction between RAS and PI3K.

Remodeling of the extracellular matrix (ECM) is a key feature of cancer regulated by cancer-associated fibroblasts (CAFs), which secrete ECM components and ECM-modifying enzymes such as matrix metalloproteinases (MMPs). The different RAS isoforms control ECM synthesis through activation of Akt and transforming growth factor β (TGF-β) expression in a model of renal fibrosis (Fuentes-Calvo et al. 2012, 2013). PI3Ks also control the interaction between cancer cells and the tumor microenvironment (TME), namely, the production of MMP in normal and cancer cells (Chen et al. 2009; Awad et al. 2010; Kim et al. 2010; Hutti et al. 2012). In particular, MMP-14/CD44 binding results in activation of epidermal growth factor receptor (EGFR) and downstream engagement of PI3K signaling pathway in cancer cells, thus promoting tumor cell migration (Zarrabi et al. 2011).

Thus, RAS and PI3K are important players in the formation of a functional stroma that supports tumor growth and dissemination through a complex signaling network. A better understanding of the many ways by which PI3K contributes to the stromal remodeling activity of oncogenic RAS may uncover new avenues for therapeutic interventions targeting tumor stromal components.

RAS/PI3K Signaling in Migration and Metastasis Promotion

The most life-threatening consequence of an evolving tumor is the acquisition of metastatic properties. Many metastatic tumors (lung, pancreas, and colon) contain RAS mutations, which also confer metastatic properties to mouse cells in culture. The PI3K pathway has also been shown to exert a fundamental role regulating some of the steps of this process.

One of the first requirements for metastasis formation is the acquisition of a migratory phenotype accompanied by extensive remodeling of the actin cytoskeleton. The involvement of oncogenic RAS and PI3K in migration and remodeling of actin cytoskeleton has been well established through their interaction with Rac and Rho proteins both in mammalian cells and in the slime mold *Dictyostelium discoideum* (Campbell and Der 2004; Giehl 2005; Pollock et al. 2005; Sasaki and Firtel 2006; Castellano and Downward 2011; Devreotes and Horwitz 2015; Di Blasio et al. 2017). However, whether all these mechanisms are directly related to RAS signaling through PI3K is not clear.

RAS activation of Rac requires PI3K, and actin rearrangement correlates with the ability of RAS mutants to activate PI3K. Inhibition of PI3K activity blocks RAS induction of membrane ruffling, whereas activated PI3K is sufficient to induce membrane ruffling, acting through Rac (Han et al. 1998; Nimnual et al. 1998). Migration of endothelial cells is necessary for the creation of new vessels and repair of lesions, a process that is dependent on nitric oxide (NO). Signal transduction mediated by NO can trigger the process of cell adhesion and migration, releasing traction forces exerted by the endothelial cells on a surface and resulting in their spreading (Chatterjee et al. 2008). A recent report suggests that NO activates RAS, localizing PI3K to the plasma membrane and facilitating the interaction between PI3K and Rac, which is essential for the migration of endothelial cells (Eller-Borges et al. 2015).

RAS, through its interaction with PI3K, regulates migration of fibroblasts in response to several growth factors by regulating Rac activation and inhibiting Reelin expression (Castellano et al. 2016). When Reelin is expressed it prevents cells from migrating and up-regulates of E-cadherin, thereby impacting on cell–cell interaction and migration. Loss of E-cadherin is associated with increased cell migration and invasion (Kardash et al. 2010; Canel et al. 2013), a key event in the acquisition of invasive capacity, because reexpression of E-cadherin suppresses the invasion of tumor cells in vitro (Kardash et al. 2010; Canel et al. 2013). Mouse lung tumors in which the RAS–PI3K interaction is blocked display an increase in Reelin and E-cadherin expression and expression analysis in human samples showed that both lung adenocarcinoma and breast carcinoma patients whose tumors express high levels of Reelin have a better prognosis. These results suggest that inhibition of Reelin expression by RAS/PI3K signaling may be part of a mechanism by which cancer cells acquire the enhanced motility necessary to metastasize.

Another component of these mechanisms may be Sur8, a RAS-Raf scaffolding protein that also binds to p110α; this interaction is important for Sur8-mediated cell migration and invasion, along with tumor metastasis. Sur8 regulates migration and invasion of cells in response to RAS/PI3K signaling through activation of Rac and MMPs (Kaduwal et al. 2015; Lee et al. 2016).

Formation of metastasis also requires tumor cells to avoid anoikis or apoptosis caused by matrix deprivation (Paoli et al. 2013). Oncogenic RAS and PI3K can promote the loss of anchorage-dependent growth. For example, in Madin–Darby canine kidney (MDCK) epithelial cells, PI3K, but not RAF, is both necessary and sufficient for the protection provided by RAS from anoikis (Frisch et al. 1996; Khwaja et al. 1997). In tumors, PI3K signaling downstream from EGFR assists cancer cells in overcoming detachment-induced metabolic stress and, in turn, prevents programmed cell death (Schafer et al. 2009) and tumor cells growing in anchorage-independent conditions are addicted to PI3K signaling, and thus more sensitive to PI3K/mTOR inhibitors than matrix adherent cells (Muranen et al. 2012).

TARGETING THE RAS/PI3K PATHWAY IN CANCER THERAPY

For many years, oncogenic RAS proteins have been considered a key but "undruggable" cancer target because of the abundance of its substrate GTP, the high-affinity with which RAS binds it (Takashima and Faller 2013), and the apparent lack of suitable surfaces in critical regions necessary for small-molecule binding.

However, recently several groups have published on small molecule inhibitors of KRAS, which has restored hope that direct targeting of this protein can be achieved. Genentech identified a compound called DCAI that binds to a pocket located between the α2 helix and the β-sheet of KRAS4B, which is the binding site for the RAS guanidine exchange factor SOS1. In live-cell experiments, DCAI prevents membrane recruitment of the RBD of c-Raf kinase (Maurer et al. 2012). Independently, a group from Vanderbilt University discovered several molecules that weakly bind to GDP-loaded KRAS (Sun et al. 2012) into the same pocket as DCAI. Shokat and colleagues (Ostrem et al. 2013) screened a library of small molecules with GDP-bound $KRAS^{G12C}$, followed by the design and synthesis of analogs. The compounds discovered were found to bind to a pocket between the α2 and α3 helices, on the other side of the SII region from the Genentech and Vanderbilt compounds. As predicted, the compounds blocked the SOS1-mediated nucleotide exchange and decreased the binding of RAS to both BRAF and CRAF. It also seemed to selectively kill cancer cells harboring the G12C mutation (Ostrem et al. 2013). However, none of these compounds are very potent.

An effective RAS inhibitor must prevent the activation by RAS of its downstream binding partners, which could be achieved through direct inhibition of RAS GTPase activity, impairment of RAS localization or trafficking (Gysin et al. 2011; Wang et al. 2012; Bhattacharya et al. 2015; Singh et al. 2015; Ostrem and Shokat 2016; Asati et al. 2017). Recently, immunological approaches to target mutant RAS have also been investigated (Meyer et al. 2007; Rahma et al. 2014; Singh et al. 2015; Gil-Bazo et al. 2016).

An alternate approach to inhibit oncogenic RAS is to directly inhibit effector pathways downstream from RAS (Athuluri-Divakar et al. 2016; Keeton et al. 2017). However, the critical downstream effector for a given tumor may vary with the tissue type and even the specific mutational variant of RAS (Ihle et al. 2012; Alagesan et al. 2015).

Although several studies showed a role for PI3K signaling in RAS-mediated tumorigenesis (Rodriguez-Viciana et al. 1996b, 1997; Sheng et al. 2001; Li et al. 2004; Gupta et al. 2007; Castellano et al. 2013), PI3K pathway inhibitors used as single agents are not effective against RAS-driven cell lines and xenotransplantation models. Resistance to these inhibitors has been ascribed to negative feedback mechanisms and reflexive activation of other downstream signaling partners of RAS (Engelman et al. 2008; Ihle et al. 2009; Sos et al. 2009; Dan et al. 2010). Conversely, treating RAS mutant cancer cells with MEK inhibitors results in increased phosphorylation of the PI3K pathway effector Akt (Hoeflich et al. 2009; Mirzoeva et al. 2009; Sos et al. 2009). Thus, dual inhibition of both pathways may be required to block the growth of RAS-driven tumors. In support of this hypothesis, several studies showed that combined inhibition of MEK and PI3K pathway in RAS mutant background is superior to single-agent inhibition several in vitro and in vivo models, resulting in a synergistic inhibition of tumor growth, cell viability, and increased apoptosis (Engelman et al. 2008; Hoeflich et al. 2009; Mirzoeva et al. 2009; Sos et al. 2009; Castellano et al. 2013; Qu et al. 2014; Jifu et al. 2015; Schrauwen et al. 2015; Talbert et al. 2017). However, in a KRAS-driven pancreatic model combined therapy showed only modest activity (Alagesan et al. 2015), suggesting that there may be tissue and/or tumor specificity of the various downstream effector pathways of RAS and stressing the obvious need to better understand signaling downstream from oncogenic RAS in different tumor types.

Data from human cancers and transgenic mouse models indicate that RAS-driven cancers have the ability to overcome host-protecting adaptive immune responses (Clark et al. 2009). There is increasing appreciation for how small molecules targeting the cancer cell affect the immune context of the tumor (Vanneman and Dranoff 2012; Zitvogel et al. 2013). Extensive studies of the PI3K/Akt network in immune cells have shown that PI3K activation is not a simple on/off switch (Fruman and Bismuth 2009; Okkenhaug 2013; Okkenhaug et al. 2016). Inhibiting the pathway can either suppress immune responses or them through effects on diverse subsets of innate and adaptive immune cells and therefore in KRAS mutant cancers it may be possible to implement treatment regimens that increase immune rejection of tumors with concomitant direct antitumor effect. Proper assessment of candidate inhibitor compounds therefore requires testing in genetically engineered mouse models, with extensive monitoring of tumor infiltration and activity of diverse immune subsets including macrophages, T cells, and natural killer cells.

Tumor-immune infiltration studies have shown that macrophage recruitment to tumors is severely impaired in animals in which RAS cannot bind and activate PI3K (Castellano et al. 2013; Murillo et al. 2014). It will be crucial to determine which agents targeting PI3K/Akt enhance or suppress the efficacy of emerging cancer immunotherapies and vaccines. In mouse models, PI3K inhibitors can enhance the efficacy of immune-directed therapies (Marshall et al. 2012b). It is relevant to consider that isoform-selective agents minimize immune suppressive effects on lymphocytes compared with pan-class I inhibitors (So et al. 2013). Thus, pan-PI3K inhibitors are more likely to enforce or accelerate the immune exhaustion state, whereas isoform-selective PI3K are likely to be better suited for use in combination with cancer immunotherapy. Matching patients to the right combinations will require knowledge of the genomic driver and the immune fingerprint of the tumor.

Interruption of RAS-effector binding is also a promising strategy for drug design against RAS-driven cancers and other diseases associated with RAS. Strong support for the concept of disrupting activated RAS/effector protein complexes as a therapeutic strategy came from studies using a membrane-directed, single immunoglobulin antibody domain (iDab#6-memb). This antibody fragment binds the switch I and II domains of active HRAS predicted to be mutually exclusive of the amino-terminal RAS-binding domain of RAF1 (RAF-RBD), RALGDS, and PI3K. In a transgenic mouse model of KRAS-driven lung cancer, tumor initiation was dramatically reduced by expressing the antibody fragment (Tanaka and Rabbitts 2010). Although RAS-effector domain antibodies provide convincing proof-of-concept, isolated antibody fragments are not currently a viable clinical approach for disrupting intracellular signaling.

However, interfering with protein–protein interactions is, by its nature, difficult. Specifically, the binding of small molecules to protein–protein contact surfaces remains a major challenge (Scott et al. 2016). In the case of RAS and other GTPases, allosteric inhibition of protein–protein interactions may be a more plausible approach. Multiple lines of evidence point to this being the case for RAS. First, switch I acts as the main protein–protein interface, and it is known to be highly dynamic, undergoing substantial conformational rearrangements between nucleotide states. Second, mutations at sites outside the interface have been shown to affect the conformation in ways that impair effector binding. Finally, structural and NMR studies have enabled the identification of multiple inactive conformations. Compounds that bind to and stabilize these inactive states have the potential to potently and effectively inhibit RAS. One example of this kind of blockade comes from small molecules that bind to the pseudokinase "kinase suppressor of RAS" (KSR) (Brennan et al. 2011). The kinase inhibitor ASC24 binds to an inactive conformation of KSR and blocks the allosteric transition that is induced on heterodimerization of KSR with BRAF.

The switch region of RAS interacts with a great number of effector proteins that contain a common RBD. Because RBD-mediated interactions are necessary for RAS signaling, blocking RBD association with small molecules makes an attractive therapeutic course. Mutational studies of the RBD in the p110α catalytic subunit of PI3K have shown the importance of its RAS-binding domain in the establishment and maintenance of KRAS-driven lung tumors, validating this protein–protein interaction as a therapeutic target for RAS-driven cancers (Gupta et al. 2007; Castellano et al. 2013). Additionally, a recent report described an unexpected interaction between the Polo-like kinase inhibitor, Rigosertib, with several RAS effector proteins. Rigosertib acts as a RAS mimetic that binds to the RBDs of the RAS effectors and interferes with their ability to bind to RAS, resulting in the inhibition of RAS-RAF/MEK and PI3K/Akt signaling pathways. The compound is able to inhibit the interactions of both mutant and wild-type isoforms of K- or NRAS isoforms (Athuluri-Divakar et al. 2016). The real test for this drug will most likely be whether or not it has efficacy in patients with KRAS mutant tumors. This question will be solved relatively soon because multiple clinical trials (some of them in phase III) for hematologic and solid malignancies are ongoing at this time.

CONCLUDING REMARKS

RAS proteins activate multiple effector pathways. Dissecting the contributions of these interactions in normal cells and in cancer is an ongoing challenge. The role of RAS proteins in regulating PI3K has been particularly complicated, because these kinases can be activated by multiple upstream signals, some of which depend on cooperation with RAS proteins, whereas others do not. However, research over the last decade has shown, in a variety of systems, that this signaling pathway plays essential roles in both normal and oncogenic signaling.

Recent data has broadened our knowledge on a RAS/PI3K signaling pathway, suggesting that this pathway is equally active in cancer cells and in the major components of the TME, controlling not only tumor growth but also tumor-associated immune responses, tumor angiogenesis, tumor-induced ECM reorganization, and tumor dissemination. The ability of RAS/PI3K signaling to promote tumorigenesis by acting both in "seed" and "soil"-dependent mechanisms indicates that therapeutic PI3K inhibition can be seen as a plausible opportunity to attack RAS-driven tumors on multiple fronts. More research is needed in this front to understand the specific molecular mechanisms by which RAS/PI3K signaling exerts control over the host responses. Targeting pathways that depend on RAS/PI3K signaling, instead of direct components of the pathway, will avoid feedback loops and could be a way forward to treat RAS-driven cancers.

ACKNOWLEDGMENTS

We thank Claudio Santos and Fran Balkwill for assistance and feedback on this review. This work is funded by Queen Mary University of London. The authors have no conflicts of interest to declare.

REFERENCES

Aksamitiene E, Kiyatkin A, Kholodenko BN. 2012. Cross-talk between mitogenic Ras/MAPK and survival PI3K/Akt pathways: A fine balance. *Biochem Soc Trans* **40:** 139–146.

Alagesan B, Contino G, Guimaraes AR, Corcoran RB, Deshpande V, Wojtkiewicz GR, Hezel AF, Wong KK, Loda M, Weissleder R, et al. 2015. Combined MEK and PI3K inhibition in a mouse model of pancreatic cancer. *Clin Cancer Res* **21:** 396–404.

Alessi DR, Saito Y, Campbell DG, Cohen P, Sithanandam G, Rapp U, Ashworth A, Marshall CJ, Cowley S. 1994. Identification of the sites in MAP kinase kinase-1 phosphorylated by p74raf-1. *EMBO J* **13:** 1610–1619.

Asati V, Mahapatra DK, Bharti SK. 2017. *K-Ras* and its inhibitors towards personalized cancer treatment: Pharmacological and structural perspectives. *Eur J Med Chem* **125:** 299–314.

Athuluri-Divakar SK, Vasquez-Del Carpio R, Dutta K, Baker SJ, Cosenza SC, Basu I, Gupta YK, Reddy MV, Ueno L, Hart JR, et al. 2016. A small molecule RAS-mimetic disrupts RAS association with effector proteins to block signaling. *Cell* **165:** 643–655.

Awad AE, Kandalam V, Chakrabarti S, Wang X, Penninger JM, Davidge ST, Oudit GY, Kassiri Z. 2010. Tumor necrosis factor induces matrix metalloproteinases in cardiomyocytes and cardiofibroblasts differentially via superoxide production in a PI3Kγ-dependent manner. *Am J Physiol Cell Physiol* **298:** C679–C692.

Barbacid M. 1987. *ras* genes. *Annu Rev Biochem* **56:** 779–827.

Berry DL, Baehrecke EH. 2007. Growth arrest and autophagy are required for salivary gland cell degradation in *Drosophila*. *Cell* **131:** 1137–1148.

Bhattacharya S, Socinski MA, Burns TF. 2015. *KRAS* mutant lung cancer: Progress thus far on an elusive therapeutic target. *Clin Transl Med* **4:** 35.

Bos JL. 1989. *ras* oncogenes in human cancer: A review. *Cancer Res* **49:** 4682–4689.

Boya P, Reggiori F, Codogno P. 2013. Emerging regulation and functions of autophagy. *Nat Cell Biol* **15:** 713–720.

Brennan DF, Dar AC, Hertz NT, Chao WC, Burlingame AL, Shokat KM, Barford D. 2011. A Raf-induced allosteric transition of KSR stimulates phosphorylation of MEK. *Nature* **472:** 366–369.

Buday L, Downward J. 2008. Many faces of Ras activation. *Biochim Biophys Acta* **1786:** 178–187.

Campbell PM, Der CJ. 2004. Oncogenic Ras and its role in tumor cell invasion and metastasis. *Semin Cancer Biol* **14:** 105–114.

Canel M, Serrels A, Frame MC, Brunton VG. 2013. E-cadherin-integrin crosstalk in cancer invasion and metastasis. *J Cell Sci* **126:** 393–401.

Castellano E, Downward J. 2010. Role of RAS in the regulation of PI 3-kinase. *Curr Top Microbiol Immunol* **346:** 143–169.

Castellano E, Downward J. 2011. RAS interaction with PI3K: More than just another effector pathway. *Genes Cancer* **2:** 261–274.

Castellano E, Sheridan C, Thin MZ, Nye E, Spencer-Dene B, Diefenbacher ME, Moore C, Kumar MS, Murillo MM, Gronroos E, et al. 2013. Requirement for interaction of PI3-kinase p110α with RAS in lung tumor maintenance. *Cancer Cell* **24:** 617–630.

Castellano E, Molina-Arcas M, Krygowska AA, East P, Warne P, Nicol A, Downward J. 2016. RAS signalling through PI3-kinase controls cell migration via modulation of reelin expression. *Nat Commun* **7:** 11245.

Cescon DW, Gorrini C, Mak TW. 2015. Breaking up is hard to do: PI3K isoforms on the rebound. *Cancer Cell* **27:** 5–7.

Chaft JE, Arcila ME, Paik PK, Lau C, Riely GJ, Pietanza MC, Zakowski MF, Rusch V, Sima CS, Ladanyi M, et al. 2012. Coexistence of *PIK3CA* and other oncogene mutations in lung adenocarcinoma—Rationale for comprehensive mutation profiling. *Mol Cancer Ther* **11:** 485–491.

Chakrabarty A, Rexer BN, Wang SE, Cook RS, Engelman JA, Arteaga CL. 2010. H1047R phosphatidylinositol 3-kinase mutant enhances HER2-mediated transformation by heregulin production and activation of HER3. *Oncogene* **29:** 5193–5203.

Chandarlapaty S, Sawai A, Scaltriti M, Rodrik-Outmezguine V, Grbovic-Huezo O, Serra V, Majumder PK, Baselga J, Rosen N. 2011. AKT inhibition relieves feedback suppression of receptor tyrosine kinase expression and activity. *Cancer Cell* **19:** 58–71.

Chatterjee A, Black SM, Catravas JD. 2008. Endothelial nitric oxide (NO) and its pathophysiologic regulation. *Vascul Pharmacol* **49:** 134–140.

Chen JS, Wang Q, Fu XH, Huang XH, Chen XL, Cao LQ, Chen LZ, Tan HX, Li W, Bi J, et al. 2009. Involvement of PI3K/PTEN/AKT/mTOR pathway in invasion and metastasis in hepatocellular carcinoma: Association with MMP-9. *Hepatol Res* **39:** 177–186.

Christoforidis S, Miaczynska M, Ashman K, Wilm M, Zhao L, Yip SC, Waterfield MD, Backer JM, Zerial M. 1999. Phosphatidylinositol-3-OH kinases are Rab5 effectors. *Nat Cell Biol* **1:** 249–252.

Clark CE, Hingorani SR, Mick R, Combs C, Tuveson DA, Vonderheide RH. 2007. Dynamics of the immune reaction to pancreatic cancer from inception to invasion. *Cancer Res* **67:** 9518–9527.

Clark CE, Beatty GL, Vonderheide RH. 2009. Immunosurveillance of pancreatic adenocarcinoma: Insights from genetically engineered mouse models of cancer. *Cancer Lett* **279:** 1–7.

Coleman ML, Marshall CJ, Olson MF. 2004. RAS and RHO GTPases in G_1-phase cell-cycle regulation. *Nat Rev Mol Cell Biol* **5:** 355–366.

Cox AD, Der CJ. 2003. The dark side of Ras: Regulation of apoptosis. *Oncogene* **22:** 8999–9006.

Cox AD, Der CJ. 2010. Ras history: The saga continues. *Small GTPases* **1:** 2–27.

Dan S, Okamura M, Seki M, Yamazaki K, Sugita H, Okui M, Mukai Y, Nishimura H, Asaka R, Nomura K, et al. 2010. Correlating phosphatidylinositol 3-kinase inhibitor efficacy with signaling pathway status: In silico and biological evaluations. *Cancer Res* **70:** 4982–4994.

Datta SR, Brunet A, Greenberg ME. 1999. Cellular survival: A play in three Akts. *Genes Dev* **13:** 2905–2927.

Dbouk HA, Vadas O, Shymanets A, Burke JE, Salamon RS, Khalil BD, Barrett MO, Waldo GL, Surve C, Hsueh C, et al. 2012. G protein–coupled receptor-mediated activation of p110β by Gβγ is required for cellular transformation and invasiveness. *Sci Signal* **5:** ra89.

Devreotes P, Horwitz AR. 2015. Signaling networks that regulate cell migration. *Cold Spring Harb Perspect Biol* **7:** a005959.

Di Blasio L, Gagliardi PA, Puliafito A, Primo L. 2017. Serine/threonine kinase 3-phosphoinositide-dependent protein kinase 1 (PDK1) as a key regulator of cell migration and cancer dissemination. *Cancers (Basel)* **9:** E25.

Djordjevic S, Driscoll PC. 2002. Structural insight into substrate specificity and regulatory mechanisms of phosphoinositide 3-kinases. *Trends Biochem Sci* **27:** 426–432.

Downward J. 1997. Role of phosphoinositide-3-OH kinase in Ras signaling. *Adv Second Messenger Phosphoprotein Res* **31:** 1–10.

Downward J. 2003. Targeting RAS signalling pathways in cancer therapy. *Nat Rev Cancer* **3:** 11–22.

Downward J. 2004. PI 3-kinase, Akt and cell survival. *Semin Cell Dev Biol* **15:** 177–182.

Dudler T, Gelb MH. 1996. Palmitoylation of Ha-Ras facilitates membrane binding, activation of downstream effectors, and meiotic maturation in *Xenopus* oocytes. *J Biol Chem* **271:** 11541–11547.

Ebi H, Corcoran RB, Singh A, Chen Z, Song Y, Lifshits E, Ryan DP, Meyerhardt JA, Benes C, Settleman J, et al. 2011. Receptor tyrosine kinases exert dominant control over PI3K signaling in human *KRAS* mutant colorectal cancers. *J Clin Invest* **121:** 4311–4321.

Elion EA. 2001. The Ste5p scaffold. *J Cell Sci* **114:** 3967–3978.

Eller-Borges R, Batista WL, da Costa PE, Tokikawa R, Curcio MF, Strumillo ST, Sartori A, Moraes MS, de Oliveira GA, Taha MO, et al. 2015. Ras, Rac1, and phosphatidylinositol-3-kinase (PI3K) signaling in nitric oxide induced endothelial cell migration. *Nitric Oxide* **47:** 40–51.

Engelman JA, Chen L, Tan X, Crosby K, Guimaraes AR, Upadhyay R, Maira M, McNamara K, Perera SA, Song Y, et al. 2008. Effective use of PI3K and MEK inhibitors to treat mutant Kras G12D and PIK3CA H1047R murine lung cancers. *Nat Med* **14:** 1351–1356.

Feig LA, Urano T, Cantor S. 1996. Evidence for a Ras/Ral signaling cascade. *Trends Biochem Sci* **21:** 438–441.

Fernandes MS, Melo S, Velho S, Carneiro P, Carneiro F, Seruca R. 2016. Specific inhibition of p110α subunit of PI3K: Putative therapeutic strategy for *KRAS* mutant colorectal cancers. *Oncotarget* **7:** 68546–68558.

Fernandez-Medarde A, Santos E. 2011. Ras in cancer and developmental diseases. *Genes Cancer* **2:** 344–358.

Filchtinski D, Sharabi O, Ruppel A, Vetter IR, Herrmann C, Shifman JM. 2010. What makes Ras an efficient molecular switch: A computational, biophysical, and structural study of Ras-GDP interactions with mutants of Raf. *J Mol Biol* **399:** 422–435.

Forbes SA, Bindal N, Bamford S, Cole C, Kok CY, Beare D, Jia M, Shepherd R, Leung K, Menzies A, et al. 2011. COSMIC: Mining complete cancer genomes in the catalogue of somatic mutations in cancer. *Nucleic Acids Res* **39:** D945–D950.

Frisch SM, Vuori K, Ruoslahti E, Chan-Hui PY. 1996. Control of adhesion-dependent cell survival by focal adhesion kinase. *J Cell Biol* **134:** 793–799.

Fritsch R, Downward J. 2013. SnapShot: Class I PI3K isoform signaling. *Cell* **154:** 940–940.e1.

Fritsch R, de Krijger I, Fritsch K, George R, Reason B, Kumar MS, Diefenbacher M, Stamp G, Downward J. 2013. RAS and RHO families of GTPases directly regulate distinct phosphoinositide 3-kinase isoforms. *Cell* **153:** 1050–1063.

Fruman DA, Bismuth G. 2009. Fine tuning the immune response with PI3K. *Immunol Rev* **228:** 253–272.

Fruman DA, Rommel C. 2014. PI3K and cancer: Lessons, challenges and opportunities. *Nat Rev Drug Discov* **13:** 140–156.

Fuentes-Calvo I, Blazquez-Medela AM, Eleno N, Santos E, Lopez-Novoa JM, Martinez-Salgado C. 2012. H-Ras isoform modulates extracellular matrix synthesis, proliferation, and migration in fibroblasts. *Am J Physiol Cell Physiol* **302:** C686–C697.

Fuentes-Calvo I, Crespo P, Santos E, Lopez-Novoa JM, Martinez-Salgado C. 2013. The small GTPase N-Ras regulates extracellular matrix synthesis, proliferation and migration in fibroblasts. *Biochim Biophys Acta* **1833:** 2734–2744.

Furuta S, Hidaka E, Ogata A, Yokota S, Kamata T. 2004. Ras is involved in the negative control of autophagy through the class I PI3-kinase. *Oncogene* **23:** 3898–3904.

Garlich JR, De P, Dey N, Su JD, Peng X, Miller A, Murali R, Lu Y, Mills GB, Kundra V, et al. 2008. A vascular targeted pan phosphoinositide 3-kinase inhibitor prodrug, SF1126, with antitumor and antiangiogenic activity. *Cancer Res* **68:** 206–215.

Giehl K. 2005. Oncogenic Ras in tumour progression and metastasis. *Biol Chem* **386:** 193–205.

Gil-Bazo I, Sweet-Cordero A, Vicent S. 2016. The present and the future of the RAS pathway: From function and genomics to inhibition. *Cancer Biol Ther* **17:** 719–722.

Graupera M, Guillermet-Guibert J, Foukas LC, Phng LK, Cain RJ, Salpekar A, Pearce W, Meek S, Millan J, Cutillas PR, et al. 2008. Angiogenesis selectively requires the p110α isoform of PI3K to control endothelial cell migration. *Nature* **453:** 662–666.

Gritsman K, Yuzugullu H, Von T, Yan H, Clayton L, Fritsch C, Maira SM, Hollingworth G, Choi C, Khandan T, et al. 2014. Hematopoiesis and RAS-driven myeloid leukemia differentially require PI3K isoform p110α. *J Clin Invest* **124:** 1794–1809.

Guillermet-Guibert J, Bjorklof K, Salpekar A, Gonella C, Ramadani F, Bilancio A, Meek S, Smith AJ, Okkenhaug K, Vanhaesebroeck B. 2008. The p110β isoform of phosphoinositide 3-kinase signals downstream of G protein-coupled receptors and is functionally redundant with p110γ. *Proc Natl Acad Sci* **105:** 8292–8297.

Gupta S, Ramjaun AR, Haiko P, Wang Y, Warne PH, Nicke B, Nye E, Stamp G, Alitalo K, Downward J. 2007. Binding of ras to phosphoinositide 3-kinase p110α is required for Ras-driven tumorigenesis in mice. *Cell* **129:** 957–968.

Gyori D, Chessa T, Hawkins PT, Stephens LR. 2017. Class (I) phosphoinositide 3-kinases in the tumor microenvironment. *Cancers (Basel)* **9:** E24.

Gysin S, Salt M, Young A, McCormick F. 2011. Therapeutic strategies for targeting ras proteins. *Genes Cancer* **2:** 359–372.

Hamad NM, Elconin JH, Karnoub AE, Bai W, Rich JN, Abraham RT, Der CJ, Counter CM. 2002. Distinct requirements for Ras oncogenesis in human versus mouse cells. *Genes Dev* **16:** 2045–2057.

Han J, Luby-Phelps K, Das B, Shu X, Xia Y, Mosteller RD, Krishna UM, Falck JR, White MA, Broek D. 1998. Role of substrates and products of PI 3-kinase in regulating activation of Rac-related guanosine triphosphatases by Vav. *Science* **279:** 558–560.

Hao Y, Wang C, Cao B, Hirsch BM, Song J, Markowitz SD, Ewing RM, Sedwick D, Liu L, Zheng W, et al. 2013. Gain of interaction with IRS1 by p110α–helical domain mutants is crucial for their oncogenic functions. *Cancer Cell* **23:** 583–593.

Hemmings BA, Restuccia DF. 2015. The PI3K-PKB/Akt pathway. *Cold Spring Harb Perspect Biol* **7:** a026609.

Herrmann C. 2003. Ras-effector interactions: After one decade. *Curr Opin Struct Biol* **13:** 122–129.

Hirakawa T, Ruley HE. 1988. Rescue of cells from *ras* oncogene-induced growth arrest by a second, complementing, oncogene. *Proc Natl Acad Sci* **85:** 1519–1523.

Hoeflich KP, O'Brien C, Boyd Z, Cavet G, Guerrero S, Jung K, Januario T, Savage H, Punnoose E, Truong T, et al. 2009. In vivo antitumor activity of MEK and phosphatidylinositol 3-kinase inhibitors in basal-like breast cancer models. *Clin Cancer Res* **15:** 4649–4664.

Hofer F, Fields S, Schneider C, Martin GS. 1994. Activated Ras interacts with the Ral guanine nucleotide dissociation stimulator. *Proc Natl Acad Sci* **91:** 11089–11093.

Hutti JE, Pfefferle AD, Russell SC, Sircar M, Perou CM, Baldwin AS. 2012. Oncogenic PI3K mutations lead to NF-κB-dependent cytokine expression following growth factor deprivation. *Cancer Res* **72:** 3260–3269.

Ihle NT, Lemos R Jr., Wipf P, Yacoub A, Mitchell C, Siwak D, Mills GB, Dent P, Kirkpatrick DL, Powis G. 2009. Mutations in the phosphatidylinositol-3-kinase pathway predict for antitumor activity of the inhibitor PX-866 whereas oncogenic Ras is a dominant predictor for resistance. *Cancer Res* **69:** 143–150.

Ihle NT, Byers LA, Kim ES, Saintigny P, Lee JJ, Blumenschein GR, Tsao A, Liu S, Larsen JE, Wang J, et al. 2012. Effect of *KRAS* oncogene substitutions on protein behavior: Implications for signaling and clinical outcome. *J Natl Cancer Inst* **104:** 228–239.

Janku F, Lee JJ, Tsimberidou AM, Hong DS, Naing A, Falchook GS, Fu S, Luthra R, Garrido-Laguna I, Kurzrock R. 2011. *PIK3CA* mutations frequently coexist with *RAS* and *BRAF* mutations in patients with advanced cancers. *PLoS ONE* **6:** e22769.

Jifu E, Xing J, Gong H, He J, Zhang W. 2015. Combine MEK inhibition with PI3K/mTOR inhibition exert inhibitory tumor growth effect on *KRAS* and *PIK3CA* mutation CRC xenografts due to reduced expression of VEGF and matrix metallopeptidase-9. *Tumour Biol* **36:** 1091–1097.

Kaduwal S, Jeong WJ, Park JC, Lee KH, Lee YM, Jeon SH, Lim YB, Min do S, Choi KY. 2015. Sur8/Shoc2 promotes cell motility and metastasis through activation of Ras-PI3K signaling. *Oncotarget* **6:** 33091–33105.

Kang S, Bader AG, Vogt PK. 2005. Phosphatidylinositol 3-kinase mutations identified in human cancer are oncogenic. *Proc Natl Acad Sci* **102:** 802–807.

Kardash E, Reichman-Fried M, Maitre JL, Boldajipour B, Papusheva E, Messerschmidt EM, Heisenberg CP, Raz E. 2010. A role for Rho GTPases and cell–cell adhesion in single-cell motility in vivo. *Nat Cell Biol* **12:** 47–53.

Karnoub AE, Weinberg RA. 2008. Ras oncogenes: Split personalities. *Nat Rev Mol Cell Biol* **9:** 517–531.

Katso R, Okkenhaug K, Ahmadi K, White S, Timms J, Waterfield MD. 2001. Cellular function of phosphoinositide 3-kinases: Implications for development, homeostasis, and cancer. *Annu Rev Cell Dev Biol* **17:** 615–675.

Keeton AB, Salter EA, Piazza GA. 2017. The RAS–effector interaction as a drug target. *Cancer Res* **77:** 221–226.

Khwaja A, Rodriguez-Viciana P, Wennstrom S, Warne PH, Downward J. 1997. Matrix adhesion and Ras transformation both activate a phosphoinositide 3-OH kinase and protein kinase B/Akt cellular survival pathway. *EMBO J* **16:** 2783–2793.

Kiel C, Wohlgemuth S, Rousseau F, Schymkowitz J, Ferkinghoff-Borg J, Wittinghofer F, Serrano L. 2005. Recognizing and defining true Ras binding domains II: In silico prediction based on homology modelling and energy calculations. *J Mol Biol* **348:** 759–775.

Kim S, Huang W, Mottillo EP, Sohail A, Ham YA, Conley-Lacomb MK, Kim CJ, Tzivion G, Kim HR, Wang S, et al. 2010. Posttranslational regulation of membrane type 1-matrix metalloproteinase (MT1-MMP) in mouse PTEN null prostate cancer cells: Enhanced surface expression and differential O-glycosylation of MT1-MMP. *Biochim Biophys Acta* **1803:** 1287–1297.

Land H, Parada LF, Weinberg RA. 1983a. Cellular oncogenes and multistep carcinogenesis. *Science* **222:** 771–778.

Land H, Parada LF, Weinberg RA. 1983b. Tumorigenic conversion of primary embryo fibroblasts requires at least two cooperating oncogenes. *Nature* **304:** 596–602.

Lane HA, Wood JM, McSheehy PM, Allegrini PR, Boulay A, Brueggen J, Littlewood-Evans A, Maira SM, Martiny-Baron G, Schnell CR, et al. 2009. mTOR inhibitor RAD001 (everolimus) has antiangiogenic/vascular properties distinct from a VEGFR tyrosine kinase inhibitor. *Clin Cancer Res* **15:** 1612–1622.

Lee YM, Kaduwal S, Lee KH, Park JC, Jeong WJ, Choi KY. 2016. Sur8 mediates tumorigenesis and metastasis in colorectal cancer. *Exp Mol Med* **48:** e249.

Li Z, Gorfe AA. 2012. What drives the clustering of membrane-bound Ras? *Small GTPases* **3:** 244–247.

Li W, Han M, Guan KL. 2000. The leucine-rich repeat protein SUR-8 enhances MAP kinase activation and forms a complex with Ras and Raf. *Genes Dev* **14:** 895–900.

Li W, Zhu T, Guan KL. 2004. Transformation potential of Ras isoforms correlates with activation of phosphatidylinositol 3-kinase but not ERK. *J Biol Chem* **279:** 37398–37406.

Ligresti G, Militello L, Steelman LS, Cavallaro A, Basile F, Nicoletti F, Stivala F, McCubrey JA, Libra M. 2009. PIK3CA mutations in human solid tumors: Role in sensitivity to various therapeutic approaches. *Cell Cycle* **8:** 1352–1358.

Liu L, Zhu S, Gong Z, Low BC. 2008. K-ras/PI3K-Akt signaling is essential for zebrafish hematopoiesis and angiogenesis. *PLoS ONE* **3:** e2850.

Marshall CB, Meiri D, Smith MJ, Mazhab-Jafari MT, Gasmi-Seabrook GM, Rottapel R, Stambolic V, Ikura M. 2012a. Probing the GTPase cycle with real-time NMR: GAP and GEF activities in cell extracts. *Methods* **57:** 473–485.

Marshall NA, Galvin KC, Corcoran AM, Boon L, Higgs R, Mills KH. 2012b. Immunotherapy with PI3K inhibitor and Toll-like receptor agonist induces IFN-γ^{+}IL-17^{+} polyfunctional T cells that mediate rejection of murine tumors. *Cancer Res* **72:** 581–591.

Marte BM, Rodriguez-Viciana P, Wennstrom S, Warne PH, Downward J. 1997. R-Ras can activate the phosphoinositide 3-kinase but not the MAP kinase arm of the Ras effector pathways. *Curr Biol* **7:** 63–70.

Martini M, De Santis MC, Braccini L, Gulluni F, Hirsch E. 2014. PI3K/AKT signaling pathway and cancer: An updated review. *Ann Med* **46:** 372–383.

Maurer T, Garrenton LS, Oh A, Pitts K, Anderson DJ, Skelton NJ, Fauber BP, Pan B, Malek S, Stokoe D, et al. 2012. Small-molecule ligands bind to a distinct pocket in Ras and inhibit SOS-mediated nucleotide exchange activity. *Proc Natl Acad Sci* **109:** 5299–5304.

Mayer IA, Arteaga CL. 2016. The PI3K/AKT pathway as a target for cancer treatment. *Annu Rev Med* **67:** 11–28.

Mendoza MC, Er EE, Blenis J. 2011. The Ras-ERK and PI3K-mTOR pathways: Cross-talk and compensation. *Trends Biochem Sci* **36:** 320–328.

Meyer RG, Korn S, Micke P, Becker K, Huber C, Wolfel T, Buhl R. 2007. An open-label, prospective phase I/II study evaluating the immunogenicity and safety of a ras peptide vaccine plus GM-CSF in patients with non-small cell lung cancer. *Lung Cancer* **58:** 88–94.

Mirzoeva OK, Das D, Heiser LM, Bhattacharya S, Siwak D, Gendelman R, Bayani N, Wang NJ, Neve RM, Guan Y, et al. 2009. Basal subtype and MAPK/ERK kinase (MEK)-phosphoinositide 3-kinase feedback signaling determine susceptibility of breast cancer cells to MEK inhibition. *Cancer Res* **69:** 565–572.

Mizushima N. 2007. Autophagy: Process and function. *Genes Dev* **21:** 2861–2873.

Mizushima N, Levine B. 2010. Autophagy in mammalian development and differentiation. *Nat Cell Biol* **12:** 823–830.

Molina-Arcas M, Hancock DC, Sheridan C, Kumar MS, Downward J. 2013. Coordinate direct input of both KRAS and IGF1 receptor to activation of PI3 kinase in *KRAS*-mutant lung cancer. *Cancer Discov* **3:** 548–563.

Moodie SA, Willumsen BM, Weber MJ, Wolfman A. 1993. Complexes of Ras·GTP with Raf-1 and mitogen-activated protein kinase kinase. *Science* **260:** 1658–1661.

Morrison DK, Davis RJ. 2003. Regulation of MAP kinase signaling modules by scaffold proteins in mammals. *Annu Rev Cell Dev Biol* **19:** 91–118.

Muranen T, Selfors LM, Worster DT, Iwanicki MP, Song L, Morales FC, Gao S, Mills GB, Brugge JS. 2012. Inhibition of PI3K/mTOR leads to adaptive resistance in matrix-attached cancer cells. *Cancer Cell* **21:** 227–239.

Murillo MM, Zelenay S, Nye E, Castellano E, Lassailly F, Stamp G, Downward J. 2014. RAS interaction with PI3K p110α is required for tumor-induced angiogenesis. *J Clin Invest* **124:** 3601–3611.

Murphy LO, Blenis J. 2006. MAPK signal specificity: The right place at the right time. *Trends Biochem Sci* **31:** 268–275.

Nezis IP, Vaccaro MI, Devenish RJ, Juhasz G. 2014. Autophagy in development, cell differentiation, and homeodynamics: From molecular mechanisms to diseases and pathophysiology. *Biomed Res Int* **2014:** 349623.

Nimnual AS, Yatsula BA, Bar-Sagi D. 1998. Coupling of Ras and Rac guanosine triphosphatases through the Ras exchanger Sos. *Science* **279:** 560–563.

Oda K, Okada J, Timmerman L, Rodriguez-Viciana P, Stokoe D, Shoji K, Taketani Y, Kuramoto H, Knight ZA, Shokat KM, et al. 2008. PIK3CA cooperates with other phosphatidylinositol 3′-kinase pathway mutations to effect oncogenic transformation. *Cancer Res* **68:** 8127–8136.

Okkenhaug K. 2013. Signaling by the phosphoinositide 3-kinase family in immune cells. *Annu Rev Immunol* **31:** 675–704.

Okkenhaug K, Graupera M, Vanhaesebroeck B. 2016. Targeting PI3K in cancer: Impact on tumor cells, their protective stroma, angiogenesis, and immunotherapy. *Cancer Discov* **6:** 1090–1105.

Ong SH, Hadari YR, Gotoh N, Guy GR, Schlessinger J, Lax I. 2001. Stimulation of phosphatidylinositol 3-kinase by fibroblast growth factor receptors is mediated by coordinated recruitment of multiple docking proteins. *Proc Natl Acad Sci* **98:** 6074–6079.

Ostrem JM, Shokat KM. 2016. Direct small-molecule inhibitors of KRAS: From structural insights to mechanism-based design. *Nat Rev Drug Discov* **15:** 771–785.

Ostrem JM, Peters U, Sos ML, Wells JA, Shokat KM. 2013. K-Ras^{G12C} inhibitors allosterically control GTP affinity and effector interactions. *Nature* **503:** 548–551.

Pacold ME, Suire S, Perisic O, Lara-Gonzalez S, Davis CT, Walker EH, Hawkins PT, Stephens L, Eccleston JF, Williams RL. 2000. Crystal structure and functional analysis of Ras binding to its effector phosphoinositide 3-kinase γ. *Cell* **103:** 931–943.

Pang H, Flinn R, Patsialou A, Wyckoff J, Roussos ET, Wu H, Pozzuto M, Goswami S, Condeelis JS, Bresnick AR, et al. 2009. Differential enhancement of breast cancer cell motility and metastasis by helical and kinase domain mutations of class IA phosphoinositide 3-kinase. *Cancer Res* **69:** 8868–8876.

Paoli P, Giannoni E, Chiarugi P. 2013. Anoikis molecular pathways and its role in cancer progression. *Biochem Biophys Acta* **1833:** 3481–3498.

Parker JA, Mattos C. 2015. The Ras-membrane interface: Isoform-specific differences in the catalytic domain. *Mol Cancer Res* **13:** 595–603.

Parsons DW, Wang TL, Samuels Y, Bardelli A, Cummins JM, DeLong L, Silliman N, Ptak J, Szabo S, Willson JK, et al. 2005. Colorectal cancer: Mutations in a signalling pathway. *Nature* **436:** 792.

Polivka J Jr, Janku F. 2014. Molecular targets for cancer therapy in the PI3K/AKT/mTOR pathway. *Pharmacol Ther* **142:** 164–175.

Pollock CB, Shirasawa S, Sasazuki T, Kolch W, Dhillon AS. 2005. Oncogenic *K-RAS* is required to maintain changes in cytoskeletal organization, adhesion, and motility in colon cancer cells. *Cancer Res* **65:** 1244–1250.

Prior IA, Lewis PD, Mattos C. 2012. A comprehensive survey of Ras mutations in cancer. *Cancer Res* **72:** 2457–2467.

Pruitt K, Der CJ. 2001. Ras and Rho regulation of the cell cycle and oncogenesis. *Cancer Lett* **171:** 1–10.

Pylayeva-Gupta Y, Grabocka E, Bar-Sagi D. 2011. RAS oncogenes: Weaving a tumorigenic web. *Nat Rev Cancer* **11:** 761–774.

Qu Y, Wu X, Yin Y, Yang Y, Ma D, Li H. 2014. Antitumor activity of selective MEK1/2 inhibitor AZD6244 in combination with PI3K/mTOR inhibitor BEZ235 in gefitinib-resistant NSCLC xenograft models. *J Exp Clin Cancer Res* **33:** 52.

Quilliam LA, Khosravi-Far R, Huff SY, Der CJ. 1995. Guanine nucleotide exchange factors: Activators of the Ras superfamily of proteins. *BioEssays* **17:** 395–404.

Rahma OE, Hamilton JM, Wojtowicz M, Dakheel O, Bernstein S, Liewehr DJ, Steinberg SM, Khleif SN. 2014. The immunological and clinical effects of mutated ras peptide vaccine in combination with IL-2, GM-CSF, or both in patients with solid tumors. *J Transl Med* **12:** 55.

Rajalingam K, Schreck R, Rapp UR, Albert S. 2007. Ras oncogenes and their downstream targets. *Biochem Biophys Res Commun* **1773:** 1177–1195.

Riely GJ, Marks J, Pao W. 2009. *KRAS* mutations in non-small cell lung cancer. *Proc Am Thoracic Soc* **6:** 201–205.

Rodriguez-Viciana P, Warne PH, Dhand R, Vanhaesebroeck B, Gout I, Fry MJ, Waterfield MD, Downward J. 1994. Phosphatidylinositol-3-OH kinase as a direct target of Ras. *Nature* **370:** 527–532.

Rodriguez-Viciana P, Marte BM, Warne PH, Downward J. 1996a. Phosphatidylinositol 3′ kinase: One of the effectors of Ras. *Phil Trans R Soc Lond B* **351:** 225–231; discussion 231–222.

Rodriguez-Viciana P, Warne PH, Vanhaesebroeck B, Waterfield MD, Downward J. 1996b. Activation of phosphoinositide 3-kinase by interaction with Ras and by point mutation. *EMBO J* **15:** 2442–2451.

Rodriguez-Viciana P, Warne PH, Khwaja A, Marte BM, Pappin D, Das P, Waterfield MD, Ridley A, Downward J. 1997. Role of phosphoinositide 3-OH kinase in cell transformation and control of the actin cytoskeleton by Ras. *Cell* **89:** 457–467.

Rodriguez-Viciana P, Sabatier C, McCormick F. 2004. Signaling specificity by Ras family GTPases is determined by the full spectrum of effectors they regulate. *Mol Cell Biol* **24:** 4943–4954.

Rodrik-Outmezguine VS, Chandarlapaty S, Pagano NC, Poulikakos PI, Scaltriti M, Moskatel E, Baselga J, Guichard S, Rosen N. 2011. mTOR kinase inhibition causes feedback-dependent biphasic regulation of AKT signaling. *Cancer Discov* **1:** 248–259.

Rojas JM, Oliva JL, Santos E. 2011. Mammalian son of sevenless guanine nucleotide exchange factors: Old concepts and new perspectives. *Genes Cancer* **2:** 298–305.

Roskoski R Jr. 2010. RAF protein-serine/threonine kinases: Structure and regulation. *Biochem Biophys Res Commun* **399:** 313–317.

Ruley HE. 1983. Adenovirus early region 1A enables viral and cellular transforming genes to transform primary cells in culture. *Nature* **304:** 602–606.

Samuels Y, Diaz LA Jr., Schmidt-Kittler O, Cummins JM, Delong L, Cheong I, Rago C, Huso DL, Lengauer C, Kinzler KW, et al. 2005. Mutant PIK3CA promotes cell growth and invasion of human cancer cells. *Cancer Cell* **7:** 561–573.

Sasaki AT, Firtel RA. 2006. Regulation of chemotaxis by the orchestrated activation of Ras, PI3K, and TOR. *Eur J Cell Biol* **85:** 873–895.

Schafer ZT, Grassian AR, Song L, Jiang Z, Gerhart-Hines Z, Irie HY, Gao S, Puigserver P, Brugge JS. 2009. Antioxidant and oncogene rescue of metabolic defects caused by loss of matrix attachment. *Nature* **461:** 109–113.

Schmukler E, Kloog Y, Pinkas-Kramarski R. 2014. Ras and autophagy in cancer development and therapy. *Oncotarget* **5:** 577–586.

Schnell CR, Stauffer F, Allegrini PR, O'Reilly T, McSheehy PM, Dartois C, Stumm M, Cozens R, Littlewood-Evans A, Garcia-Echeverria C, et al. 2008. Effects of the dual phosphatidylinositol 3-kinase/mammalian target of rapamycin inhibitor NVP-BEZ235 on the tumor vasculature: Implications for clinical imaging. *Cancer Res* **68:** 6598–6607.

Schrauwen S, Depreeuw J, Coenegrachts L, Hermans E, Lambrechts D, Amant F. 2015. Dual blockade of PI3K/AKT/mTOR (NVP-BEZ235) and Ras/Raf/MEK (AZD6244) pathways synergistically inhibit growth of primary endometrioid endometrial carcinoma cultures, whereas NVP-BEZ235 reduces tumor growth in the corresponding xenograft models. *Gynecol Oncol* **138:** 165–173.

Scott DE, Bayly AR, Abell C, Skidmore J. 2016. Small molecules, big targets: Drug discovery faces the protein-protein interaction challenge. *Nat Rev Drug Discov* **15:** 533–550.

Shao J, Yang X, Liu T, Zhang T, Xie QR, Xia W. 2016. Autophagy induction by SIRT6 is involved in oxidative stress-induced neuronal damage. *Protein Cell* **7:** 281–290.

Sheng H, Shao J, DuBois RN. 2001. Akt/PKB activity is required for Ha-Ras-mediated transformation of intestinal epithelial cells. *J Biol Chem* **276:** 14498–14504.

Shieh A, Ward AF, Donlan KL, Harding-Theobald ER, Xu J, Mullighan CG, Zhang C, Chen SC, Su X, Downing JR, et al. 2013. Defective K-Ras oncoproteins overcome impaired effector activation to initiate leukemia in vivo. *Blood* **121:** 4884–4893.

Singh H, Longo DL, Chabner BA. 2015. Improving prospects for targeting RAS. *J Clin Oncol* **33:** 3650–3659.

Smith MJ, Ikura M. 2014. Integrated RAS signaling defined by parallel NMR detection of effectors and regulators. *Nat Chem Biol* **10:** 223–230.

So L, Yea SS, Oak JS, Lu M, Manmadhan A, Ke QH, Janes MR, Kessler LV, Kucharski JM, Li LS, et al. 2013. Selective inhibition of phosphoinositide 3-kinase p110α preserves lymphocyte function. *J Biol Chem* **288:** 5718–5731.

Soler A, Serra H, Pearce W, Angulo A, Guillermet-Guibert J, Friedman LS, Vinals F, Gerhardt H, Casanovas O, Graupera M, et al. 2013. Inhibition of the p110α isoform of PI 3-kinase stimulates nonfunctional tumor angiogenesis. *J Exp Med* **210:** 1937–1945.

Soler A, Angulo-Urarte A, Graupera M. 2015. PI3K at the crossroads of tumor angiogenesis signaling pathways. *Mol Cell Oncol* **2:** e975624.

Sos ML, Fischer S, Ullrich R, Peifer M, Heuckmann JM, Koker M, Heynck S, Stuckrath I, Weiss J, Fischer F, et al. 2009. Identifying genotype-dependent efficacy of single and combined PI3K- and MAPK-pathway inhibition in cancer. *Proc Natl Acad Sci* **106:** 18351–18356.

Stephen AG, Esposito D, Bagni RK, McCormick F. 2014. Dragging Ras back in the ring. *Cancer Cell* **25:** 272–281.

Stites EC, Ravichandran KS. 2009. A systems perspective of Ras signaling in cancer. *Clin Cancer Res* **15:** 1510–1513.

Suire S, Condliffe AM, Ferguson GJ, Ellson CD, Guillou H, Davidson K, Welch H, Coadwell J, Turner M, Chilvers ER, et al. 2006. Gβγs and the Ras binding domain of p110γ are both important regulators of $PI_3K\gamma$ signalling in neutrophils. *Nat Cell Biol* **8:** 1303–1309.

Sun Q, Burke JP, Phan J, Burns MC, Olejniczak ET, Waterson AG, Lee T, Rossanese OW, Fesik SW. 2012. Discovery of small molecules that bind to K-Ras and inhibit Sos-mediated activation. *Angew Chem Int Ed Engl* **51:** 6140–6143.

Takashima A, Faller DV. 2013. Targeting the RAS oncogene. *Expert Opin Ther Targets* **17:** 507–531.

Takaya A, Kamio T, Masuda M, Mochizuki N, Sawa H, Sato M, Nagashima K, Mizutani A, Matsuno A, Kiyokawa E, et al. 2007. R-Ras regulates exocytosis by Rgl2/Rlf-mediated activation of RalA on endosomes. *Mol Biol Cell* **18:** 1850–1860.

Talbert EE, Yang J, Mace TA, Farren MR, Farris AB, Young GS, Elnaggar O, Che Z, Timmers CD, Rajasekera P, et al. 2017. Dual inhibition of MEK and PI3K/Akt rescues cancer cachexia through both tumor-extrinsic and -intrinsic activities. *Mol Cancer Ther* **16:** 344–356.

Tamanoi F, Hsueh EC, Goodman LE, Cobitz AR, Detrick RJ, Brown WR, Fujiyama A. 1988. Posttranslational modification of *ras* proteins: Detection of a modification prior to fatty acid acylation and cloning of a gene responsible for the modification. *J Cell Biochem* **36:** 261–273.

Tan J, Yu Q. 2013. Molecular mechanisms of tumor resistance to PI3K-mTOR-targeted therapy. *Chin J Cancer* **32:** 376–379.

Tanaka T, Rabbitts TH. 2010. Interfering with RAS-effector protein interactions prevent RAS-dependent tumour initiation and causes stop-start control of cancer growth. *Oncogene* **29:** 6064–6070.

Thorpe LM, Yuzugullu H, Zhao JJ. 2015. PI3K in cancer: Divergent roles of isoforms, modes of activation and therapeutic targeting. *Nat Rev Cancer* **15:** 7–24.

Trahey M, McCormick F. 1987. A cytoplasmic protein stimulates normal N-*ras* p21 GTPase, but does not affect oncogenic mutants. *Science* **238:** 542–545.

Tran Thang NN, Derouazi M, Philippin G, Arcidiaco S, Di Berardino-Besson W, Masson F, Hoepner S, Riccadonna C, Burkhardt K, Guha A, et al. 2010. Immune infiltration of spontaneous mouse astrocytomas is dominated by immunosuppressive cells from early stages of tumor development. *Cancer Res* **70:** 4829–4839.

Vanhaesebroeck B, Guillermet-Guibert J, Graupera M, Bilanges B. 2010a. The emerging mechanisms of isoform-specific PI3K signalling. *Nat Rev Mol Cell Biol* **11:** 329–341.

Vanhaesebroeck B, Vogt PK, Rommel C. 2010b. PI3K: From the bench to the clinic and back. *Curr Top Microbiol Immunol* **347:** 1–19.

Vanhaesebroeck B, Stephens L, Hawkins P. 2012. PI3K signalling: The path to discovery and understanding. *Nat Rev Mol Cell Biol* **13:** 195–203.

Vanneman M, Dranoff G. 2012. Combining immunotherapy and targeted therapies in cancer treatment. *Nat Rev Cancer* **12:** 237–251.

Vigil D, Cherfils J, Rossman KL, Der CJ. 2010. Ras superfamily GEFs and GAPs: Validated and tractable targets for cancer therapy? *Nat Rev Cancer* **10:** 842–857.

Walker EH, Perisic O, Ried C, Stephens L, Williams RL. 1999. Structural insights into phosphoinositide 3-kinase catalysis and signalling. *Nature* **402:** 313–320.

Wang C, Wang Y, McNutt MA, Zhu WG. 2011. Autophagy process is associated with anti-neoplastic function. *Acta Biochim Biophys Sin (Shanghai)* **43:** 425–432.

Wang W, Fang G, Rudolph J. 2012. Ras inhibition via direct Ras binding—Is there a path forward? *Bioorg Med Chem Lett* **22:** 5766–5776.

Wang L, Hu H, Pan Y, Wang R, Li Y, Shen L, Yu Y, Li H, Cai D, Sun Y, et al. 2014. *PIK3CA* mutations frequently coexist with *EGFR/KRAS* mutations in non-small cell lung cancer and suggest poor prognosis in *EGFR/KRAS* wildtype subgroup. *PLoS ONE* **9:** e88291.

Warne PH, Viciana PR, Downward J. 1993. Direct interaction of Ras and the amino-terminal region of Raf-1 in vitro. *Nature* **364:** 352–355.

Wohlgemuth S, Kiel C, Kramer A, Serrano L, Wittinghofer F, Herrmann C. 2005. Recognizing and defining true Ras binding domains. I: Biochemical analysis. *J Mol Biol* **348:** 741–758.

Wood KC. 2015. Mapping the pathways of resistance to targeted therapies. *Cancer Res* **75:** 4247–4251.

Yang HW, Shin MG, Lee S, Kim JR, Park WS, Cho KH, Meyer T, Heo WD. 2012. Cooperative activation of PI3K by Ras and Rho family small GTPases. *Mol Cell* **47:** 281–290.

Yoon S, Seger R. 2006. The extracellular signal-regulated kinase: Multiple substrates regulate diverse cellular functions. *Growth Factors* **24:** 21–44.

Yuan TL, Cantley LC. 2008. PI3K pathway alterations in cancer: Variations on a theme. *Oncogene* **27:** 5497–5510.

Yuan TL, Choi HS, Matsui A, Benes C, Lifshits E, Luo J, Frangioni JV, Cantley LC. 2008. Class 1A PI3K regulates vessel integrity during development and tumorigenesis. *Proc Natl Acad Sci* **105:** 9739–9744.

Zarrabi K, Dufour A, Li J, Kuscu C, Pulkoski-Gross A, Zhi J, Hu Y, Sampson NS, Zucker S, Cao J. 2011. Inhibition of matrix metalloproteinase 14 (MMP-14)-mediated cancer cell migration. *J Biol Chem* **286:** 33167–33177.

Zhang J, Lodish HF. 2005. Identification of K-ras as the major regulator for cytokine-dependent Akt activation in erythroid progenitors in vivo. *Proc Natl Acad Sci* **102:** 14605–14610.

Zhang X, Tang N, Hadden TJ, Rishi AK. 2011. Akt, FoxO and regulation of apoptosis. *Biochim Biophys Acta* **1813:** 1978–1986.

Zhang J, Zheng J, Yang Y, Lu J, Gao J, Lu T, Sun J, Jiang H, Zhu Y, Zheng Y, et al. 2015. Molecular spectrum of KRAS, NRAS, BRAF and PIK3CA mutations in Chinese colorectal cancer patients: Analysis of 1,110 cases. *Sci Rep* **5:** 18678.

Zhao L, Vogt PK. 2008. Class I PI3K in oncogenic cellular transformation. *Oncogene* **27:** 5486–5496.

Zhao L, Vogt PK. 2010. Hot-spot mutations in p110α of phosphatidylinositol 3-kinase (pI3K): Differential interactions with the regulatory subunit p85 and with RAS. *Cell Cycle* **9:** 596–600.

Zhao JJ, Cheng H, Jia S, Wang L, Gjoerup OV, Mikami A, Roberts TM. 2006. The p110α isoform of PI3K is essential for proper growth factor signaling and oncogenic transformation. *Proc Natl Acad Sci* **103:** 16296–16300.

Zhou Y, Liang H, Rodkey T, Ariotti N, Parton RG, Hancock JF. 2014. Signal integration by lipid-mediated spatial cross talk between Ras nanoclusters. *Mol Cell Biol* **34:** 862–876.

Zitvogel L, Galluzzi L, Smyth MJ, Kroemer G. 2013. Mechanism of action of conventional and targeted anticancer therapies: Reinstating immunosurveillance. *Immunity* **39:** 74–88.

The Interdependent Activation of Son-of-Sevenless and Ras

Pradeep Bandaru,[1] Yasushi Kondo,[1] and John Kuriyan[2]

[1]Department of Molecular and Cell Biology, California Institute for Quantitative Biosciences, Howard Hughes Medical Institute, University of California, Berkeley, California 94720

[2]Departments of Molecular and Cell Biology and of Chemistry, California Institute for Quantitative Biosciences, Molecular Biophysics and Integrated Bioimaging Division, Lawrence Berkeley National Laboratory, Howard Hughes Medical Institute, University of California, Berkeley, California 94720

Correspondence: kuriyan@berkeley.edu

The guanine-nucleotide exchange factor (GEF) Son-of-Sevenless (SOS) plays a critical role in metazoan signaling by converting Ras•GDP (guanosine diphosphate) to Ras•GTP (guanosine triphosphate) in response to tyrosine kinase activation. Structural studies have shown that SOS differs from other Ras-specific GEFs in that SOS is itself activated by Ras•GTP binding to an allosteric site, distal to the site of nucleotide exchange. The activation of SOS involves membrane recruitment and conformational changes, triggered by lipid binding, that open the allosteric binding site for Ras•GTP. This is in contrast to other Ras-specific GEFs, which are activated by second messengers that more directly affect the active site. Allosteric Ras•GTP binding stabilizes SOS at the membrane, where it can turn over other Ras molecules processively, leading to an ultrasensitive response that is distinct from that of other Ras-specific GEFs.

An evolutionary innovation in metazoans is the coupling of Ras activation to the phosphorylation of tyrosine residues on cell-surface receptors. This occurs through the SH2-mediated recruitment to these receptors of a Ras-specific guanine nucleotide exchange factor (GEF), Son-of-Sevenless (SOS) (Fig. 1A) (Lowenstein et al. 1992; Buday and Downward 1993; Egan et al. 1993; Gale et al. 1993; Rozakis-Adcock et al. 1993). There are several classes of Ras-specific GEFs in mammalian cells, and a key difference between SOS and other Ras-specific GEFs is that the latter are regulated by soluble second messengers, such as calcium, cyclic AMP (cAMP), and diacylglycerol, whereas SOS is not (Bos et al. 2007). Instead, SOS is recruited to the membrane by tyrosine phosphorylation of receptors or scaffold proteins and, in addition, the activation of SOS requires the binding of Ras•GTP to an allosteric site (Margarit et al. 2003).

All Ras-specific GEFs share a catalytic module, consisting of a Ras exchanger motif (REM) domain and a Cdc25 domain (Boguski and McCormick 1993). The Cdc25 domain contains the active site, where nucleotide-free Ras is bound during the exchange process. In SOS, allosteric activation occurs by the binding of Ras•GTP to a distal site bracketed by the REM and Cdc25 domain (Margarit et al. 2003; Sondermann et al. 2004) but, as far as is known, the principal role of

Cite this article as *Cold Spring Harb Perspect Med* doi: 10.1101/cshperspect.a031534

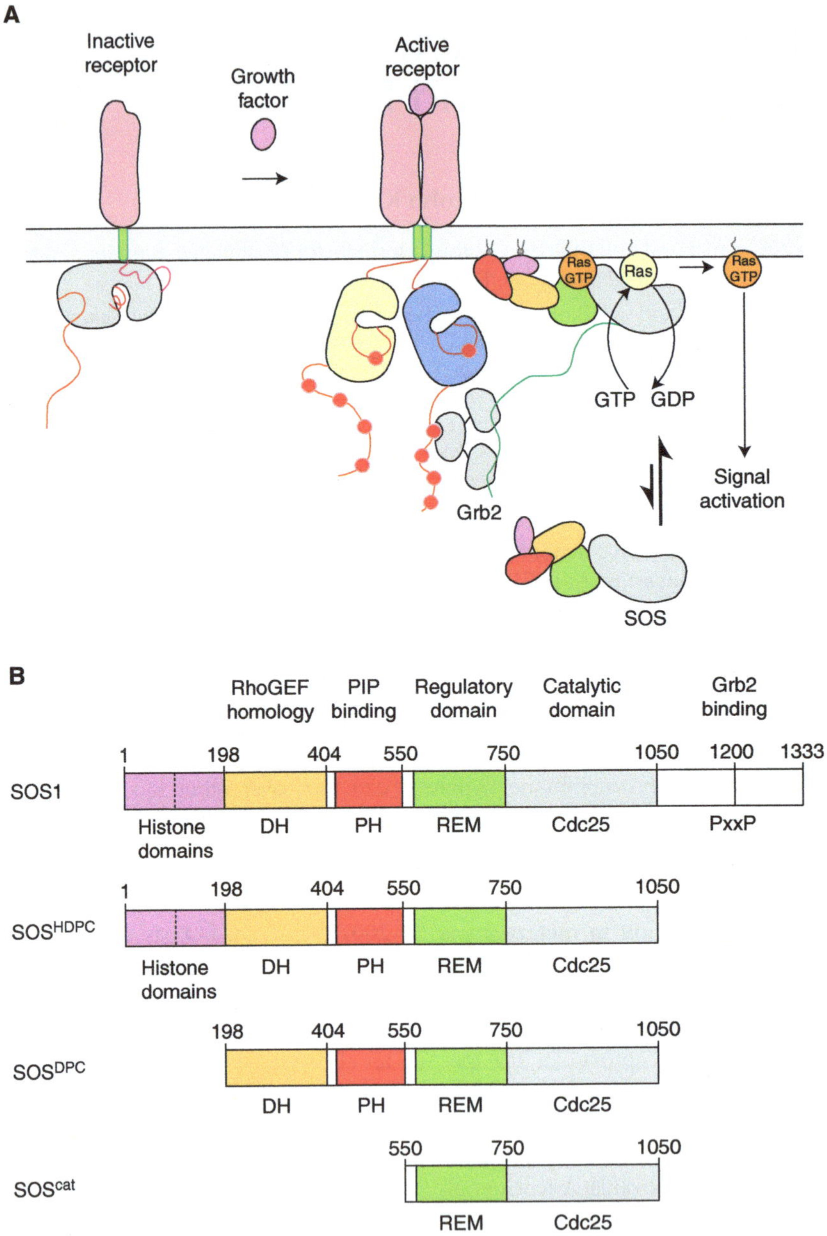

Figure 1. The activation and domain organization of Son-of-Sevenless (SOS). (*A*) Ras activation is coupled to the phosphorylation of tyrosine residues on cell-surface receptors. This primarily occurs through the SH2-mediated recruitment of SOS to these receptors by adaptor proteins such as Grb2. Ras activation by SOS subsequently triggers downstream signaling responses in cells. (*B*) Schematic representation of the various constructs of SOS discussed in this review. DH, Dbl homology; PH, Pleckstrin homology; REM, Ras exchanger motif.

Cite this article as *Cold Spring Harb Perspect Med* doi: 10.1101/cshperspect.a031534

the REM domain in other Ras-specific GEFs appears to be structural. Each class of Ras-specific GEFs has a distinct regulatory mechanism, unrelated to that of SOS. For example, Epac proteins are autoinhibited by a cAMP-binding domain, which prevents Ras binding to the Cdc25 domain in the absence of cAMP (de Rooij et al. 2000; Rehmann et al. 2006, 2008; Bos 2017). Similarly, calcium and diacylglycerol binding to C1 and EF-hand domains, respectively, relieves the autoinhibition of RasGRP proteins at the membrane (Iwig et al. 2013; Jun et al. 2013a; Vercoulen et al. 2017).

In common with other GEFs, SOS activates Ras by causing dissociation of the otherwise tightly bound nucleotide, thereby facilitating the binding of guanosine triphosphate (GTP) (Bos et al. 2007; Cherfils and Zeghouf 2013). SOS is not normally membrane associated, and it was thought that SOS was activated simply by recruitment to the membrane, where it can more efficiently encounter membrane-anchored Ras. SOS is bound to the adaptor protein Grb2, and the SH2 domain of Grb2 binds to activated receptors, either directly or through the bridging action of the adaptor protein Shc (Rozakis-Adcock et al. 1992; Buday et al. 1994). In this way, receptor activation results in the membrane localization of SOS. Consistent with this, the constitutive membrane localization of the catalytic module of SOS activates Ras without receptor input (Aronheim et al. 1994). Subsequent studies showed, however, that multiple layers of inhibition are imposed on SOS by its regulatory domains, and these need to be relieved by interaction with lipids in the membrane, such as PIP_2 and PA, for SOS to be fully active at the membrane (Corbalan-Garcia et al. 1998; Hall et al. 2002; Sondermann et al. 2004; Gureasko et al. 2008; Yadav and Bar-Sagi 2010). The end point of these actions is the binding of Ras•GTP to the allosteric site on SOS, which is blocked prior to activation by the regulatory domains of SOS (Sondermann et al. 2004; Gureasko et al. 2010). Thus, the activation of SOS requires its recruitment to the membrane as well as the release of blockage of the allosteric site of SOS.

The binding of Ras to the allosteric site tethers SOS to the membrane, and stimulates the nucleotide exchange activity of SOS strongly, in a positive feedback loop (Gureasko et al. 2008). The fact that activated SOS can be anchored to the membrane by Ras•GTP has important consequences. Upon release of autoinhibition, SOS molecules that are anchored to the membrane by Ras•GTP can activate large numbers of Ras molecules processively (Iversen et al. 2014). In cells, the pool of Ras•GTP that is required to trigger SOS activity is provided, in part, by the activity of the RasGRP family of GEFs (Roose et al. 2007). This interplay between SOS and other Ras-specific GEFs, such as RasGRP1, can result in a switch-like response to the activation of cell-surface receptors (Das et al. 2009). The allosteric activation of SOS by Ras is important for sustained Erk activation (Boykevisch et al. 2006), and for T-cell signaling (Roose et al. 2007; Das et al. 2009; Prasad et al. 2009). The allosteric stimulation of SOS by oncogenic Ras can result in the activation of wild-type alleles of Ras, resulting in stronger growth stimulation (Jeng et al. 2012).

In this review, we focus on the structural basis for the interplay between Ras and SOS that results in their mutual activation. We describe the mechanism of nucleotide exchange by SOS, and the nature of the allosteric stimulation of SOS by Ras. We then discuss how SOS is autoinhibited. Finally, we discuss some of the functional consequences of the allosteric activation of SOS at the membrane.

THE DOMAIN ORGANIZATION OF SOS

There are two isoforms of SOS in humans (SOS1 and SOS2) that have the same domain organization (Fig. 1B). SOS1 has 1333 residues and is 74% identical to SOS2, with a high degree of conservation distributed evenly throughout the sequence of both proteins. Most studies of human SOS have focused on SOS1, which we will refer to simply as "SOS." SOS acts upon all three principal isoforms of Ras in vertebrates (Buday and Downward 1993; Jaumot et al. 2002; Ehrhardt et al. 2004). The ~500-residue minimal segment of SOS that promotes efficient nucleotide exchange is the catalytic module, denoted SOS^{cat}, consisting of the REM domain and the

Cdc25 domain. The Cdc25 domain engages the substrate Ras molecule, causing it to release nucleotide (Boriack-Sjodin et al. 1998). A REM domain is usually found associated with the Cdc25 domain in Ras-specific GEFs (Boguski and McCormick 1993). The sequence of the Cdc25 domain is conserved, and is readily identifiable in fungi and animals, whereas the sequence of the REM domain is more variable across organisms (van Dam et al. 2009).

The amino-terminal ∼550 residues of SOS consist of three well-defined structural domains, located just before the catalytic module: a histone-fold domain (Sondermann et al. 2003), a Pleckstrin homology (PH) domain, and a Dbl homology (DH) domain (Soisson et al. 1998). Together, these domains cooperate to inhibit SOS activity by blocking Ras•GTP binding to the allosteric site (Corbalan-Garcia et al. 1998; Hall et al. 2002; Sondermann et al. 2004). Both the histone domain and the PH domain have been shown to bind to specific lipids in membranes, an important step in the release of autoinhibition (Gureasko et al. 2008; Yadav and Bar-Sagi 2010). DH domains are typically found in GEFs that are specific for the Rac, Rho, and cdc42 family of GTPases (Soisson et al. 1998; Aghazadeh et al. 2000; Worthylake et al. 2000), and SOS has been implicated in coupling Ras activation to the activation of Rac (Nimnual et al. 1998; Jun et al. 2013b), although it is unclear whether this coupling is direct. The structure of autoinhibited SOS, including the amino-terminal regulatory domains and the catalytic module, but excluding the carboxy-terminal tail, has been determined (Sondermann et al. 2004; Gureasko et al. 2010). These structures have revealed multiple ways in which the regulatory domains exert control on SOS activity, highlighting the exceptionally tight control of Ras activation, and its release by membrane interactions (Findlay and Pawson 2008).

The carboxy-terminal ∼280 residues of SOS are disordered, and this segment contains the proline-rich motifs that are binding sites for the SH3 domain of Grb2 (Chardin et al. 1993; Li et al. 1993). Mutagenesis and nuclear magnetic resonance (NMR) studies have provided insights into the interaction of Grb2 with SOS (Lim et al. 1994; Terasawa et al. 1994; Wittekind et al. 1994). The carboxy-terminal tail also inhibits SOS (Corbalan-Garcia et al. 1998), and its presence weakens the affinity of Ras•GTP for the allosteric site, although the precise structural details of this mechanism are unknown (Lee et al. 2017).

SOS INDUCES A WIDE OPENING OF THE ACTIVE SITE OF Ras, DISPLACING NUCLEOTIDE

Ras binds with very high affinity to both guanosine diphosphate (GDP) and GTP, with dissociation constants in the picomolar range (Goody et al. 1991), and Ras is unstable in the absence of nucleotide. The structure of a 1:1 Ras:SOS^{cat} complex showed that SOS stabilizes the nucleotide-free form of Ras by forming an extensive interface involving the Switch I and Switch II elements of Ras (Fig. 2A) (Boriack-Sjodin et al. 1998; Hall et al. 2001). Switch II provides the main anchor point for SOS on Ras. A helical hairpin element that protrudes from the main body of the Cdc25 domain of SOS is inserted between the Switch I and Switch II elements of Ras, acting as a molecular "tongue depressor" that causes the active site of Ras to open widely and expel the nucleotide (Fig. 2B). The principal conformational change in Ras is a peeling away of Switch I from the rest of Ras, caused by the insertion of the helical hairpin of SOS (Fig. 2C). The insertion of the helical hairpin into the Switch I region does not occlude the guanine and ribose binding sites in Ras, thereby facilitating the rebinding of nucleotide and the dislodging of SOS. Instead, residues within the helical hairpin, and the associated conformational changes in Ras, prevent the coordination of the phosphate groups of the nucleotide and the associated magnesium ion. Key features of this mechanism are preserved in other Ras-specific GEFs (Rehmann et al. 2008; Vercoulen et al. 2017).

Studies on the Ras-specific GEFs Cdc25 and RCC1 have shown that the GEF does not impose a preference on whether Ras is reloaded with GTP or GDP (Lenzen et al. 1998; Klebe et al. 1995). Once the nucleotide bound to Ras is re-

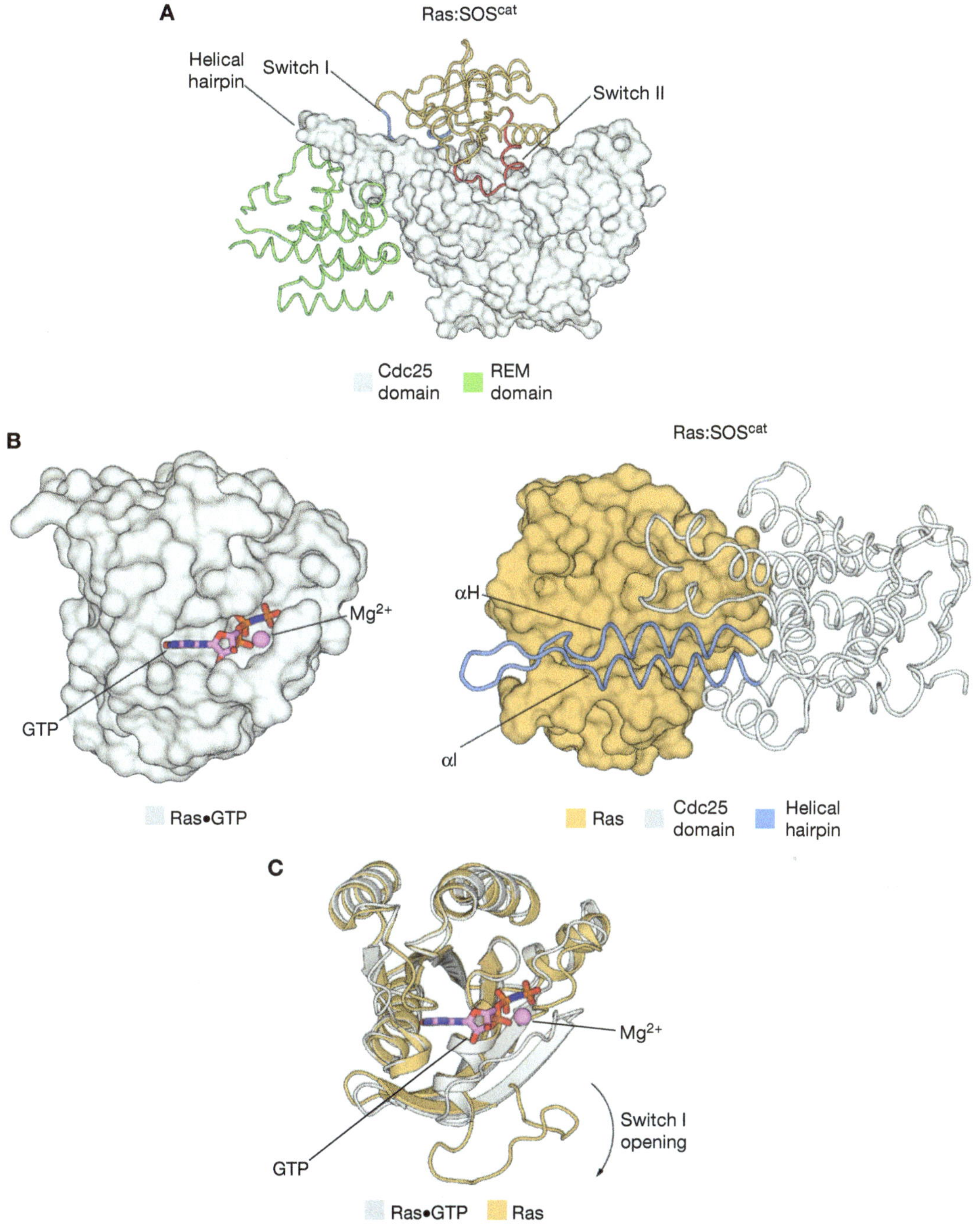

Figure 2. The structural mechanism of Ras activation by Son-of-Sevenless (SOS). (*A*) The structure of the Ras:SOS^{cat} complex (Protein Data Bank [PDB] code: 1BKD) showed that SOS stabilizes the nucleotide-free form of Ras by forming an extensive interface involving Switch I and Switch II of Ras. (*B*) (*Left*) The structure of Ras•GTP (PDB code: 5P21). (*Right*) A helical hairpin element that protrudes from the main body of the Cdc25 domain of SOS is inserted between the Switch I and Switch II elements of Ras, and causes the active site of Ras to open widely and expel the nucleotide. (*C*) After the insertion of the helical hairpin of SOS into Ras, the principal conformational change in Ras that leads to nucleotide release is a peeling away of Switch I from the rest of the protein. REM, Ras exchanger motif; GTP, guanosine triphosphate.

leased by the GEF, Ras is reset by the binding of GTP, driven by the higher cellular concentrations of GTP over GDP (Bennett et al. 2009). Although the data for other GEFs indicate that SOS has no intrinsic preference for loading Ras with either GDP or GTP, it has been suggested recently that SOS is more active when binding and stimulating the exchange of Ras•GDP instead of Ras•GTP (Vo et al. 2016). This conclusion was not, however, based on a direct comparison of the rates of nucleotide exchange for Ras•GDP and Ras•GTP by SOS, and so this question needs further investigation.

Peptide-based inhibitors of Ras have been developed that mimic the action of the helical hairpin of SOS (Patgiri et al. 2011). Brefeldin-A is a natural product that inhibits the action of Arf-specific GEFs (Klausner et al. 1992). It does so by binding to the interface between Arf and the GEF, slowing down the release of the complex (Mossessova et al. 2003; Renault et al. 2003). A Brefeldin-A-like molecule has not yet been developed for SOS, but compounds have been found that bind to Ras near the Ras:SOS interface, and enhance the rate of nucleotide exchange by SOS (Burns et al. 2014).

THE ACTIVATION OF SOS BY RAS•GTP

The discovery that SOS is allosterically activated by Ras•GTP was unexpected, and it was brought about by the serendipitous crystallization of a 2:1 complex of Ras and SOS^{cat}, by Margarit and colleagues (2003). Analysis of the resulting structure revealed that nucleotide-free Ras was bound to the active site in the Cdc25 domain, as seen earlier for the structure of the 1:1 Ras:SOS^{cat} complex, but that a second Ras•GTP molecule was bound to a distal site, wedged between the REM and Cdc25 domains (Fig. 3A) (Margarit et al. 2003). This site is specific for Ras•GTP and, based on comparison to the structures of complexes of Ras•GTP bound to effectors such as PI3-kinase (Pacold et al. 2000), the structure suggested that SOS is an "effector" for Ras•GTP. Biochemical studies showed that SOS^{cat} and Ras form a positive feedback loop, wherein Ras•GTP produced during the nucleotide exchange process binds to SOS^{cat} at the distal binding site, resulting in a dramatic increase in the rate of nucleotide exchange (Margarit et al. 2003; Gureasko et al. 2008) and potentiation of Ras activation by the epidermal growth factor receptor (Boykevisch et al. 2006).

The mechanism by which the binding of Ras•GTP to the allosteric site activates SOS was revealed by comparing the structure of isolated SOS^{cat}, with no Ras bound to the structures of either the 1:1 Ras:SOS complex or the ternary Ras•GTP:SOS^{cat}:Ras complex (Freedman et al. 2006). A prominent difference between the two structures is in the conformation of the helical hairpin of the cdc25 domain of SOS. The Ras-free structure of SOS^{cat} shows that the helical hairpin is tilted toward the active site of SOS, thereby constricting the site where SOS engages Switch II of nucleotide-free Ras (Fig. 3B). This suggests that Ras binding to the active site of SOS requires the helical hairpin to be pulled back. Indeed, the allosteric binding of Ras•GTP to SOS facilitates the rotation and opening of the helical hairpin, which frees the catalytic site to bind Ras. This structural observation was validated by mutating residues in the helical hairpin so as to stabilize the catalytically competent open conformation. These mutations led to an increase in the rate of nucleotide exchange (Freedman et al. 2009; Vo et al. 2016).

THE AMINO-TERMINAL REGULATORY DOMAINS OF SOS BLOCK THE BINDING OF RAS•GTP AT THE ALLOSTERIC SITE

SOS contains a ~350-residue DH-PH module located immediately before the catalytic module. DH domains, which are usually associated with PH domains, are GEFs for Rho-family GTPases (Cerione and Zheng 1996; Whitehead et al. 1997; Bishop and Hall 2000; Rossman et al. 2005). PH domains bind to specific phospholipids in the membrane (Lemmon and Ferguson 2000). The structure of the DH-PH module of SOS was the first to reveal how the two domains are integrated into a structural module, but, ironically, the DH domain of SOS is not active as a Rho-family GEF (Soisson et al. 1998). Comparison of the crystal structures of the DH-PH modules of SOS with the structure of DH-PH

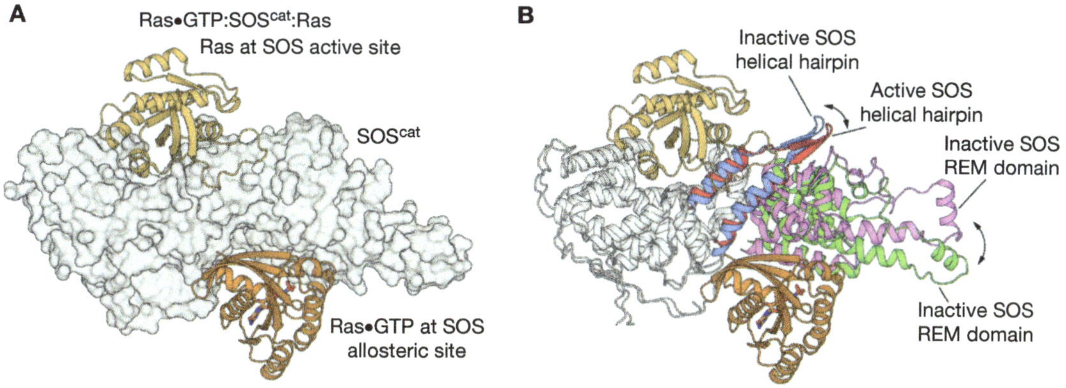

Figure 3. The activation of Son-of-Sevenless (SOS) by Ras•GTP (guanosine triphosphate). (*A*) The structure of the 2:1 Ras:SOS^{cat} complex (Protein Data Bank [PDB] code: 1NVV) revealed that nucleotide-free Ras was bound to the active site of the Cdc25 domain, and that a second Ras•GTP molecule is bound to a distal allosteric site, wedged between the Ras exchanger motif (REM) and Cdc25 domains (*B*) Comparison of the Ras-free structure of SOS^{cat} (PDB code: 2II0) to the 2:1 Ras:SOS^{cat} complex shows that the helical hairpin is tilted toward the active site of SOS in the Ras-free state, thereby constricting the site where SOS engages Switch II of nucleotide-free Ras. Allosteric binding of Ras•GTP to SOS leads to the rotation and opening of the helical hairpin, accompanied by the outward rotation of the REM domain, which frees the catalytic site to bind Ras.

modules that act as Rho-family GEFs indicated that the conformation of the DH-PH module of SOS is not compatible with binding to small GTPases, unless a large conformational change occurs (Soisson et al. 1998; Aghazadeh et al. 2000; Worthylake et al. 2000). Although it is possible that such a conformational change could occur when the PH domain engages the membrane, sequence features of the DH-PH module of SOS also suggest that it does not act directly as a nucleotide exchange factor (Soisson et al. 1998).

The role of the DH-PH domain in SOS is to regulate binding of Ras•GTP to the allosteric site of SOS (Fig. 4A). The structure of a construct of SOS containing the DH-PH module and the catalytic module (SOS^{DPC}) showed that the location of the DH-PH module is such that it sterically occludes the binding of Ras•GTP at the allosteric site (Fig. 4B). Biochemical measurements show that SOS^{DPC} has lower activity than SOS^{cat} (Fig. 4C) (Sondermann et al. 2004; Gureasko et al. 2010). Relief of autoinhibition occurs when the PH domain binds to negatively charged membrane phospholipids. This binding event triggers a conformational change in the DH domain that ultimately reveals the allosteric Ras•GTP binding site on SOS (Gureasko et al. 2008).

The first ~200 residues of SOS comprise an amino-terminal tandem histone-fold domain, the presence of which in a cytoplasmic domain is unusual (Sondermann et al. 2003). Reflecting the organization of the nucleosome, most non-histone proteins that contain histone-like domains are present in the nucleus as dimers, and play a role in nucleic acid metabolism (Baxevanis et al. 1995). The two histone domains of SOS are also arranged similarly to two histone proteins in the dimeric units that make up the nucleosome. The DNA-binding face of the "histone dimer" of SOS retains the pattern of positive charge that, in histones, binds DNA. This suggested that the histone domain of SOS may play a role in lipid binding, by using these positively charged residues to interact with the phosphate headgroups of lipids, instead of with the phosphate backbone of DNA. Indeed, mutations in this region affect the affinity of the histone fold for membrane lipids (Yadav and Bar-Sagi 2010; Saliba et al. 2014).

The crystal structure of SOS^{HDPC} shows that the histone domain binds to the linker connecting the DH-PH module to the REM domain,

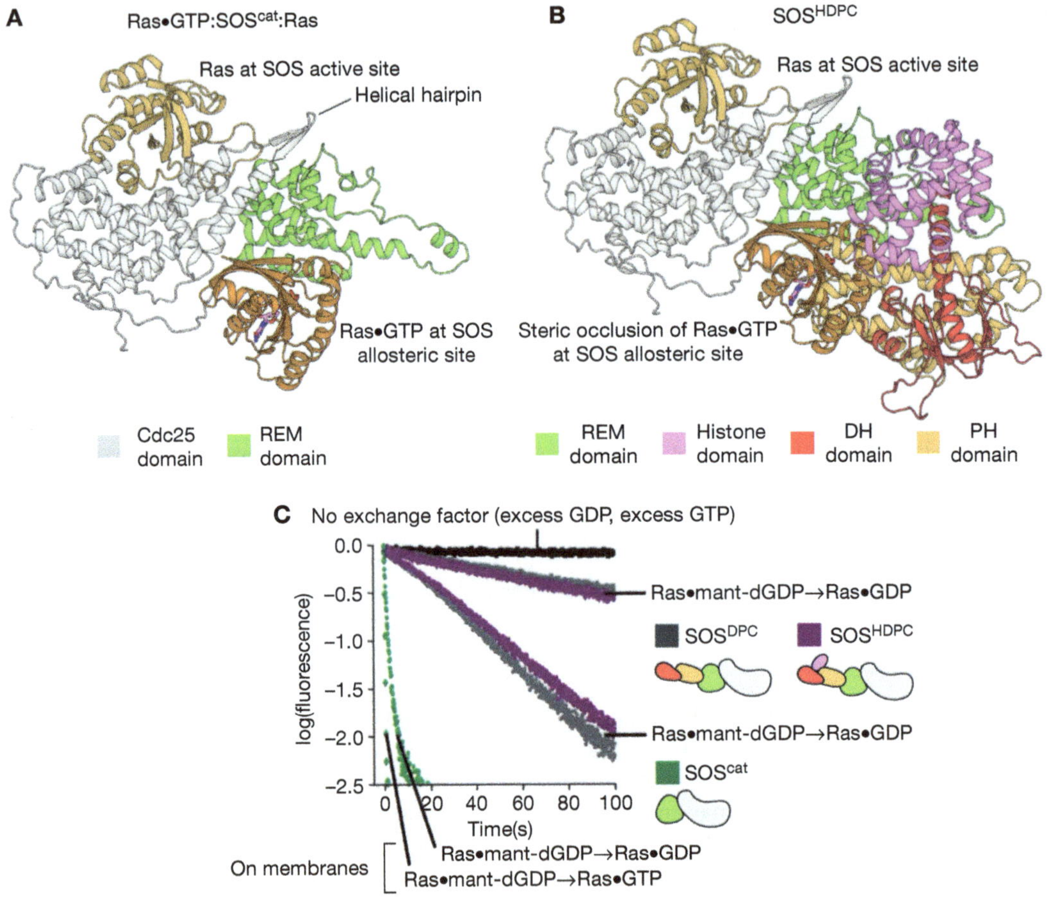

Figure 4. The autoinhibition of Son-of-Sevenless (SOS) by regulatory domains. (*A*) The structure of the 2:1 Ras: SOS^{cat} complex, depicting Ras bound to the Cdc25 domain and Ras•GTP (guanosine triphosphate) bound to the allosteric site of SOS. (*B*) The structure of SOS containing the Dbl homology (DH)-Pleckstrin homology (PH) module and the catalytic module (SOS^{DPC}) (Protein Data Bank [PDB] code: 3KSY) show that the DH-PH module sterically occludes the binding of Ras•GTP at the allosteric site. (*C*) Biochemical measurements show that SOS^{DPC} has lower activity than SOS^{cat}. The fastest traces are for membrane-bound Ras (Gureasko et al. 2008). REM, Ras exchanger motif; GDP, guanosine diphosphate; dGDP, deoxyguanosine diphosphate.

thereby stabilizing the autoinhibited state of SOS (Sondermann et al. 2004; Gureasko et al. 2010; Lee et al. 2017). Notably, the presence of the histone domain suppresses the ability of membrane lipids to release the inhibitory action of the DH-PH module on SOS activity. Arg 552, located in the linker connecting the DH-PH module to the REM domain, stabilizes the autoinhibited conformation of SOS by interacting with the histone domain (Sondermann et al. 2005), and mutation of this arginine residue activates SOS. Such gain-of-function mutations in SOS have been associated with Noonan syndrome (Roberts et al. 2007; Tartaglia et al. 2007).

ACTIVATION OF SOS AT THE MEMBRANE

Although it was known that the constitutive membrane localization of SOS activates Ras (Aronheim et al. 1994), a series of experiments showed that the activity of SOS is greatly enhanced by allosteric binding of Ras•GTP at the membrane (Gureasko et al. 2008). In these experiments, Ras•GTP was either covalently at-

 Cite this article as *Cold Spring Harb Perspect Med* doi: 10.1101/cshperspect.a031534

tached to supported lipid bilayers or tethered to lipid vesicles, and the rate of nucleotide exchange was measured upon membrane recruitment and allosteric activation of SOS^{cat}. On vesicles, the nucleotide exchange rate was enhanced by approximately 500-fold compared to a solution measurement, and the nucleotide exchange rate on lipid bilayers increased as a function of Ras•GTP surface density, in a manner that is dependent on Ras•GTP binding to the allosteric site of SOS. Mutations that weaken the allosteric binding of Ras•GTP fail to recruit SOS^{cat} to the membrane, and therefore prevent the stimulation of nucleotide exchange.

Once a single molecule of SOS is activated allosterically by Ras•GTP at the membrane, hundreds of Ras molecules can then be activated processively by that SOS molecule (Fig. 5). This was shown by integrating structural and bulk biochemical observations with single-molecule measurements of SOS on membranes (Iversen et al. 2014). Solution experiments have shown that Ras•GDP binds about 10-fold less tightly to the allosteric site than Ras•GTP, and a crystal structure revealed that Ras with hydrolyzed GTP ($GDP+P_i$) can be bound to the allosteric site (Sondermann et al. 2004). The feedback activation of SOS by Ras occurs at two levels: first, Ras•GTP is more efficient at recruiting SOS to the membrane, by virtue of having a higher binding affinity to the allosteric site than Ras•GDP. Second, on the membrane, Ras•GTP is more effective at enabling SOS to transition from states of lower GEF activity to states of much higher activity, as demonstrated by single-molecule measurements of the nucleotide exchange rates of SOS^{cat} corralled within micrometer-size partitions on a supported lipid bilayer (Iversen et al. 2014). In these experiments, a spectrum of SOS activity was observed, and a minority of highly active states of SOS dominated the nucleotide exchange reaction. Once SOS is anchored to the membrane by either Ras•GDP or Ras•GTP, the difference between the nucleotide-loading

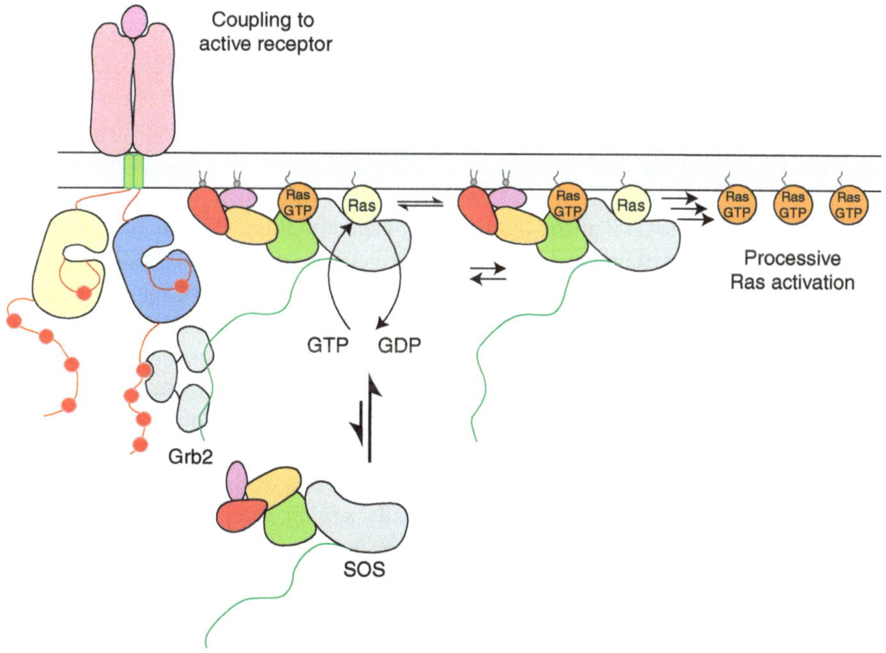

Figure 5. The activation of Son-of-Sevenless (SOS) at the membrane. SOS is initially recruited to the membrane by adaptor proteins, such as Grb2, that bind to activated cell-surface receptors. Once a single molecule of SOS is activated allosterically by Ras•GTP (guanosine triphosphate) at the membrane, many more Ras molecules can then be activated processively by that SOS molecule, even if it were to disengage from the receptor. GDP, Guanosine diphosphate.

state of Ras bound at the allosteric site is only manifested in the efficiency of accessing the most highly active states. The structural basis for the interconversion of highly and lowly active states of SOS is not yet understood.

The slow off-rate of Ras•GTP at the allosteric site contributes to SOS^{cat} retention on the membrane in the absence of regulatory domains (Iversen et al. 2014). However, once SOS^{HDPC} is recruited to the membrane, it is retained there much longer than SOS^{cat} (Christensen et al. 2016), largely as a result of interactions between lipid-binding regulatory domains with membrane lipids. In cells, SOS is also tethered to transmembrane receptors via interactions with Grb2, and SOS activation is balanced by the reversible binding kinetics of the SOS:Grb2 interaction (Christensen et al. 2016). Furthermore, SOS remains at the membrane until actively being removed by membrane endocytosis, thus preventing new Ras molecules from being activated (Christensen et al. 2016). Recent work has pointed to the existence of phase-separated signaling clusters on membranes, where multivalent interactions between the T-cell receptor, costimulatory receptors, SOS, Grb2, and other proteins lead to downstream signal transduction in T cells (Huang et al. 2016; Su et al. 2016). The ability of SOS to make multiple interactions to scaffold proteins via Grb2 and Shc, as well as its ability to stay engaged to Ras•GTP at the membrane, are critical properties that are likely to facilitate these phase transitions.

BISTABILITY IN RAS ACTIVATION BY SOS IN CELLS

In lymphocytes, RasGRP exchange factors produce Ras•GTP that subsequently allosterically activates SOS, potentiating the direct activation of SOS via Grb2 (Roose et al. 2007). The subsequent production of Ras•GTP by SOS leads to sharp responses in signaling pathways that control the generation of self-tolerant immune cells (Das et al. 2009). RasGRP proteins are activated by binding to membrane-localized diacylglycerol or phosphorylation by protein kinase C (Ebinu et al. 1998, 2000), and the strength of the external stimulus can further tune the activity of SOS and RasGRP proteins. Therefore, the interplay between SOS and RasGRP proteins, and their ability to be activated by distinct stimuli, leads to dynamic downstream signaling responses in cells.

In the case of SOS, low levels of receptor stimulation, and consequently low levels of SOS activation, lead to a predominant population of cells containing a low amount of Ras•GTP. Conversely, high levels of receptor stimulation led to a predominant population of cells containing a high amount of Ras•GTP (Das et al. 2009). Intermediate levels of receptor stimulation lead to two distinct cell populations, containing low and high amounts of Ras•GTP, respectively. Therefore, an intermediate level of SOS activity introduces a bistability in the amount of Ras•GTP produced in response to receptor stimulation (Das et al. 2009). This bistability disappears past a threshold amount of SOS activity, suggesting that T lymphocytes occupy either one of two binary "on" or "off" states with regard to Ras activation, in a "digital" manner (Das et al. 2009). For RasGRP proteins, increasing levels of receptor stimulation lead to an increasing number of cells containing Ras•GTP, and bistability in Ras•GTP levels is not observed. This suggests a graded, or "analog," response of Ras•GTP production to receptor stimulation.

Once the allosteric site of SOS is primed and engaged by Ras•GTP, the bistable nature of SOS activation emerges, and might give rise to a molecular memory of past signals. This "hysteresis" in the Ras response was predicted by computational simulations, and was validated in T cells subject to repeated stimulation over the course of tens of minutes (Das et al. 2009). Hysteresis in Ras activation allows T cells to produce sustained levels of Ras•GTP in response to potentially weaker receptor stimulation, in a manner that is dependent on the feedback interaction between Ras and SOS.

SOS MAY HAVE CONSTRAINED Ras EVOLUTION IN METAZOANS

SOS is the only Ras-specific GEF that is known to be allosterically activated by Ras•GTP, pointing to an evolutionary divergence between SOS

 Cite this article as *Cold Spring Harb Perspect Med* doi: 10.1101/cshperspect.a031534

and other Ras-specific GEFs. Insight into the unique dependence of SOS on Ras may be gained from analyzing the sequence conservation of Ras in metazoans, relative to other Ras family GTPases. The three principal isoforms of Ras (H-Ras, K-Ras, and N-Ras) are generally highly conserved in metazoans (Rojas et al. 2012; Johnson et al. 2017), though less so in invertebrates compared to vertebrates (Bandaru et al. 2017). The sequence of Ras from the choanoflagellate *Salpingoeca rosetta*, thought to be the closest living relative of metazoans, is 72% identical to human Ras, whereas Ras is nearly invariant (>90% identical to human) in vertebrates. In comparison, the sequence of the Ras-family GTPase, Rap, in *S. rosetta* is 92% identical to that of human Rap, indicating that the selective pressures on Rap did not change as much in the transition into the vertebrate lineage. It is intriguing to note that the high conservation of Ras in vertebrates can possibly be traced to interactions with SOS, which may have been a factor that constrained the evolution of Ras (Bandaru et al. 2017).

The mutational sensitivity of human H-Ras was determined by screening a library consisting of all possible point mutations to the protein sequence, in the presence of a GTPase-activating protein (GAP) and a GEF (Bandaru et al. 2017). It was found that the overall conservation of Ras was a consequence of most residues being involved in binding the nucleotide, maintaining protein stability, interacting with the GAP and the GEF, as well as suppressing spontaneous activation. Interestingly, four regions in the sequence of Ras are more tolerant of mutation, and these contain residues that differed between invertebrate and human Ras, yet are nearly invariant in vertebrates. Residues in two of the four variable regions of Ras are directly involved in allosteric binding to SOS, and residues in the remaining two variable regions are involved in a two-state conformational transition that might affect the affinity of Ras for the allosteric site of SOS. This conformational transition may be triggered by the binding of an allosteric effector between the α3 and α4 helices of Ras (Buhrman et al. 2010), leading to the outward rotation of the Tyr 71 residue, which appears inconsistent with Ras binding to the allosteric site of SOS. The structure of *S. rosetta* Ras adopts the outward conformation of Tyr 71, whereas the structure human Ras does not, and *S. rosetta* Ras also fails to allosterically activate SOS (Bandaru et al. 2017). These observations suggest that SOS may have coevolved with Ras in vertebrates.

CONCLUDING REMARKS

Ras is a highly conserved and central molecular switch in metazoan cell signaling. In vertebrates, Ras displays the unique property of being able to activate its own activator, SOS. Structural and functional analysis has revealed the remarkable complexity underlying the interdependent activation of Ras and SOS. The culmination of the release of autoinhibition of SOS by regulatory domains is the allosteric binding of Ras•GTP to SOS. This binding event sets up a positive feedback loop, which allows receptor tyrosine kinases to generate highly nonlinear responses in response to extracellular stimuli. This has profound consequences on cell signaling, most importantly in generating a sharp response in pathways that ultimately control cellular differentiation and growth. It is intriguing that the structural element in SOS that promotes the allosteric activation by Ras is the REM domain, which is also present in other Ras-specific GEFs. An intriguing question for further investigation is whether the REM domains of other Ras-specific GEFs modulate activity by interacting with as-of-yet unidentified allosteric regulators.

REFERENCES

**Reference is also in this collection.*

Aghazadeh B, Lowry WE, Huang XY, Rosen MK. 2000. Structural basis for relief of autoinhibition of the Dbl homology domain of proto-oncogene Vav by tyrosine phosphorylation. *Cell* **102:** 625–633.

Aronheim A, Engelberg D, Li N, al-Alawi N, Schlessinger J, Karin M. 1994. Membrane targeting of the nucleotide exchange factor Sos is sufficient for activating the Ras signaling pathway. *Cell* **78:** 949–961.

Bandaru P, Shah NH, Bhattacharyya M, Barton JP, Kondo Y, Cofsky JC, Gee CL, Chakraborty AK, Kortemme T, Ranganathan R, Kuriyan, J. 2017. Deconstruction of the Ras switching cycle through saturation mutagenesis. *eLife* **6: e27810**.

Baxevanis AD, Arents G, Moudrianakis EN, Landsman D. 1995. A variety of DNA-binding and multimeric proteins contain the histone fold motif. *Nucleic Acids Res* **23:** 2685–2691.

Bennett BD, Kimball EH, Gao M, Osterhout R, Van Dien SJ, Rabinowitz JD. 2009. Absolute metabolite concentrations and implied enzyme active site occupancy in *Escherichia coli*. *Nat Chem Biol* **5:** 593–599.

Bishop AL, Hall A. 2000. Rho GTPases and their effector proteins. *Biochem J* **348:** 241–255.

Boguski MS, McCormick F. 1993. Proteins regulating Ras and its relatives. *Nature* **366:** 643–654.

Boriack-Sjodin PA, Margarit SM, Bar-Sagi D, Kuriyan J. 1998. The structural basis of the activation of Ras by Sos. *Nature* **394:** 337–343.

* Bos JL. 2017. From ras to Rap and back, a journey of 35 years. *Cold Spring Harb Perspect Med* doi: 10.1101/cshperspect.a031468.

Bos JL, Rehmann H, Wittinghofer A. 2007. GEFs and GAPs: Critical elements in the control of small G proteins. *Cell* **129:** 865–877.

Boykevisch S, Zhao C, Sondermann H, Philippidou P, Halegoua S, Kuriyan J, Bar-Sagi D. 2006. Regulation of ras signaling dynamics by Sos-mediated positive feedback. *Curr Biol* **16:** 2173–2179.

Buday L, Downward J. 1993. Epidermal growth factor regulates p21ras through the formation of a complex of receptor, Grb2 adapter protein, and Sos nucleotide exchange factor. *Cell* **73:** 611–620.

Buday L, Egan SE, Rodriguez Viciana P, Cantrell DA, Downward J. 1994. A complex of Grb2 adaptor protein, Sos exchange factor, and a 36-kDa membrane-bound tyrosine phosphoprotein is implicated in ras activation in T cells. *J Biol Chem* **269:** 9019–9023.

Buhrman G, Holzapfel G, Fetics S, Mattos C. 2010. Allosteric modulation of Ras positions Q61 for a direct role in catalysis. *Proc Natl Acad Sci* **107:** 4931–4936.

Burns MC, Sun Q, Daniels RN, Camper D, Kennedy JP, Phan J, Olejniczak ET, Lee T, Waterson AG, Rossanese OW, et al. 2014. Approach for targeting Ras with small molecules that activate SOS-mediated nucleotide exchange. *Proc Natl Acad Sci* **111:** 3401–3406.

Cerione RA, Zheng Y. 1996. The Dbl family of oncogenes. *Curr Opin Cell Biol* **8:** 216–222.

Chardin P, Camonis JH, Gale NW, van Aelst L, Schlessinger J, Wigler MH, Bar-Sagi D. 1993. Human Sos1: A guanine nucleotide exchange factor for Ras that binds to GRB2. *Science* **260:** 1338–1343.

Cherfils J, Zeghouf M. 2013. Regulation of small GTPases by GEFs, GAPs, and GDIs. *Physiol Rev* **93:** 269–309.

Christensen SM, Tu HL, Jun JE, Alvarez S, Triplet MG, Iwig JS, Yadav KK, Bar-Sagi D, Roose JP, Groves JT. 2016. One-way membrane trafficking of SOS in receptor-triggered Ras activation. *Nat Struct Mol Biol* **23:** 838–846.

Corbalan-Garcia S, Margarit SM, Galron D, Yang SS, Bar-Sagi D. 1998. Regulation of Sos activity by intramolecular interactions. *Mol Cell Biol* **18:** 880–886.

Das J, Ho M, Zikherman J, Govern C, Yang M, Weiss A, Chakraborty AK, Roose JP. 2009. Digital signaling and hysteresis characterize ras activation in lymphoid cells. *Cell* **136:** 337–351.

de Rooij J, Rehmann H, van Triest M, Cool RH, Wittinghofer A, Bos JL. 2000. Mechanism of regulation of the Epac family of cAMP-dependent RapGEFs. *J Biol Chem* **275:** 20829–20836.

Ebinu JO, Bottorff DA, Chan EY, Stang SL, Dunn RJ, Stone JC. 1998. RasGRP, a Ras guanyl nucleotide- releasing protein with calcium- and diacylglycerol-binding motifs. *Science* **280:** 1082–1086.

Ebinu JO, Stang SL, Teixeira C, Bottorff DA, Hooton J, Blumberg PM, Barry M, Bleakley RC, Ostergaard HL, Stone JC. 2000. RasGRP links T-cell receptor signaling to Ras. *Blood* **95:** 3199–3203.

Egan SE, Giddings BW, Brooks MW, Buday L, Sizeland AM, Weinberg RA. 1993. Association of Sos Ras exchange protein with Grb2 is implicated in tyrosine kinase signal transduction and transformation. *Nature* **363:** 45–51.

Ehrhardt A, David MD, Ehrhardt GRA, Schrader JW. 2004. Distinct mechanisms determine the patterns of differential activation of H-Ras, N-Ras, K-Ras 4B, and M-Ras by receptors for growth factors or antigen. *Mol Cell Biol* **24:** 6311–6323.

Findlay GM, Pawson T. 2008. How is SOS activated? Let us count the ways. *Nat Struct Mol Biol* **15:** 538–540.

Freedman TS, Sondermann H, Friedland GD, Kortemme T, Bar-Sagi D, Marqusee S, Kuriyan J. 2006. A Ras-induced conformational switch in the Ras activator Son of sevenless. *Proc Natl Acad Sci* **103:** 16692–16697.

Freedman TS, Sondermann H, Kuchment O, Friedland GD, Kortemme T, Kuriyan J. 2009. Differences in flexibility underlie functional differences in the Ras activators son of sevenless and Ras guanine nucleotide releasing factor 1. *Structure* **17:** 41–53.

Gale NW, Kaplan S, Lowenstein EJ, Schlessinger J, Bar-Sagi D. 1993. Grb2 mediates the EGF-dependent activation of guanine nucleotide exchange on Ras. *Nature* **363:** 88–92.

Goody RS, Frech M, Wittinghofer A. 1991. Affinity of guanine nucleotide binding proteins for their ligands: Facts and artefacts. *Trends Biochem Sci* **16:** 327–328.

Gureasko J, Galush WJ, Boykevisch S, Sondermann H, Bar-Sagi D, Groves JT, Kuriyan J. 2008. Membrane-dependent signal integration by the Ras activator Son of sevenless. *Nat Struct Mol Biol* **15:** 452–461.

Gureasko J, Kuchment O, Makino DL, Sondermann H, Bar-Sagi D, Kuriyan J. 2010. Role of the histone domain in the autoinhibition and activation of the Ras activator Son of Sevenless. *Proc Natl Acad Sci* **107:** 3430–3435.

Hall BE, Yang SS, Boriack-Sjodin PA, Kuriyan J, Bar-Sagi D. 2001. Structure-based mutagenesis reveals distinct functions for Ras switch 1 and switch 2 in Sos-catalyzed guanine nucleotide exchange. *J Biol Chem* **276:** 27629–27637.

Hall BE, Yang SS, Bar-Sagi D. 2002. Autoinhibition of Sos by intramolecular interactions. *Front Biosci* **7:** d288–d294.

Huang WYC, Yan Q, Lin WC, Chung JK, Hansen SD, Christensen SM, Tu HL, Kuriyan J, Groves JT. 2016. Phosphotyrosine-mediated LAT assembly on membranes drives kinetic bifurcation in recruitment dynamics of the Ras activator SOS. *Proc Natl Acad Sci* **113:** 8218–8223.

Iversen L, Tu HL, Lin WC, Christensen SM, Abel SM, Iwig J, Wu HJ, Gureasko J, Rhodes C, Petit RS, et al. 2014. Molecular kinetics. Ras activation by SOS: Allosteric regulation by altered fluctuation dynamics. *Science* 345: 50–54.

Iwig JS, Vercoulen Y, Das R, Barros T, Limnander A, Che Y, Pelton JG, Wemmer DE, Roose JP, Kuriyan J. 2013. Structural analysis of autoinhibition in the Ras-specific exchange factor RasGRP1. *eLife* **2:** e00813.

Jaumot M, Yan J, Clyde-Smith J, Sluimer J, Hancock JF. 2002. The linker domain of the Ha-Ras hypervariable region regulates interactions with exchange factors, Raf-1 and phosphoinositide 3-kinase. *J Biol Chem* **277:** 272–278.

Jeng HH, Taylor LJ, Bar-Sagi D. 2012. Sos-mediated cross-activation of wild-type Ras by oncogenic Ras is essential for tumorigenesis. *Nat Commun* **3:** 1168.

Johnson CW, Reid D, Parker JA, Salter S, Knihtila R, Kuzmic P, Mattos C. 2017. The small GTPases K-Ras, N-Ras and H-Ras have distinct biochemical properties determined by allosteric effects. *J Biol Chem* **292:** 12981–12993.

Jun JE, Rubio I, Roose JP. 2013a. Regulation of ras exchange factors and cellular localization of ras activation by lipid messengers in T cells. *Front Immunol* **4:** 239.

Jun JE, Yang M, Chen H, Chakraborty AK, Roose JP. 2013b. Activation of extracellular signal-regulated kinase but not of p38 mitogen-activated protein kinase pathways in lymphocytes requires allosteric activation of SOS. *Mol Cell Biol* **33:** 2470–2484.

Klausner RD, Donaldson JG, Lippincott-Schwartz J. 1992. Brefeldin A: Insights into the control of membrane traffic and organelle structure. *J Cell Biol* **116:** 1071–1080.

Klebe C, Prinz H, Wittinghofer A, Goody RS. 1995. The kinetic mechanism of Ran–nucleotide exchange catalyzed by RCC1. *Biochemistry* **34:** 12543–12552.

Lee YK, Low-Nam ST, Chung JK, Hansen SD, Lam HYM, Alvarez S, Groves JT. 2017. Mechanism of SOS PR-domain autoinhibition revealed by single-molecule assays on native protein from lysate. *Nat Commun* **8:** 15061.

Lemmon MA, Ferguson KM. 2000. Signal-dependent membrane targeting by Pleckstrin homology (PH) domains. *Biochem J* **350:** 1–18.

Lenzen C, Cool RH, Prinz H, Kuhlmann J, Wittinghofer A. 1998. Kinetic analysis by fluorescence of the interaction between Ras and the catalytic domain of the guanine nucleotide exchange factor Cdc25Mm. *Biochemistry* **37:** 7420–7430.

Li N, Batzer A, Daly R, Yajnik V, Skolnik E, Chardin P, Bar-Sagi D, Margolis B, Schlessinger J. 1993. Guanine-nucleotide-releasing factor hSos1 binds to Grb2 and links receptor tyrosine kinases to Ras signalling. *Nature* **363:** 85–88.

Lim WA, Richards FM, Fox RO. 1994. Structural determinants of peptide-binding orientation and of sequence specificity in SH3 domains. *Nature* **372:** 375–379.

Lowenstein EJ, Daly RJ, Batzer AG, Li W, Margolis B, Lammers R, Ullrich A, Skolnik EY, Bar-Sagi D, Schlessinger J. 1992. The SH2 and SH3 domain-containing protein GRB2 links receptor tyrosine kinases to ras signaling. *Cell* **70:** 431–442.

Margarit SM, Sondermann H, Hall BE, Nagar B, Hoelz A, Pirruccello M, Bar-Sagi D, Kuriyan J. 2003. Structural evidence for feedback activation by Ras•GTP of the Ras-specific nucleotide exchange factor SOS. *Cell* **112:** 685–695.

Mossessova E, Corpina RA, Goldberg J. 2003. Crystal structure of ARF1*Sec7 complexed with Brefeldin A and its implications for the guanine nucleotide exchange mechanism. *Mol Cell* **12:** 1403–1411.

Nimnual AS, Yatsula BA, Bar-Sagi D. 1998. Coupling of Ras and Rac guanosine triphosphatases through the Ras exchanger Sos. *Science* **279:** 560–563.

Pacold ME, Suire S, Perisic O, Lara-Gonzalez S, Davis CT, Walker EH, Hawkins PT, Stephens L, Eccleston JF, Williams RL. 2000. Crystal structure and functional analysis of Ras binding to its effector phosphoinositide 3-kinase γ. *Cell* **103:** 931–943.

Patgiri A, Yadav KK, Arora PS, Bar-Sagi D. 2011. An orthosteric inhibitor of the Ras-Sos interaction. *Nat Chem Biol* **7:** 585–587.

Prasad A, Zikherman J, Das J, Roose JP, Weiss A, Chakraborty AK. 2009. Origin of the sharp boundary that discriminates positive and negative selection of thymocytes. *Proc Natl Acad Sci* **106:** 528–533.

Rehmann H, Das J, Knipscheer P, Wittinghofer A, Bos JL. 2006. Structure of the cyclic-AMP-responsive exchange factor Epac2 in its auto-inhibited state. *Nature* **439:** 625–628.

Rehmann H, Arias-Palomo E, Hadders MA, Schwede F, Llorca O, Bos JL. 2008. Structure of Epac2 in complex with a cyclic AMP analogue and RAP1B. *Nature* **455:** 124–127.

Renault L, Guibert B, Cherfils J. 2003. Structural snapshots of the mechanism and inhibition of a guanine nucleotide exchange factor. *Nature* **426:** 525–530.

Roberts AE, Araki T, Swanson KD, Montgomery KT, Schiripo TA, Joshi VA, Li L, Yassin Y, Tamburino AM, Neel BG, et al. 2007. Germline gain-of-function mutations in SOS1 cause Noonan syndrome. *Nat Genet* **39:** 70–74.

Rojas AM, Fuentes G, Rausell A, Valencia A. 2012. The Ras protein superfamily: Evolutionary tree and role of conserved amino acids. *J Cell Biol* **196:** 189–201.

Roose JP, Mollenauer M, Ho M, Kurosaki T, Weiss A. 2007. Unusual interplay of two types of Ras activators, RasGRP and SOS, establishes sensitive and robust Ras activation in lymphocytes. *Mol Cell Biol* **27:** 2732–2745.

Rossman KL, Der CJ, Sondek J. 2005. GEF means go: Turning on RHO GTPases with guanine nucleotide-exchange factors. *Nat Rev Mol Cell Biol* **6:** 167–180.

Rozakis-Adcock M, McGlade J, Mbamalu G, Pelicci G, Daly R, Li W, Batzer A, Thomas S, Brugge J, Pelicci PG, et al. 1992. Association of the Shc and Grb2/Sem5 SH2-containing proteins is implicated in activation of the Ras pathway by tyrosine kinases. *Nature* **360:** 689–692.

Rozakis-Adcock M, Fernley R, Wade J, Pawson T, Bowtell D. 1993. The SH2 and SH3 domains of mammalian Grb2 couple the EGF receptor to the Ras activator mSos1. *Nature* **363:** 83–85.

Saliba A-E, Vonkova I, Ceschia S, Findlay GM, Maeda K, Tischer C, Deghou S, van Noort V, Bork P, Pawson T, et al. 2014. A quantitative liposome microarray to systematically characterize protein-lipid interactions. *Nat Methods* **11:** 47–50.

Soisson SM, Nimnual AS, Uy M, Bar-Sagi D, Kuriyan J. 1998. Crystal structure of the Dbl and Pleckstrin homology domains from the human Son of sevenless protein. *Cell* **95:** 259–268.

Sondermann H, Soisson SM, Bar-Sagi D, Kuriyan J. 2003. Tandem histone folds in the structure of the N-terminal segment of the ras activator Son of Sevenless. *Structure* **11:** 1583–1593.

Sondermann H, Soisson SM, Boykevisch S, Yang S-S, Bar-Sagi D, Kuriyan J. 2004. Structural analysis of autoinhibition in the Ras activator Son of sevenless. *Cell* **119:** 393–405.

Sondermann H, Nagar B, Bar-Sagi D, Kuriyan J. 2005. Computational docking and solution x-ray scattering predict a membrane-interacting role for the histone domain of the Ras activator son of sevenless. *Proc Natl Acad Sci* **102:** 16632–16637.

Su X, Ditlev JA, Hui E, Xing W, Banjade S, Okrut J, King DS, Taunton J, Rosen MK, Vale RD. 2016. Phase separation of signaling molecules promotes T cell receptor signal transduction. *Science* **352:** 595–599.

Tartaglia M, Pennacchio LA, Zhao C, Yadav KK, Fodale V, Sarkozy A, Pandit B, Oishi K, Martinelli S, Schackwitz W, et al. 2007. Gain-of-function SOS1 mutations cause a distinctive form of Noonan syndrome. *Nat Genet* **39:** 75–79.

Terasawa H, Kohda D, Hatanaka H, Tsuchiya S, Ogura K, Nagata K, Ishii S, Mandiyan V, Ullrich A, Schlessinger J, et al. 1994. Structure of the N-terminal SH3 domain of GRB2 complexed with a peptide from the guanine nucleotide releasing factor Sos. *Nat Struct Mol Biol* **1:** 891–897.

van Dam TJP, Rehmann H, Bos JL, Snel B. 2009. Phylogeny of the CDC25 homology domain reveals rapid differentiation of Ras pathways between early animals and fungi. *Cell Signal* **21:** 1579–1585.

Vercoulen Y, Kondo Y, Iwig JS, Janssen AB, White KA, Amini M, Barber DL, Kuriyan J, Roose JP. 2017. A histidine pH sensor regulates activation of the Ras-specific guanine nucleotide exchange factor RasGRP1. *eLife* doi: 10.7554/eLife.209002.

Vo U, Vajpai N, Flavell L, Bobby R, Breeze AL, Embrey KJ, Golovanov AP. 2016. Monitoring Ras interactions with the nucleotide exchange factor Son of Sevenless (Sos) using site-specific NMR reporter signals and intrinsic fluorescence. *J Biol Chem* **291:** 1703–1718.

Whitehead IP, Campbell S, Rossman KL, Der CJ. 1997. Dbl family proteins. *Biochim Biophys Acta* **1332:** F1–F23.

Wittekind M, Mapelli C, Farmer BT, Suen KL, Goldfarb V, Tsao J, Lavoie T, Barbacid M, Meyers CA, Mueller L. 1994. Orientation of peptide fragments from Sos proteins bound to the N-terminal SH3 domain of Grb2 determined by NMR spectroscopy. *Biochemistry* **33:** 13531–13539.

Worthylake DK, Rossman KL, Sondek J. 2000. Crystal structure of Rac1 in complex with the guanine nucleotide exchange region of Tiam1. *Nature* **408:** 682–688.

Yadav KK, Bar-Sagi D. 2010. Allosteric gating of Son of sevenless activity by the histone domain. *Proc Natl Acad Sci* **107:** 3436–3440.

Posttranslational Modifications of RAS Proteins

Ian Ahearn, Mo Zhou, and Mark R. Philips

Department of Medicine, Perlmutter Cancer Center, New York University School of Medicine, New York, New York 10016

Correspondence: philim01@nyu.edu

The three human *RAS* genes encode four proteins that play central roles in oncogenesis by acting as binary molecular switches that regulate signaling pathways for growth and differentiation. Each is subject to a set of posttranslational modifications (PTMs) that modify their activity or are required for membrane targeting. The enzymes that catalyze the various PTMs are potential targets for anti-RAS drug discovery. The PTMs of RAS proteins are the focus of this review.

Activating mutations of *RAS* genes have been implicated in human cancer for more than three decades. Modern cancer genome analysis has confirmed that the three human *RAS* genes, *KRAS*, *NRAS*, and *HRAS*, are mutated in cancer more frequently than any other oncogene (Forbes et al. 2011; Prior et al. 2012). Thus, the biochemical and cell biological properties of the products of *RAS* genes are a major focus for cancer biologists who hope to uncover vulnerabilities that can be exploited for drug discovery (Cox et al. 2014).

The three *RAS* genes give rise to four proteins because transcripts from the *KRAS* locus are alternately spliced to encode either KRAS4A or KRAS4B, which use alternate fourth exons. Although activating mutations of each RAS gene transforms rodent fibroblasts with similar efficiency (Maher et al. 1995), there is a strong and unexplained isoform difference in driving human cancer with the hierarchy *KRAS* (85%) >*NRAS* (12%) >*HRAS* (3%) (Hobbs et al. 2016). Each of the four RAS proteins is a member of a superfamily of small GTPases that includes the RAS, RHO, RAB, ARF, and RAN families. RAS, like the other members of this superfamily, is a guanine nucleotide-binding protein that functions as a binary molecular switch that is interconverted from an inactive to an active state by exchange of guanosine triphosphate (GTP) for guanosine diphosphate (GDP). GTP binding reorders two regions known as Switch I and Switch II on the surface of RAS, providing the structural basis for the activated state (Vetter and Wittinghofer 2001). The activation portion of the cycle is catalyzed by guanine nucleotide exchange factors (GEFs) that promote loss of GDP such that GTP, 10-fold more abundant in cells, can bind. Several proteins function as GEFs for RAS, including SOS, which transmits the signal from tyrosine kinase growth factor receptors to RAS (Bos et al. 2007). Signaling downstream from RAS is initiated by GTP-bound RAS binding to any of a dozen effectors that harbor RAS-binding domains (RBDs) that bind only GTP-bound RAS. The best-studied RAS effectors are RAF1 and PIK3CA that mediate mitogen-activated protein kinase (MAPK)

Cite this article as *Cold Spring Harb Perspect Med* doi: 10.1101/cshperspect.a031484

and AKT/mechanistic target of rapamycin (mTOR) signaling, respectively (Marshall 1996; Cox and Der 2011). The inactivation cycle is mediated by the GTPase activity of RAS and other small GTPases, which hydrolyzes bound GTP to GDP and thereby returns the molecule to the off state. However, the intrinsic GTPase activity of RAS is quite low (Gibbs et al. 1984; Chung et al. 1993) such that an accessory protein, GTPase-activating protein (GAP), is required, which accelerates catalysis up to 10^5-fold (Ahmadian et al. 1997). Oncogenic mutations of RAS render it insensitive to GAP, allowing the GTPase to accumulate in the on state and sustain signaling (Trahey and McCormick 1987; Scheffzek et al. 1997).

The most straightforward approach to developing RAS inhibitors is to target the RAS protein directly. Efforts to inhibit RAS by interfering with GTP/GDP binding have proven fruitless because of the binding affinity of guanine nucleotides (John et al. 1990), although a thiol-reactive derivative of GTP has shown some in vitro efficacy in oncogenic RAS proteins with G12C mutations that afford a reactive cysteine in the guanine nucleotide-binding pocket (Lim et al. 2014). The crystal structure of RAS, first solved in 1989 (Santos and Nebreda 1989), revealed no pockets other than the guanine nucleotide-binding site that might be targeted by small molecules that could interrupt the switch function of RAS. However, recent, renewed efforts using this approach have yielded some fruit. Shokat and colleagues have exploited a shallow binding pocket under the Switch II region of RAS that is in proximity to the cysteine of G12C mutants (which account for 12% of RAS-driven tumors) and can be targeted with thiol-reactive compounds (Ostrem et al. 2013). These compounds have been shown to lock Switch II in a GDP-bound conformation thus abolishing signaling (Lito et al. 2016; Patricelli et al. 2016). Recently, Stockwell and colleagues reported on a small molecule that binds directly to all RAS proteins and interferes with signaling (Welsch et al. 2017), and Reddy and colleagues characterized rigosertib as a molecule that can bind to RBDs thereby interrupting RAS signaling (Athuluri-Divakar et al. 2016), although another study casts doubt on this mechanism of action (Ritt et al. 2016).

The difficulties in directly targeting RAS have led cancer biologists to look for alternate approaches. Among these have been synthetic lethal screens that have yet to bear fruit (Downward 2015; Wang et al. 2017) and targeting the kinases downstream from RAS, an approach that has been very successful, although has yet to produce durable clinical responses (Samatar and Poulikakos 2014). Another approach has been to target the many posttranslational modifications (PTMs) of RAS, among them those that remodel the carboxyl terminus of RAS proteins and thereby allow RAS to associate with membranes, a feature that is required for signaling. It is the PTMs of RAS proteins that are the subject of this review.

CONSTITUTIVE CARBOXY-TERMINAL MODIFICATIONS: MEMBRANE ASSOCIATION AND TRAFFICKING

CAAX Processing

RAS and RHO family small GTPases are members of a larger family of proteins that end with a CAAX sequence in which C is cysteine, A is usually aliphatic, and X is any amino acid. A CAAX motif signals for sequential modifications by three enzymes (Wright and Philips 2006). Primary translation products that terminate with a CAAX sequence immediately become substrates for one of two cytosolic prenyltransferases, farnesyltransferase (FTase) or geranylgeranyltransferase 1 (GGTase1) that add a 15- or 20-carbon, respectively, polyisoprene lipid via a stable thioether linkage to the sulphydryl group of the cysteine (Fu and Casey 1999). When a CAAX sequence ends in S or M, the protein is a primary substrate for FTase, whereas those CAAX motifs that end in L are modified by GGTase1 (Casey et al. 1989; Seabra et al. 1991). All four RAS proteins are modified by FTase (Fig. 1). Although CAAX prenylation has been considered immediately posttranslational and unregulated, recently, Williams and colleagues have found that RAS binding to splice variants of a promiscuous GEF, SmgGDS, can regulate entry into the CAAX pro-

 Cite this article as *Cold Spring Harb Perspect Med* doi: 10.1101/cshperspect.a031484

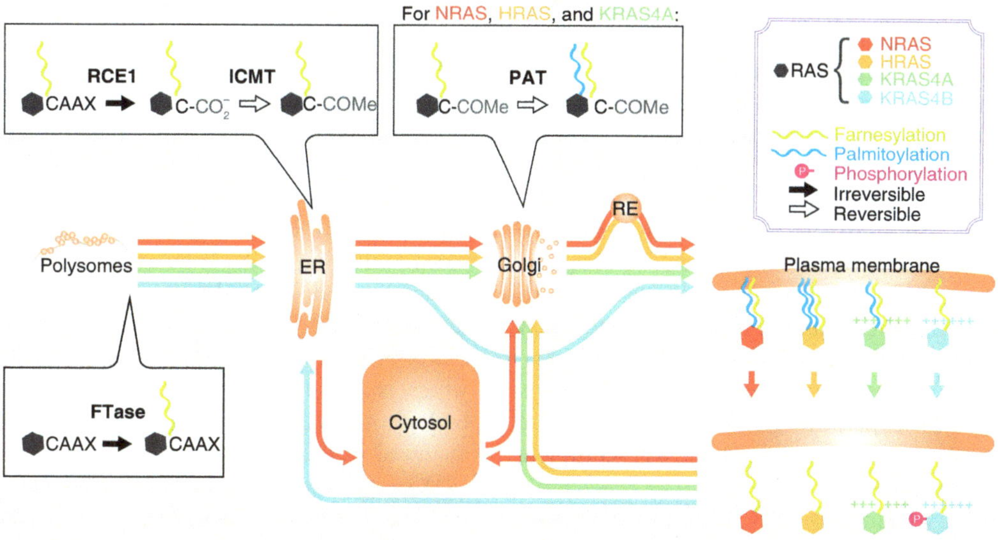

Figure 1. Differential posttranslational processing regulates RAS isoform subcellular trafficking. The trafficking of RAS isoforms HRAS, NRAS, KRAS4A, and KRAS4B (color coded, see legend) are schematized for the nascent proteins following messenger RNA (mRNA) translation on polysomes in the cytosol. The enzymes that catalyze the posttranslational modifications (PTMs) that direct trafficking are highlighted in dialog boxes (FTase, farnesyltransferase; RCE1, Ras-converting enzyme 1 protease; ICMT, isoprenylcysteine carboxylmethyltransferase; PATs, DHHC family of palmitoyl acyl transferases) with irreversible modifications (farnesylation and proteolysis) indicated by solid arrows, and reversible modifications (carboxyl methylation, palmitoylation, and phosphorylation) indicated by outlined arrows (see legend). Lipid modifications (farnesylation and palmitoylation) are color coded (see legend). The subcellular locations at which these modifications occur (ER, endoplasmic reticulum; Golgi, Golgi apparatus; RE, recycling endosome; cytosol, plasma membrane) are also indicated. Between the indicated subcellular compartments, the color-coded arrows denote both the RAS isoform undergoing transit and the direction of trafficking. For KRAS4A and KRAS4B, (+) signs indicate polybasic residues that confer electrostatic affinity for phospholipid headgroups, which, in the case of KRAS4B, is interrupted by phosphorylation (via protein kinase C [PKC] and GMP-dependent protein kinase 2 [PKG2] as discussed in the text but not depicted here). For NRAS, arrows indicate transit to the cytosol, a compartment in which this isoform accumulates when unpalmitoylated, from both the ER and plasma membrane. Prenyl-dependent chaperone interactions for N/KRAS with PDE6δ or for NRAS with VPS35 are not depicted but are discussed in the text.

cessing pathway (Berg et al. 2010). More recently, regulation of KRAS prenylation via interactions with the noncoding, small nucleolar RNAs, SNORD50A and SNORD50B, was reported. However, regulation of Ras prenylation was shown indirectly using proximity ligation analysis and the identification of a presumptive prenylation-dependent electrophoretic mobility shift via western blotting (Siprashvili et al. 2016).

Prenylated CAAX sequences become substrates for the second enzyme—RAS converting enzyme 1 (RCE1)—an endoprotease that removes the AAX amino acids converting the prenylcysteine into the new carboxyl terminus (Boyartchuk et al. 1997; Freije et al. 1999; Otto et al. 1999). Carboxy-terminal prenylcysteines then become substrates for the final CAAX-processing enzyme—isoprenylcysteine carboxylmethyltransferase (ICMT)—that methylesterifies the α carboxyl group, thereby neutralizing the negative charge at the carboxyl terminus (Clarke et al. 1988; Gutierrez et al. 1989; Hancock et al. 1991b; Hrycyna et al. 1991; Pillinger et al. 1994). Whereas carboxylmethylation is reversible (although no specific esterase had been identified), prenylation is not and persists through the lifetime of the protein. The end result of CAAX processing is to remodel the hydrophilic carboxyl terminus of a globular hydrophilic protein and render it hydrophobic such that it has affinity for membranes (Fig. 1).

Proteins that are modified only with a prenyl lipid have affinity for the endoplasmic reticulum (ER) (Choy et al. 1999). The basis for this preference is not established but likely relates to the biophysical properties of a prenyl lipid and the ER membrane. Both RCE1 and ICMT are polytopic enzymes that are restricted to the ER with their active sites disposed toward the cytoplasm (Dai et al. 1998; Schmidt et al. 1998; Wright et al. 2009). Thus, the affinity of prenylated proteins for the ER is part and parcel of directing these proteins to the enzymes that will complete CAAX processing.

The elucidation of CAAX processing by 1990 and the demonstration that it is required for RAS transformation of fibroblasts (Willumsen et al. 1984; Bergo et al. 2002, 2004) pointed the way to the first rational design of anti-RAS drugs. FTase, the enzyme that catalyzes the first and rate-limiting step was targeted. Numerous FTase inhibitors (FTIs) were developed that proved relatively nontoxic, could be made orally available, and hit their target in vivo. Some of these went to phase III trials (Appels et al. 2005). Unfortunately, these trials, mostly involving tumors that are driven by mutant *KRAS*, failed. It was later observed that although under physiologic conditions all RAS proteins are modified by FTase, in the presence of an FTI, KRAS and NRAS but not HRAS were capable of modification by GGTase1, which allowed for full biological function (Whyte et al. 1997). Recently, an interest in FTIs for *HRAS*-driven tumors has been rekindled and clinical trials for head and neck tumors (ClinicalTrials.gov identifier: NCT02383927) and urothelial carcinoma (ClinicalTrials.gov identifier: NCT02535650) are underway. In principle, dual FTase/GGTase1 inhibitors would be expected to have effective anti-RAS activity but this combination, which should inhibit all RAS and RHO family proteins and some RAB proteins, is predicted to be too toxic. Some preclinical animal studies support this idea (Lobell et al. 2002; Liu et al. 2010) but the possibility of a therapeutic window has not been explored. It is important to keep in mind that FTIs did not fail because blocking RAS association with membranes is a poor strategy, but rather that FTIs did not block KRAS and NRAS association with membranes.

Although RCE1 and ICMT are required for the transformation of rodent fibroblasts (Bergo et al. 2002, 2004), studies of transgenic mice with conditional RCE1 and ICMT alleles have provided conflicting results. Whereas ICMT deficiency ameliorated *KRAS*-driven myeloproliferative disease (Wahlstrom et al. 2008), RCE1 deficiency exacerbated the disease (Wahlstrom et al. 2007), which is a result difficult to reconcile with the biology of CAAX processing. Surprisingly, when analyzed in the context of the *LSL-KRAS*^{G12D}; *PDX1-CRE* model of pancreatic adenocarcinoma, ICMT deficiency exacerbated the neoplastic disease (Court et al. 2013), and follow-up studies have implicated CAAX processing of RAB proteins that are required for NOTCH signaling (NOTCH is a tumor suppressor in this model) as the mechanism (Court et al. 2017). Thus, it is clear that the utility of targeting RCE1 or ICMT will be context dependent in cancer.

RCE1 and ICMT were recently reported to be synthetic lethal with oncogenic *NRAS* in Ba/F3 cells subjected to a genome-wide CRISPR/Cas9 screen (Wang et al. 2017), renewing interest in these postprenylation CAAX-processing enzymes. Like FTase, RCE1 and ICMT have been targeted for anti-RAS drug discovery. RCE1 inhibitors remain under development (Mohammed et al. 2016). Although they have been shown to mislocalize RAS proteins, preclinical evidence for inhibition of tumor growth has not been reported. The most potent ICMT inhibitor described to date also showed a strong effect on membrane association of RAS but relatively weak inhibition of tumor cell proliferation, although, in contrast to previous studies, these investigators found that *KRAS*-transformed cell lines were exquisitely sensitive to FTIs (Judd et al. 2011). Importantly, Casey and Wang have recently established a requirement for *ICMT* in human tumor cell lines and showed that an ICMT inhibitor inhibits tumor growth and induces cancer cell death (Lau et al. 2014, 2017; Sun et al. 2016; Manu et al. 2017). Even if therapeutics targeted to RCE1 or ICMT prove ineffective as single agents, the possibility that they might synergize with FTIs or other RAS therapeutics remains a scientifically sound possibility.

Palmitoylation/Depalmitoylation

CAAX processing is necessary but not sufficient for delivery of RAS proteins to the plasma membrane (PM) (Fig. 1). Also required is a so-called "second signal" immediately upstream of the CAAX motif (Hancock et al. 1991a). For KRAS4B, this consists of a polybasic region with eight lysines that allow for an electrostatic interaction with the negatively charged headgroups of phospholipids on the cytoplasmic face of the PM. The second signal for NRAS and HRAS consists of one or two, respectively, additional cysteines immediately upstream of the CAAX cysteine that become modified by palmitic acid (Hancock et al. 1990). KRAS4A is also palmitoylated but it has a hybrid second signal consisting of the palmitate that works in conjunction with a weaker polybasic region (Laude and Prior 2008; Tsai et al. 2015). Whereas the second signal of KRAS4B is constitutive and requires no further PTM, palmitoylation of NRAS, HRAS, and KRAS4A is regulated and reversible and therefore affords another opportunity for therapeutic intervention.

RAS proteins are among hundreds encoded in mammalian genomes that are modified by acylation, which includes both myristoylation and palmitoylation (Veit and Schmidt 2006). The study of protein acylation has exploded during the past 10 years as innovative, nonradioactive methods have been developed for detecting these modifications (Drisdel and Green 2004; Drisdel et al. 2006; Charron et al. 2009; Yang et al. 2010). The combination of acyl-exchange chemistry with alkyne and azido derivatives of fatty acids that permit covalent linkage to affinity tags through click chemistry has permitted exploration of the "palmitoylome," which has proven to include hundreds and perhaps thousands of gene products (Sanders et al. 2015).

RAS palmitoylation occurs on cysteines at the carboxyl terminus via thioester linkages, referred to *S*-palmitoylation. For many other signaling molecules, including src family kinases, palmitoylation occurs at the amino terminus often in association with myristoylation of glycine. Many other proteins are palmitoylated at sites remote from the amino or carboxyl terminus, including transmembrane proteins that, by definition, do not require acylation for membrane association (Resh 1999). The protein palmitoylation machinery is comprised of a family of 25 protein acyltransferases (PATs) that each possess a DHHC sequence in their catalytic sites (Mitchell et al. 2006). Unlike other lipid modifications such as prenylation or myristoylation, no consensus signal has emerged that accounts for which surface-exposed cysteines can become substrates for PATs (Rocks et al. 2010). Genetic screens in yeast first identified Erf2 and Erf4 as the heterodimeric enzyme responsible for RAS palmitoylation in *Saccharomyces cerevesiae* (Lobo et al. 2002). In yeast, Erf2 is the DHHC-containing PAT subunit, whereas Erf4 regulates both the ubiquitin-mediated degradation of Erf2 and the stability of the palmitate-Erf2 catalytic intermediate (Mitchell et al. 2006). The mammalian orthologs of this complex are DDHC9 and GCP16 (Swarthout et al. 2005). Although HRAS is a substrate for DHHC9/GCP16, knockdown of DHHC9 does not block HRAS palmitoylation, suggesting that multiple PATs have activity toward the GTPase (Rocks et al. 2010). Indeed, the substrate promiscuity of PATs has made it difficult to assign enzyme–substrate pairs. Although no PAT has been characterized with regard to a full substrate repertoire, it is clear that each assumes a specific subcellular distribution, predominantly on the Golgi apparatus or PM. DHHC9 is restricted to the Golgi and may account for the localization of NRAS and HRAS on this compartment in addition to the PM (Swarthout et al. 2005). Interestingly, KRAS4A does not accumulate on Golgi, suggesting that this RAS protein is a substrate for another PAT (Tsai et al. 2014).

Upon palmitoylation, RAS acquires a 100-fold greater affinity for membranes than that of the prenylated-only protein (Shahinian and Silvius 1995; Schroeder et al. 1997). This increased affinity creates a kinetic trap that enriches NRAS and HRAS at the cytoplasmic face of Golgi membranes from where these RAS proteins can engage vesicular transport and gain access to other membranes, including the PM (Goodwin et al. 2005). Palmitoylation occurs on Cys 181 of NRAS, Cys 181 and 184 of HRAS, and on Cys 180 of KRAS4A. For both NRAS and HRAS, palmitoylation at Cys 181 directs sorting from the Golgi to recycling endosomes, and this is necessary for subsequent trafficking on the cytoplasmic face of exocytic vesicles destined for the PM (Choy et al. 1999; Apolloni et al. 2000). In contrast, palmitoylation of Cys 184 on HRAS is not absolutely required for delivery to the PM (Roy et al. 2005). Similarly, although palmitoylation of Cys 180 on KRAS4A dramatically increases the efficiency of PM targeting, it is not required because the polybasic regions can function without acylation (Tsai et al. 2015).

The requirement for palmitoylation in Ras-driven tumors has been tested in models of NRAS-driven leukemia, in which palmitoylation of $NRAS^{G12D}$ was necessary for leukemogenesis (Cuiffo and Ren 2010; Liu et al. 2016). A similar approach to investigate palmitoylation of $KRAS4A^{G12D}$ in leukemogenesis revealed that even in the absence of palmitoylation, the polybasic region cooperates sufficiently with prenylation to promote disease, albeit with a notable delay in disease onset (Zhao et al. 2015).

In addition to prenylation and palmitoylation, the distribution of RAS molecules at the PM and the Golgi is also regulated by depalmitoylation. Although the thioester linkage of protein *S*-acylation is labile at physiologic pH, suggesting that at least some depalmitoylation could be noncatalytic, current evidence suggests that palmitoylation is reversed in a regulated manner by cellular thioesterases. The first nonlysosomal proteins found to have thioesterase activity toward RAS were acylprotein thioesterase 1 and 2 (APT1 and 2) (Duncan and Gilman 1998; Tomatis et al. 2010; Rusch et al. 2011). Studies of APT1/2 have been aided by the development of inhibitors of thioesterases, including palmostatin B and M (Dekker et al. 2010; Dekker and Hedberg 2011; Hedberg et al. 2011). However, these compounds have proven to possess broad inhibitory activity against serine hydrolases and cannot be used alone to implicate APT1. Some doubt exists as to whether RAS proteins are physiologic substrates for APT1/2. The yeast ortholog of APT1 has activity toward $Gi_{\alpha 1}$ but not RAS (Duncan and Gilman 2002). Knockdown or inhibition of APT1/2 blocks depalmitoylation of Huntingtin but not NRAS or PSD95 (Lin and Conibear 2015).

APT1 and 2 are serine hydrolases that are themselves palmitoylated (Kong et al. 2013) and this PTM affords some affinity for membranes, suggesting a mechanism whereby the esterase colocalizes with its substrates. However, it is the depalmitoylated, cytosolic pool of APT1 that manifests depalmitoylating activity (Vartak et al. 2014). Recently, activity profiling using palmostatin B identified the ABHD17 family of serine hydrolases as physiologically important thioesterases. Unlike APT1 or 2, ABHD17a shows robust NRAS depalmitoylating activity (Lin and Conibear 2015). Like APT1/2, ABHD17A is also palmitoylated (Kang et al. 2008; Martin and Cravatt 2009), but unlike APT1, this modification is required for both its membrane association and NRAS esterase activity. The full repertoire, specificity, and activity of cellular thioesterases that depalmitoylate RAS remains largely uncharacterized (Tabaczar et al. 2017). Nonetheless, disruption of the acylation cycle using inhibitors of depalmitoylation has been shown to limit cytokine-independent colony formation of $Nras^{G12D}$ but not $Kras4B^{G12D}$, expressing fetal liver and bone marrow cells from mice (Xu et al. 2012).

To establish a physiologic role for the RAS palmitoylation/depalmitoylation cycle and to study the effect of it perturbation one must accurately measure the $t_{1/2}$ of palmitate on RAS proteins in living cells and tissues. However, accurate measurement of the palmitate half-life on endogenous RAS has been limited by the poor sensitivity of detecting radiolabeled [^{3}H] palmitoyl modification (Tsai et al. 2014). Iodinated derivatives of palmitate have been used to enhance sensitivity (Liang et al. 2001), but these reagents are not commercially available. Most reports have used a classical pulse-chase methodology with a relatively long (up to 1 h) pulse followed by a chase in the presence of cycloheximide to inhibit new protein synthesis (Lu and Hofmann 1995; Baker et al. 2000, 2003). This creates two problems. First, a long "pulse" leads

 Cite this article as *Cold Spring Harb Perspect Med* doi: 10.1101/cshperspect.a031484

to equilibration of cellular pools of palmitate, palmitoyl-CoA, and palmitoyl proteins. Because [^{3}H]palmitoyl-CoA can be generated from both [^{3}H]palmitate and [^{3}H]palmitoyl proteins, it is not feasible to abruptly shut down [^{3}H]palmitoylation by adding a great excess of unlabeled palmitate to the medium. Second, we have shown that cycloheximide markedly increases HRAS palmitoylation because it inhibits FKBP12, a prolyl isomerase that acts on a carboxy-terminal proline of HRAS and thereby promotes depalmitoylation (Ahearn et al. 2011). Baker et al. reported a $t_{1/2}$ of 2.4 h of palmitate on HRAS in cells "pulsed" for 3 h in the presence of cycloheximide (Baker et al. 2000, 2003). Subsequently, Lu and Hofmann (1995) measured a $t_{1/2}$ for HRAS of 90 min using a 30 min pulse and cycloheximide. Interestingly, singly palmitoylated HRAS mutants had a dramatically reduced $t_{1/2}$ of 20 min. Although dual palmitoylation of HRAS could account for the discrepancy, it is more likely that a relatively long pulse and addition of cycloheximide increased the observed $t_{1/2}$ because a subsequent study with a 5 min pulse without cycloheximide found the $t_{1/2}$ of palmitate on HRAS to be less than 5 min (Ahearn et al. 2011). Using a pulse of 5 min without cycloheximide, Magee et al. reported that the $t_{1/2}$ of palmitate on singly palmitoyated NRAS was 20 min (Magee et al. 1987). Fluorescence recovery after photobleaching studies of NRAS transport to and from the Golgi, a process regulated by the palmitotylation/depalmitlylation cycle, are consistent with a rapid turnover of seconds to a few minutes (Goodwin et al. 2005; Rocks et al. 2005). Thus, the kinetics of turnover of palmitate on RAS proteins is uncertain but is likely to be measured in a few minutes, whereas the $t_{1/2}$ of the proteins is on the order of 24 h. It is unclear whether the newer techniques of measuring protein palmitoylation by click chemistry and acyl-exchange chemistry will improve the accuracy of measuring palmitate turnover, and there is reason to suspect that the additional steps and manipulations involved may complicate the analysis. Using the 17-ODYA palmitate analog and click chemistry, Lin and Conibear (2015) reported a $t_{1/2}$ for the palmitate on NRAS of <0.5 h.

It is clear that the hypervariable regions (HVRs) of RAS proteins and the PTMs thereof encode information with regard to targeting to phospholipid bilayers and trafficking within the cell. The palmitate modification has long been known to promote association of both peripheral and intrinsic membrane proteins, such as HRAS and LAT, respectively, to liquid-ordered membrane domains, in the past referred to as lipid rafts (Webb et al. 2000; Liang et al. 2001; Lorent and Levental 2015). More recently, the "code" of the KRAS4B polybasic HVR has been read. Using gold particle labeling and electron microscopy spatial mapping, Zhou et al. have described how anionic lipid sorting of KRAS4B is encoded in combination by prenylation, the polybasic stretch of lysines, and by tertiary conformations revealed by molecular dynamic simulations (Zhou et al. 2017). In this model, these chemical and structural codes promote partitioning to PtdSer or PtdInositol bisphosphate-enriched domains (Maher et al. 1995; Cox et al. 2014). In turn, nanoclustering and differential effector pathway activation is established. The investigators also show how phosphorylation at serine 181 of KRAS4B (see below) operates to regulate this partitioning.

CONDITIONAL POSTTRANSLATIONAL MODIFICATIONS

Phosphorylation of KRAS4B and the Farnesyl Electrostatic Switch

Although KRAS4B requires no modification beyond CAAX processing to associate with membranes, the strength of the interaction is regulated by PTM. Phosphorylation of serine 181 within the polybasic region can diminish membrane affinity by partially neutralizing the positive charge (Fig. 1). Although phosphorylation of serine 181 does not completely inhibit association with the PM, the distribution of the GTPase between PM and the endomembrane is shifted in favor of the endomembrane (Bivona et al. 2006). In this way, the prenylated carboxyl terminus of KRAS4B mimics the myristoylated amino terminus of the MARKS protein that also requires a polybasic region in conjunction with the acyl lipid to effect membrane binding and harbors serines within the polybasic region that, when phosphorylated by protein kinase C (PKC), reverses membrane binding through a mechanism known as a myristoyl electrostatic switch (McLaughlin and Aderem 1995). By analogy, KRAS4B harbors a farnesyl electrostatic switch. Serine 181 of KRAS4B is a substrate for both PKCs (Bivona et al. 2006; Alvarez-Moya et al. 2010) and cyclic GMP-dependent protein kinase 2 (PKG2) (Cho et al. 2016). PKG2 is activated to phosphorylate KRAS4B downstream from AMP-activated protein kinase (AMPK) and requires endothelial nitric oxide synthase and soluble guanylyl cyclase. Phosphorylation of KRAS4B on serine 181 regulates association of the HVR with membrane microdomains (Barceló et al. 2013; Jang et al. 2015). Bryostatin, a potent agonist of PKC, was found to inhibit the growth of oncogenic KRAS4B-driven xenografts in an S181-dependent fashion, suggesting that phosphorylation is a negative regulatory event in tumor maintenance (Mohammad et al. 1998; Bivona et al. 2006; Kollar et al. 2014). Consistent with this observation, oncogenic KRAS4B with a phosphomimetic group at position 181, $KRAS^{12V181E}$, was found to diminish cell survival (Bivona et al. 2006). One mechanism for decreased survival is the ability of $KRAS^{12V181E}$ to interfere with one of the prosurvival actions of Bcl-X_L (Sung et al. 2013). A constant low level of calcium flux from ER stores to mitochondria is required for metabolic health and suppression of autophagy (White et al. 2005; Cárdenas et al. 2010). This calcium flux is delivered through inositol-*tris*-phosphate (IP_3) receptors (IP3Rs) and their IP_3-stimulated conductance is potentiated by Bcl-X_L, thus accounting for one of the prosurvival actions of Bcl-X_L. KRAS12V181E blocks this action of Bcl-X_L (Sung et al. 2013). Wang et al. (2015) recently showed that phosphorylation of KRAS4B on serine 181 inhibits tumor initiation by blocking interaction with calmodulin and thereby abrogating suppression of noncanonical Wnt signaling. Thus, there are multiple independent lines of evidence suggesting that phosphorylation of KRAS4B is a negative regulatory event with regard to oncogenesis. One contrary report in the literature suggests that phosphorylation of KRAS4B on S181 is required for transformed rodent fibroblasts and colon cancer cell lines to grow subcutaneously in nude mice (Barceló et al. 2014). However, we have observed that LSL-$KRAS^{12D181A}$ double knockin mice that harbor an inducible KRAS4B oncogene that cannot be phosphorylated produce the same number of lung tumors on tracheal administration of adeno-Cre than do LSL-$KRAS^{12D181S}$ mice (unpubl.), suggesting that in vivo phosphorylation of S181 is not required for oncogenesis.

Ubiquitination and Sumoylation

Despite intensive study of RAS proteins for three decades, little is known about their catabolism. HRAS has been shown to be polyubiquitinated when co-overexpressed with FLAG-Ub (Jeong et al. 2012). The E3 ligase was identified as β-TrCP (Kim et al. 2009). Zeng et al. reported interactions of H,N and KRAS4B with the E3 ligase Nedd4-1 and showed polyubiquitination of KRAS4B on Lys5 using HA-Ub. Signaling via growth factor receptors and oncogenic RAS both impaired Nedd4-1-mediated RAS degradation and in turn enhanced Nedd4-1-mediated PTEN degradation to drive cancer growth (Zeng et al. 2014). Mono- and diubiquitination of RAS proteins is also established. HRAS, NRAS, and KRAS4B have all been shown to be either mono- or diubiquitinated (Jura et al. 2006; Sasaki et al. 2010). These modifications were first reported on NRAS and HRAS and modification of the latter promoted association with endosomes. The E3 ligase for HRAS was identified as RABEX5 (Xu et al. 2010), a protein that also has exchange activity for RAB5. Interestingly, the RAS effector RIN1 was required for the ubiquitination reaction. Later mass spectrometry analysis identified Lys147 on both HRAS and KRAS as a site of mono-/diubiquitination (Sasaki et al. 2010). The same study identified Lys117 and Lys170 as additional sites on HRAS and Campbell's group added Lys104 of KRAS as an additional site (Baker et al. 2013). Importantly, Sasaki et al. (2010) reported that mono- and diubiquitination of KRAS on Lys 147 enhanced GTP loading and interaction with effectors, and Baker et

 Cite this article as *Cold Spring Harb Perspect Med* doi: 10.1101/cshperspect.a031484

al. (2013) reported that mono- or diubiquitination of KRAS on Lys117 accelerated nucleotide exchange through a mechanism distinct from that of modified Lys147. More recently, we found that HRAS, NRAS, and KRAS4B are modified on Lys42 by the ubiquitin-like protein SUMO3 (BH Choi, C Chen, M Philips, and W Dai, unpubl.). We have identified the E3 ligase as PIASγ and found that sumoylation promotes MAPK signaling and tumor progression. The weakness of all of these studies is that the stoichiometry of modification has not been determined and the physiologic significance is uncertain.

Nitrosylation

The concept of redox regulation of cellular signaling prompted investigation into direct modification of RAS proteins by reactive oxygen species. The most surface-exposed cysteine, Cy118, conserved in all RAS isoforms, was shown to be nitrosylated (Lander et al. 1995a). Two mechanisms have been described for this modification, one involving direct modification with $\bullet NO_2$, a product of NO and O_2, and one involving a glutathionyl radical intermediate (Lander et al. 1996; Heo and Campbell 2004). *S*-nitrosylation of RAS on cys118 has been shown to increase GDP/GTP exchange (Lander et al. 1995b; Williams et al. 2003) and has been implicated in breast cancer in which nitric oxide synthase 2 has been shown to be up-regulated (Switzer et al. 2012). Most compelling are studies by Counter and colleagues who showed that activation of eNOS by AKT is required for xenograft initiation and maintenance, a pathway mediated by *S*-nitrosylation of wild-type RAS in human tumor cells harboring an oncogenic RAS allele (Lim et al. 2008). Interestingly, eNOS is palmitoylated, highlighting how nistrosylation of RAS may be partially regulated by compartmentalization (Garcia-Cardena et al. 1996).

Acetylation

KRAS can be acetylated on Lys104, which acts as a negative regulatory modification by interfering with GEF-catalyzed nucleotide exchange and thereby reducing transforming capacity (Yang et al. 2012). Acetylation of Lys104 of KRAS is reversed by HDAC6 or SIRT2 (Yang et al. 2013). Thus, among the antineoplastic activities of inhibitors of these deacetylating enzymes may be removal of the acetyl group from KRAS.

Bacterial Toxins and Exoenzymes

Pathogenic bacteria have evolved a plethora of virulence factors, many of which are enzymes that modify signaling molecules, including members of the RAS superfamily of small GTPases (Aktories 2011). Although the majority of these target RHO family proteins, some are active against RAS. Exoenzyme S (ExoS) is an ADP-ribosyl transferase produced by *Pseudomonas aeruginosa* and injected directly into eukaryotic cells by type III secretion. ExoS ADP ribosylates Arg41 and Arg 128 on RAS, which attenuates GTP loading and inhibits interaction with effectors (Ganesan et al. 1999). Lethal toxin in *Clostridium sordeli* is a monoglucosyltranferase closely related to toxin B in *Clostridium difficile*, which uses UDP-glucose to modify Thr35 and thereby inhibits MAPK signaling (Just et al. 1996; Herrmann et al. 1998).

CONSEQUENCES OF RAS PTMs

PTMs of RAS proteins can be divided into those that modify amino acids in the guanine nucleotide binding (G) domain and those that modify the carboxy-terminal HVR (Fig. 2). Not surprisingly, modifications of the G domain (mono- or diubiquitination, sumoylation, acetylation, glucosylation, and nitrosylation) are conditional and modulate GDP/GTP exchange and engagement of effectors. One exception to this rule is mono- and diubiquitination of HRAS that promotes trafficking to endosomes (Jura et al. 2006). Ubiquitination and sumoylation should markedly alter the electrophoretic mobility of RAS proteins. The fact that immunodetectable endogenous cellular RAS proteins run at 21 kDa suggests that the stoichiometry of these modifications is relatively low. The stoichiometries of RAS acetylation and nitrosylation are unknown but mass spectrometry analysis suggests that they too are relatively low

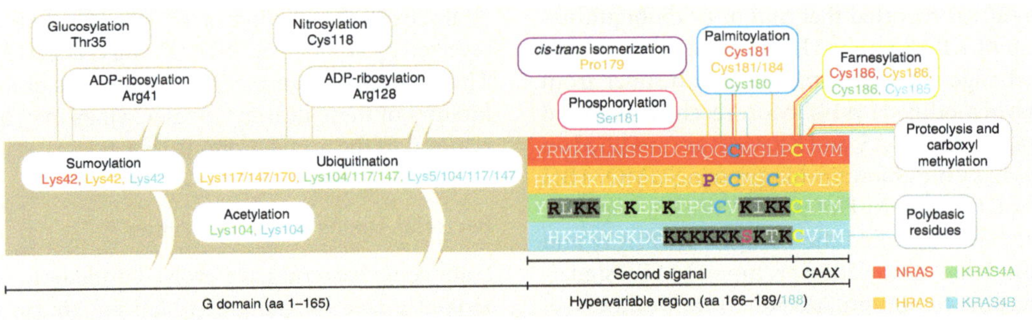

Figure 2. Posttranslational modifications (PTMs) of RAS isoforms. The full length of RAS is depicted schematically and includes partitioning of the G domain (aa1–165) and the hypervariable region (HVR) (aa166–188/189). Color coding of the HVR distinguishes each RAS isoform (HRAS, NRAS, KRAS4A, and KRAS4B). PTMs occurring within the G domain are indicated (glucosylation, sumolyation, ADP-ribosylation, acetylation, ubiquitination, and nitrosylation). For the HVR, unique color coding highlights residues modified by the various PTMs (phosphorylation, *cis-trans* [prolyl] isomerization, palmitoylation, farnesylation, proteolysis, and carboxyl methylation). Polybasic residues in the HVRs of KRAS4A and KRAS4B are indicated by black text. HVR modifications are correlated to the RAS isoforms on which they occur via isoform color-coded linkers, and also via the isoform color-coded text for the modified amino acids listed within the modifications dialog boxes.

(Lander et al. 1997; Yang et al. 2012). Accordingly, although small pools of modified RAS might have significant biological consequences, it is not clear what are the physiologic roles of these modifications in the biology of RAS. Clarification of this issue is of great importance in that the enzymes that catalyze the PTMs become rational targets for anti-RAS drug discovery once a physiologic role is established.

In contrast to modifications of the G domain, the PTMs of the CaaX sequence (prenylation, proteolysis, and carboxyl methylation) are constitutive and immediately posttranslational such that the stoichiometry is high and the vast majority of endogenous RAS proteins are fully processed. Whereas CaaX processing is irreversible, palmitoylation of NRAS, HRAS, and KRAS4A is dynamic and reversible (Bijlmakers and Marsh 2003). Although it appears that initial palmitoylation is constitutive in that it occurs immediately posttranslationally and stoichiometrically, sending a first wave of nascent RAS to the PM, the lifetimes of palmitoylated RAS proteins are characterized by many cycles of palmitoylation/depalmitoylation. Likewise, phosphorylation of KRAS4B on ser181 is reversible and the stoichiometry is unknown.

It is clear that the carboxy-terminal PTMs control affinity for membranes in general and trafficking of the proteins between cellular compartments. For example, whereas palmitoylated HRAS is efficiently targeted to the PM, a depalmitoylated pool accumulates on the ER and Golgi (Choy et al. 1999). In addition, the addition and subtraction of palmitate influences the microdomains into which RAS proteins partition (Plowman and Hancock 2005). This realization has given rise to the idea of compartmentalized signaling of RAS, whereby one set of pathways might be engaged most efficiently form one intracellular compartment and another set from another location. Compartmentalized signaling increases the complexity of RAS signaling and greatly increases the repertoire of outcomes, a feature consistent with the requirement for RAS signaling in a very large number of cellular processes that must be independently regulated (Fehrenbacher et al. 2009). Differential signaling outcomes of stringent, artificial targeting of activated RAS proteins to different cellular compartments supports this idea (Chiu et al. 2002) as does evidence for compartmentalized RAS signaling in *Schizosaccharomyces pombe* (Onken et al. 2006). Perhaps the most compelling example of compartmentalized RAS signaling is in lymphocyte development in which RAS signaling is required for both positive and negative selection of nascent lymphocytes in the thymus,

 Cite this article as *Cold Spring Harb Perspect Med* doi: 10.1101/cshperspect.a031484

two diametrically opposed cellular outcomes (proliferation vs. apoptosis). Palmer and colleagues showed that whereas negative selection was associated with RAS/MAPK signaling from the PM, positive selection was associated with signaling from internal membranes (Daniels et al. 2006).

The carboxy terminal PTMs of RAS control affinity not only for membranes and membrane microdomains but also for cytosolic prenyl-binding proteins. RHO family small GTPases undergo the same PTMs at their carboxyl termini as do RAS proteins. Because they are modified by a 20-carbon geranylgeranyl polyisoprene lipid rather than the 15-carbon farnesyl lipid of RAS proteins, they have even greater intrinsic affinity for membranes. Nevertheless, the bulk of endogenous RAC, RHO, and CDC42 are present in the aqueous environment of the cytosol (Michaelson et al. 2001). This phenomenon is a consequence of a prenyl-binding protein knows as RhoGDI. RhoGDI sequesters the geranylgeranyl moiety of RHO proteins and allows delivery of the GTPase to the relevant membrane comportment on which signaling takes place (Hoffman et al. 2000). The subcellular trafficking of RAB GTPases is accomplished in the same manner through the action of RabGDI (Ullrich et al. 1993). Importantly, carboxylmethylation of RAB proteins has been shown to inhibit binding to RabGDI (Leung et al. 2007), demonstrating how PTMs can regulate the interactions with prenyl-binding proteins.

This ability of GDP dissociation inhibitors (GDIs) to regulate RHO and RAB proteins has caused RAS biologists to explore the possibility that cytosolic farnesyl-binding proteins are required for trafficking of RAS. Several such proteins have been reported. We reported that the vast majority of endogenous, farnesylated NRAS resides in the cytosol in a high molecular weight pool and that at least a portion of that pool includes NRAS bound to VPS35, a component of the coat of the retromer (Zhou et al. 2016). Other proteins, including PRA1 (Figueroa et al. 2001) and SmgGDS (Berg et al. 2010), have been described as prenyl-binding proteins capable of binding RAS. However, the prenyl-binding protein that has attracted the most attention is the δ subunit of phosphodiesterase type 6 (PDE6δ). PDE6δ was first described in photoreceptors as a component of the PDE6 complex that bound the prenylated α and β subunits (Cook et al. 2000; Zhang et al. 2004). The crystal structure of PDE6δ revealed structural homology with RhoGDI with a prenyl-binding grove between β sheets (Hanzal-Bayer et al. 2002). PDE6δ is a promiscuous carrier protein accepting a wide variety of prenylated proteins as cargo. In an elegant series of studies, Wittinghoffer and colleagues showed that active ARL2/3, a nonprenylated small GTPase, acts as a releasing factor for farnesylated RHEB bound to PDE6δ (Ismail et al. 2011). Overexpression of PDE6δ can extract green fluorescent protein (GFP)-tagged NRAS and KRAS4B, but not HRAS or KRAS4A, from the PM (Nancy et al. 2002; Tsai et al. 2015). Bastiaens and colleagues have proposed that PDE6δ is absolutely required for proper trafficking of KRAS4B (Chandra et al. 2012; Schmick et al. 2014), a conclusion at odds with the fact that, whereas *KRAS* is essential for mouse development (Johnson et al. 1997), PDE6δ null mice are small but otherwise normal (Zhang et al. 2007). Nevertheless, the importance of PDE6δ in KRAS biology was confirmed by Waldmann's group that developed a series of inhibitors of RAS binding to PDE6δ (e.g., deltarasin), and showed that these compounds inhibit the growth of *KRAS*-dependent pancreatic adenocarcinoma cells (Zimmermann et al. 2013, 2014). A recent crystal structure of PDE6δ in complex with KRAS4B reveals that binding requires, in addition to the protein–lipid interaction that can accommodate either a farnesyl or geranylgeranyl lipid, some protein–protein interaction with the HVR of KRAS4B (Dharmaiah et al. 2016). Moreover, this study revealed that carboxylmethylation of KRAS4B was required for efficient binding to PDE6δ. The latter observation makes sense because the prenylcysteine enters the hydrophobic pocket of PDE6δ but is surprisingly contrary to the results with RAB proteins and RabGDI in which carboxylmethylation inhibits the interaction. Either way, these are examples of PTMs regulating interactions of small

GTPases not only with membranes but also with cytosolic chaperones.

CONCLUSION

Because *RAS* mutations are so frequently drivers in human cancer, there is a great desire to develop direct inhibitors of the small GTPases that are the products of the *RAS* oncogenes. Because RAS proteins bind guanine nucleotides with picomolar affinity inhibitors of GTP binding are not practical, other approaches have been sought. RAS proteins are substrates for a wide variety of PTMs discussed above. Because RAS PTMs are catalyzed by specific enzymes, each one represents a potential target for anticancer drug discovery. The G domain of RAS proteins is modified by mono- and diubiquitination, sumoylation, acetylation, and nitrosylation, each of which is reversible and, in principle, susceptible to pharmacological manipulation. However, it is not clear what is the physiologic role of these modifications in RAS signaling, either downstream from growth factor receptors or as a consequence of oncogenic mutation. In contrast, the constitutive PTMs of the carboxy-terminal HVRs of RAS proteins that regulate association with membrane and prenyl binding proteins are clearly essential for RAS protein function and remain rational targets for drug discovery.

REFERENCES

Ahearn IM, Tsai FD, Court H, Zhou M, Jennings BC, Ahmed M, Fehrenbacher N, Linder ME, Philips MR. 2011. FKBP12 binds to acylated H-Ras and promotes depalmitoylation. *Mol Cell* **41:** 173–185.

Ahmadian MR, Stege P, Scheffzek K, Wittinghofer A. 1997. Confirmation of the arginine-finger hypothesis for the GAP-stimulated GTP-hydrolysis reaction of Ras. *Nat Struct Biol* **4:** 686–689.

Aktories K. 2011. Bacterial protein toxins that modify host regulatory GTPases. *Nat Rev Microbiol* **9:** 487–498.

Alvarez-Moya B, Lopez-Alcala C, Drosten M, Bachs O, Agell N. 2010. K-Ras4B phosphorylation at Ser181 is inhibited by calmodulin and modulates K-Ras activity and function. *Oncogene* **29:** 5911–5922.

Apolloni A, Prior IA, Lindsay M, Parton RG, Hancock JF. 2000. H-Ras but not K-Ras traffics to the plasma membrane through the exocytic pathway. *Mol Cell Biol* **20:** 2475–2487.

Appels NM, Beijnen JH, Schellens JH. 2005. Development of farnesyl transferase inhibitors: A review. *Oncologist* **10:** 565–578.

Athuluri-Divakar SK, Vasquez-Del Carpio R, Dutta K, Baker SJ, Cosenza SC, Basu I, Gupta YK, Reddy MV, Ueno L, Hart JR, et al. 2016. A small molecule RAS-mimetic disrupts RAS association with effector proteins to block signaling. *Cell* **165:** 643–655.

Baker TL, Booden MA, Buss JE. 2000. *S*-Nitrosocysteine increases palmitate turnover on Ha-Ras in NIH 3T3 cells. *J Biol Chem* **275:** 22037–22047.

Baker TL, Zheng H, Walker J, Coloff JL, Buss JE. 2003. Distinct rates of palmitate turnover on membrane-bound cellular and oncogenic H-Ras. *J Biol Chem* **278:** 19292–19300.

Baker R, Wilkerson EM, Sumita K, Isom DG, Sasaki AT, Dohlman HG, Campbell SL. 2013. Differences in the regulation of K-Ras and H-Ras isoforms by monoubiquitination. *J Biol Chem* **288:** 36856–36862.

Barceló C, Paco N, Beckett AJ, Alvarez-Moya B, Garrido E, Gelabert M, Tebar F, Jaumot M, Prior I, Agell N. 2013. Oncogenic K-Ras segregates at spatially distinct plasma membrane signaling platforms according to its phosphorylation status. *J Cell Sci* **126:** 4553–4559.

Barceló C, Paco N, Morell M, Alvarez-Moya B, Bota-Rabassedas N, Jaumot M, Vilardell F, Capella G, Agell N. 2014. Phosphorylation at Ser-181 of oncogenic KRAS is required for tumor growth. *Cancer Res* **74:** 1190–1199.

Berg TJ, Gastonguay AJ, Lorimer EL, Kuhnmuench JR, Li R, Fields AP, Williams CL. 2010. Splice variants of SmgGDS control small GTPase prenylation and membrane localization. *J Biol Chem* **285:** 35255–35266.

Bergo MO, Ambroziak P, Gregory C, George A, Otto JC, Kim E, Nagasae H, Casey PJ, Balmain A, Young SG. 2002. Absence of the CAAX endoprotease Rce1: Effects on cell growth and transformation. *Mol Cell Biol* **22:** 171–181.

Bergo MO, Gavino BJ, Hong C, Beigneux AP, McMahon M, Casey PJ, Young SG. 2004. Inactivation of *Icmt* inhibits transformation by oncogenic K-Ras and B-Raf. *J Clin Invest* **113:** 539–550.

Bijlmakers MJ, Marsh M. 2003. The on–off story of protein palmitoylation. *Trends Cell Biol* **13:** 32–42.

Bivona TG, Quatela SE, Bodemann BO, Ahearn IO, Soskis MJ, Mor A, Miura J, Wiener HH, Wright L, Saba SG, et al. 2006. PKC regulates a farnesyl-electrostatic switch on K-Ras that promotes its association with Bcl-X_L on mitochondria and induces apoptosis. *Mol Cell* **21:** 481–493.

Bos JL, Rehmann H, Wittinghofer A. 2007. GEFs and GAPs: Critical elements in the control of small G proteins. *Cell* **129:** 865–877.

Boyartchuk VL, Ashby MN, Rine J. 1997. Modulation of Ras and a-factor function by carboxyl-terminal proteolysis. *Science* **275:** 1796–1800.

Cárdenas C, Miller RA, Smith I, Bui T, Molgó J, Müller M, Vais H, Cheung KH, Yang J, Parker I, et al. 2010. Essential regulation of cell bioenergetics by constitutive InsP3 receptor Ca^{2+} transfer to mitochondria. *Cell* **142:** 270–283.

Casey PJ, Solski PA, Der CJ, Buss JE. 1989. p21ras is modified by a farnesyl isoprenoid. *Proc Natl Acad Sci* **86:** 8323–8327.

Chandra A, Grecco HE, Pisupati V, Perera D, Cassidy L, Skoulidis F, Ismail SA, Hedberg C, Hanzal-Bayer M, Venkitaraman AR, et al. 2012. The GDI-like solubilizing factor PDEδ sustains the spatial organization and signalling of Ras family proteins. *Nat Cell Biol* **14:** 148–158.

Charron G, Zhang MM, Yount JS, Wilson J, Raghavan AS, Shamir E, Hang HC. 2009. Robust fluorescent detection of protein fatty-acylation with chemical reporters. *J Am Chem Soc* **131:** 4967–4975.

Chiu VK, Bivona T, Hach A, Sajous JB, Silletti J, Wiener H, Johnson RL II, Cox AD, Philips MR. 2002. Ras signalling on the endoplasmic reticulum and the Golgi. *Nat Cell Biol* **4:** 343–350.

Cho KJ, Casteel DE, Prakash P, Tan L, van der Hoeven D, Salim AA, Kim C, Capon RJ, Lacey E, Cunha SR, et al. 2016. AMPK and endothelial nitric oxide synthase signaling regulates K-Ras plasma membrane interactions via cyclic GMP-dependent protein kinase 2. *Mol Cell Biol* **36:** 3086–3099.

Choy E, Chiu VK, Silletti J, Feoktistov M, Morimoto T, Michaelson D, Ivanov IE, Philips MR. 1999. Endomembrane trafficking of Ras: The CAAX motif targets proteins to the ER and Golgi. *Cell* **98:** 69–80.

Chung HH, Benson DR, Cornish VW, Schultz PG. 1993. Probing the role of loop 2 in Ras function with unnatural amino acids. *Proc Natl Acad Sci* **90:** 10145–10149.

Clarke S, Vogel JP, Deschenes RJ, Stock J. 1988. Posttranslational modification of the Ha-Ras oncogene protein: Evidence for a third class of protein carboxyl methyltransferases. *Proc Natl Acad Sci* **85:** 4643–4647.

Cook TA, Ghomashchi F, Gelb MH, Florio SK, Beavo JA. 2000. Binding of the δ subunit to rod phosphodiesterase catalytic subunits requires methylated, prenylated C-termini of the catalytic subunits. *Biochemistry* **39:** 13516–13523.

Court H, Amoyel M, Hackman M, Lee KE, Xu R, Miller G, Bar-Sagi D, Bach EA, Bergö MO, Philips MR. 2013. Isoprenylcysteine carboxylmethyltransferase deficiency exacerbates KRAS-driven pancreatic neoplasia via Notch suppression. *J Clin Invest* **123:** 4681–4694.

Court H, Ahearn IM, Amoyel M, Bach E, Philips MR. 2017. Regulation of NOTCH signaling by RAB7 and RAB8 require carboxyl methylation by ICMT. *J Cell Biol* DOI: 10.1083/jcb.201701053.

Cox AD, Der CJ. 2011. Ras history: The saga continues. *Small GTPases* **1:** 2–27.

Cox AD, Fesik SW, Kimmelman AC, Luo J, Der CJ. 2014. Drugging the undruggable RAS: Mission possible? *Nat Rev Drug Discov* **13:** 828–851.

Cuiffo B, Ren R. 2010. Palmitoylation of oncogenic NRAS is essential for leukemogenesis. *Blood* **115:** 3598–3605.

Dai Q, Choy E, Chiu V, Romano J, Slivka SR, Steitz SA, Michaelis S, Philips MR. 1998. Mammalian prenylcysteine carboxyl methyltransferase is in the endoplasmic reticulum. *J Biol Chem* **273:** 15030–15034.

Daniels MA, Teixeiro E, Gill J, Hausmann B, Roubaty D, Holmberg K, Werlen G, Holländer GA, Gascoigne NR, Palmer E. 2006. Thymic selection threshold defined by compartmentalization of Ras/MAPK signalling. *Nature* **444:** 724–729.

Dekker FJ, Hedberg C. 2011. Small molecule inhibition of protein depalmitoylation as a new approach towards downregulation of oncogenic Ras signalling. *Bioorg Med Chem* **19:** 1376–1380.

Dekker FJ, Rocks O, Vartak N, Menninger S, Hedberg C, Balamurugan R, Wetzel S, Renner S, Gerauer M, Schölermann B, et al. 2010. Small-molecule inhibition of APT1 affects Ras localization and signaling. *Nat Chem Biol* **6:** 449–456.

Dharmaiah S, Bindu L, Tran TH, Gillette WK, Frank PH, Ghirlando R, Nissley DV, Esposito D, McCormick F, Stephen AG, et al. 2016. Structural basis of recognition of farnesylated and methylated KRAS4b by PDEδ. *Proc Natl Acad Sci* **113:** E6766–E6775.

Downward J. 2015. RAS synthetic lethal screens revisited: Still seeking the elusive prize? *Clin Cancer Res* **21:** 1802–1809.

Drisdel RC, Green WN. 2004. Labeling and quantifying sites of protein palmitoylation. *BioTechniques* **36:** 276–285.

Drisdel RC, Alexander JK, Sayeed A, Green WN. 2006. Assays of protein palmitoylation. *Methods* **40:** 127–134.

Duncan JA, Gilman AG. 1998. A cytoplasmic acyl-protein thioesterase that removes palmitate from G protein α subunits and $p21^{RAS}$. *J Biol Chem* **273:** 15830–15837.

Duncan JA, Gilman AG. 2002. Characterization of *Saccharomyces cerevisiae* acyl-protein thioesterase 1, the enzyme responsible for G protein α subunit deacylation in vivo. *J Biol Chem* **277:** 31740–31752.

Fehrenbacher N, Bar-Sagi D, Philips M. 2009. Ras/MAPK signaling from endomembranes. *Mol Oncol* **3:** 297–307.

Figueroa C, Taylor J, Vojtek AB. 2001. Prenylated Rab acceptor protein is a receptor for prenylated small GTPases. *J Biol Chem* **276:** 28219–28225.

Forbes SA, Bindal N, Bamford S, Cole C, Kok CY, Beare D, Jia M, Shepherd R, Leung K, Menzies A, et al. 2011. COSMIC: Mining complete cancer genomes in the Catalogue of Somatic Mutations in Cancer. *Nucleic Acids Res* **39** (Database issue): D945–D950.

Freije JM, Blay P, Pendás AM, Cadiñanos J, Crespo P, López-Otín C. 1999. Identification and chromosomal location of two human genes encoding enzymes potentially involved in proteolytic maturation of farnesylated proteins. *Genomics* **58:** 270–280.

Fu HW, Casey PJ. 1999. Enzymology and biology of CaaX protein prenylation. *Recent Prog Horm Res* **54:** 315–342; discussion 42–43.

Ganesan AK, Vincent TS, Olson JC, Barbieri JT. 1999. *Pseudomonas aeruginosa* exoenzyme S disrupts Ras-mediated signal transduction by inhibiting guanine nucleotide exchange factor-catalyzed nucleotide exchange. *J Biol Chem* **274:** 21823–21829.

Garcia-Cardena G, Oh P, Liu J, Schnitzer JE, Sessa WC. 1996. Targeting of nitric oxide synthase to endothelial cell caveolae via palmitoylation: Implications for nitric oxide signaling. *Proc Natl Acad Sci* **93:** 6448–6453.

Gibbs JB, Sigal IS, Poe M, Scolnick EM. 1984. Intrinsic GTPase activity distinguishes normal and oncogenic *ras* p21 molecules. *Proc Natl Acad Sci* **81:** 5704–5708.

Goodwin JS, Drake KR, Rogers C, Wright L, Lippincott-Schwartz J, Philips MR, Kenworthy AK. 2005. Depalmitoylated Ras traffics to and from the Golgi complex via a nonvesicular pathway. *J Cell Biol* **170:** 261–272.

Gutierrez L, Magee AI, Marshall CJ, Hancock JF. 1989. Post-translational processing of p21ras is two-step and involves carboxyl-methylation and carboxyl terminal proteolysis. *EMBO J* **8:** 1093–1098.

Hancock JF, Paterson H, Marshall CJ. 1990. A polybasic domain or palmitoylation is required in addition to the CAAX motif to loacalize $p21^{ras}$ to the plasma membrane. *Cell* **63:** 133–139.

Hancock JF, Cadwallader K, Paterson H, Marshall CJ. 1991a. A CAAX or a CAAL motif and a second signal are sufficient for plasma membrane targeting of ras proteins. *EMBO J* **10:** 4033–4039.

Hancock JF, Cadwallader K, Marshall CJ. 1991b. Methylation and proteolysis are essential for efficient membrane binding of prenylated $p21^{K\text{-}ras(B)}$. *EMBO J* **10:** 641–646.

Hanzal-Bayer M, Renault L, Roversi P, Wittinghofer A, Hillig RC. 2002. The complex of Arl2-GTP and PDEδ: From structure to function. *EMBO J* **21:** 2095–2106.

Hedberg C, Dekker FJ, Rusch M, Renner S, Wetzel S, Vartak N, Gerding-Reimers C, Bon RS, Bastiaens PI, Waldmann H. 2011. Development of highly potent inhibitors of the Ras-targeting human acyl protein thioesterases based on substrate similarity design. *Angew Chem Int Ed Engl* **50:** 9832–9837.

Heo J, Campbell SL. 2004. Mechanism of $p21^{Ras}$ S-nitrosylation and kinetics of nitric oxide-mediated guanine nucleotide exchange. *Biochemistry* **43:** 2314–2322.

Herrmann C, Ahmadian MR, Hofmann F, Just I. 1998. Functional consequences of monoglucosylation of Ha-Ras at effector domain amino acid threonine 35. *J Biol Chem* **273:** 16134–16139.

Hobbs GA, Der CJ, Rossman KL. 2016. RAS isoforms and mutations in cancer at a glance. *J Cell Sci* **129:** 1287–1292.

Hoffman GR, Nassar N, Cerione RA. 2000. Structure of the Rho family GTP-binding protein Cdc42 in complex with the multifunctional regulator RhoGDI. *Cell* **100:** 345–356.

Hrycyna CA, Sapperstein SK, Clarke S, Michaelis S. 1991. The *Saccharomyces cerevisiae STE14* gene encodes a methyltransferase that mediates C-terminal methylation of a-factor and RAS proteins. *EMBO J* **10:** 1699–1709.

Ismail SA, Chen YX, Rusinova A, Chandra A, Bierbaum M, Gremer L, Triola G, Waldmann H, Bastiaens PI, Wittinghofer A. 2011. Arl2-GTP and Arl3-GTP regulate a GDI-like transport system for farnesylated cargo. *Nat Chem Biol* **7:** 942–949.

Jang H, Abraham SJ, Chavan TS, Hitchinson B, Khavrutskii L, Tarasova NI, Nussinov R, Gaponenko V. 2015. Mechanisms of membrane binding of small GTPase K-Ras4B farnesylated hypervariable region. *J Biol Chem* **290:** 9465–9477.

Jeong WJ, Yoon J, Park JC, Lee SH, Lee SH, Kaduwal S, Kim H, Yoon JB, Choi KY. 2012. Ras stabilization through aberrant activation of Wnt/β-catenin signaling promotes intestinal tumorigenesis. *Sci Signal* **5:** ra30.

John J, Sohmen R, Feuerstein J, Linke R, Wittinghofer A, Goody RS. 1990. Kinetics of interaction of nucleotides with nucleotide-free H-Ras p21. *Biochemistry* **29:** 6058–6065.

Johnson L, Greenbaum D, Cichowski K, Mercer K, Murphy E, Schmitt E, Bonson RT, Umanoff H, Edelmann W, Kucherlapati R, et al. 1997. K-*ras* is an essential gene in the mouse with partial functional overlap with N-*ras*. *Genes Dev* **11:** 2468–2481.

Judd WR, Slattum PM, Hoang KC, Bhoite L, Valppu L, Alberts G, Brown B, Roth B, Ostanin K, Huang L, et al. 2011. Discovery and SAR of methylated tetrahydropyranyl derivatives as inhibitors of isoprenylcysteine carboxyl methyltransferase (ICMT). *J Med Chem* **54:** 5031–5047.

Jura N, Scotto-Lavino E, Sobczyk A, Bar-Sagi D. 2006. Differential modification of Ras proteins by ubiquitination. *Mol Cell* **21:** 679–687.

Just I, Selzer J, Hofmann F, Green GA, Aktories K. 1996. Inactivation of Ras by *Clostridium sordellii* lethal toxin-catalyzed glucosylation. *J Biol Chem* **271:** 10149–10153.

Kang R, Wan J, Arstikaitis P, Takahashi H, Huang K, Bailey AO, Thompson JX, Roth AF, Drisdel RC, Mastro R, et al. 2008. Neural palmitoyl-proteomics reveals dynamic synaptic palmitoylation. *Nature* **456:** 904–909.

Kim SE, Yoon JY, Jeong WJ, Jeon SH, Park Y, Yoon JB, Park YN, Kim H, Choi KY. 2009. H-Ras is degraded by Wnt/β-catenin signaling via β-TrCP-mediated polyubiquitylation. *J Cell Sci* **122:** 842–848.

Kollar P, Rajchard J, Balounova Z, Pazourek J. 2014. Marine natural products: Bryostatins in preclinical and clinical studies. *Pharm Biol* **52:** 237–242.

Kong E, Peng S, Chandra G, Sarkar C, Zhang Z, Bagh MB, Mukherjee AB. 2013. Dynamic palmitoylation links cytosol-membrane shuttling of acyl-protein thioesterase-1 and acyl-protein thioesterase-2 with that of proto-oncogene H-Ras product and growth-associated protein-43. *J Biol Chem* **288:** 9112–9125.

Lander HM, Ogiste JS, Teng KK, Novogrodsky A. 1995a. p21ras as a common signaling target of reactive free radicals and cellular redox stress. *J Biol Chem* **270:** 21195–21198.

Lander HM, Ogiste JS, Pearce SF, Levi R, Novogrodsky A. 1995b. Nitric oxide-stimulated guanine nucleotide exchange on p21ras. *J Biol Chem* **270:** 7017–7020.

Lander HM, Milbank AJ, Tauras JM, Hajjar DP, Hempstead BL, Schwartz GD, Kraemer RT, Mirza UA, Chait BT, Burk SC, et al. 1996. Redox regulation of cell signalling. *Nature* **381:** 380–381.

Lander HM, Hajjar DP, Hempstead BL, Mirza UA, Chait BT, Campbell S, Quilliam LA. 1997. A molecular redox switch on $p21^{ras}$. Structural basis for the nitric oxide-$p21^{ras}$ interaction. *J Biol Chem* **272:** 4323–4326.

Lau HY, Ramanujulu PM, Guo D, Yang T, Wirawan M, Casey PJ, Go ML, Wang M. 2014. An improved isoprenylcysteine carboxylmethyltransferase inhibitor induces cancer cell death and attenuates tumor growth in vivo. *Cancer Biol Ther* **15:** 1280–1291.

Lau HY, Tang J, Casey PJ, Wang M. 2017. Isoprenylcysteine carboxylmethyltransferase is critical for malignant transformation by all RAS isoforms. *Oncogene* **36:** 3934–3942.

Laude AJ, Prior IA. 2008. Palmitoylation and localisation of RAS isoforms are modulated by the hypervariable linker domain. *J Cell Sci* **121:** 421–427.

Leung KF, Baron R, Ali BR, Magee AI, Seabra MC. 2007. Rab GTPases containing a CAAX motif are processed post-geranylgeranylation by proteolysis and methylation. *J Biol Chem* **282:** 1487–1497.

Liang X, Nazarian A, Erdjument-Bromage H, Bornmann W, Tempst P, Resh MD. 2001. Heterogeneous fatty acylation of Src family kinases with polyunsaturated fatty acids regulates raft localization and signal transduction. *J Biol Chem* **276:** 30987–30994.

Lim KH, Ancrile BB, Kashatus DF, Counter CM. 2008. Tumour maintenance is mediated by eNOS. *Nature* **452:** 646–649.

Lim S, Westover K, Ficarro S, Harrison R, Choi H, Pacold M, Carrasco M, Hunter J, Kim ND, Xie T, et al. 2014. Therapeutic targeting of oncogenic K-Ras by a covalent catalytic site inhibitor. *Angew Chem Int Ed Engl* **53:** 199–204.

Lin DT, Conibear E. 2015. ABHD17 proteins are novel protein depalmitoylases that regulate N-Ras palmitate turnover and subcellular localization. *eLife* **4:** e11306.

Lito P, Solomon M, Li LS, Hansen R, Rosen N. 2016. Allele-specific inhibitors inactivate mutant KRAS G12C by a trapping mechanism. *Science* **351:** 604–608.

Liu M, Sjogren AK, Karlsson C, Ibrahim MX, Andersson KM, Olofsson FJ, Wahlstrom AM, Dalin M, Yu H, Chen Z, et al. 2010. Targeting the protein prenyltransferases efficiently reduces tumor development in mice with K-RAS-induced lung cancer. *Proc Natl Acad Sci* **107:** 6471–6476.

Liu P, Jiao B, Zhang R, Zhao H, Zhang C, Wu M, Li D, Zhao X, Qiu Q, Li J, et al. 2016. Palmitoylacyltransferase Zdhhc9 inactivation mitigates leukemogenic potential of oncogenic Nras. *Leukemia* **30:** 1225–1228.

Lobell RB, Liu D, Buser CA, Davide JP, DePuy E, Hamilton K, et al. 2002. Preclinical and clinical pharmacodynamic assessment of L-778,123, a dual inhibitor of farnesyl:protein transferase and geranylgeranyl:protein transferase type-I. *Mol Cancer Ther* **1:** 747–758.

Lobo S, Greentree WK, Linder ME, Deschenes RJ. 2002. Identification of a Ras palmitoyltransferase in *Saccharomyces cerevisiae*. *J Biol Chem* **21:** 21.

Lorent JH, Levental I. 2015. Structural determinants of protein partitioning into ordered membrane domains and lipid rafts. *Chem Phys Lipids* **192:** 23–32.

Lu JY, Hofmann SL. 1995. Depalmitoylation of CAAX motif proteins. Protein structural determinants of palmitate turnover rate. *J Biol Chem* **270:** 7251–7256.

Magee AI, Gutierrez L, McKay IA, Marshall CJ, Hall A. 1987. Dynamic fatty acylation of p21N-ras. *EMBO J* **6:** 3353–3357.

Maher J, Baker DA, Manning M, Dibb NJ, Roberts IA. 1995. Evidence for cell-specific differences in transformation by N-, H- and K-Ras. *Oncogene* **11:** 1639–1647.

Manu KA, Chai TF, Teh JT, Zhu WL, Casey PJ, Wang M. 2017. Inhibition of isoprenylcysteine carboxylmethyltransferase induces cell-cycle arrest and apoptosis through p21 and p21-regulated BNIP3 induction in pancreatic cancer. *Mol Cancer Ther* **16:** 914–923.

Marshall CJ. 1996. Ras effectors. *Curr Opin Cell Biol* **8:** 197–204.

Martin BR, Cravatt BF. 2009. Large-scale profiling of protein palmitoylation in mammalian cells. *Nat Methods* **6:** 135–138.

McLaughlin S, Aderem A. 1995. The myristoyl-electrostatic switch: A modulator of reversible protein–membrane interactions. *Trends Biochem Sci* **20:** 272–276.

Michaelson D, Silletti J, Murphy G, D'Eustachio P, Rush M, Philips MR. 2001. Differential localization of Rho GTPases in live cells. Regulation by hypervariable regions and rhogdi binding. *J Cell Biol* **152:** 111–126.

Mitchell DA, Vasudevan A, Linder ME, Deschenes RJ. 2006. Protein palmitoylation by a family of DHHC protein S-acyltransferases. *J Lipid Res* **47:** 1118–1127.

Mohammad RM, Dugan MC, Mohamed AN, Almatchy VP, Flake TM, Dergham ST, Sheilds AF, Al-Katib AA, Vaitkevicius VK, Sarkar FH. 1998. Establishment of a human pancreatic tumor xenograft model: Potential application for preclinical evaluation of novel therapeutic agents. *Pancreas* **16:** 19–25.

Mohammed I, Hampton SE, Ashall L, Hildebrandt ER, Kutlik RA, Manandhar SP, Floyd BJ, Smith HE, Dozier JK, Distefano MD, et al. 2016. 8-Hydroxyquinoline-based inhibitors of the Rce1 protease disrupt Ras membrane localization in human cells. *Bioorg Med Chem* **24:** 160–178.

Nancy V, Callebaut I, El Marjou A, de Gunzburg J. 2002. The δ subunit of retinal rod cGMP phosphodiesterase regulates the membrane association of Ras and Rap GTPases. *J Biol Chem* **277:** 15076–15084.

Onken B, Wiener H, Philips M, Chang EC. 2006. Compartmentalized signaling of Ras in fission yeast. *Proc Natl Acad Sci* **103:** 9045–9050.

Ostrem JM, Peters U, Sos ML, Wells JA, Shokat KM. 2013. K-Ras^{G12C} inhibitors allosterically control GTP affinity and effector interactions. *Nature* **503:** 548–551.

Otto JC, Kim E, Young SG, Casey PJ. 1999. Cloning and characterization of a mammalian prenyl protein-specific protease. *J Biol Chem* **274:** 8379–8382.

Patricelli MP, Janes MR, Li LS, Hansen R, Peters U, Kessler LV, Chen Y, Kucharski JM, Feng J, Ely T, et al. 2016. Selective inhibition of oncogenic KRAS output with small molecules targeting the inactive state. *Cancer Discov* **6:** 316–329.

Pillinger MH, Volker C, Stock JB, Weissmann G, Philips MR. 1994. Characterization of a plasma membrane-associated prenylcysteine-directed α carboxyl methyltransferase in human neutrophils. *J Biol Chem* **269:** 1486–1492.

Plowman SJ, Hancock JF. 2005. Ras signaling from plasma membrane and endomembrane microdomains. *Biochim Biophys Acta* **1746:** 274–283.

Prior IA, Lewis PD, Mattos C. 2012. A comprehensive survey of Ras mutations in cancer. *Cancer Res* **72:** 2457–2467.

Resh MD. 1999. Fatty acylation of proteins: New insights into membrane targeting of myristoylated and palmitoylated proteins. *Biochim Biophys Acta* **1451:** 1–16.

Ritt DA, Abreu-Blanco MT, Bindu L, Durrant DE, Zhou M, Specht SI, Stephen AG, Holderfield M, Morrison DK. 2016. Inhibition of Ras/Raf/MEK/ERK pathway signaling by a stress-induced phospho-regulatory circuit. *Mol Cell* **64:** 875–887.

Rocks O, Peyker A, Kahms M, Verveer PJ, Koerner C, Lumbierres M, Kuhlmann J, Waldmann H, Wittinghofer A, Bastiaens PI. 2005. An acylation cycle regulates localization and activity of palmitoylated Ras isoforms. *Science* **307:** 1746–1752.

Rocks O, Gerauer M, Vartak N, Koch S, Huang ZP, Pechlivanis M, Kuhlmann J, Brunsveld L, Chandra A, Ellinger B, et al. 2010. The palmitoylation machinery is a spatially organizing system for peripheral membrane proteins. *Cell* **141:** 458–471.

Roy S, Plowman S, Rotblat B, Prior IA, Muncke C, Grainger S, Parton RG, Henis YI, Kloog Y, Hancock JF. 2005. Individual palmitoyl residues serve distinct roles in H-Ras trafficking, microlocalization, and signaling. *Mol Cell Biol* **25:** 6722–6733.

Rusch M, Zimmermann TJ, Bürger M, Dekker FJ, Görmer K, Triola G, Brockmeyer A, Janning P, Böttcher T, Sieber SA, et al. 2011. Identification of acyl protein thioesterases 1 and 2 as the cellular targets of the Ras-signaling modulators palmostatin B and M. *Angew Chem Int Ed Engl* **50:** 9838–9842.

Samatar AA, Poulikakos PI. 2014. Targeting RAS-ERK signalling in cancer: Promises and challenges. *Nat Rev Drug Discov* **13:** 928–942.

Sanders SS, Martin DD, Butland SL, Lavallée-Adam M, Calzolari D, Kay C, Yates JR III, Hayden MR. 2015. Curation of the mammalian palmitoylome indicates a pivotal role for palmitoylation in diseases and disorders of the nervous system and cancers. *PLoS Comput Biol* **11:** e1004405.

Santos E, Nebreda AR. 1989. Structural and functional properties of ras proteins. *FASEB J* **3:** 2151–2163.

Sasaki AT, Carracedo A, Locasale JW, Anastasiou D, Takeuchi K, Kahoud ER, Haviv S, Asara JM, Pandolfi PP, Cantley LC. 2010. Ubiquitination of K-Ras enhances activation and facilitates binding to select downstream effectors. *Sci Signal* **4:** ra13.

Scheffzek K, Ahmadian MR, Kabsch W, Wiesmüller L, Lautwein A, Schmitz F, Wittinghofer A. 1997. The Ras-RasGAP complex: Structural basis for GTPase activation and its loss in oncogenic mutants. *Science* **277:** 5324.

Schmick M, Vartak N, Papke B, Kovacevic M, Truxius DC, Rossmannek L, Bastiaens PIH. 2014. KRas localizes to the plasma membrane by spatial cycles of solubilization, trapping and vesicular transport. *Cell* **157:** 459–471.

Schmidt WK, Tam A, Fujimura-Kamada K, Michaelis S. 1998. Endoplasmic reticulum membrane localization of Rce1p and Ste24p, yeast proteases involved in carboxyl-terminal CAAX protein processing and amino-terminal a-factor cleavage. *Proc Natl Acad Sci* **95:** 11175–11180.

Schroeder H, Leventis R, Rex S, Schelhaas M, Nägele E, Waldmann H, Silvius JR. 1997. *S*-Acylation and plasma membrane targeting of the farnesylated carboxyl-terminal peptide of N-*ras* in mammalian fibroblasts. *Biochemistry* **36:** 13102–13109.

Seabra MC, Reiss Y, Casey PJ, Brown MS, Goldstein JL. 1991. Protein farnesyltransferase and geranylgeranyltransferase share a common α subunit. *Cell* **65:** 429–434.

Shahinian S, Silvius JR. 1995. Doubly-lipid-modified protein sequence motifs exhibit long-lived anchorage to lipid bilayer membranes. *Biochemistry* **34:** 3813–3822.

Siprashvili Z, Webster DE, Johnston D, Shenoy RM, Ungewickell AJ, Bhaduri A, Flockhart R, Zarnegar BJ, Che Y, Meschi F, et al. 2016. The noncoding RNAs SNORD50A and SNORD50B bind K-Ras and are recurrently deleted in human cancer. *Nat Genet* **48:** 53–58.

Sun WT, Xiang W, Ng BL, Asari K, Bunte RM, Casey PJ, Wang M, Chuah C. 2016. Inhibition of isoprenylcysteine carboxylmethyltransferase augments BCR-ABL1 tyrosine kinase inhibition-induced apoptosis in chronic myeloid leukemia. *Exp Hematol* **44:** 189–193 e2.

Sung PJ, Tsai FD, Vais H, Court H, Yang J, Fehrenbacher N, Foskett JK, Philips MR. 2013. Phosphorylated K-Ras limits cell survival by blocking Bcl-x_L sensitization of inositol trisphosphate receptors. *Proc Natl Acad Sci* **110:** 20593–20598.

Swarthout JT, Lobo S, Farh L, Croke MR, Greentree WK, Deschenes RJ, Linder ME. 2005. DHHC9 and GCP16 constitute a human protein fatty acyltransferase with specificity for H- and N-Ras. *J Biol Chem* **280:** 31141–31148.

Switzer CH, Cheng RY, Ridnour LA, Glynn SA, Ambs S, Wink DA. 2012. Ets-1 is a transcriptional mediator of oncogenic nitric oxide signaling in estrogen receptor-negative breast cancer. *Breast Cancer Res* **14:** R125.

Tabaczar S, Czogalla A, Podkalicka J, Biernatowska A, Sikorski AF. 2017. Protein palmitoylation: Palmitoyltransferases and their specificity. *Exp Biol Med (Maywood)* **242:** 1150–1157.

Tomatis VM, Trenchi A, Gomez GA, Daniotti JL. 2010. Acyl-protein thioesterase 2 catalyzes the deacylation of peripheral membrane-associated GAP-43. *PLoS ONE* **5:** e15045.

Trahey M, McCormick F. 1987. A cytoplasmic protein stimulates normal N-ras p21 GTPase, but does not affect oncogenic mutants. *Science* **238:** 542–545.

Tsai FD, Wynne JP, Ahearn IM, Philips MR. 2014. Metabolic labeling of Ras with tritiated palmitate to monitor palmitoylation and depalmitoylation. *Methods Mol Biol* **1120:** 33–41.

Tsai FD, Lopes MS, Zhou M, Court H, Ponce O, Fiordalisi JJ, Gierut JJ, Cox AD, Haigis KM, Philips MR. 2015. K-Ras4A splice variant is widely expressed in cancer and uses a hybrid membrane-targeting motif. *Proc Natl Acad Sci* **112:** 779–784.

Ullrich O, Stenmark H, Alexandrov K, Huber LA, Kaibuchi K, Sasaki T, Takai Y, Zerial M. 1993. Rab GDP dissociation inhibitor as a general regulator for the membrane association of rab proteins. *J Biol Chem* **268:** 18143–18150.

Vartak N, Papke B, Grecco HE, Rossmannek L, Waldmann H, Hedberg C, Bastiaens PI. 2014. The autodepalmitoylating activity of APT maintains the spatial organization of palmitoylated membrane proteins. *Biophysical J* **106:** 93–105.

Veit M, Schmidt MFG. 2006. *Encyclopedic reference of genomics and proteomics in molecular medicine.* Springer, Berlin.

Vetter IR, Wittinghofer A. 2001. The guanine nucleotide-binding switch in three dimensions. *Science* **294:** 1299–1304.

Wahlstrom AM, Cutts BA, Karlsson C, Andersson KM, Liu M, Sjogren AK, Swolin B, Young SG, Bergo MO. 2007. Rce1 deficiency accelerates the development of K-RAS-induced myeloproliferative disease. *Blood* **109:** 763–78.

Wahlstrom AM, Cutts BA, Liu M, Lindskog A, Karlsson C, Sjogren AK, Andersson KM, Young SG, Bergo MO. 2008. Inactivating *Icmt* ameliorates K-RAS-induced myeloproliferative disease. *Blood* **112:** 1357–1365.

Wang MT, Holderfield M, Galeas J, Delrosario R, To MD, Balmain A, McCormick F. 2015. K-Ras promotes tumorigenicity through suppression of non-canonical Wnt signaling. *Cell* **163:** 1237–1251.

Wang T, Yu H, Hughes NW, Liu B, Kendirli A, Klein K, Chen WW, Lander ES, Sabatini DM. 2017. Gene essentiality profiling reveals gene networks and synthetic lethal interactions with oncogenic Ras. *Cell* **168:** 890–903.e15.

Webb Y, Hermida-Matsumoto L, Resh MD. 2000. Inhibition of protein palmitoylation, raft localization, and T-cell signaling by 2-bromopalmitate and polyunsaturated fatty acids. *J Biol Chem* **275:** 261–270.

Welsch ME, Kaplan A, Chambers JM, Stokes ME, Bos PH, Zask A, Zhang Y, Sanchez-Martin M, Badgley MA, Huang CS, et al. 2017. Multivalent small-molecule pan-RAS inhibitors. *Cell* **168:** 878–89 e29.

White C, Li C, Yang J, Petrenko NB, Madesh M, Thompson CB, Foskett JK. 2005. The endoplasmic reticulum gateway to apoptosis by Bcl-X_L modulation of the $InsP_3R$. *Nat Cell Biol* **7:** 1021–1028.

Whyte DB, Kirschmeier P, Hockenberry TN, Nunez-Oliva I, James L, Catino JJ, Bishop WR, Pai JK. 1997. K- and N-Ras are geranylgeranylated in cells treated with farnesyl protein transferase inhibitors. *J Biol Chem* **272:** 14459–14464.

Williams JG, Pappu K, Campbell SL. 2003. Structural and biochemical studies of p21Ras *S*-nitrosylation and nitric oxide-mediated guanine nucleotide exchange. *Proc Natl Acad Sci* **100:** 6376–6381.

Willumsen BM, Norris K, Papageorge AG, Hubbert NL, Lowy DR. 1984. Harvey murine sarcoma virus p21 ras protein: Biological and biochemical significance of the cysteine nearest the carboxy terminus. *EMBO J* **3:** 2581–2585.

Wright LP, Philips MR. 2006. Thematic review series: Lipid posttranslational modifications. CAAX modification and membrane targeting of Ras. *J Lipid Res* **47:** 883–891.

Wright LP, Court H, Mor A, Ahearn IM, Casey PJ, Philips MR. 2009. Topology of mammalian isoprenylcysteine carboxyl methyltransferase determined in live cells with a fluorescent probe. *Mol Cell Biol* **29:** 1826–1833.

Xu L, Lubkov V, Taylor LJ, Bar-Sagi D. 2010. Feedback regulation of Ras signaling by Rabex-5-mediated ubiquitination. *Curr Biol* **20:** 1372–1377.

Xu J, Hedberg C, Dekker FJ, Li Q, Haigis KM, Hwang E, Waldmann H, Shannon K. 2012. Inhibiting the palmitoylation/depalmitoylation cycle selectively reduces the growth of hematopoietic cells expressing oncogenic Nras. *Blood* **119:** 1032–1035.

Yang YY, Ascano JM, Hang HC. 2010. Bioorthogonal chemical reporters for monitoring protein acetylation. *J Am Chem Soc* **132:** 3640–3641.

Yang MH, Nickerson S, Kim ET, Liot C, Laurent G, Spang R, Philips MR, Shan Y, Shaw DE, Bar-Sagi D, et al. 2012. Regulation of RAS oncogenicity by acetylation. *Proc Natl Acad Sci* **109:** 10843–10888.

Yang MH, Laurent G, Bause AS, Spang R, German N, Haigis MC, Haigis KM. 2013. HDAC6 and SIRT2 regulate the acetylation state and oncogenic activity of mutant K-RAS. *Mol Cancer Res* **11:** 1072–1077.

Zeng T, Wang Q, Fu J, Lin Q, Bi J, Ding W, Qiao Y, Zhang S, Zhao W, Lin H, et al. 2014. Impeded Nedd4-1-mediated Ras degradation underlies Ras-driven tumorigenesis. *Cell Rep* **7:** 871–882.

Zhang H, Liu XH, Zhang K, Chen CK, Frederick JM, Prestwich GD, Baehr W. 2004. Photoreceptor cGMP phosphodiesterase δ subunit (PDEδ) functions as a prenyl-binding protein. *J Biol Chem* **279:** 407–413.

Zhang H, Li S, Doan T, Rieke F, Detwiler PB, Frederick JM, Baehr W. 2007. Deletion of PrBP/δ impedes transport of GRK1 and PDE6 catalytic subunits to photoreceptor outer segments. *Proc Natl Acad Sci* **104:** 8857–8862.

Zhao H, Liu P, Zhang R, Wu M, Li D, Zhao X, Zhang C, Jiao B, Chen B, Chen Z, et al. 2015. Roles of palmitoylation and the KIKK membrane-targeting motif in leukemogenesis by oncogenic KRAS4A. *J Hematol Oncol* **8:** 132.

Zhou M, Wiener H, Su W, Zhou Y, Liot C, Ahearn I, Hancock JF, Philips MR. 2016. VPS35 binds farnesylated N-Ras in the cytosol to regulate N-Ras trafficking. *J Cell Biol* **214:** 445–458.

Zhou Y, Prakash P, Liang H, Cho KJ, Gorfe AA, Hancock JF. 2017. Lipid-sorting specificity encoded in K-Ras membrane anchor regulates signal output. *Cell* **168:** 239–251 e16.

Zimmermann G, Papke B, Ismail S, Vartak N, Chandra A, Hoffmann M, Hahn SA, Triola G, Wittinghofer A, Bastiaens PI, et al. 2013. Small molecule inhibition of the KRAS-PDEδ interaction impairs oncogenic KRAS signalling. *Nature* **497:** 638–642.

Zimmermann G, Schultz-Fademrecht C, Küchler P, Murarka S, Ismail S, Triola G, Nussbaumer P, Wittinghofer A, Waldmann H. 2014. Structure guided design and kinetic analysis of highly potent benzimidazole inhibitors targeting the PDEδ prenyl binding site. *J Med Chem* **57:** 5435–5448.

Ras and the Plasma Membrane: A Complicated Relationship

Yong Zhou, Priyanka Prakash, Alemayehu A. Gorfe, and John F. Hancock

Department of Integrative Biology and Pharmacology, University of Texas Health Science Center at Houston, Houston, Texas 77030

Correspondence: john.f.hancock@uth.tmc.edu

The primary site of Ras signal transduction is the plasma membrane (PM). On the PM, the ubiquitously expressed Ras isoforms, H-, N-, and K-Ras, spatially segregate to nonoverlapping nanometer-sized domains, called nanoclusters, with further lateral segregation into nonoverlapping guanosine triphosphate (GTP)-bound and guanosine diphosphate (GDP)-bound nanoclusters. Effector binding and activation is restricted to GTP nanoclusters, rendering the underlying assembly mechanism essential to Ras signaling. Ras nanoclusters have distinct lipid compositions as a result of lipid-sorting specificity encoded in each Ras carboxy-terminal membrane anchor. The role of the G-domain in regulating anchor–membrane interactions is becoming clearer. Ras G-domains undergo significant conformational orientation changes on guanine nucleotide switch, leading to differential direct contacts between the G-domain and reorganization of the membrane anchor. Ras G-domains also contain weak dimer interfaces, resulting in homodimerization, which is an obligate step of nanoclustering. Modulating the formation of Ras dimers, the lipid composition of the PM or lateral dynamics of key PM phospholipids represent novel mechanisms whereby the extent of Ras nanoclustering can be regulated to tune the gain in Ras signaling circuits.

In this review, we will discuss the spatiotemporal organization of Ras proteins on the cell plasma membrane (PM) and the emergent signaling consequences of the attendant spatial distribution of Ras signaling complexes. Specifically, we will discuss how the posttranslationally modified carboxy-terminal anchoring domains of Ras isoforms selectively sort PM lipids and drive isoform-specific nanoclustering. We will then explore the latest findings on G-domain-mediated Ras homodimerization and how Ras dimers potentially contribute to the formation of higher order oligomers. We will conclude the review by discussing the potential biological implications of Ras nanoclusters and how altering lipid composition within Ras nanoclusters influences signal transduction.

RAS ANCHORS AND PLASMA MEMBRANE TARGETING

Ras superfamily proteins oscillate between inactive guanosine diphosphate (GDP)-bound and active guanosine triphosphate (GTP)-bound states to control signaling cascades that regulate multiple aspects of cell biology, including cell

Cite this article as *Cold Spring Harb Perspect Med* doi: 10.1101/cshperspect.a031831

survival, migration, and proliferation. Prototypical examples are the three Ras isoforms that are ubiquitously expressed in human cells: K-, H-, and N-Ras, of which K-Ras and N-Ras are frequently mutated in human cancers (Prior et al. 2012). Ras proteins contain nearly identical G-domains that bind guanine nucleotides and interact with a common set of activators and effectors but show significant sequence variation in their carboxy-terminal hypervariable regions (Prior and Hancock 2001; Hancock 2003). For biological activity, Ras proteins must localize to the PM (Willumsen et al. 1984; Hancock et al. 1989) using a carboxy-terminal membrane anchor. This follows because Ras effectors must be recruited to the PM for activation. A Ras PM anchor consists of two components. The first anchor component, common to all Ras isoforms, is a carboxy-terminal *S*-farnesyl cysteine carboxylmethyl ester, which is generated by a sequential set of irreversible posttranslational modifications of a conserved carboxy-terminal CAAX motif present in the nascent Ras protein (Fig. 1). The modifications are carried out by farnesyl protein transferase in the cytosol, and two endoplasmic reticulum (ER)-localized enzymes, the Rce1 protease and isoprenyl-carboxymethyl transferase (Reiss et al. 1990; Dai et al. 1998; Bergo et al. 2000, 2001, 2002). The second anchor component comprises monopalmitoylation of N-Ras, duopalmitoylation of H-Ras, and a polybasic domain (PBD) of six contiguous lysines in K-Ras4B, the more prevalent splice form of K-Ras, hereafter called just K-Ras (Casey et al. 1989; Hancock et al. 1989, 1990). Palmitoylation is accomplished by ER- and Golgi-localized palmitoyltransferases. Of note, this type of bipartite membrane anchor is found in many other Ras superfamily proteins (Hancock et al. 1990; Michaelson et al. 2001). After anchor assembly is completed on the cytosolic surface of the ER, H-Ras, and N-Ras access the PM via the exocytic pathway through the Golgi (Choy et al. 1999; Apolloni et al. 2000), whereas K-Ras

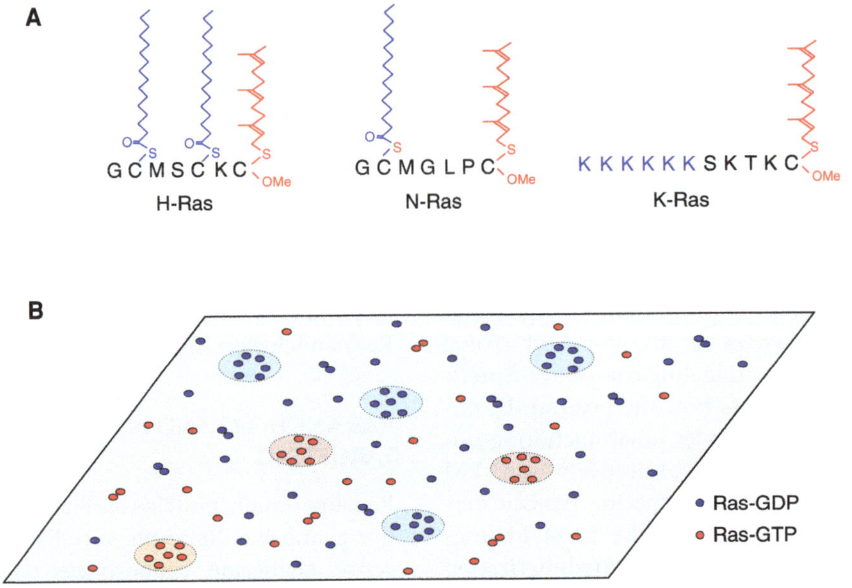

Figure 1. Ras targeting and spatial organization on the plasma membrane (PM). (*A*) Ras minimal membrane anchors. All Ras anchors have a farnesyl group, whereas H-Ras has two additional palmitoyl chains and N-Ras has one additional palmitoyl chain. K-Ras has a contiguous hexalysine polybasic domain (PBD). (*B*) Ras isoforms spatially segregate into nonoverlapping nanoclusters on cell PM. Guanine nucleotide binding states also drive further spatial separation for each isoform. Each Ras nanocluster is ~18 nm in diameter and contains about five to six Ras molecules. ~56% of Ras proteins are found as monomers, whereas the remaining 44% form higher-order oligomers and nanoclusters.

 Cite this article as *Cold Spring Harb Perspect Med* doi: 10.1101/cshperspect.a031831

is delivered to the PM bypassing the Golgi via the recycling endosome (RE) (Choy et al. 1999; Apolloni et al. 2000; Schmick et al. 2014).

After initial delivery to the PM, each Ras protein must be actively maintained on the PM to preserve the fidelity of PM targeting and prevent entropic equilibration across all cellular membranes. To this end, H- and N-Ras undergo cycles of depalmitoylation, by as-yet poorly characterized thioesterases after internalization on the cytosolic surface of endosomes, followed by repalmitoylation in the ER and Golgi for vesicular transport back to the PM (Goodwin et al. 2005; Rocks et al. 2005). Several putative chaperones that bind nonpalmitoylated H- and N-Ras have been identified, including VPS35 and PDEδ, which facilitate cytosolic diffusion of prenylated, nonpalmitoylated Ras and contribute to engagement with membrane-bound palmitoyl transferases (Schmick et al. 2014; Zhou et al. 2016). Similarly, K-Ras is released from endosomes immediately after internalization, caused by the loss of anionic lipid asymmetry, and is captured by the chaperone protein PDEδ. The K-Ras PDEδ complex is, in turn, dissociated by binding the small GTPase Arl2 in the spatial vicinity of the RE, allowing K-Ras to bind to the cytosolic surface of RE vesicles for forward transport back to the PM (Chandra et al. 2012; Zimmermann et al. 2013; Schmick et al. 2014).

SPATIAL ORGANIZATION ON THE PM: DIMERS AND NANOCLUSTERS

The major Ras signaling platform, the PM, is a complex, dynamic structure with heterogeneity on many length and time scales (Mayor and Rao 2004; Simons and Vaz 2004; Kusumi et al. 2005; Hancock 2006; Lingwood et al. 2009; Prior and Hancock 2012). The origins of this heterogeneity include several thousand species of lipid, compartmentalization by a submembrane actin cytoskeleton, and the transient lateral assembly of different subsets of lipids driven by protein–lipid and lipid–lipid interactions (Mayor and Rao 2004; Simons and Vaz 2004; Kusumi et al. 2005; Hancock 2006; Lingwood et al. 2009; Prior and Hancock 2012). On the PM, Ras proteins diffuse laterally as monomers and dimers and assemble into higher-order oligomers that lead to the formation of nanoclusters (Prior et al. 2003; Plowman et al. 2005; Nan et al. 2013; Lin et al. 2014; Muratcioglu et al. 2015; Zhou and Hancock 2015; Jang et al. 2016; Prakash et al. 2016). A nanocluster contains about six Ras proteins, has a radius of ~9 nm and a lifetime of <1 s (Prior et al. 2003; Murakoshi et al. 2004; Hancock and Parton 2005; Plowman et al. 2005; Abankwa et al. 2010; Zhou et al. 2014; Zhou and Hancock 2015). H-, N-, and K-Ras proteins assemble into spatially nonoverlapping nanoclusters with further lateral segregation between nanoclusters of GTP-loaded and GDP-loaded Ras (Prior et al. 2003; Gorfe et al. 2007b; Abankwa et al. 2008, 2010; Weise et al. 2011; Zhou et al. 2014; Prakash et al. 2016). Ras-GTP nanoclusters act as signaling platforms for effector binding and signal transmission and are therefore essential to Ras signaling (Plowman et al. 2005, 2008; Tian et al. 2007; Nan et al. 2013; Zhou and Hancock 2015) as will be discussed later.

The spatial distribution of Ras on the cell PM has been extensively quantified using electron microscopy (EM) combined with spatial analysis. EM estimates of the relative distributions of fractions, coupled with earlier single-fluorophore video tracking (SFVT) experiments, suggest that ~50% of Ras proteins diffuse as monomers, with the remainder present as mobile dimers and immobile nanoclusters. This is a nonequilibrium state in that the fraction of a Ras protein population assembling into nanoclusters is constant over a multilog range of expression levels; that is, the clustered fraction is the same at endogenous levels of expression as well as ectopic expression levels reaching 20× over endogenous (Plowman et al. 2005). Interestingly, glycophosphatidylinositol (GPI)-anchored proteins show an identical nonequilibrium distribution (Mayor and Rao 2004; Goswami et al. 2008; Gowrishankar et al. 2012; Raghupathy et al. 2015). These and other considerations suggest that an intrinsic property of lipid-anchored proteins on a lipid bilayer may be a propensity to form clusters. Indeed, the clustering phenomenon is seen when multiple anchors are simulated on a bilayer using molecular dynamics (MD) and on supported bilayers (Weise et al.

2011; Li and Gorfe 2013; Lin et al. 2015). Our current understanding of the basic molecular mechanism revolves around oligomerization of the G-domain, the role of the anchor and its interactions with the lipid bilayer, and how the G-domain can modulate the behavior of the anchor (Fig. 2). These are interrelated but will be initially considered here as individual determinants. Other contributing factors that will not be considered further are thermodynamic fluctuations in lipid diffusion that result from actin turnover and trundling in the submembrane actin cytoskeleton (Goswami et al. 2008; Gowrishankar et al. 2012), which have been shown theoretically and experimentally to play a role in maintaining the nonequilibrium clustered distribution of GPI-anchored proteins and which may also therefore contribute to maintaining the similar nonequilibrium clustered distribution of Ras proteins. Connectivity between the inner and outer leaflets mediated by long-chain PtdSer and actin interactions contributes to nanoclustering of outer leaflet GPI-anchored proteins (Raghupathy et al. 2015), although relevance here to Ras is unclear.

Much recent work has focused on characterizing K-Ras dimers and putative interaction surfaces that may be relevant to PM-bound K-Ras. Dimers have been observed using super-resolution microscopy, dimers, and nanoclusters by EM of intact PM, and the existence of diverse higher-order oligomers inferred by SFVT (Murakoshi et al. 2004; Nan et al. 2015; Spencer-Smith et al. 2017). Using a combination of computational and experimental techniques, we and others recently identified two G-domain surfaces involved in the assembly of K-Ras monomers into multiple types of dimer (Muratcioglu et al. 2015; Prakash et al. 2016). The surfaces are low affinity and are stable only when the Ras protein is attached to the membrane bilayer. Both surfaces are stabilized by ionic interactions

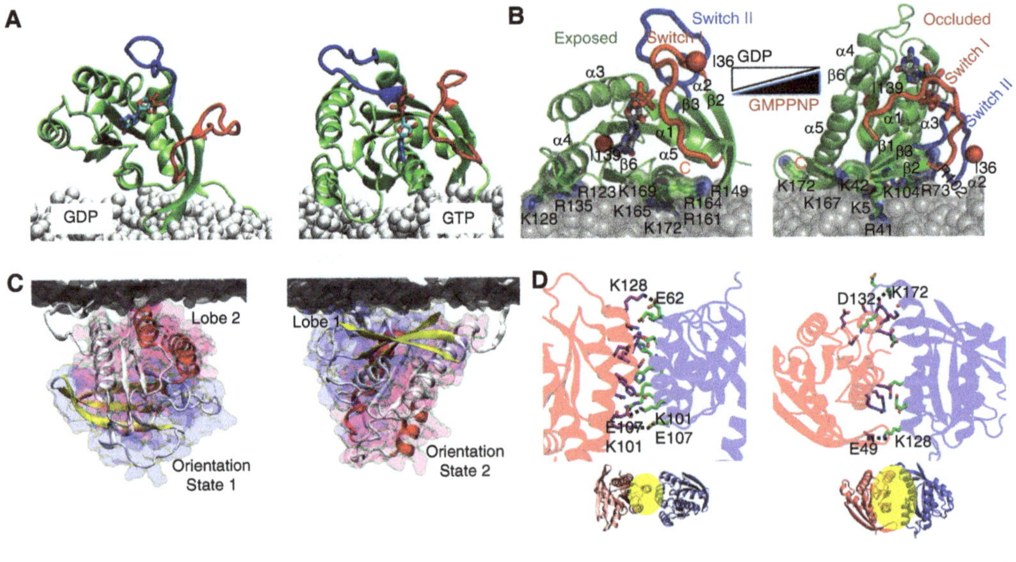

Figure 2. Ras G-domain conformational orientations drive spatial segregation on the plasma membrane. (*A*) H-Ras-guanosine diphosphate (GDP) favors a G-domain orientation that is perpendicular to the membrane. Guanosine triphosphate (GTP)-bound H-Ras favors a G-domain orientation that is rotated through 100° such that the helix α4 now interacts with lipid head groups. (*B*) The K-Ras G-domain is more dynamic, but samples more extensively orientate aligning parallel with the bilayer when GDP-bound and semiperpendicular when GTP-bound, with different G-domain and hypervariable region (HVR) residue hydrogen bonding to the bilayer in each case. (From Mazhab-Jafari et al. 2015; reprinted, with permission, from the National Academy of Sciences © 2015.) (*C*) The GTP-bound K-Ras G-domain further samples two distinct conformations, OS1 and OS2 as described in the text (Prakash et al. 2016). (*D*) Two partially overlapping dimer surfaces that have been validated experimentally are located on helix α3/α4 and helix α4/α5.

between two Ras molecules: the h3/h4 (h3 = α-helix3/h4 = α-helix4) surface contains reciprocal K101-E107 and E107-K101 salt bridges and the h4/h5 surface contains ion pairs E98-K165 and D105-K172. The role of additional K-Ras dimerization surfaces observed in MD simulations remains uncertain (Muratcioglu et al. 2015; Sayyed-Ahmad et al. 2016; Prakash et al. 2017).

Taken together with the known spatial organization of K-Ras, these dimerization data lead to the hypothesis that the formation of K-Ras dimers on the PM is a prerequisite for the formation of higher-order oligomers that are, in turn, a prelude to the assembly of a nanocluster. Several lines of evidence are concordant with this hypothesis: first is the identification of multiple dimer surfaces that can be used to assemble a series of higher-order oligomers, including a pentamer (Sarkar-Banerjee et al. 2017). Of special note, the computed diameter of this pentameric structure matches exactly the lower bound diameter of a K-Ras nanocluster visualized previously by EM (Plowman et al. 2005; Sarkar-Banerjee et al. 2017). Second, if dimerization is indeed an obligate step in nanocluster assembly then perturbing dimer formation should in turn affect nanocluster formation. This has been shown experimentally in two different studies. Two sets of ionic interactions that stabilize the h3/h4 dimer or the h4/h5 dimer can be abrogated by replacing one of the partners with an amino acid of the opposite charge. Such mutations, K101E and E107K, significantly reduce K-Ras nanoclustering, whereas nanoclustering is unaffected in a double mutant K101E + E107K, in which the opposed charges are simply swapped (Prakash et al. 2017). Similarly, charge reversal mutagenesis to perturb the h4/h5 surface (E98K/D105K) impairs nanoclustering, whereas swapping the charges so that salt bridge formation is still possible (E98K/D105K + K165E/K127D) has no effect on nanoclustering (Prakash et al. 2017). Concordantly, ectopic expression of a novel monobody that targets the h3/h4 interface profoundly inhibits K-Ras dimerization, K-Ras nanoclustering, and oncogenic K-Ras signal transmission (Spencer-Smith et al. 2017). Conversely, enhancing dimerization enhances nanoclustering, for example, introducing two cysteine mutations at K101 and E107 allows the opposed cysteine side chains to form a disulfide bond, which stabilizes the h3/h4 dimer and increases K-Ras nanocluster formation (Prakash et al. 2017). Similarly, K-Ras^{G12V} dimerization can be enhanced extrinsically by treating cells with BRAF kinase inhibitors to generate stable BRAF–CRAF dimers (Heidorn et al. 2010; Poulikakos et al. 2010), or by ectopically expressing a Ras-binding domain (RBD), RBD–RBD dimer (Cho et al. 2012b). In both cases, the bipartite RBDs enhance K-Ras^{G12V} dimerization and concomitantly enhance K-Ras nanoclustering (Cho et al. 2012b). Artificially induced dimerization of mCherry-K-Ras^{G12V} also leads to significant activation of the mitogen-activated protein kinase (MAPK) signaling pathway, concordant with enhanced dimerization driving enhanced nanoclustering and, hence, increased signal output (Nan et al. 2015).

Dimers of H-Ras and N-Ras have also been observed in MD simulations and verified experimentally (Guldenhaupt et al. 2012; Lin et al. 2014). The dimerization surfaces identified are not exactly congruous with those of K-Ras. Experiments using time-resolved fluorescence anisotropy, fluorescence correlation spectroscopy, and single-particle tracking of lipidated H-Ras on a supported lipid bilayer (Lin et al. 2014) showed that alanine mutation of Tyr64 (Y64A) significantly increases H-Ras lateral diffusion by eliminating dimers (Lin et al. 2014). A dimerization interface has been identified comprising helix α5 and loops β2/3 region of N-Ras (Guldenhaupt et al. 2012). Fluorescence lifetime measurements show that N-Ras forms extensive dimers when attached to a supported bilayer but not in solution (Guldenhaupt et al. 2012). MD simulations further propose that a dimer interface exists in the region of helices α4, α5 and β4, α5 (Guldenhaupt et al. 2012). Key residues include intermolecular salt bridges between R161 and D154 of helix α5 on one N-Ras molecule with K135 in helix α4 and E49 in loop β2/β3 on the adjacent N-Ras molecule. Intermolecular salt bridges between D47 (loops β2/β3) and R161/R164 of helix α5 on the adjacent N-Ras molecule are also prominent. There is a general consensus among these and other studies that dimerization is restricted to membrane-bound Ras molecules because the binding affinities are too low to be detectable in solution (Dementiev 2012; Guldenhaupt et al. 2012; Lin et al. 2014; Muratcioglu et al. 2015; Nan et al. 2015; Prakash et al. 2016). Taken together, these studies also suggest that heterodimerization between different Ras isoforms would be unlikely and that homodimerization is strongly favored. A recent study examined this directly and verified that, at least for K-Ras and H-Ras, the respective dimers are isoform specific (Jang et al. 2016). This is an important result because if dimers are the building blocks of higher-order oligomers and nanoclusters, then fidelity of dimer assembly will, in turn, lead to fidelity of nanocluster assembly and provide a molecular basis for isoform-specific lateral segregation that is observed experimentally.

Complementary MD simulations and fluorescence lifetime imaging–fluorescence resonance energy transfer (FLIM–FRET) experiments have revealed that the likely basis of guanine nucleotide-dependent lateral segregation, whereby the GTP-bound and GDP-bound forms of each Ras isoform assemble into spatially nonoverlapping nanoclusters, involves conformational orientations of the G-domain with respect to the PM. For example, the H-Ras G-domain operates like a balance. In the GDP-bound state, most helices of the G-domain orient parallel to the membrane, with only the hypervariable region (HVR) and carboxy-terminal anchor engaging the membrane (Abankwa et al. 2008, 2010). This conformation allows extensive interaction of R169 and K170 with charged lipids in the membrane and the dual palmitoyl chains fully extended into the bilayer. This anchor structure can readily account for the segregation of H-Ras-GDP to ordered cholesterol-enriched domains (Prior et al. 2003; Abankwa et al. 2008, 2010). GTP binding leads to conformational reorientation and new salt bridges between D47/E49 in the β2-β3 loops and R161/R164 in helix α5 (Abankwa et al. 2008). The G-domain, in turn, swings ~100°, resulting in extensive interaction of R128 and R135 in helix α4 with polar lipids in the membrane and with R169 and K170 being disengaged from the membrane (Abankwa et al. 2008, 2010). The dual palmitoyl chains also become disordered again, accounting for the segregation of H-Ras-GTP away from cholesterol-rich domains (Gorfe et al. 2007a,b; Abankwa et al. 2008, 2010).

The behavior of the K-Ras G-domain differs from that of H-Ras. Liquid nuclear magnetic resonance (NMR) analyses of K-Ras anchored to lipid nanodiscs (Mazhab-Jafari et al. 2015) and infrared spectroscopy of K-Ras on synthetic model membranes (Weise et al. 2011; Kapoor et al. 2012) show that the G-domain helices of GDP-bound K-Ras align parallel to the membrane, with the surface of β1 and β6, helices α4 and α5, as well as β2-β3 loops contacting acidic lipids in the bilayer. GTP loading leads to a semiperpendicular alignment of the G-domain with the membrane, with β1–3 and helices α2 and α3 now in contact. The same alignment was predicted in all-atom MD simulations. The computational study, however, showed that K-Ras-GTP actually samples several distinct orientation states (OSs). A combination of MD simulations, EM-univariate spatial mapping, and FLIM–FRET revealed that K-Ras-GTP shifts between two predominant OSs (Prakash et al. 2016): OS1 presents helices α3/α4 to the membrane while OS2 uses β1–3 and helix α2. OS1 and OS2 are mutually exclusive in terms of interacting with the membrane and use R102, which is persistently attached to the membrane, as a hinge during switching. Detailed structures of the K-Ras minimal anchor under each nucleotide-bound state have not been reported, but the different lipid-sorting capacities discussed in the following section would suggest that they are indeed different

LIPID SORTING AND LIPID COMPOSITION OF RAS NANOCLUSTERS

Early studies showed that the cholesterol content of the PM was differentially important for H-Ras but not K-Ras signaling and subsequently that the integrity of H-Ras-GDP and N-Ras-GTP clusters, but not K-Ras-GTP or K-Ras-GDP clusters, was cholesterol dependent (Roy et al. 1999, 2005; Prior et al. 2001, 2003;

Ariotti et al. 2014). Such results suggested that each Ras nanocluster may have a different lipid composition, likely reflecting different lipid-binding preferences of the individual anchors that when, aggregated in a cluster, would in turn sort or assemble different cohorts of PM lipids. In turn, different lipid compositions of H-, N-, and K-Ras nanoclusters could account for signaling differences among the three isoforms. Subsequent work has validated each of these basic concepts.

The lipid composition of Ras nanoclusters in intact cell PM was recently determined using EM to quantify the extent of colocalization of different Ras isoforms with a set of lipid-binding probes (Figs. 3 and 4) (Zhou et al. 2014, 2015, 2017). For example, H-Ras-GDP nanoclusters were highly enriched with PIP_2, whereas those of K-Ras-GTP had a high phosphatidic acid (PA) content but minimal PIP_2. All Ras nanoclusters contain PtdSer, but respond to changing PtdSer levels in distinct manners. Although enhancing or decreasing PM PtdSer content increases or decreases, respectively, K-Ras^{G12V} PM localization and clustering, such changes in PM PtdSer content have no effect on H-Ras^{G12V} (Zhou et al. 2014, 2015), leading to the conclusion that PtdSer is a critical structural component of K-Ras-GTP nanoclusters, but not H-Ras-GTP nanoclusters. The K-Ras PBD is a key structural element for associating with anionic lipids in the PM (Hancock et al. 1990; McLaughlin et al. 2002). Traditionally, PBDs have been thought to operate largely as charge sensors with the number of basic residues within a PBD, determining the strength of electrostatic interaction with acidic lipids and similarly interacting more strongly with acidic lipids that have a high charge density. However, K-Ras nanoclusters are enriched with monovalent PtdSer but largely de-

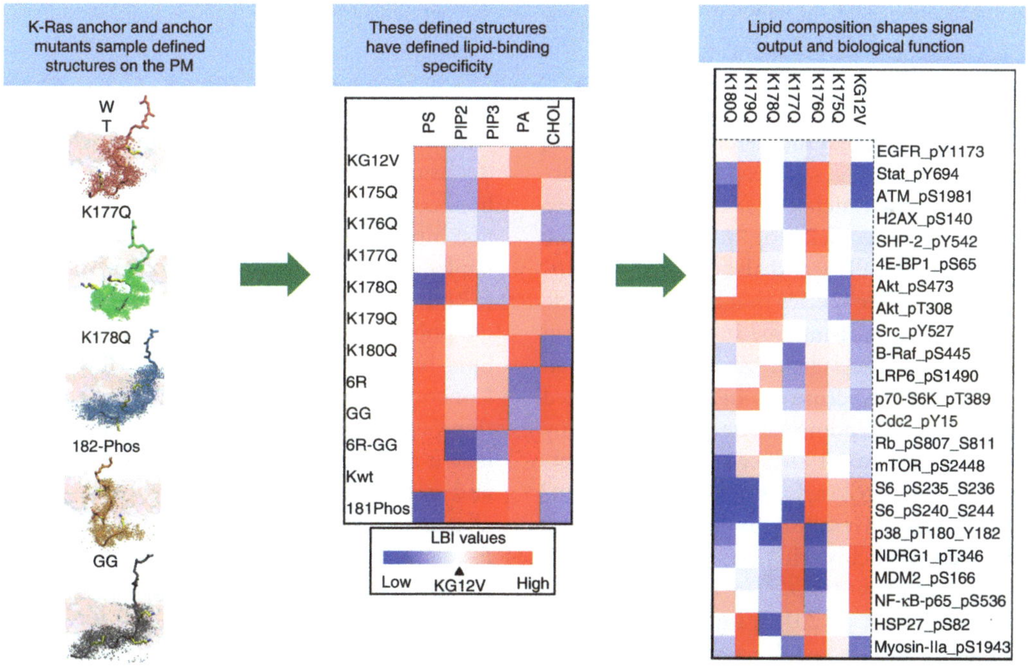

Figure 3. The K-Ras membrane anchor encodes a highly specific capacity for lipid binding. The precise amino acid sequence of the polybasic domain (PBD) and prenyl group encodes lipid-binding specificity that is realized as a defined anchor structure. Aggregation of a set of like anchors leads to assembly of nanoclusters with defined lipid compositions (shown as a heat map). The lipid composition of the resulting nanocluster shapes signal output (shown as a heat map of signaling pathway activity) (Zhou et al. 2017). LBI, Lipopolysaccharide-binding index.

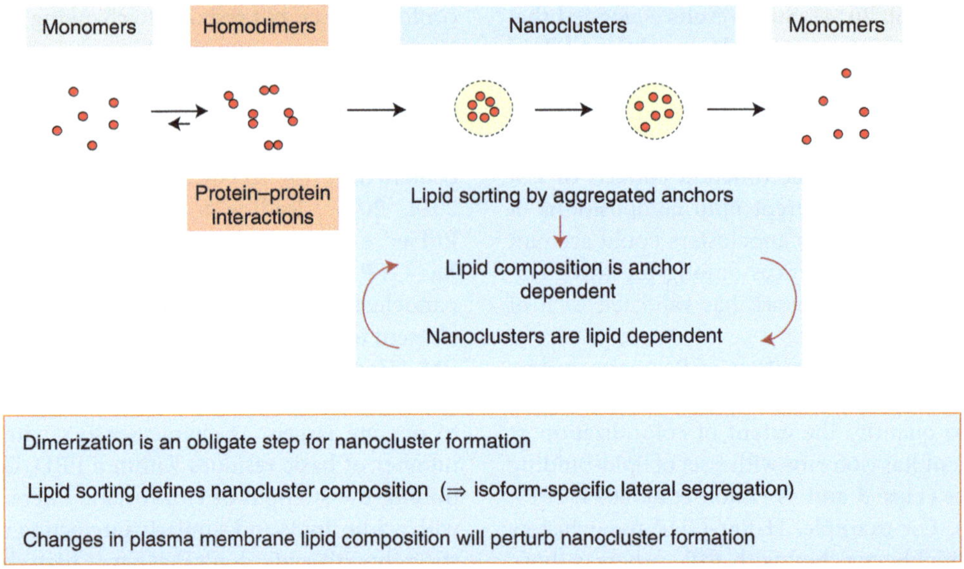

Figure 4. Putative molecular mechanism for nanocluster formation. Ras homodimer formation is an obligate step of Ras nanoclustering. Homodimers form between Ras monomers, mediated by their weak G-domain interfaces. Ras homodimers further oligomerize into nanoclusters composed of five to six proteins. Lipid sorting by the higher-order Ras oligomers generates a lipid-based nanocluster concordantly weakening G-domain interactions. The purely lipid-stabilized nanocluster then disintegrates, releasing Ras monomers.

pleted of multivalent PIP_2 (Zhou et al. 2014, 2017), suggesting that the K-Ras PBD is sensing more than just charges. Indeed, the K-Ras bipartite PBD–farnesyl anchor has now been shown to have exquisite specificity for PtdSer over anionic lipids, including PIP_2, and also to selectively interact with PtdSer species that have asymmetric lipid chains comprising one fully saturated and one unsaturated acyl chain (Zhou et al. 2017). In lipid add-back experiments in PtdSer-depleted cells, K-Ras PM binding can be recovered by any PtdSer species, except PtdSer with two fully saturated chains. Strikingly, however, only asymmetric PtdSer can rescue both K-Ras PM binding and K-Ras nanoclustering, and in turn restore the capacity of K-Ras^{G12V} to recruit CRAF to the PM (Zhou et al. 2017), implying that only asymmetric PtdSer can be assembled into K-Ras-GTP nanoclusters.

EM bivariate colocalization analysis was then used to map lipid composition of K-Ras PBD mutants, each of which contains a single alanine point mutation at each of six different positions in the PBD. Each K-Ras PBD mutant still retains five lysines, thus equivalent to the amount of charges among all PBD mutants. Yet, each K-Ras PBD mutant possesses distinct lipid composition (Zhou et al. 2017). Of particular interest, K-RasK177Q and K-RasK178Q are depleted of PtdSer and enriched with PIP_2 (Zhou et al. 2017); remarkably then, these single-point mutations are sufficient to change the lipid-binding specificity of the K-Ras anchor. Concordant with their different lipid compositions, each different PBD sequence anchor generated a different cellular signal output. Lysine and arginine each contains only one positive charge and are thus electrostatically equivalent to each other. However, replacing the lysines of the PBD with arginines fundamentally changed the lipid-binding specificity of a K-Ras anchor as does changing the farnesyl group to a geranylgeranyl group (Zhou et al. 2017). These results show that the precise amino acid sequence of the K-Ras PBD together with the prenyl group define a combinatorial code for lipid binding that extends beyond simple electrostatics. Within the code, lysine and arginine residues are nonequivalent and prenyl chain length modifies nascent PBD lipid preferences (Zhou et al. 2017). All-atom MD simulations show that this newly discovered PBD-prenyl code embedded within the K-Ras anchor is realized by distinct dynamic tertiary structures of the anchor on the PM that govern amino acid side-chain–lipid interactions (Zhou et al. 2017).

POSSIBLE MOLECULAR MODEL FOR NANOCLUSTERING

Collecting together the data and ideas described above, we can generate a working model for nanocluster assembly. We propose that G-domain-mediated Ras homodimerization is an obligate step for Ras nanocluster formation. Freely diffusing Ras homodimers assemble as a result of molecular collisions into higher-order oligomers up to and including pentamers. A key consequence of the assembly of these higher-order structures is an aggregation of a set of identical carboxy-terminal anchors with defined lipid-binding specificity that can sort or retain specific cohorts of PM lipids. Each nanocluster therefore acquires a defined, nonrandom assembly of lipids. We speculate that as lipid composition of the nanocluster is established, protein–protein interactions are destabilized, leaving the nanocluster as a lipid-based assembly of Ras monomers. This is an intrinsically unstable entity with a finite lifetime that disassembles as lipids diffuse away, releasing a set of freely diffusing Ras monomers.

SIGNAL PROCESSING AND EMERGENT SYSTEM PROPERTIES OF THE RAS NANOCLUSTER SYSTEM

RAF, MEK, and ERK are recruited and activated in Ras nanoclusters not by freely diffusing Ras monomers (Tian et al. 2007; Plowman et al. 2008). This spatiotemporal restriction of MAPK cascade activation results in key emergent properties, including high-fidelity and low-noise signal transmission within the Ras-MAPK signaling circuit (Tian et al. 2007; Harding and Hancock 2008b; Kholodenko et al. 2010). In brief, the fixed nanoclustered fraction (ϕ) at all Ras expression levels results in a linear relationship between the amount of Ras-GTP on the PM, generated, for example, in response to epidermal growth factor receptor (EGFR) activation and the number of Ras-GTP nanoclusters assembled (Fig. 5). Each nanocluster then operates as a transient digital switch in activating MAPK, generating a fixed quantal output of ERKpp before the nanocluster disassembles and terminates MAPK activation (Harding et al. 2005; Tian et al. 2007, 2010; Harding and Hancock 2008a,b; Inder and Hancock 2008; Inder et al. 2008; Kholodenko et al. 2010). The whole system operates as an analog–digital–analog converter, in which the assembly of nanoclusters digitizes the incoming analog signal (acting as an analog-to-digital converter [ADC]). The cytosolic summed output of all the ERKpp generated by the nanoclusters then regenerates the analog input signal (acting as a digital-to-analog converter [DAC]). These features of K-Ras nanoclusters allow cells to use digital biochemistry to build high-fidelity analog signaling circuits (Tian et al. 2007; Harding and Hancock 2008b; Kholodenko et al. 2010). The dynamics of nanocluster assembly and turnover are thus critical for Ras function (Plowman et al. 2005; Tian et al. 2007; Kholodenko et al. 2010) such that the nanoclustered fraction (ϕ) becomes a critical system parameter that can set signal gain. Increasing ϕ results in increases in signal gain, whereby there is increased MAPK activation per unit of Ras-GTP, whereas reducing ϕ has the opposite effect; at the limit, completely blocking nanocluster formation abrogates Ras signaling even if Ras is GTP loaded and PM localized.

In intact cells, ϕ can be regulated by multiple mechanisms. Extensive early studies implicated galectin-1 and galectin-3 as H-Ras-GTP and K-Ras-GTP nanocluster scaffolds, respectively (Belanis et al. 2008; Shalom-Feuerstein et al. 2008). Increasing the cytosolic concentration of these galectins enhanced H-Ras and K-Ras nanoclustering and MAPK signaling concordantly. Intriguingly, cytosolic levels of nucleophosmin and nucleolin also enhance K-Ras clustering (Inder et al. 2009). Caveolae remotely control Ras nanoclustering in part by sequestering PtdSer (Ariotti et al. 2014; Zhou et al. 2014). Disruption of caveolae by expressing

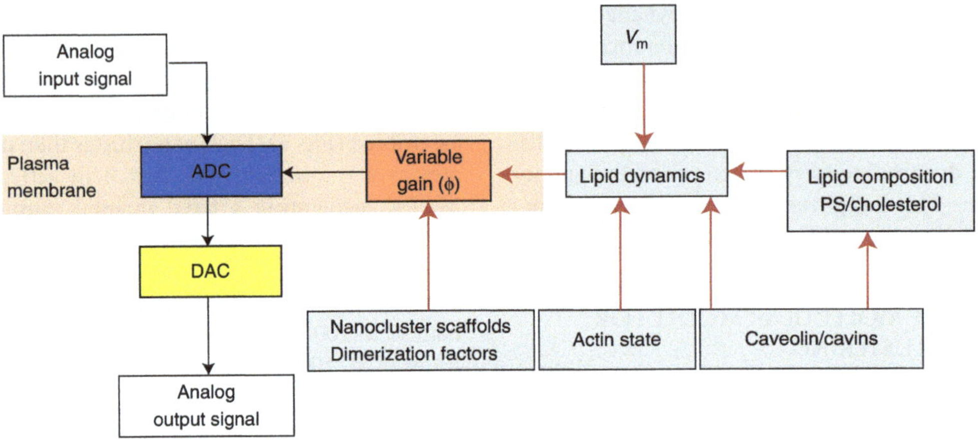

Figure 5. Multiple plasma membrane (PM) mechanisms regulate the gain in Ras signaling circuits. Ras nanoclusters set up the PM as an analog–digital–analog converter for high-fidelity mitogen-activated protein kinase (MAPK) signal transmission. The Ras clustered fraction (ϕ) is a key system parameter that sets the gain in the signaling circuit and that can, in turn, be controlled by multiple mechanisms as indicated on the diagram. Of particular interest are mechanisms that use PM lipid dynamics or effective lipid composition to regulate ϕ, including V_m (transmembrane potential). In this context, the PM can operate as a high-level controller of signal transmission. PS, Phosphatidylserine.

dominant-negative caveolin mutants, or knockdown of caveolin or cavin, alters PM PtdSer and cholesterol content and distribution to enhance K-Ras-GTP clustering, but abolish GTP-dependent lateral segregation of H-Ras (Ariotti et al. 2014; Zhou et al. 2014). PM depolarization also induces nanoscale reorganization PtdSer and PIP_2 but not other anionic phospholipids (Zhou et al. 2015). The reorganization of PtdSer, in turn, enhances K-Ras nanoclustering, which robustly amplifies K-Ras-dependent MAPK signaling. Conversely, PM repolarization disrupts K-Ras nanoclustering and inhibits MAPK signaling. Therefore, by responding to voltage-induced changes in PtdSer spatiotemporal dynamics, K-Ras nanoclusters set up the PM as a biological field-effect transistor, allowing membrane potential to control the gain in mitogenic signaling circuits. Together, these results show that ϕ is an important parameter, subject to multiple levels of control that directly control the strength of Ras signal transduction.

The important role of PtdSer in regulating K-Ras spatiotemporal organization on cell PM and signaling is further shown by the pharmacology of two K-Ras inhibitors: fendiline and staurosporine (STS). Both of these compounds perturb sphingomyelin metabolism, fendiline, by inhibiting acid sphingomyelinase, and STS by inhibiting the ORMDL complex that inhibits the first step in sphingomyelin biosynthesis (Cho et al. 2012a, 2015; van der Hoeven et al. 2013; Maekawa et al. 2016). Both compounds cause sphingomyelin loading of the endolysomal system, leading to aberrant lipid trafficking and loss of PtdSer and cholesterol from the PM. Both compounds in consequence therefore potently suppress K-Ras signaling, arguing that pharmacological targeting of PM lipid content may have some merit as an approach to block oncogenic K-Ras function.

CONCLUDING REMARKS

Ras proteins spatially segregate into distinct nanoclusters on the PM in an isoform- and guanine nucleotide-dependent manner. Nanocluster formation is essential to Ras-dependent signal transmission because these transient lipid-dependent assemblies operate as platforms for effector binding and activation. The molecular mechanisms that lead to nanocluster formation are becoming clearer and involve contributions from protein–protein interactions, largely confined to the G-domain, and protein–lipid interactions involving both the G-domain and the carboxy-terminal lipid anchor. Other small GTPases also share similar G-domain structures and analogous bipartite lipid anchors, and we speculate that these proteins, including Rho, Rac, and CDC42, will all undergo lateral segregation into distinct nanoclusters with specific lipid compositions, using the same basic molecular and biophysical mechanisms as Ras. These mechanisms can, on the one hand, be used by the cell to tune Ras signal transmission, and on the other hand may offer novel therapeutic approaches to block oncogenic Ras function.

ACKNOWLEDGMENTS

Work in the authors' laboratories is supported by the Cancer Research and Prevention Institute of Texas (CPRIT: RP170233 and DP150065) to J.F.H. and the National Institutes of Health (NIH: P30 DK056338) to Y.Z.

REFERENCES

Abankwa D, Hanzal-Bayer M, Ariotti N, Plowman SJ, Gorfe AA, Parton RG, McCammon JA, Hancock JF. 2008. A novel switch region regulates H-Ras membrane orientation and signal output. *EMBO J* **27:** 727–735.

Abankwa D, Gorfe AA, Inder K, Hancock JF. 2010. Ras membrane orientation and nanodomain localization generate isoform diversity. *Proc Natl Acad Sci* **107:** 1130–1135.

Apolloni A, Prior IA, Lindsay M, Parton RG, Hancock JF. 2000. H-Ras but not K-Ras traffics to the plasma membrane through the exocytic pathway. *Mol Cell Biol* **20:** 2475–2487.

Ariotti N, Fernández-Rojo MA, Zhou Y, Hill MM, Rodkey T, Inder K, Hancock JF, Parton RG. 2014. Caveolae regulate the nanoscale organization of the plasma membrane to remotely control Ras signaling. *J Cell Biol* **204:** 777–792.

Belanis L, Plowman SJ, Rotblat B, Hancock JF, Kloog Y. 2008. Galectin-1 is a novel structural component and a major regulator of H-Ras nanoclusters. *Mol Biol Cell* **19:** 1404–1414.

Bergo MO, Leung GK, Ambroziak P, Otto JC, Casey PJ, Young SG. 2000. Targeted inactivation of the isoprenylcysteine carboxyl methyltransferase gene causes mislocalization of K-Ras in mammalian cells. *J Biol Chem* **23:** 17605–17610.

Bergo MO, Leung GK, Ambroziak P, Otto JC, Casey PJ, Gomes AQ, Seabra MC, Young SG. 2001. Isoprenylcysteine carboxyl methyltransferase deficiency in mice. *J Biol Chem* **276:** 5841–5845.

Bergo MO, Ambroziak P, Gregory C, George A, Otto JC, Kim E, Nagase H, Casey PJ, Balmain A, Young SG. 2002. Absence of the CAAX endoprotease Rce1: Effects on cell growth and transformation. *Mol Cell Biol* **22:** 171–181.

Casey PJ, Solski PA, Der CJ, Buss JE. 1989. p21ras is modified by a farnesyl isoprenoid. *Proc Natl Acad Sci* **86:** 8323–8327.

Chandra A, Grecco HE, Pisupati V, Perera D, Cassidy L, Skoulidis F, Ismail SA, Hedberg C, Hanzal-Bayer M, Venkitaraman AR, et al. 2012. The GDI-like solubilizing factor PDEδ sustains the spatial organization and signalling of Ras family proteins. *Nat Cell Biol* **14:** 148–158.

Cho K-j, Park J-H, Piggott AM, Salim A, Gorfe AA, Parton RG, Capon RJ, Lacey E, Hancock JF. 2012a. Staurosporines disrupt phosphatidylserine trafficking and mislocalize Ras proteins. *J Biol Chem* **287:** 43573–43584.

Cho KJ, Kasai RS, Park JH, Chigurupati S, Heidorn SJ, van der Hoeven D, Plowman SJ, Kusumi A, Marais R, Hancock JF. 2012b. Raf inhibitors target Ras spatiotemporal dynamics. *Curr Biol* **22:** 945–955.

Cho KJ, van der Hoeven D, Zhou Y, Maekawa M, Ma X, Chen W, Fairn GD, Hancock JF. 2015. Inhibition of acid sphingomyelinase depletes cellular phosphatidylserine and mislocalizes K-Ras from the plasma membrane. *Mol Cell Biol* **36:** 363–374.

Choy E, Chiu VK, Silletti J, Feoktisitov M, Morimoto T, Michaelson D, Ivanov IE, Philips MR. 1999. Endomembrane trafficking of Ras: The CAAX motif targets proteins to the ER and Golgi. *Cell* **98:** 69–80.

Dai Q, Choy E, Chiu V, Romano J, Slivka SR, Seitz SA, Michaelis S, Philips MR. 1998. Mammalian prenylcysteine carboxyl methyltransferase is in the endoplasmic reticulum. *J Biol Chem* **273:** 15030–15034.

Dementiev A. 2012. K-Ras4B lipoprotein synthesis: Biochemical characterization, functional properties, and dimer formation. *Protein Expr Purif* **84:** 86–93.

Goodwin JS, Drake KR, Rogers C, Wright L, Lippincott-Schwartz J, Philips MR, Kenworthy AK. 2005. Depalmitoylated Ras traffics to and from the Golgi complex via a nonvesicular pathway. *J Cell Biol* **170:** 261–272.

Gorfe AA, Babakhani A, McCammon JA. 2007a. H-Ras protein in a bilayer: Interaction and structure perturbation. *J Am Chem Soc* **129:** 12280–12286.

Gorfe AA, Hanzal-Bayer M, Abankwa D, Hancock JF, McCammon JA. 2007b. Structure and dynamics of the full-length lipid-modified H-Ras protein in a 1,2-dimyristoylglycero-3-phosphocholine bilayer. *J Med Chem* **50:** 674–684.

Goswami D, Gowrishankar K, Bilgrami S, Ghosh S, Raghupathy R, Chadda R, Vishwakarma R, Rao M, Mayor S. 2008. Nanoclusters of GPI-anchored proteins are formed by cortical actin-driven activity. *Cell* **135:** 1085–1097.

Gowrishankar K, Ghosh S, Saha S, C R, Mayor S, Rao M. 2012. Active remodeling of cortical actin regulates spatiotemporal organization of cell surface molecules. *Cell* **149:** 1353–1367.

Guldenhaupt J, Rudack T, Bachler P, Mann D, Triola G, Waldmann H, Kotting C, Gerwert K. 2012. N-Ras forms dimers at POPC membranes. *Biophys J* **103:** 1585–1593.

Hancock JF. 2003. Ras proteins: Different signals from different locations. *Nat Rev Mol Cell Biol* **4:** 373–384.

Hancock JF. 2006. Lipid rafts: Contentious only from simplistic standpoints. *Nat Rev Mol Cell Biol* **7:** 456–462.

Hancock JF, Parton RG. 2005. Ras plasma membrane signalling platforms. *Biochem J* **389:** 1–11.

Hancock JF, Magee AI, Childs JE, Marshall CJ. 1989. All *ras* proteins are polyisoprenylated but only some are palmitoylated. *Cell* **57:** 1167–1177.

Hancock JF, Paterson H, Marshall CJ. 1990. A polybasic domain or palmitoylation is required in addition to the CAAX motif to localize p21ras to the plasma membrane. *Cell* **63:** 133–139.

Harding A, Hancock JF. 2008a. Ras nanoclusters: Combining digital and analog signaling. *Cell Cycle* **7:** 127–134.

Harding AS, Hancock JF. 2008b. Using plasma membrane nanoclusters to build better signaling circuits. *Trends Cell Biol* **18:** 364–371.

Harding A, Tian T, Westbury E, Frische E, Hancock JF. 2005. Subcellular localization determines MAP kinase signal output. *Curr Biol* **15:** 869–873.

Heidorn SJ, Milagre C, Whittaker S, Nourry A, Niculescu-Duvas I, Dhomen N, Hussain J, Reis-Filho JS, Springer CJ, Pritchard C, et al. 2010. Kinase-dead BRAF and oncogenic RAS cooperate to drive tumor progression through CRAF. *Cell* **140:** 209–221.

Inder K, Hancock JF. 2008. System output of the MAPK module is spatially regulated. *Commun Integr Biol* **1:** 178–179.

Inder K, Harding A, Plowman SJ, Philips MR, Parton RG, Hancock JF. 2008. Activation of the MAPK module from different spatial locations generates distinct system outputs. *Mol Biol Cell* **19:** 4776–4784.

Inder KL, Lau C, Loo D, Chaudhary N, Goodall A, Martin S, Jones A, van der Hoeven D, Parton RG, Hill MM, et al. 2009. Nucleophosmin and nucleolin regulate K-Ras plasma membrane interactions and MAPK signal transduction. *J Biol Chem* **284:** 28410–28419.

Jang H, Muratcioglu S, Gursoy A, Keskin O, Nussinov R. 2016. Membrane-associated Ras dimers are isoform-specific: K-Ras dimers differ from H-Ras dimers. *Biochem J* **473:** 1719–1732.

Kapoor S, Weise K, Erlkamp M, Triola G, Waldmann H, Winter R. 2012. The role of G-domain orientation and nucleotide state on the Ras isoform-specific membrane interaction. *Eur Biophys J* **41:** 801–813.

Kholodenko BN, Hancock JF, Kolch W. 2010. Signalling ballet in space and time. *Nat Rev Mol Cell Biol* **11:** 414–426.

Kusumi A, Nakada C, Ritchie K, Murase K, Suzuki K, Murakoshi H, Kasai RS, Kondo J, Fujiwara T. 2005. Paradigm shift of the plasma membrane concept from the two-dimensional continuum fluid to the partitioned fluid: High-speed single-molecule tracking of membrane molecules. *Annu Rev Biophys Biomol Struct* **34:** 351–378.

Li Z, Gorfe AA. 2013. Deformation of a two-domain lipid bilayer due to asymmetric insertion of lipid-modified Ras peptides. *Soft Matter* doi: 10.1039/C3SM51388B.

Lin WC, Iversen L, Tu HL, Rhodes C, Christensen SM, Iwig JS, Hansen SD, Huang WY, Groves JT. 2014. H-Ras forms dimers on membrane surfaces via a protein–protein interface. *Proc Natl Acad Sci* **111:** 2996–3001.

Lin X, Li Z, Gorfe AA. 2015. Reversible effects of peptide concentration and lipid composition on H-Ras lipid anchor clustering. *Biophys J* **109:** 2467–2470.

Lingwood D, Kaiser HJ, Levental I, Simons K. 2009. Lipid rafts as functional heterogeneity in cell membranes. *Biochem Soc Trans* **37:** 955–960.

Maekawa M, Lee M, Wei K, Ridgway ND, Fairn GD. 2016. Staurosporines decrease ORMDL proteins and enhance sphingomyelin synthesis resulting in depletion of plasmalemmal phosphatidylserine. *Sci Rep* **6:** 35762.

Mayor S, Rao M. 2004. Rafts: Scale-dependent, active lipid organization at the cell surface. *Traffic* **5:** 231–240.

Mazhab-Jafari MT, Marshall CB, Smith MJ, Gasmi-Seabrook GM, Stathopulos PB, Inagaki F, Kay LE, Neel BG, Ikura M. 2015. Oncogenic and RASopathy-associated K-RAS mutations relieve membrane-dependent occlusion of the effector-binding site. *Proc Natl Acad Sci* **112:** 6625–6630.

McLaughlin S, Wang J, Gambhir A, Murray D. 2002. PIP_2 and proteins: Interactions, organization, and information flow. *Annu Rev Biophys Biomol Struct* **31:** 151–175.

Michaelson D, Silletti J, Murphy G, D'Eustachio P, Rush M, Philips MR. 2001. Differential localization of Rho GTPases in live cells: Regulation by hypervariable regions and RhoGDI binding. *J Cell Biol* **152:** 111–126.

Murakoshi H, Iino R, Kobayashi T, Fujiwara T, Ohshima C, Yoshimura A, Kusumi A. 2004. Single-molecule imaging analysis of Ras activation in living cells. *Proc Natl Acad Sci* **101:** 7317–7322.

Muratcioglu S, Chavan TS, Freed BC, Jang H, Khavrutskii L, Freed RN, Dyba MA, Stefanisko K, Tarasov SG, Gursoy A, et al. 2015. GTP-dependent K-Ras dimerization. *Structure* **23:** 1325–1335.

Nan X, Collisson EA, Lewis S, Huang J, Tamguney TM, Liphardt JT, McCormick F, Gray JW, Chu S. 2013. Single-molecule superresolution imaging allows quantitative analysis of RAF multimer formation and signaling. *Proc Natl Acad Sci* **110:** 18519–18524.

Nan X, Tamguney TM, Collisson EA, Lin LJ, Pitt C, Galeas J, Lewis S, Gray JW, McCormick F, Chu S. 2015. Ras-GTP dimers activate the mitogen-activated protein kinase (MAPK) pathway. *Proc Natl Acad Sci* **112:** 7996–8001.

Plowman SJ, Muncke C, Parton RG, Hancock JF. 2005. H-Ras, K-Ras and inner plasma membrane raft proteins operate in nanoclusters that exhibit differential dependence on the actin cytoskeleton. *Proc Natl Acad Sci* **102:** 15500–15505.

Plowman SJ, Ariotti N, Goodall A, Parton RG, Hancock JF. 2008. Electrostatic interactions positively regulate K-Ras nanocluster formation and function. *Mol Cell Biol* **28:** 4377–4385.

Poulikakos PI, Zhang C, Bollag G, Shokat KM, Rosen N. 2010. RAF inhibitors transactivate RAF dimers and ERK signalling in cells with wild-type BRAF. *Nature* **464:** 427–430.

Prakash P, Zhou Y, Liang H, Hancock JF, Gorfe AA. 2016. Oncogenic K-Ras binds to an anionic membrane in two distinct orientations: A molecular dynamics analysis. *Biophys J* **110:** 1125–1138.

Prakash P, Sayyed-Ahmad A, Cho KJ, Dolino DM, Chen W, Li H, Grant BJ, Hancock JF, Gorfe AA. 2017. Computational and biochemical characterization of two partially overlapping interfaces and multiple weak-affinity K-Ras dimers. *Sci Rep* **7:** 40109.

Prior IA, Hancock JF. 2001. Compartmentalization of Ras proteins. *J Cell Sci* **114:** 1603–1608.

Prior IA, Hancock JF. 2012. Ras trafficking, localization and compartmentalized signalling. *Semin Cell Dev Biol* **23:** 145–153.

Prior IA, Harding A, Yan J, Sluimer J, Parton RG, Hancock JF. 2001. GTP-dependent segregation of H-Ras from lipid rafts is required for biological activity. *Nat Cell Biol* **3:** 368–375.

Prior IA, Muncke C, Parton RG, Hancock JF. 2003. Direct visualization of Ras proteins in spatially distinct cell surface microdomains. *J Cell Biol* **160:** 165–170.

Prior IA, Lewis PD, Mattos C. 2012. A comprehensive survey of Ras mutations in cancer. *Cancer Res* **72:** 2457–2467.

Raghupathy R, Anilkumar AA, Polley A, Singh PP, Yadav M, Johnson C, Suryawanshi S, Saikam V, Sawant SD, Panda A, et al. 2015. Transbilayer lipid interactions mediate nanoclustering of lipid-anchored proteins. *Cell* **161:** 581–594.

Reiss Y, Goldstein JL, Seabra MC, Casey PJ, Brown MS. 1990. Inhibition of purified p21^{ras} farnesyl:protein transferase by Cys-AAX tetrapeptides. *Cell* **62:** 81–88.

Rocks O, Peyker A, Kahms M, Verveer PJ, Koerner C, Lumbierres M, Kuhlmann J, Waldmann H, Wittinghofer A, Bastiaens PI. 2005. An acylation cycle regulates localization and activity of palmitoylated Ras isoforms. *Science* **307:** 1746–1752.

Roy S, Luetterforst R, Harding A, Apolloni A, Etheridge M, Stang E, Rolls B, Hancock JF, Parton RG. 1999. Dominant-negative caveolin inhibits H-Ras function by disrupting cholesterol-rich plasma membrane domains. *Nat Cell Biol* **1:** 98–105.

Roy S, Plowman S, Rotblat B, Prior IA, Muncke C, Grainger S, Parton RG, Henis YI, Kloog Y, Hancock JF. 2005. Individual palmitoyl residues serve distinct roles in H-Ras trafficking, microlocalization, and signaling. *Mol Cell Biol* **25:** 6722–6733.

Sarkar-Banerjee S, Sayyed-Ahmad A, Prakash P, Cho K-J, Waxham MN, Hancock JF, Gorfe AA. 2017. Spatiotemporal analysis of K-Ras plasma membrane interactions reveals multiple high order homo-oligomeric complexes. *J Am Chem Soc* **139:** 13466–13475.

Sayyed-Ahmad A, Cho KJ, Hancock JF, Gorfe AA. 2016. Computational equilibrium thermodynamic and kinetic analysis of K-Ras dimerization through an effector binding surface suggests limited functional role. *J Phys Chem B* **120:** 8547–8556.

Schmick M, Vartak N, Papke B, Kovacevic M, Truxius DC, Rossmannek L, Bastiaens PI. 2014. KRas localizes to the plasma membrane by spatial cycles of solubilization, trapping and vesicular transport. *Cell* **157:** 459–471.

Shalom-Feuerstein R, Plowman SJ, Rotblat B, Ariotti N, Tian T, Hancock JF, Kloog Y. 2008. K-Ras nanoclustering is subverted by overexpression of the scaffold protein galectin-3. *Cancer Res* **68:** 6608–6616.

Simons K, Vaz WL. 2004. Model systems, lipid rafts, and cell membranes. *Annu Rev Biophys Biomol Struct* **33:** 269–295.

Spencer-Smith R, Koide A, Zhou Y, Eguchi RR, Sha F, Gajwani P, Santana D, Gupta A, Jacobs M, Herrero-Garcia E, et al. 2017. Inhibition of RAS function through targeting an allosteric regulatory site. *Nat Chem Biol* **13:** 62–68.

Tian T, Harding A, Inder K, Plowman S, Parton RG, Hancock JF. 2007. Plasma membrane nanoswitches generate high-fidelity Ras signal transduction. *Nat Cell Biol* **9:** 905–914.

Tian T, Plowman SJ, Parton RG, Kloog Y, Hancock JF. 2010. Mathematical modeling of K-Ras nanocluster formation on the plasma membrane. *Biophys J* **99:** 534–543.

van der Hoeven D, Cho KJ, Ma X, Chigurupati S, Parton RG, Hancock JF. 2013. Fendiline inhibits K-Ras plasma membrane localization and blocks K-Ras signal transmission. *Mol Cell Biol* **33:** 237–251.

Weise K, Kapoor S, Denter C, Nikolaus J, Opitz N, Koch S, Triola G, Herrmann A, Waldmann H, Winter R. 2011. Membrane-mediated induction and sorting of K-Ras microdomain signaling platforms. *J Am Chem Soc* **133:** 880–887.

Willumsen BM, Christensen A, Hubbert NL, Papageorge AG, Lowy DR. 1984. The p21 ras C-terminus is required for transformation and membrane association. *Nature* **310:** 583–586.

Zhou Y, Hancock JF. 2015. Ras nanoclusters: Versatile lipid-based signaling platforms. *Biochim Biophys Acta* **1853:** 841–849.

Zhou Y, Liang H, Rodkey T, Ariotti N, Parton RG, Hancock JF. 2014. Signal integration by lipid-mediated spatial cross talk between Ras nanoclusters. *Mol Cell Biol* **34:** 862–876.

Zhou Y, Wong CO, Cho KJ, van der Hoeven D, Liang H, Thakur DP, Luo J, Babic M, Zinsmaier KE, Zhu MX, et al. 2015. Membrane potential modulates plasma membrane phospholipid dynamics and K-Ras signaling. *Science* **349:** 873–876.

Zhou M, Wiener H, Su W, Zhou Y, Liot C, Ahearn I, Hancock JF, Philips MR. 2016. VPS35 binds farnesylated N-Ras in the cytosol to regulate N-Ras trafficking. *J Cell Biol* **214:** 445–458.

Zhou Y, Prakash P, Liang H, Cho KJ, Gorfe AA, Hancock JF. 2017. Lipid-sorting specificity encoded in K-Ras membrane anchor regulates signal output. *Cell* **168:** 239–251.

Zimmermann G, Papke B, Ismail S, Vartak N, Chandra A, Hoffmann M, Hahn SA, Triola G, Wittinghofer A, Bastiaens PI, et al. 2013. Small molecule inhibition of the KRAS-PDEδ interaction impairs oncogenic KRAS signalling. *Nature* **497:** 638–642.

MRAS: A Close but Understudied Member of the RAS Family

Lucy C. Young[1] and Pablo Rodriguez-Viciana[2]

[1]UCSF Helen Diller Family Comprehensive Cancer Center, San Francisco, California 94158

[2]UCL Cancer Institute, Paul O'Gorman Building, University College London, London WC1E 6BT, United Kingdom

Correspondence: p.rodriguez-viciana@ucl.ac.uk

MRAS is the closest relative to the classical RAS oncoproteins and shares most regulatory and effector interactions. However, it also has unique functions, including its ability to function as a phosphatase regulatory subunit when in complex with SHOC2 and protein phosphatase 1 (PP1). This phosphatase complex regulates a crucial step in the activation cycle of RAF kinases and provides a key coordinate input required for efficient ERK pathway activation and transformation by RAS. MRAS mutations rarely occur in cancer but deregulated expression may play a role in tumorigenesis in some settings. Activating mutations in MRAS (as well as SHOC2 and PP1) do occur in the RASopathy Noonan syndrome, underscoring a key role for MRAS within the RAS-ERK pathway. MRAS also has unique roles in cell migration and differentiation and has properties consistent with a key role in the regulation of cell polarity. Further investigations should shed light on what remains a relatively understudied RAS family member.

The RRAS subgroup (RRAS, TC21/RRAS2, and MRAS/RRAS3) of the RAS family GTPases (RFGs) are the closest relatives to the classical RAS oncogenes (H/N/KRAS, hereafter referred to collectively as RAS). These GTPases share many regulatory and effector proteins with RAS as well as transforming abilities (Chan et al. 1994; Saez et al. 1994; Kimmelman et al. 1997; Rodriguez-Viciana et al. 2006). Members of this group have distinct functions, and MRAS has been shown to play a number of roles in cellular processes such as differentiation, cytoskeletal remodeling, and cell migration. Uniquely among RFGs, MRAS is part of a phosphatase complex that positively regulates RAF kinase activation and is required to cooperate with RAS proteins for efficient ERK pathway activation.

SEQUENCE FEATURES OF MRAS

The RRAS subgroup lies within a distinct branch of the tree of all small GTPases with classical RAS and ERAS (Fig. 1A). MRAS is highly conserved between vertebrates and has considerable similarity to the *ras-2* gene in *Caenorhabditis elegans* (Fig. 1B). Interestingly, although classical RAS orthologs exist in fly, fish, and nematode, they are absent in ascidian, which does have orthologs to MRAS and RRAS (Fig. 1B). Therefore, MRAS evolved independently of RRAS in metazoans and has been suggested to compensate for the lack of classical RAS in ascidian (Keduka et al. 2009).

The G domain of MRAS is similar in sequence to the classical RAS proteins (it shares

Cite this article as *Cold Spring Harb Perspect Med* doi: 10.1101/cshperspect.a033621

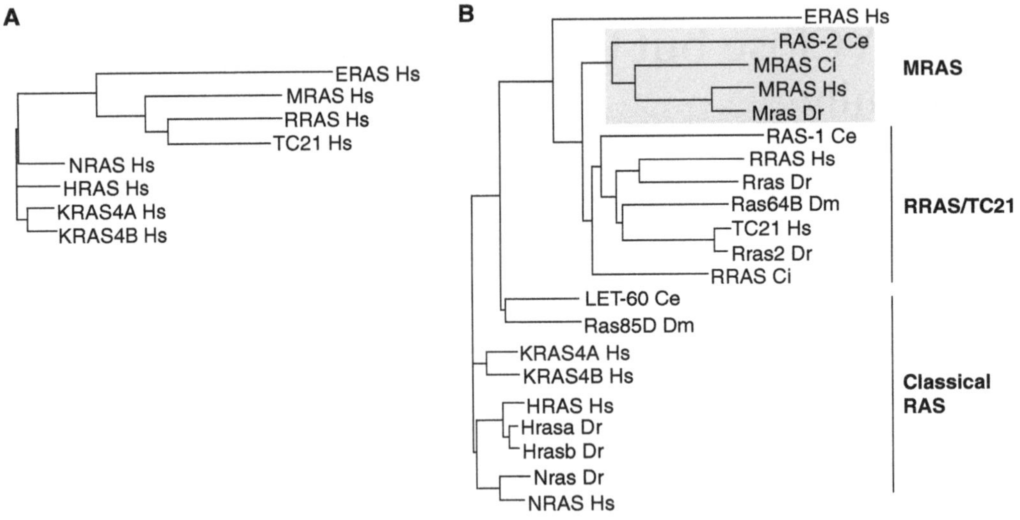

Figure 1. MRAS phylogeny. (*A*) Phylogenetic relationship of human RAS family proteins. (*B*) Phylogenetic analysis of RAS family proteins in human (Hs), zebrafish (Dr), fruitfly (Dm), nematode (Ce), and ascidian *Ciona intestinalis* (Ci). Sequence alignment was performed using MUSCLE (see ebi.ac.uk/Tools/msa/muscle) and tree generation with iTOL (see itol.embl.de).

53% amino acid identity with KRAS), and like RRAS and TC21, it has a short amino-terminal extension. The terminal residue of MRAS is lysine, which differs from the terminal residues of RAS (Fig. 2A), and consequently rather than being farnesylated, MRAS is predicted to be posttranslationally modified instead by geranylgeranylation (Zhang and Casey 1996). The hypervariable region (HVR) of MRAS has similarities to KRAS4B in that carboxy-terminal cysteines capable of palmytoylation are absent and instead a polybasic region is found (Fig. 2A). Like KRAS4B, MRAS is found in disordered membrane domains rather than organized lipid rafts (Ehrhardt et al. 2002), which suggests these proteins may signal in similar pathways and/or be similarly regulated.

MRAS REGULATION AND CONTROL OF DOWNSTREAM PATHWAYS

MRAS shares many regulatory proteins with other RFGs; it can be activated by SOS1, RASGRF1, RASGRP1, and RASGRP3 guanine nucleotide exchange factors (GEFs) and inactivated by p120 RASGAP, neurofibromin, and GAP1m GTPase-activating proteins (GAPs) (Mitin et al. 2005). Although some of these proteins also regulate RRAS/TC21, full analysis of this regulation links MRAS closely to classical RAS in terms of GEF/GAP specificity (Ohba et al. 2000). This implies that most physiological signals that activate RAS will also activate MRAS simultaneously and is consistent with MRAS functioning together with RAS to provide coordinate inputs for efficient RAF activation (see below).

Given its sequence similarity and identical effector domain (Fig. 2B), it is unsurprising that MRAS can bind many of the same effectors as RAS such as A-, B-, and CRAF, AFDN/AF6, RASSF5, RalGEFs, and PI3K (Quilliam et al. 2001; Ortiz-Vega et al. 2002; Rodriguez-Viciana et al. 2004). Through binding RGL2/RLF, MRAS activates RAL and ELK1 in MCF-7 cells in an ERK-independent manner (Ehrhardt et al. 2001; Castro et al. 2012). MRAS also controls activation of RAP activity through binding MR-GEF/RAPGEF5 (Rebhun et al. 2000) and RA-GEF2/RAPGEF6 (Gao et al. 2001), the latter being specifically linked to control of cell adhesion through tumor necrosis factor (TNF)-α-triggered integrin activation in haematopoietic cells (Yoshikawa et al. 2007).

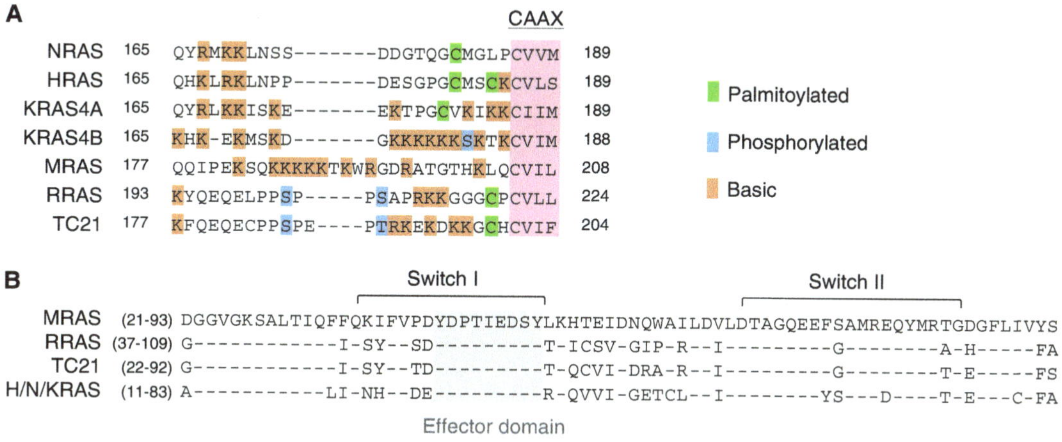

Figure 2. Comparison of MRAS amino acid sequence with other RAS family GTPases (RFGs). (*A*) Alignment of RAS protein hypervariable regions with features shaded according to property. (*B*) Sequence alignment (of Switch I and II containing regions) of MRAS with H/N/KRAS, RRAS, and TC21.

MRAS is a weak activator of the ERK pathway compared with RAS (Kimmelman et al. 1997; Rodriguez-Viciana et al. 2004), which, at least in part, could be attributed to the lower affinity of MRAS for RAF compared with RAS. One reason for this may be the differences in sequence of MRAS, particularly in the Switch I region (Fig. 2B). Both the GppNHp- and guanosine diphosphate (GDP)-bound forms of MRAS have an open Switch I conformation, and mutation of these residues to those of RAS result in a higher proportion of MRAS in the closed state. In addition, these mutations increase the affinity of MRAS for the RAS-binding domain (RBD) of RAF (Ye et al. 2005), which implies that there are regions unique to MRAS (with others yet to be identified) that govern its affinity and specificity for various effectors.

MRAS REGULATION OF RAF ACTIVITY THROUGH SHOC2/PROTEIN PHOSPHATASE 1 (PP1)

In addition to sharing interactions with many of RAS effectors, MRAS also has its own unique effector interactions with RAPGEFs (Gao et al. 2001) as well as SHOC2 and PP1 (Rodriguez-Viciana et al. 2006). SHOC2 is a ubiquitously expressed protein comprised almost exclusively of leucine rich repeats that was originally identified in *C. elegans* as a positive modulator of the ERK pathway (Selfors et al. 1998; Sieburth et al. 1998). Active MRAS forms a ternary complex with SHOC2 and PP1 to form a phosphatase holoenzyme that specifically dephosphorylates a conserved inhibitory site in RAF kinases (S259 in CRAF, S365 in BRAF, and S214 in ARAF, hereby referred to as S259) that functions as a 14-3-3 binding site and plays a key role in the RAF-activation cycle (Rodriguez-Viciana et al. 2006).

The consensus model of RAF activation stipulates that RAF is maintained in an autoinhibited inactive state in the cytosol by an intramolecular interaction between the amino-terminal region and the catalytic domain, which is in part mediated by a 14-3-3 dimer bound to two phosphorylated residues (S259 and S621 in CRAF) (Fig. 3A) (Tzivion et al. 1998; Matallanas et al. 2011; Lavoie and Therrien 2015). RAS-guanosine triphosphate (GTP) binding to the RBD of RAF results in RAF translocation to the plasma membrane in which other activating steps then take place. Chief among these is the dephoshphorylation of the "S259" site, which leads to 14-3-3 displacement from this site, destabilizes the closed conformation of RAF, allows the cysteine-rich domain (CRD) to further anchor RAF to the membrane, and facilitates RAF dimerization (Fig. 3B) (Lavoie and Therrien 2015). The MRAS–

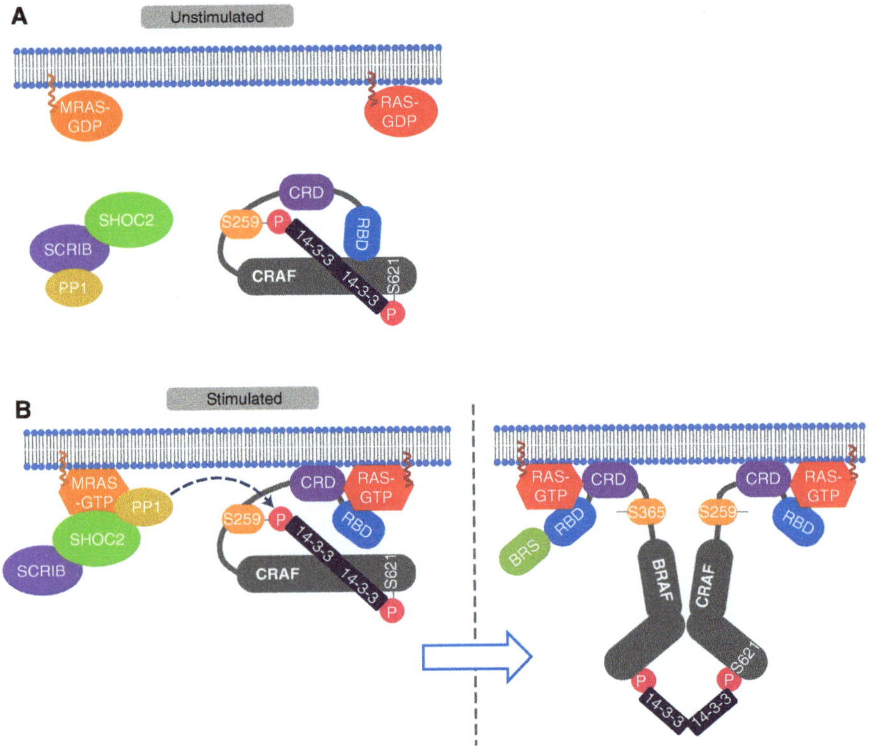

Figure 3. Coordinate inputs from RAS and MRAS–SHOC2–protein phosphatase 1 (PP1) are required for effective RAF activation. (*A*) Under basal conditions, in which RAS and MRAS are guanosine diphosphate (GDP)-bound and inactive, SHOC2 interacts with PP1 via SCRIB. BRAF and CRAF exist as autoinhibited monomers, which are phosphorylated at S259. (*B*) After pathway stimulation by a mitogen such as epidermal growth factor (EGF), guanosine triphosphate (GTP)-bound RAS recruits BRAF/CRAF to the membrane, and SHOC2 forms a complex with MRAS-GTP and PP1. Here, the complex dephosphorylates the S259/S365 site on RAF and the 14-3-3 dimer can now mediate RAF dimerization (shown here as a heterodimer of BRAF and CRAF). RBD, RAS-binding domain; CRD, cysteine-rich domain; BRS, BRAF-specific region.

SHOC2–PP1 complex functions as a key phosphatase promoting this dephosphorylation step, which occurs preferentially on the RAF that has been recruited to the plasma membrane by RAS proteins (Fig. 3B) (Rodriguez-Viciana et al. 2006).

PP1 is known to interact with hundreds of regulatory proteins that confer substrate specificity and unique properties to each resulting holoenzyme. By analogy with other PP1 holoenzymes as well as the heterotrimeric PP2A complex (Cho and Xu 2007; Shi 2009; Peti et al. 2013), MRAS and SHOC2 are predicted to "remodel" the substrate recognition surface within the complex, altering the physiochemical landscape to create a surface specific for recognition of P-S259 RAF but not other phosphorylation sites, even on the same target (Rodriguez-Viciana et al. 2006).

A ROLE FOR MRAS IN POLARITY

Active MRAS can also associate with the polarity protein SCRIB, although this interaction is indirect and mediated by SHOC2 (Young et al. 2013). SCRIB is also a PP1-interacting protein and MRAS has the potential to "rearrange" the PP1 molecules within the SCRIB–SHOC2 complex. In the absence of MRAS, SHOC2 has very low affinity for PP1, and the PP1 in the complex is bound to SCRIB (Fig. 3A). In the presence of active MRAS, SHOC2 and PP1 form a ternary complex with high affinity and the PP1 interaction is now independent of SCRIB (Fig. 3B) (Young et al. 2013).

Through its interaction with SHOC2, MRAS is expected to recruit SCRIB and its associated interactome to sites of activation (Richier et al. 2010; Anastas et al. 2012; Young et al. 2013). SCRIB antagonizes SHOC2-mediated RAF "S259" dephosphorylation, at least partly, through a mechanism involving competition for PP1 molecules within the same macromolecular complex. SCRIB recruitment, through its interaction with the PIX–GIT complex, SGEF, and VANGL proteins, would also allow for the regulation of RAC/CDC42, ARF, RHOG, and the planar cell polarity pathways, respectively (Ellerbroek et al. 2004; Frank and Hansen 2008; Tada and Kai 2012). Furthermore, MRAS impairs the association of NOS1AP and the exchange factor βPIX with the SHOC2–SCRIB complex and thus has the ability to regulate the SCRIB interactome. MRAS is also expected to recruit to the same signaling platforms as other effectors, such as RAPGEFs, RALGEFs, and PI3K, which are also known to be involved in polarity adhesion and migration (Kimmelman et al. 2000; Ehrhardt et al. 2002; Rodriguez-Viciana et al. 2004). MRAS is therefore an excellent candidate to behave as a master regulator of polarity (Fig. 4).

MRAS IN HUMAN DISEASE

Despite being the most closely related to RAS by sequence similarity, and given its shared interaction with many of the same effector proteins (as well as GAPs and GEFs), it is somewhat surprising that activating mutations in MRAS are rarely found in cancer, in clear contrast to RAS genes (Catalog of Somatic Mutations in Cancer [COSMIC]). Considering the key role of the RAF-ERK pathway in mediating RAS oncogenic properties, MRAS's lower affinity for RAF and its considerably weaker activation of the ERK pathway may at least partly account for this observation. Additionally, it is also possible that MRAS's unique role in polarity may make constitutive activation disfavored in some contexts (Young et al. 2013). However, MRAS up-regulation may be linked to cancer in other ways.

MRAS expression is up-regulated in estrogen receptor (ER)-negative breast carcinomas compared with ER-positive in three independent studies (van de Vijver et al. 2002; Chin et al. 2006; Hess et al. 2006), and overexpression of constitutively active MRAS enables MCF-7 breast carcinoma cells to proliferate in the absence of estrogen (Castro et al. 2012). MRAS is part of the epithelial-to-mesenchymal (EMT) signature (Huang et al. 2012) and expression of active mutants causes EMT and oncogenic

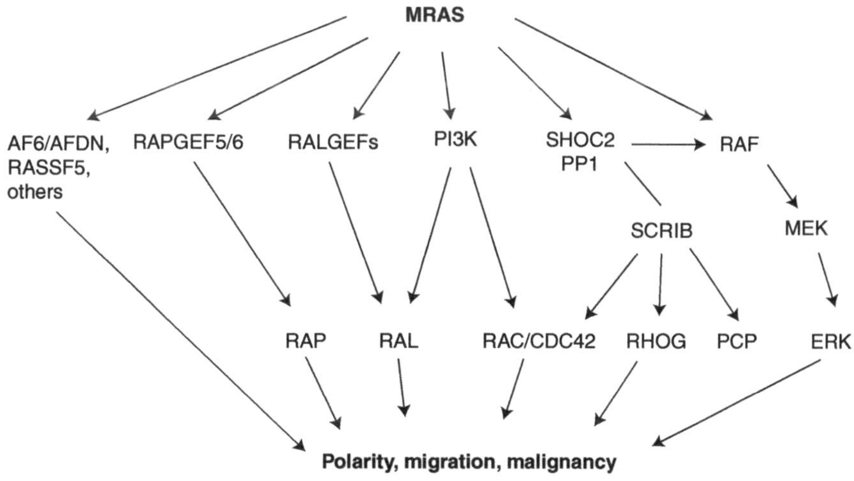

Figure 4. MRAS regulates cell polarity and migration through coordination of multiple signaling pathways (see text for details).

transformation in mouse scp-2 cells. These cells can grow in the absence of serum, lose contact inhibition, gain the ability to grow in an anchorage-independent manner, and form tumors in mice that have elevated levels of P-ERK and P-AKT (Ward et al. 2004). Many of these MRAS-induced characteristics are dependent on an Hepatocyte growth factor (HGF) autocrine mechanism (Zhang et al. 2004), which is interesting given that invasive growth and metastasis of mammary tumors correlates with HGF secretion (which itself occurs in the majority of tumors of that type) (Jeffers et al. 1996; Nagy et al. 1996).

The prominence of MRAS overexpression in cancer may also be context specific. For example, *MRAS* (along with *MET*) is overexpressed in multiple tumor types in a cytokine-driven/signal transducers and activators of transcription (STAT)3 activity-dependent manner (Yang et al. 2005). The Cancer Genome Atlas (TCGA) studies show that MRAS is amplified in 17% of lung squamous cell carcinomas but very rarely in lung adenocarcinoma (Fig. 5), which correlates inversely with the frequency of RAS mutations, which is high in adenocarcinomas but low in squamous cell carcinomas (cBioportal) (Cerami et al. 2012; Gao et al. 2013). MRAS was also overexpressed and/or amplified in 11% of ovarian serous cystadenocarcinoma and in 10% of head and neck squamous cell carcinoma (cBioportal). MRAS mutation frequency may be linked to specific subsets of cancers, for example, Borrmann type IV gastric cancer, which carries particularly poor prognoses and has a higher frequency of nonsynonymous MRAS mutations (17%) compared with overall gastric cancers (0.7% MRAS mutations) (Yasumoto et al. 2017). However, functional assays are still required to show that these mutations are activating.

Two mutations in MRAS (p.Gly23Val and p.Thr68Ile) (Fig. 6) have been identified in patients with Noonan syndrome ([NS] MIM 163950), a developmental disorder that is part of the RASopathies family of related syndromes, which are driven by mutations in components of the RAS-ERK pathway. NS typically features facial dysmorphisms, slow growth rates, skeletal anomalies, mental retardation, predisposition to malignancies, and, often, cardiac defects. Of

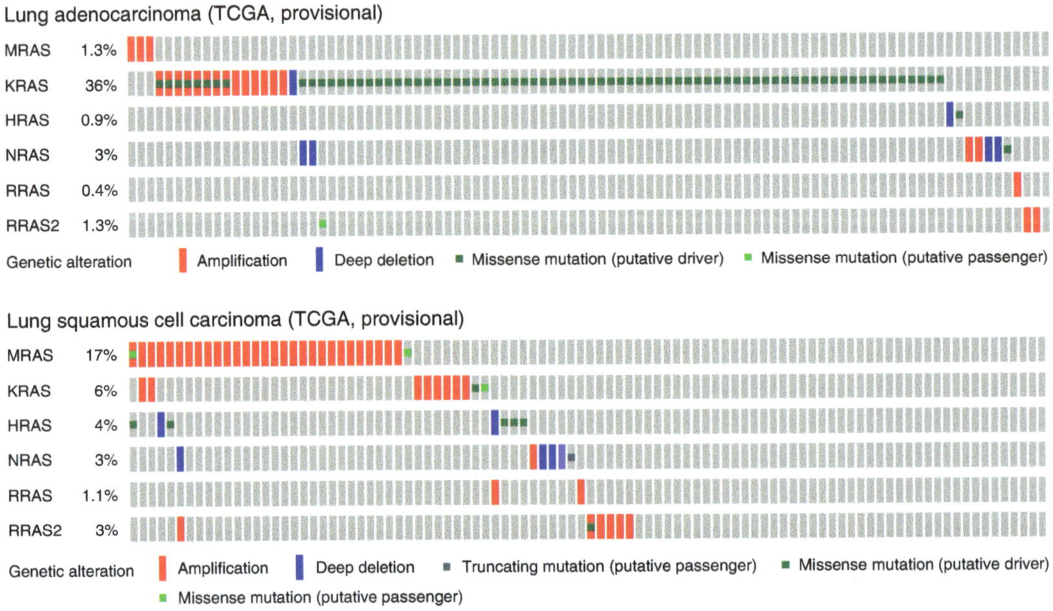

Figure 5. cBioportal oncoprints depicting genetic alterations in the classical RAS and RRAS subgroups of the RAS family of GTPases. TCGA, The Cancer Genome Atlas.

 Cite this article as *Cold Spring Harb Perspect Med* doi: 10.1101/cshperspect.a033621

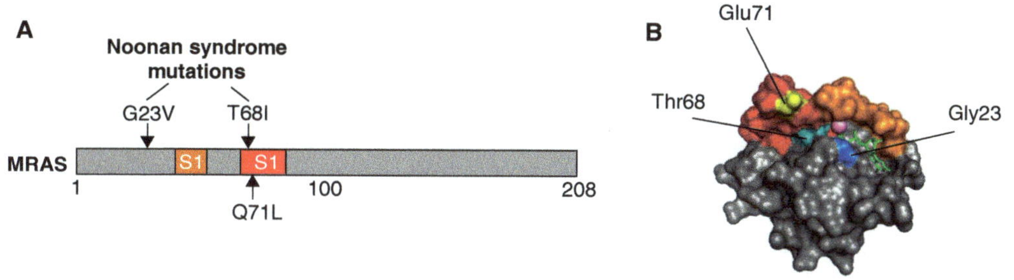

Figure 6. Activating mutations in MRAS. (*A*) Position of Noonan syndrome mutations (*upper*) and activating mutation equivalent to codon 61 in RAS (*lower*). (*B*) Structure of MRAS bound to guanosine diphosphate (GDP) (green sticks) (PDB 1X1R). Switch I (orange), Switch II (red), and Mg^{2+} (pink sphere) are shown, as well as positions of key residues Gly23 (blue), Thr68 (turquoise), and Glu71 (yellow).

the two *MRAS* mutation-positive patients so far described, both had developmental delays, facial dysmorphisms, and cardiac hypertrophy. Similar clinical phenotypes are observed in patients with NS with loose anagen hair (NS-LAH) driven by mutations in MRAS-binding partners—SHOC2 and PP1 (Cordeddu et al. 2009; Gripp et al. 2016). MRAS p.Gly23Val corresponds to Gly13 of H/N/KRAS—a known oncogenic mutation. G23V-MRAS is primarily GTP-bound and activates ERK in cells (Higgins et al. 2017), and although not functionally tested as yet, mutation of position T68 is predicted to be activating given that it is within the Switch II region and lies within the GTP/GDP binding pocket (Fig. 6). Based on the observed phenotypes and what is known about their role in ERK pathway regulation, MRAS-driven NS is likely to be functioning through SHOC2–PP1 complex activity.

Gain-of-function *CRAF* mutations are also found in NS and cluster around S259 to disrupt 14-3-3 binding (Pandit et al. 2007; Razzaque et al. 2007; Kobayashi et al. 2010; Molzan et al. 2010), Of note, S259F as well as others in this region (S257L, 261S, and V263A), have been identified in cancer as well as NS. These results underscore the important role of MRAS–SHOC2–PP1 and RAF S259 dephosphorylation on RAS-ERK pathway activation.

Large-scale genome-wide association studies have identified the MRAS locus as a risk factor in cardiovascular disease (Erdmann et al. 2009; Schunkert et al. 2011; Liu et al. 2013). Intriguingly, although MRAS is widely expressed, it has particularly high expression in the heart (see biogps.org). Although a role for MAP kinase pathways in the heart is known (Rose et al. 2010), the precise role of MRAS remains to be elucidated.

OTHER FUNCTIONS OF MRAS

MRAS signaling is linked to differentiation and development in a variety of mammalian cell types. It is activated by nerve growth factor (NGF) and is required for ERK-dependent neuronal differentiation of rat pheochromocytoma PC12 cells (Kimmelman et al. 2002; Sun et al. 2006) and, in developing mouse bone, BMP2 treatment not only activates MRAS, but also increases its expression levels. Active MRAS stimulates transdifferentiation of C2C12 mouse myoblasts into osteoblasts, and causes osteoblast differentiation in a p38- and c-Jun amino-terminal kinase (JNK)-dependent manner (Watanabe-Takano et al. 2010). MRAS is up-regulated during dendrite development and contributes to dendrite growth via the ERK pathway. In these cells, semaphorin-4D (or Sema4D, a repulsive guidance molecule in the developing nervous system) binds to its receptor PLEXIN-B1, the cytoplasmic domain of which acts as a GAP on MRAS (and RRAS, but not TC21 or RAS). When MRAS activity is suppressed in this way, it results in growth cone collapse and remodeling of dendrite morphology (Saito et al. 2009). In the absence of Sema4D, active MRAS binds to an effector, LPD (lamellopodin), and recruits it to the membrane of growing dendrites where it participates in actin cytoskeleton remodeling via Ena/VASP proteins (Tasaka et al. 2012). However, other studies have suggested that sempahorins function as GAPs for RAPs not MRAS (Wang et al. 2012).

These ties to differentiation have also been studied in mouse embryonic stem cells (mESCs). Of all the *Ras* family members, *Mras* is the only gene whose expression is limited to undifferentiated mESCs. Before differentiation, expression is controlled by the cytokine LIF, and, as such, *Mras* is described as a marker of stemness (Mathieu et al. 2013). Persistent *Mras* down-regulation by removal of LIF affects the normal balance of expression of core pluripotency markers such as OCT4 and CAECAM1, and overexpression of MRAS leads to sustained expression of OCT4 and NANOG. In *Xenopus*, Mras is expressed throughout the embryo, maintaining pluripotency until the blastula stage and, as in mice, it is required for neuronal differentiation, which indicates that this protein has conserved functions across vertebrates (Mathieu et al. 2013).

At the organismal level, *Mras* null mice were initially described as phenotypically normal (Nunez Rodriguez et al. 2006), but further studies since then link *Mras* to the control of normal urinary function (Ehrhardt et al. 2015b) and normal olfactory and/or social processes in adult male mice because *Mras*$^{-/-}$ males are phenotypically more aggressive and show increased sexual behavior (Ehrhardt et al. 2015a).

CONCLUDING REMARKS

Although *MRAS* mutations are rarely found in cancer, it is likely that dysregulation of *MRAS* at the expression level may be a contributing factor to tumorigenesis in some contexts. Considering this and the studies of MRAS in cell-fate determination and stemness, understanding how MRAS expression is regulated in different cell types is of interest. The recent identification of activating mutations in *MRAS* in NS (as well as *SHOC2* and *PPP1CB*) highlights a key role for the MRAS–SHOC2–PP1 complex in the regulation of the ERK pathway. MRAS has the ability to regulate multiple signaling pathways, including many that directly or indirectly regulate other GTPases of the RAS superfamily, and that suggests a role for MRAS as a master regulator of polarity. More comprehensive biochemical and structural studies will deepen our understanding of how MRAS-specific signaling contributes to cell behavior and will help to explain the phenotypes observed at the organismal level in which MRAS function is modified.

REFERENCES

Anastas JN, Biechele TL, Robitaille M, Muster J, Allison KH, Angers S, Moon RT. 2012. A protein complex of SCRIB, NOS1AP and VANGL1 regulates cell polarity and migration, and is associated with breast cancer progression. *Oncogene* **31:** 3696–3708.

Castro AF, Campos T, Babcock JT, Armijo ME, Martinez-Conde A, Pincheira R, Quilliam LA. 2012. M-Ras induces Ral and JNK activation to regulate MEK/ERK-independent gene expression in MCF-7 breast cancer cells. *J Cell Biochem* **113:** 1253–1264.

Cerami E, Gao J, Dogrusoz U, Gross BE, Sumer SO, Aksoy BA, Jacobsen A, Byrne CJ, Heuer ML, Larsson E, et al. 2012. The cBio cancer genomics portal: An open platform for exploring multidimensional cancer genomics data. *Cancer Discov* **2:** 401–404.

Chan AM, Miki T, Meyers KA, Aaronson SA. 1994. A human oncogene of the RAS superfamily unmasked by expression cDNA cloning. *Proc Natl Acad Sci* **91:** 7558–7562.

Chin K, DeVries S, Fridlyand J, Spellman PT, Roydasgupta R, Kuo WL, Lapuk A, Neve RM, Qian Z, Ryder T, et al. 2006. Genomic and transcriptional aberrations linked to breast cancer pathophysiologies. *Cancer Cell* **10:** 529–541.

Cho US, Xu W. 2007. Crystal structure of a protein phosphatase 2A heterotrimeric holoenzyme. *Nature* **445:** 53–57.

Cordeddu V, Di Schiavi E, Pennacchio LA, Ma'ayan A, Sarkozy A, Fodale V, Cecchetti S, Cardinale A, Martin J, Schackwitz W, et al. 2009. Mutation of *SHOC2* promotes aberrant protein *N*-myristoylation and causes Noonan-like syndrome with loose anagen hair. *Nat Genet* **41:** 1022–1026.

Ehrhardt GR, Korherr C, Wieler JS, Knaus M, Schrader JW. 2001. A novel potential effector of M-Ras and p21 Ras negatively regulates p21 Ras-mediated gene induction and cell growth. *Oncogene* **20:** 188–197.

Ehrhardt A, Ehrhardt GR, Guo X, Schrader JW. 2002. Ras and relatives—Job sharing and networking keep an old family together. *Exp Hematol* **30:** 1089–1106.

Ehrhardt A, Wang B, Leung MJ, Schrader JW. 2015a. Absence of M-Ras modulates social behavior in mice. *BMC Neurosci* **16:** 68.

Ehrhardt A, Wang B, Yung AC, Wang Y, Kozlowski P, van Breemen C, Schrader JW. 2015b. Urinary retention, incontinence, and dysregulation of muscarinic receptors in male mice lacking Mras. *PLoS ONE* **10:** e0141493.

Ellerbroek SM, Wennerberg K, Arthur WT, Dunty JM, Bowman DR, DeMali KA, Der C, Burridge K. 2004. SGEF, a RhoG guanine nucleotide exchange factor that stimulates macropinocytosis. *Mol Biol Cell* **15:** 3309–3319.

Erdmann J, Grosshennig A, Braund PS, Konig IR, Hengstenberg C, Hall AS, Linsel-Nitschke P, Kathiresan S, Wright B, Tregouet DA, et al. 2009. New susceptibility locus for coronary artery disease on chromosome 3q22.3. *Nat Genet* **41:** 280–282.

Frank SR, Hansen SH. 2008. The PIX-GIT complex: A G protein signaling cassette in control of cell shape. *Semin Cell Dev Biol* **19:** 234–244.

Gao X, Satoh T, Liao Y, Song C, Hu CD, Kariya Ki K, Kataoka T. 2001. Identification and characterization of RA-GEF-2, a Rap guanine nucleotide exchange factor that serves as a downstream target of M-Ras. *J Biol Chem* **276:** 42219–42225.

Gao J, Aksoy BA, Dogrusoz U, Dresdner G, Gross B, Sumer SO, Sun Y, Jacobsen A, Sinha R, Larsson E, et al. 2013. Integrative analysis of complex cancer genomics and clinical profiles using the cBioPortal. *Sci Signal* **6:** pl1.

Gripp KW, Aldinger KA, Bennett JT, Baker L, Tusi J, Powell-Hamilton N, Stabley D, Sol-Church K, Timms AE, Dobyns WB. 2016. A novel rasopathy caused by recurrent de novo missense mutations in *PPP1CB* closely resembles Noonan syndrome with loose anagen hair. *Am J Med Genet A* **170:** 2237–2247.

Hess KR, Anderson K, Symmans WF, Valero V, Ibrahim N, Mejia JA, Booser D, Theriault RL, Buzdar AU, Dempsey PJ, et al. 2006. Pharmacogenomic predictor of sensitivity to preoperative chemotherapy with paclitaxel and fluorouracil, doxorubicin, and cyclophosphamide in breast cancer. *J Clin Oncol* **24:** 4236–4244.

Higgins EM, Bos JM, Mason-Suares H, Tester DJ, Ackerman JP, MacRae CA, Sol-Church K, Gripp KW, Urrutia R, Ackerman MJ. 2017. Elucidation of MRAS-mediated Noonan syndrome with cardiac hypertrophy. *JCI Insight* **2:** e91225.

Huang S, Holzel M, Knijnenburg T, Schlicker A, Roepman P, McDermott U, Garnett M, Grernrum W, Sun C, Prahallad A, et al. 2012. MED12 controls the response to multiple cancer drugs through regulation of TGF-β receptor signaling. *Cell* **151:** 937–950.

Jeffers M, Rong S, Vande Woude GF. 1996. Enhanced tumorigenicity and invasion-metastasis by hepatocyte growth factor/scatter factor-met signalling in human cells concomitant with induction of the urokinase proteolysis network. *Mol Cell Biol* **16:** 1115–1125.

Keduka E, Kaiho A, Hamada M, Watanabe-Takano H, Takano K, Ogasawara M, Satou Y, Satoh N, Endo T. 2009. M-Ras evolved independently of R-Ras and its neural function is conserved between mammalian and ascidian, which lacks classical Ras. *Gene* **429:** 49–58.

Kimmelman A, Tolkacheva T, Lorenzi MV, Osada M, Chan AM. 1997. Identification and characterization of R-ras3: A novel member of the RAS gene family with a non-ubiquitous pattern of tissue distribution. *Oncogene* **15:** 2675–2685.

Kimmelman AC, Osada M, Chan AM. 2000. R-Ras3, a brain-specific Ras-related protein, activates Akt and promotes cell survival in PC12 cells. *Oncogene* **19:** 2014–2022.

Kimmelman AC, Nunez Rodriguez N, Chan AM. 2002. R-Ras3/M-Ras induces neuronal differentiation of PC12 cells through cell-type-specific activation of the mitogen-activated protein kinase cascade. *Mol Cell Biol* **22:** 5946–5961.

Kobayashi T, Aoki Y, Niihori T, Cave H, Verloes A, Okamoto N, Kawame H, Fujiwara I, Takada F, Ohata T, et al. 2010. Molecular and clinical analysis of *RAF1* in Noonan syndrome and related disorders: Dephosphorylation of serine 259 as the essential mechanism for mutant activation. *Hum Mutat* **31:** 284–294.

Lavoie H, Therrien M. 2015. Regulation of RAF protein kinases in ERK signalling. *Nat Rev Mol Cell Biol* **16:** 281–298.

Liu H, Huang XQ, Yang M, Ji XM, Du X, Zheng J. 2013. MRAS genetic variation is associated with atherothrombotic stroke in the Han Chinese population. *J Clin Neurol* **9:** 223–230.

Matallanas D, Birtwistle M, Romano D, Zebisch A, Rauch J, von Kriegsheim A, Kolch W. 2011. Raf family kinases: Old dogs have learned new tricks. *Genes Cancer* **2:** 232–260.

Mathieu ME, Faucheux C, Saucourt C, Soulet F, Gauthereau X, Fedou S, Trouillas M, Theze N, Thiebaud P, Boeuf H. 2013. MRAS GTPase is a novel stemness marker that impacts mouse embryonic stem cell plasticity and *Xenopus* embryonic cell fate. *Development* **140:** 3311–3322.

Mitin N, Rossman KL, Der CJ. 2005. Signaling interplay in Ras superfamily function. *Curr Biol* **15:** R563–R574.

Molzan M, Schumacher B, Ottmann C, Baljuls A, Polzien L, Weyand M, Thiel P, Rose R, Rose M, Kuhenne P, et al. 2010. Impaired binding of 14-3-3 to C-RAF in Noonan syndrome suggests new approaches in diseases with increased Ras signaling. *Mol Cell Biol* **30:** 4698–4711.

Nagy J, Curry GW, Hillan KJ, McKay IC, Mallon E, Purushotham AD, George WD. 1996. Hepatocyte growth factor/scatter factor expression and c-met in primary breast cancer. *Surg Oncol* **5:** 15–21.

Nunez Rodriguez N, Lee IN, Banno A, Qiao HF, Qiao RF, Yao Z, Hoang T, Kimmelman AC, Chan AM. 2006. Characterization of R-Ras3/M-Ras null mice reveals a potential role in trophic factor signaling. *Mol Cell Biol* **26:** 7145–7154.

Ohba Y, Mochizuki N, Yamashita S, Chan AM, Schrader JW, Hattori S, Nagashima K, Matsuda M. 2000. Regulatory proteins of R-Ras, TC21/R-Ras2, and M-Ras/R-Ras3. *J Biol Chem* **275:** 20020–20026.

Ortiz-Vega S, Khokhlatchev A, Nedwidek M, Zhang XF, Dammann R, Pfeifer GP, Avruch J. 2002. The putative tumor suppressor RASSF1A homodimerizes and heterodimerizes with the Ras-GTP binding protein Nore1. *Oncogene* **21:** 1381–1390.

Pandit B, Sarkozy A, Pennacchio LA, Carta C, Oishi K, Martinelli S, Pogna EA, Schackwitz W, Ustaszewska A, Landstrom A, et al. 2007. Gain-of-function RAF1 mutations cause Noonan and LEOPARD syndromes with hypertrophic cardiomyopathy. *Nat Genet* **39:** 1007–1012.

Peti W, Nairn AC, Page R. 2013. Structural basis for protein phosphatase 1 regulation and specificity. *FEBS J* **280:** 596–611.

Quilliam LA, Rebhun JF, Zong H, Castro AF. 2001. Analyses of M-Ras/R-Ras3 signaling and biology. *Methods Enzymol* **333:** 187–202.

Razzaque MA, Nishizawa T, Komoike Y, Yagi H, Furutani M, Amo R, Kamisago M, Momma K, Katayama H, Nakagawa M, et al. 2007. Germline gain-of-function mutations in *RAF1* cause Noonan syndrome. *Nat Genet* **39:** 1013–1017.

Rebhun JF, Castro AF, Quilliam LA. 2000. Identification of guanine nucleotide exchange factors (GEFs) for the Rap1 GTPase. Regulation of MR-GEF by M-Ras-GTP interaction. *J Biol Chem* **275:** 34901–34908.

Richier L, Williton K, Clattenburg L, Colwill K, O'Brien M, Tsang C, Kolar A, Zinck N, Metalnikov P, Trimble WS, et al. 2010. NOS1AP associates with Scribble and regulates dendritic spine development. *J Neurosci* **30:** 4796–4805.

Rodriguez-Viciana P, Sabatier C, McCormick F. 2004. Signaling specificity by Ras family GTPases is determined by the full spectrum of effectors they regulate. *Mol Cell Biol* **24:** 4943–4954.

Rodriguez-Viciana P, Oses-Prieto J, Burlingame A, Fried M, McCormick F. 2006. A phosphatase holoenzyme comprised of Shoc2/Sur8 and the catalytic subunit of PP1 functions as an M-Ras effector to modulate Raf activity. *Mol Cell* **22:** 217–230.

Rose BA, Force T, Wang Y. 2010. Mitogen-activated protein kinase signaling in the heart: Angels versus demons in a heart-breaking tale. *Physiol Rev* **90:** 1507–1546.

Saez R, Chan AM, Miki T, Aaronson SA. 1994. Oncogenic activation of human R-Ras by point mutations analogous to those of prototype H-Ras oncogenes. *Oncogene* **9:** 2977–2982.

Saito Y, Oinuma I, Fujimoto S, Negishi M. 2009. Plexin-B1 is a GTPase activating protein for M-Ras, remodelling dendrite morphology. *EMBO Rep* **10:** 614–621.

Schunkert H, Konig IR, Kathiresan S, Reilly MP, Assimes TL, Holm H, Preuss M, Stewart AF, Barbalic M, Gieger C, et al. 2011. Large-scale association analysis identifies 13 new susceptibility loci for coronary artery disease. *Nat Genet* **43:** 333–338.

Selfors LM, Schutzman JL, Borland CZ, Stern MJ. 1998. *soc-2* encodes a leucine-rich repeat protein implicated in fibroblast growth factor receptor signaling. *Proc Natl Acad Sci* **95:** 6903–6908.

Shi Y. 2009. Serine/threonine phosphatases: Mechanism through structure. *Cell* **139:** 468–484.

Sieburth DS, Sun Q, Han M. 1998. SUR-8, a conserved Ras-binding protein with leucine-rich repeats, positively regulates Ras-mediated signaling in *C. elegans*. *Cell* **94:** 119–130.

Sun P, Watanabe H, Takano K, Yokoyama T, Fujisawa J, Endo T. 2006. Sustained activation of M-Ras induced by nerve growth factor is essential for neuronal differentiation of PC12 cells. *Genes Cells* **11:** 1097–1113.

Tada M, Kai M. 2012. Planar cell polarity in coordinated and directed movements. *Curr Top Dev Biol* **101:** 77–110.

Tasaka G, Negishi M, Oinuma I. 2012. Semaphorin 4D/plexin-B1-mediated M-Ras GAP activity regulates actin-based dendrite remodeling through lamellipodin. *J Neurosci* **32:** 8293–8305.

Tzivion G, Luo Z, Avruch J. 1998. A dimeric 14-3-3 protein is an essential cofactor for Raf kinase activity. *Nature* **394:** 88–92.

van de Vijver MJ, He YD, van't Veer LJ, Dai H, Hart AA, Voskuil DW, Schreiber GJ, Peterse JL, Roberts C, Marton MJ, et al. 2002. A gene-expression signature as a predictor of survival in breast cancer. *N Engl J Med* **347:** 1999–2009.

Wang Y, He H, Srivastava N, Vikarunnessa S, Chen YB, Jiang J, Cowan CW, Zhang X. 2012. Plexins are GTPase-activating proteins for Rap and are activated by induced dimerization. *Sci Signal* **5:** ra6.

Ward KR, Zhang KX, Somasiri AM, Roskelley CD, Schrader JW. 2004. Expression of activated M-Ras in a murine mammary epithelial cell line induces epithelial–mesenchymal transition and tumorigenesis. *Oncogene* **23:** 1187–1196.

Watanabe-Takano H, Takano K, Keduka E, Endo T. 2010. M-Ras is activated by bone morphogenetic protein-2 and participates in osteoblastic determination, differentiation, and transdifferentiation. *Exp Cell Res* **316:** 477–490.

Yang J, Chatterjee-Kishore M, Staugaitis SM, Nguyen H, Schlessinger K, Levy DE, Stark GR. 2005. Novel roles of unphosphorylated STAT3 in oncogenesis and transcriptional regulation. *Cancer Res* **65:** 939–947.

Yasumoto M, Sakamoto E, Ogasawara S, Isobe T, Kizaki J, Sumi A, Kusano H, Akiba J, Torimura T, Akagi Y, et al. 2017. Muscle RAS oncogene homolog (MRAS) recurrent mutation in Borrmann type IV gastric cancer. *Cancer Med* **6:** 235–244.

Ye M, Shima F, Muraoka S, Liao J, Okamoto H, Yamamoto M, Tamura A, Yagi N, Ueki T, Kataoka T. 2005. Crystal structure of M-Ras reveals a GTP-bound "off" state conformation of Ras family small GTPases. *J Biol Chem* **280:** 31267–31275.

Yoshikawa Y, Satoh T, Tamura T, Wei P, Bilasy SE, Edamatsu H, Aiba A, Katagiri K, Kinashi T, Nakao K, et al. 2007. The M-Ras-RA-GEF-2-Rap1 pathway mediates tumor necrosis factor-α dependent regulation of integrin activation in splenocytes. *Mol Biol Cell* **18:** 2949–2959.

Young LC, Hartig N, Munoz-Alegre M, Oses-Prieto JA, Durdu S, Bender S, Vijayakumar V, Vietri Rudan M, Gewinner C, Henderson S, et al. 2013. An MRAS, SHOC2, and SCRIB complex coordinates ERK pathway activation with polarity and tumorigenic growth. *Mol Cell* **52:** 679–692.

Zhang FL, Casey PJ. 1996. Protein prenylation: Molecular mechanisms and functional consequences. *Annu Rev Biochem* **65:** 241–269.

Zhang KX, Ward KR, Schrader JW. 2004. Multiple aspects of the phenotype of mammary epithelial cells transformed by expression of activated M-Ras depend on an autocrine mechanism mediated by hepatocyte growth factor/scatter factor. *Mol Cancer Res* **2:** 242–255.

The K-Ras, N-Ras, and H-Ras Isoforms: Unique Conformational Preferences and Implications for Targeting Oncogenic Mutants

Jillian A. Parker and Carla Mattos

Department of Chemistry and Chemical Biology, Northeastern University, Boston, Massachusetts 02115

Correspondence: c.mattos@northeastern.edu

Ras controls a multitude of cellular signaling processes, including cell proliferation, differentiation, and apoptosis. Deregulation of Ras cycling often promotes tumorigenesis and various other developmental disorders, termed RASopothies. Although the structure of Ras has been known for many decades, it is still one of the most highly sought-after drug targets today, and is often referred to as "undruggable." At the center of this paradoxical protein is a lack of understanding of fundamental differences in the G domains between the highly similar Ras isoforms and common oncogenic mutations, despite the immense wealth of knowledge accumulated about this protein to date. A shift in the field during the past few years toward a high-resolution understanding of the structure confirms the hypothesis that each isoform and oncogenic mutation must be considered individually, and that not all Ras mutations are created equal. For the first time in Ras history, we have the ability to directly compare the structures of each wild-type isoform to construct a "base-line" understanding, which can then be used as a springboard for analyzing the effects of oncogenic mutations on the structure–function relationship in Ras. This is a fundamental and large step toward the goal of developing personalized therapies for patients with Ras-driven cancers and diseases.

The small GTPase Ras was discovered more than 30 years ago because of its oncogenic potential, and is frequently mutated in ~20% of all human cancers. Classically, Ras is referred to as a monomeric switch protein that is signaling active when bound to guanosine triphosphate (GTP) and inactive when bound to guanosine diphosphate (GDP). This active/inactive cycle is tightly controlled by regulatory proteins knows as guanine nucleotide exchange factors (GEFs) and GTPase-activating proteins (GAPs). In the context of cellular signaling, Ras requires the aid of these regulatory proteins to overcome its high affinity for the bound nucleotide and intrinsically slow hydrolysis rate (Ahmadian et al. 1997; Boriack-Sjodin et al. 1998; Bos et al. 2007). Changes in the nucleotide bound in the active site are accompanied by dynamic conformational changes in the switch I (residues 30–40) and switch II (residues 60–76) motifs of Ras. Mutations at residues G12, G13, and Q61 lead to GAP-insensitivity and disruption of this tightly controlled cycling, in which constitutively active Ras promotes tumorigenesis.

Cite this article as *Cold Spring Harb Perspect Med* doi: 10.1101/cshperspect.a031427

The structures of Ras proteins contain a common central 6-stranded β-sheet flanked by five α-helices in a Rossmann-type fold (Vetter 2014). The G domain of the proteins (residues 1–166) is generally divided into two halves, the effector lobe (residues 1–86) and the allosteric lobe (87–166), based on function (Buhrman et al. 2011a). All GTPases contain conserved sequence motifs throughout the protein that impart nucleotide-binding specificity and serve specific roles central to the function of these molecular switches (Bourne et al. 1991). Residues in the NKxD (^{116}NKCD^{119}) and ExSAK (^{143}ETSAK^{147}) motifs, as well as F28, are imperative for recognizing and positioning the guanine nucleotide within the active site, while the P-loop (GxxxxGK[S/T]; ^{10}GAGGVGKS^{17} in Ras) provides the appropriate charge and solvent environment for the phosphate groups and Mg^{2+} (Valencia et al. 1991). The neutron crystal structure of H-Ras (Protein Data Bank [PDB] ID 4RSG) shows that the γ-phosphate of the bound GTP analog is protonated (overall charge of −3) in the crystal, providing an active site that is amenable to both intrinsic and GAP-catalyzed hydrolysis of GTP through a dissociative-like mechanism (Knihtila et al. 2015). Residues in the switch I, specifically T35, and DxxG (^{57}DTAG^{60}) motifs recognize and stabilize the Mg^{2+}-bound GTP in the active site when Ras is signaling active. The Mg^{2+} ion is a crucial cofactor in GTPases and is involved in all aspects of GTPase function, including binding to effector and regulatory protein partners and in the hydrolysis of GTP to GDP (Pan and Wessling-Resnick 1998; Rudack et al. 2012). When Ras is bound to GTP, the active site Mg^{2+} is coordinated by the β- and γ-phosphates and the side chains of switch I residues S17 and T35, while two water molecules complete the octahedral coordination sphere. This switch I conformation is referred to as state 2, in which T35 interacts with the Mg^{2+} and Y32 is often located over the nucleotide to create a "closed" active site (Shima et al. 2010). When T35 is not interacting with the Mg^{2+}, switch I samples a range of more "open" switch I conformations, leading to greater solvent accessibility of the active site (Araki et al. 2011). The state 2 switch I conformation is observed in structures of Ras bound to effectors, whereas the state 1 conformation results in Ras that is "effector-binding-deficient" and more susceptible to nucleotide exchange (Spoerner et al. 2001).

The three isoforms of Ras, H-, N-, and K-Ras, share 90% identity in their G domain, with conserved structural and biochemical properties. There are two alternative splice variants of K-Ras: K-Ras4A and K-Ras4B. In most tissues, the K-Ras4A isoform is less predominantly expressed, although recent evidence suggests a role in some cancers (Tsai et al. 2015). Throughout this review, K-Ras will refer to the 4B splice variant, as it has been the most studied of the two. Importantly, the Ras isoforms have unique carboxy-terminal hypervariable regions (HVRs) that contain sites for posttranslational modification (PTM) imperative for correct function and localization of each isoform (Henis et al. 2009). All Ras proteins are farnesylated at C186 (C185 in K-Ras4B) to promote binding to the plasma membrane, where Ras interacts with regulatory and effector protein partners. Although the plasma membrane is the major site of Ras functional output, signaling from endomembranes has also been observed, particularly for N-Ras (Fehrenbacher et al. 2009). This may be a result, in part, of the presence of a singular palmitoyl group at C181 for this isoform, whereas H-Ras is doubly palmitoylated at C181 and C184 (Ahearn et al. 2012). K-Ras4B contains a hexalysine polybasic region (residues 175–180), which uniquely targets this isoform to areas of the plasma membrane enriched in acidic phospholipids (Hancock et al. 1990). Proteins such as PDEδ and calmodulin (CaM) interact specifically with the carboxy-terminal HVR on Ras to regulate trafficking and function on the membrane (Abraham et al. 2009; Dharmaiah et al. 2016). Isoform-specific localization patterns have been proposed as a mechanism for organizing Ras signaling clusters within the crowded cellular membrane environment (Hancock and Parton 2005). Nanoclusters containing around five to eight Ras proteins are essential for high-fidelity signal output, although the exact protein and lipid composition of these nanoclusters has been debated (Abankwa et al. 2007; Tian et al.

 Cite this article as *Cold Spring Harb Perspect Med* doi: 10.1101/cshperspect.a031427

2007; Zhou et al. 2014). Recently, Ras dimerization and higher order oligomerization has become an area of intense research (Thompson 2013). However, the details of Ras dimerization are highly contested and unclear, such as the residues involved in the dimerization interface, the importance of isoform-specific homo- and heterodimerization, the mechanism of dimer formation, the role of the membrane, and the signaling relevance of dimers (Santos 2014; Holderfield and Morrison 2017). It is becoming increasingly clear that targeting Ras dimerization could abrogate nanocluster formation and signaling output (Chen et al. 2016; Spencer-Smith et al. 2016), although there is still confusion in the literature as to the location of the dimer interface (Jang et al. 2016; Prakash et al. 2017).

Although the majority of the G domain is highly conserved across the three Ras isoforms, residue differences in the allosteric lobe are hypothesized to play a role in isoform-specific membrane localization and orientation, which in turn has been linked to regulation of Ras–effector interactions and downstream signaling (Abankwa et al. 2010; Parker and Mattos 2015). Importantly, these allosteric lobe residues can influence intraprotein communication, connecting the membrane-interacting allosteric lobe residues to those in the active site of the effector lobe (Kearney et al. 2014). The predominant assumption in the field has been that the 100% sequence conserved effector lobe, where Ras interacts with a wide array of effector proteins such as Raf, PI3K, and RalGEF, promoted identical functions in the isoforms. However, recent evidence suggests that the structure, biochemistry, and dynamics of the effector lobe of each isoform are fundamentally different and require individual attention (Johnson et al. 2017; JA Parker, AY Volmar, S Pavlopoulos, and C Mattos, in prep.). Importantly, the balance between the more "open" state 1 and "closed" state 2 conformation can potentially be exploited in the design of isoform- and mutant-specific Ras inhibitors (Kauke et al. 2017). For the first time, both the structures and biochemical properties of the GTP-bound form of wild-type (WT) Ras isoforms are available, allowing for unprecedented comparison and novel insights into the important differences that contribute to the complicated field of Ras biology and cancer. Insights gleaned from the careful study of the WT isoforms can then be applied to oncogenic mutants, which are linked to isoform- and mutant-specific cancer and tissue-type frequencies (Prior et al. 2012; Lu et al. 2016a).

GLOBAL INTRAMOLECULAR COMMUNICATION NETWORKS IN Ras

The signaling active, GTP-bound form of Ras relies on intricate interplay between the effector and allosteric lobes. A highly conserved interswitch region (β2–β3 loop, L3) connects the switch I and II regions of the effector lobe with the allosteric lobe. Specifically, D47 and E49 engage in stable salt bridge interactions with helix 5 residues R161 and R164, respectively. These residues have been dubbed "nucleotide-sensing residues" because of their ability to influence orientation of Ras with respect to the plasma membrane to ensure correct interactions with effector or regulatory proteins based on whether GDP or GTP is bound to Ras (Abankwa et al. 2008). In addition to the interswitch, water-mediated hydrogen bond networks connect membrane-interacting residues in the allosteric lobe to the effector lobe, whereas information based on the nucleotide bound in the active site is also relayed to the allosteric lobe (Kearney et al. 2014). One such network, termed the "helix 5 network," spans nearly 15 Å across Ras to connect R161 and R164 with active site residue Y32 via the catalytically important bridging water molecule (Kearney et al. 2014). This network, which is abolished in inactive GDP-bound Ras, can therefore directly sense the presence of the γ-phosphate and relay this information to membrane-interacting residues in the allosteric lobe. Molecular dynamics (MD) simulations support these nucleotide-dependent shifts in the conformational ensemble of Ras (Grant et al. 2009; Raimondi et al. 2011). Oncogenic mutations, such as Q61L and G12V, result in a severed connection between the active site and the helix 5 network because of a direct H-bond between Y32 and the γ-phosphate in the absence of the active site water molecule that bridges Y32 to the γ-phosphate (the bridging water molecule) (Marcus and Mattos 2015). It is important to point out that this communication pathway uses residues in the allosteric lobe, particularly in helix 5, helix 4, and loop 8 that are highly sequence divergent among the isoforms. Therefore, it is possible that these isoform-specific residues differentially regulate communication in the isoforms, contributing to the fact that each isoform can promote overlapping yet distinct signaling outputs (Parker and Mattos 2015). The contribution of the G domain to functional specificity in the isoforms has been largely overlooked. This review focuses on recent advances linking biochemical and dynamics features of the G domain to isoform- specific differences in the allosteric lobe.

A ligand-binding site in the allosteric lobe has been identified as an important modulator of switch II conformations in the effector lobe (Buhrman et al. 2010). Calcium and acetate, which could be acting as a surrogate for a negatively charged membrane component in H-Ras crystals, bind in the allosteric site of Ras and promote a shift in the helix 3/loop 7 motif toward the allosteric lobe to allow ordering of switch II in a fully helical conformation, as seen in the crystal structure with PDB ID 3K8Y. A water-mediated hydrogen bond network from the allosteric site through helix 3 to switch II leads to the placement of the catalytic Q61 residue in the active site, where it interacts with the bridging water molecule (Fig. 1A–C) (Buhrman et al. 2010). This conformation is referred to as the R state, whereas an empty allosteric site and consequently disordered switch II is referred to as the T state. When switch I is present in the state 2 conformation and Y32 closed over the nucleotide, similar to that seen in the H-Ras/Raf complex (PDB ID 4G0N), the active site achieves a conformation to promote intrinsic hydrolysis. While Raf orders switch I, stabilizing the catalytically competent state 2 conformation, switch II, with catalytic residue Q61, remains disordered in the T state unless ligand binds in the allosteric site to promote the R state in which switch II is ordered and Ras catalytically competent. This promotes the allosteric network associated with the R state Ras, prompting the hypothesis that these networks could regulate signaling output for the Ras/Raf/Mek/Erk, through a GAP-independent hydrolysis mechanism, by promoting intrinsic hydrolysis of GTP (Buhrman et al. 2011b). Specifically, this level of regulation may be important in the context of Raf binding, which can outcompete GAP binding because of its nanomolar affinity for Ras-GTP (Smith and Ikura 2014). Evidence of the importance of Ras regulation via the allosteric network is seen in oncogenic G12D structures of Ras, where the connection to the allosteric site is cut off because of steric hindrance from the aspartate side chain in the active site, in effect hindering both GAP-assisted and intrinsic hydrolysis (Marcus and Mattos 2015).

Unlike all other known Ras effector proteins, Raf binds solely to switch I of Ras, leaving switch II free to attain multiple conformations. This point is crucial to the intrinsic hydrolysis mechanism and has driven research to focus intensely on the Ras/Raf/Mek/Erk pathway, a major regulator of cell proliferation and, consequently, tumorigenesis (McCubrey et al. 2007). We have previously shown in H-Ras that Q61 is critical to signaling through Raf, in which oncogenic mutants such as Q61L promote an anti-catalytic conformation associated with the T state and prevent hydrolysis in the presence of Raf (Buhrman et al. 2007). On the other hand, the oncogenic G12V mutant promotes a relatively minor increase in mitogen-activated protein kinase (MAPK) phosphorylation in cells (Buhrman et al. 2011b), further confirming the growing realization that oncogenic mutants have differential effects on the signaling output of Ras. A combination of crystallographic and computational results revealed previously unrealized characteristics of the Ras/Raf complex. Raf binding to WT H-Ras induces an increase in conformational flexibility of switch II, while rigidifying residues in the allosteric site that directly bind Ca^{2+}, consistent with the intrinsic hydrolysis mechanism (Fetics et al. 2015). Most interestingly, a similar study of the H-RasQ61L/Raf complex (PDB ID 4G3X) shows how the Q61L mutation rigidifies switch II while

 Cite this article as *Cold Spring Harb Perspect Med* doi: 10.1101/cshperspect.a031427

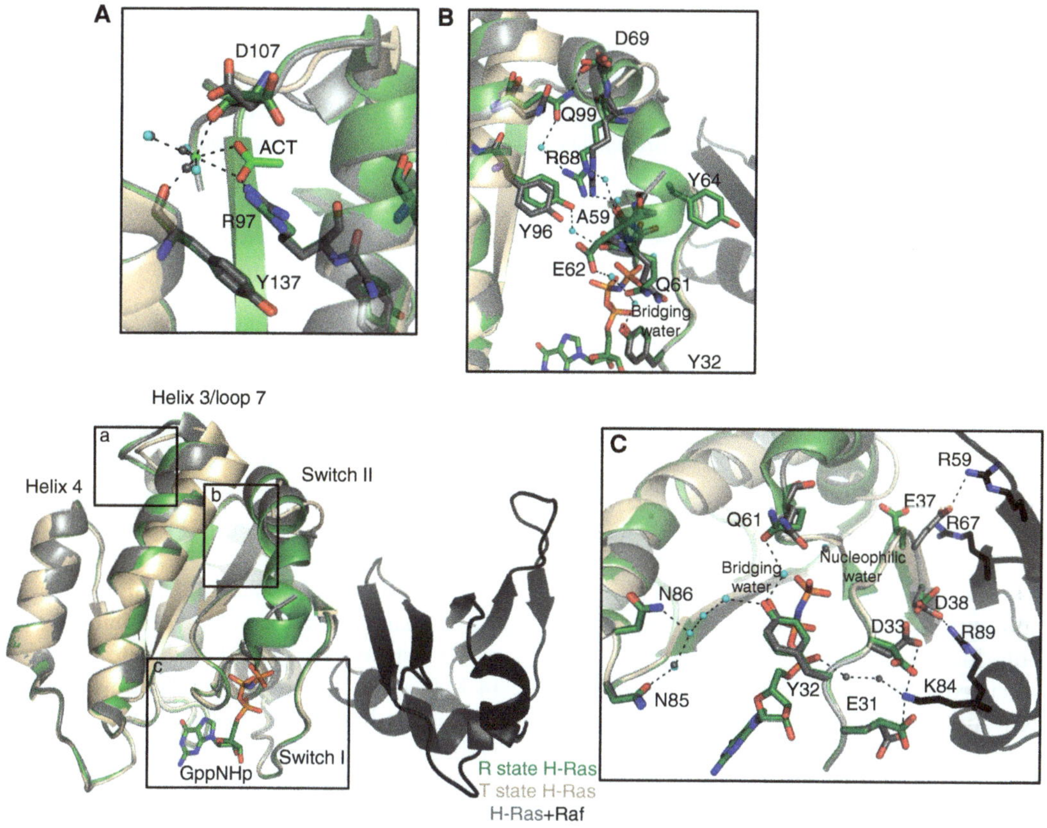

Figure 1. Allosteric network communication pathway in H-Ras. The T state of H-Ras, exemplified by the Protein Data Bank (PDB) ID 2RGE (wheat), shows a disordered switch II motif. On binding of an allosteric ligand as shown by PDB ID 3K8Y (green), the helix 3/loop 7 motif shifts toward helix 4, promoting formation of a water-mediated hydrogen bond network from the allosteric site (*A*) through an ordered switch II motif (*B*) to place Q61 in the active site (*C*). This allosteric mechanism is important in signaling through Raf, which binds at switch I as shown by the H-Ras/Raf complex (PDB ID 4G0N, H-Ras in gray and Raf in black).

increasing the flexibility of the Ca^{2+}-binding residues in the allosteric site (Fetics et al. 2015). Unlike previous assumptions that oncogenic mutants perturb only their local environment and binding with GAP, these results show that a single mutation can have global effects on the communication pathways between Ras and effector proteins, and that single-residue mutations are able to alter the balance of conformational states. These changes in global communication pathways throughout Ras raise the possibility of allosteric effects on Ras dimerization and the Ras–membrane interaction as a result of oncogenic mutations.

DIFFERENCES IN THE G DOMAIN OF THE Ras ISOFORMS

Until recently, only crystal structures of WT H-Ras bound to GTP-analogs were available in the PDB, and these structures were assumed to be good models for the K-Ras and N-Ras isoforms because of their 95% conserved G domains. Despite the fact that the isoforms share 100% sequence identity in areas where they interact with effector proteins, the networks linking the two G domain lobes led us to expect significant differences in the kinetics, biochemistry, and signaling output for H-, K-, and N-Ras. Our recent structures of both WT N-Ras and K-Ras bound to GTP analogs gives us the unprecedented ability to directly compare the three isoforms (Johnson et al. 2017; JA Parker, AY Volmar, S Pavlopoulos, and C Mattos, in prep.). The observed structural differences can be linked to functional differences associated with distinct balance in the conformational states and dynamics in each of the isoforms (Gorfe et al. 2008; Harrison et al. 2016).

In terms of the G domain, there are 17 residue differences across the isoforms (Fig. 2A). When mapped onto the structure, these isoform-specific residues cluster in the helix 3, loop 8, and helix 4 areas of Ras, sandwiching highly conserved residues in the ^{116}NKCD^{119} and ^{143}ETSAK^{147} motifs, as well as the allosteric site (Fig. 2B). The superposed structures of WT H-Ras, K-Ras, and N-Ras bound to GTP analogs show differences in the state 1 versus state 2 conformations in switch I and R versus T states associated with switch II (Fig. 3A). It is interesting to note that, among the numerous WT H-Ras structures in the GTP-bound state found in the PDB, a single one shows a state 1 conformation and that was obtained only by seeding crystals with the constitutive state 1 mutant H-RasT35S-GppNHp (Araki et al. 2011). The predominance of state 2 in WT H-Ras found both in the crystal structures and by ^{31}P nuclear magnetic resonance (NMR) (Spoerner et al. 2010) contrasts with the WT K-Ras crystal structure with an open switch I conformation consistent with a more exposed P-loop as determined by recently published hydrogen/deuterium exchange mass spectrometry (HDX-MS) experiments on both isoforms. As discussed above, the positioning of the helix 3/loop 7 motif is crucial in mediating the T-to-R state transition and formation of the allosteric network that connects membrane-interacting residues in Ras to the active site. In the R state H-Ras structure (PDB ID 3K8Y), a hydrogen bond between helix 4 residue Y137 and helix 3 residue H94 stabilizes the R-state helix 3 conformation (Fig. 3B). Both residues 94 and 95 are sites of isoform sequence diversity, in which position 95 is one of two where the residue identity is completely different across the isoforms (L95 in N-Ras, H95 in K-Ras, and Q95 in H-Ras). Differences in the hydrogen-bonding network of the allosteric site propagate across helix 3 toward switch II, where an altered set of interactions in K-Ras promotes a Y71 conformation that pushes switch I away from the active site (Fig. 3C). Although the T-state conformation is preferred in both N-Ras and K-Ras, this conformation is more extreme in the WT K-Ras structure (PDB ID 5UK9) (JA Parker, AY Volmar, S Pavlopoulos, and C Mattos, in prep.) and more intermediate in the WT N-Ras structure (PDB ID 5UHV) (Fig. 3A) (Johnson et al. 2017). Single-turnover hydrolysis experiments show that the hydrolysis rate constants for WT K- and N-Ras are indeed lower than that of H-Ras, consistent with a more prominent state 1 and/or T state conformation in solution (Johnson et al. 2017). Given that the binding affinity of K-Ras and H-Ras with Raf is essentially the same, we suspect that the energy barrier between state 1 and state 2 is lower in K-Ras than in N-Ras, which shows about a twofold reduction in binding affinity for Raf (Johnson et al. 2017). Because the effector lobe is 100% identical in terms of sequence among the three isoforms, the differences in biochemical properties of the isoforms can be attributed to differences in the allosteric lobe and differences in the balance of conformational states in solution.

Isoform-specific residues in the allosteric lobe tend to cluster around structurally important regions of Ras, such as motifs crucial for interacting with the nucleotide in the active site or the membrane. One example is the R123-E143 salt bridge, which connects the conserved ^{116}NKCD^{119} and ^{143}ETSAK^{147} motifs to position these residues for optimal interactions with the nucleotide. While the salt bridge is a head-on interaction in both H- and K-Ras, it is offset in N-Ras such that one nitrogen atom from R123 and one oxygen atom from E143 can interact (Fig. 3D) (Johnson et al. 2017). Isoform-specific residues A/P121, S/T127, and Y/F141 (H-/K-Ras, respectively) surround this important salt bridge, providing unique interactions in H-Ras compared with K-Ras in their GTP-bound forms. In addition to this salt bridge, the conserved motifs are in close prox-

A

Residue #	Residue ID HRas	Residue ID NRas	Residue ID KRas4B	Residue relationship	Structural position
87	Thr	Ser	Thr	Polar	Helix 3
91	Glu	Ala	Glu	-	Helix 3
94	His	Asn	His	-	Helix 3
95	Gln	Leu	His	-	Helix 3
107	Asp	Asp	Glu	Acidic	Loop 7
121	Ala	Pro	Pro	Hydrophobic	Loop 8
122	Ala	Thr	Ser	-	Loop 8
126	Glu	Asp	Asp	Acidic	Loop 8/helix 4
127	Ser	Thr	Thr	Polar	Helix 4
128	Arg	Lys	Arg	Basic	Helix 4
131	Gln	His	Gln	-	Helix 4
132	Asp	Glu	Asp	Acidic	Helix 4
135	Arg	Lys	Arg	Basic	Helix 4
141	Tyr	Phe	Phe	Hydrophobic	Loop 9/β6
153	Glu	Glu	Asp	Acidic	Helix 5
165	Gln	Gln	Lys	-	Carboxyl terminus
166	His	Tyr	His	-	Carboxyl terminus

B

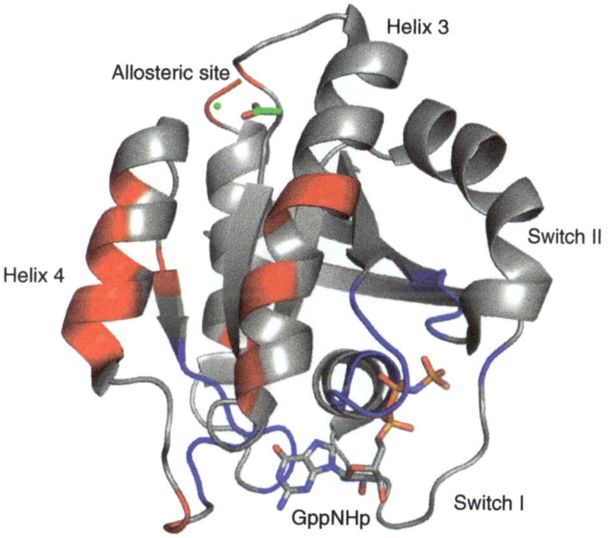

Figure 2. Residue differences in the three Ras isoforms. (*A*) H-Ras contains four unique residues (green), K-Ras contains three (blue), and N-Ras contains eight (orange). Positions 95 and 122 are different among the three isoforms (red). (*B*) Residue differences (red) mapped onto H-Ras (Protein Data Bank [PDB] ID 3K8Y, gray) cluster around 100% conserved motifs in Ras (blue).

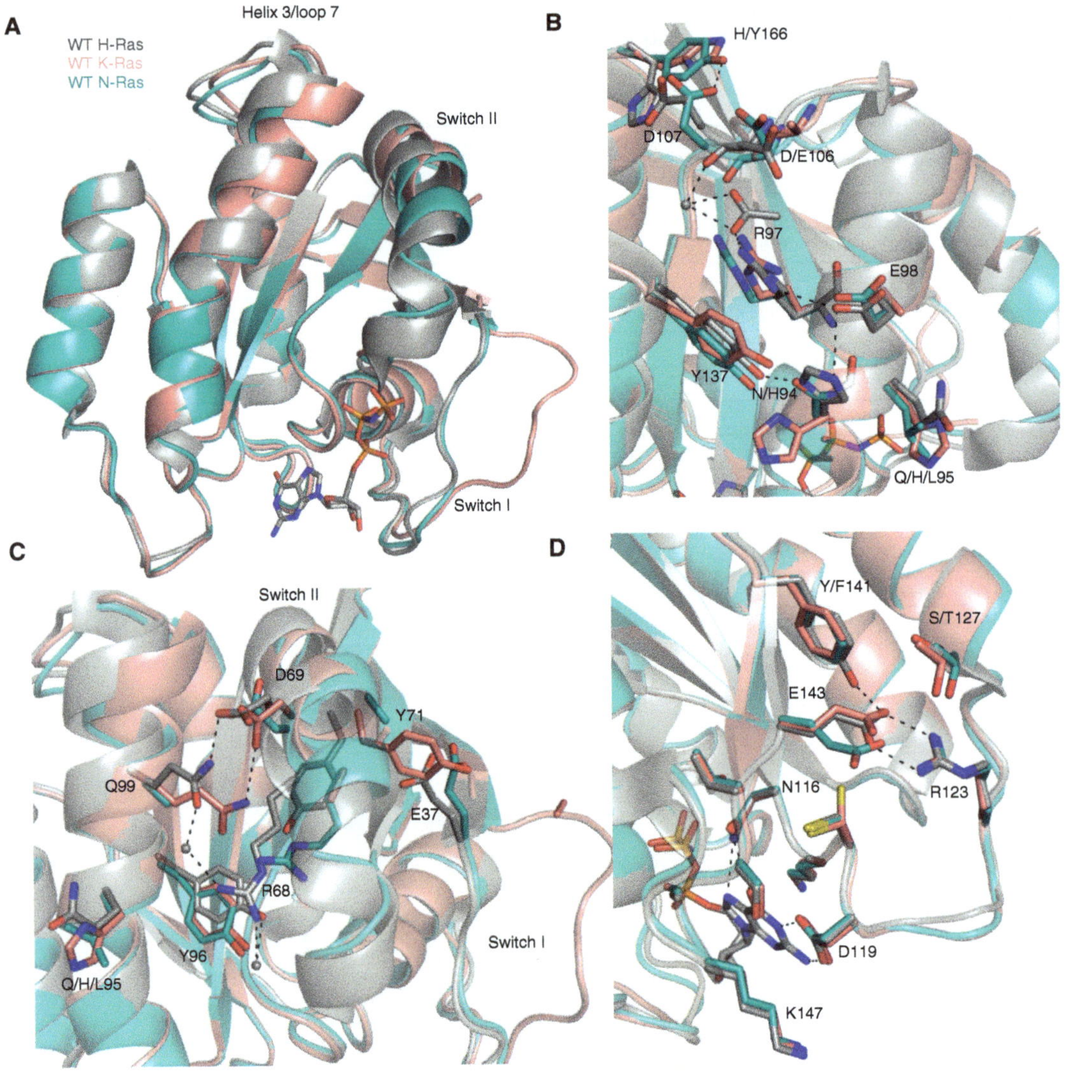

Figure 3. Isoform-specific residues promote structural differences in Ras isoforms. (*A*) Alignment of N-Ras (GppNHp) (Protein Data Bank [PDB] IL 5UHV, cyan), K-Ras (GppCH$_2$p) (PDB ID 5UK9, salmon), and H-Ras (GppNHp) (PDB ID 3K8Y, gray). Isoform-specific residues in the allosteric site (*B*) promote differences in switch II conformations in the isoforms (*C*). The altered salt bridge interaction between E123 and R143 in N-Ras is shown in *D*.

imity to residue pockets on Ras that are proposed to modulate Ras–membrane interactions in an isoform-specific manner. The helix 4 R128 and R135 hot spots and a pocket formed by H94, L133, S126, and Y137 in H-Ras, may have evolved to accommodate an isoform-specific set of lipid head groups, depending on where each isoform microlocalizes within the plasma membrane (Parker and Mattos 2015). Importantly, these isoform-specific allosteric residues are positioned along the helix 5 water-mediated hydrogen bond network that connects the active site to the membrane-interacting regions of Ras (Kearney et al. 2014). Therefore, residue differences in H-, N-, and K-Ras could lead to altered mechanisms of intraprotein communication, contributing to isoform-specific structure, biochemistry, and functional output.

 Cite this article as *Cold Spring Harb Perspect Med* doi: 10.1101/cshperspect.a031427

K-Ras AND DYNAMICS PROVIDE INSIGHT INTO THE MOST ONCOGENIC ISOFORM

Despite their similarities in global structure and the effector lobe where they interact with a similar set of effector proteins, H-, N-, and K-Ras each promote distinct signaling outputs (Ehrhardt et al. 2002; Abankwa et al. 2010). While the HVR is implicated as the major driving force for these differences in the membrane environment (Henis et al. 2009), differences in the G domain of the isoforms could also drive isoform-specific functions. Unique structural and dynamical features of the Ras isoforms have thus become an area of intense research toward understanding the fundamental divergence among the isoforms, especially in the search for isoform-specific Ras therapies. In fact, MD simulations of the three isoforms find that K-Ras is significantly more flexible than N- or H-Ras (Gorfe et al. 2008). Importantly, both switch I and allosteric lobe regions, including helix 4, helix 5, and nucleotide bind loops, of K-Ras display the most conformational flexibility, indicating that this isoform has both unique nucleotide-binding characteristics and effector-allosteric lobe communication pathways (Gorfe et al. 2008; Kapoor and Travesset 2015). Over the course of simulations, K-Ras samples more "open" state 1 conformations, while N-Ras samples more intermediate conformations between state 2 ("closed") and state 1 (Gorfe et al. 2008). It is interesting to note that in the inactive GDP-bound structures switch I is the most rigid in K-Ras when compared with the other two isoforms (Kapoor and Travesset 2015), suggesting a stable interaction mostly mediated by the canonical Y32-Y40 hydrogen bond observed in all GDP-bound Ras structures. These dynamical observations correlate directly to the open state 1 conformation of switch I observed in our recent crystal structure of WT K-Ras-GppCH$_2$p (PDB ID 5UK9) (JA Parker, AY Volmar, S Pavlopoulos, and C Mattos, in prep.). ^{1}H NMR experiments of WT K-Ras bound to nucleotide analogs indicate that the state 1 conformation is sampled in solution much more prominently than is the case for H-Ras, consistent with the crystal structures of the two isoforms (JA Parker, AY Volmar, S Pavlopoulos, and C Mattos, in prep.). The highly dynamic nature of K-Ras may pose additional challenges for targeting this isoform, which displays the highest mutation profile in human cancers (Hobbs et al. 2016).

Unlike the structure of activated WT K-Ras, the crystal structure of K-RasG12D-GppNHp (PDB ID 4DSN) shows a state 2 conformation for switch I, with T35 interacting with the Mg^{2+}-nucleotide complex in the active site and Y32 placed over the nucleotide (Maurer et al. 2012). In this structure, the bridging water molecule is also present, interacting with both Y32 and the mutated D12 side chains. The negatively charged D12 side chain would clash with both the arginine finger and catalytically important residue Q61 and thus interfere with both GAP-mediated and intrinsic hydrolysis, respectively, of GTP to GDP on Ras (Franken et al. 1993; Scheffzek et al. 1997). Consistently, the ^{1}H NMR signatures of K-RasG12D are markedly different from those of WT K-Ras, in which K-RasG12D appears to favor a state 2 conformation that is not sensitive to temperature changes (JA Parker, AY Volmar, S Pavlopoulos, and C Mattos, in prep.). These data indicate that oncogenic K-RasG12D obtains a more stable effector-competent binding conformation to propagate signaling through interactions with effector proteins while simultaneously resisting hydrolysis to lock the protein in this active conformation. The addition of solution NMR data provides novel insight into how specific oncogenic mutants can alter the balance of conformational states to increase their oncogenic propensity. These data can be used to direct more informed targeting of oncogenic Ras, especially in moving toward the goal of targeting the most prevalent K-Ras oncogenic amino acid substitution at codon 12 (Pylayeva-Gupta et al. 2011; Prior et al. 2012). The recent findings for WT K-Ras and K-RasG12D, driven by structural and biophysical characterization, advance the goal of obtaining an atomic level understanding of each isoform and oncogenic mutant, especially given the mounting evidence for isoform and mutant-specific functions and signaling outcomes (Cammarata et al. 2016; Hobbs et al. 2016).

CHALLENGES, SUCCESSES, AND FUTURE DIRECTIONS FOR TARGETING Ras

Since its discovery more than three decades ago, Ras has been one of the most highly sought-after drug targets in the fight against cancer. The failure of the blockbuster farnesyltransferase inhibitors (FTIs) was disappointing, and placed a spotlight on how little Ras structure, function, and biology was understood (Whyte et al. 1997). As discussed, when H-Ras is considered the model for the other isoforms, many key details of each proteins' individual structure, including the balance of conformational states, biochemistry, and function, are overlooked. Today, although Ras is still considered "undruggable" in the traditional sense, the field has made major strides forward.

The picomolar affinity of Ras for its nucleotide, along with the millimolar cellular concentrations of GDP and GTP, make designing a small molecule that can directly compete with nucleotide binding a daunting task (John et al. 1990). Yet, there has been moderate success for small molecules designed to covalently target the G12C oncogenic mutation in Ras, especially important for the goal of selectively targeting prominent mutations linked to specific cancer types, in this case lung adenocarcinoma (Prior et al. 2012). These molecules bind near the active site, either directly in place of the GDP nucleotide or extending into the space between switch II and helix 3 (Ostrem et al. 2013; Lim et al. 2014). One drawback to these molecules is their limitation of targeting the reactive G12C side chain, a technique that will not work on other more prominent oncogenic side chains at residues 12, 13, and 61 (Hobbs et al. 2016). Other small molecules have been shown to bind to the outer switch II region of GDP-bound Ras, in which they interfere with the GEF-mediated nucleotide exchange to keep Ras in its signaling-inactive conformation (Maurer et al. 2012; Sun et al. 2012). Although blocking the GEF-mediated exchange may be a successful approach to abrogating signaling through WT Ras, many of the oncogenic mutants that maintain Ras in the GTP-bound state would be unaffected by this type of intervention.

Aside from the active site, the surface of Ras is relatively flat and lacks well-defined pockets that can be exploited for drug design. Targeting the allosteric site between helix 3, loop 7, and helix 4, which is key in the T-to-R state transition to form an active site poised for intrinsic hydrolysis, could be a useful tactic for targeting Ras with oncogenic mutations that impair GAP-mediated hydrolysis (Buhrman et al. 2010). Additionally, because intrinsic hydrolysis is proposed to be a major regulatory mechanism in the Ras/Raf/Mek/Erk signaling cascade, this approach could target Ras in a pathway-selective manner (Buhrman et al. 2011b). To date, no allosteric site inhibitor has been identified, possibly because of the small volume of the allosteric site, although design of inhibitors that mimic the size, shape, and chemical functionality of membrane lipid head group molecules should be investigated. A zinc cyclen, the only molecule that has been shown by crystallography to bind at the carboxyl terminus and loop 7 close to the allosteric site, appears to promote the effector-binding-deficient state 1 conformation of switch I (Rosnizeck et al. 2010). This is one example of an inhibitor that takes advantage of stabilizing a conformational state known to impair Ras-effector binding. The allosteric lobe contains other hot spots that have been identified experimentally and computationally as sites of protein–ligand interactions (Buhrman et al. 2011a). These sites have been largely unexplored, yet could provide opportunities to target Ras in an isoform-specific manner (Marcus and Mattos 2015). For example, many of these sites include membrane-binding residues that could be targeted to modulate the Ras–membrane interaction in a way that disrupts the functionally critical intramolecular communication networks between the allosteric and effector lobes (Parker and Mattos 2015). Another exciting area of interest is targeting Ras dimerization at the membrane, which may be a crucial mediator of correct Ras nanocluster formation and signaling conformations at the membrane (Holderfield and Morrison 2017).

Blocking the interaction between Ras and its multitude of effector proteins, including Raf, RalGDS, and PI3K, is an area of intense interest given the potential for one therapeutic to block multiple Ras–effector interactions and shut down signaling through Ras. However, targeting the protein–protein interaction (PPI) interface on Ras, which includes both dynamic switch I and switch II regions and is large and relatively flat, presents challenges for traditional small molecules (Lu et al. 2016b). Larger cyclic peptides and stapled SOS1 peptides have shown moderate success in binding Ras with mid-nanomolar affinities to block signaling through inhibition of the Ras–Raf interaction (Upadhyaya et al. 2014; Leshchiner et al. 2015). Other attempts at designing larger molecules to target the effector-binding sites of Ras include single-antibody fragments and monobodies (Tanaka et al. 2007; Spencer-Smith et al. 2016; Keeton et al. 2017).

With the recent advances in understanding the structural and conformational ensemble differences among the Ras isoforms, we are beginning to see that these isoform-specific properties could be exploited in design of potent and specific inhibitors. The first example of an inhibitor that takes advantage of differing conformational states between WT and mutant-activated K-Ras has been illustrated by the crystallization of a small engineered binding protein identified by yeast surface display in complex with K-RasG12D-GppNHp and WT K-Ras-GppNHp (PDB ID 5UFQ and 5UFE, respectively) (Kauke et al. 2017). This small protein binds Ras at switch II, in which a high number of hydrophobic residues selected during an affinity maturation process intercalate with hydrophobic residues on Ras. Although this highly complementary interaction contributes to the single-digit nanomolar affinity to K-Ras, moderate selectivity for the G12D mutant is achieved by taking advantage of the differences in conformational states in the switch I region observed in K-Ras. Although the binder protein does not directly interact with the oncogenic G12D mutation in the active site, the state 2 conformation of switch I in this mutant leads to a unique "lysine claw" interaction between residues in switch I and II and binder residues K32 and K40. This interaction is weakened in the WT K-Ras complex, indicating that the residue selection process directed the formation of a binder protein that can take advantage of the differences in conformational states in switch I, despite binding at switch II. Excitingly, the low nanomolar affinity and larger size of this binder can directly compete with Ras/Raf complex formation and has been shown to directly interfere with signaling through the Ras/Raf/Mek/Erk pathway (Kauke et al. 2017).

The future of cancer therapies, including those directed toward Ras-driven cancers, needs to focus on patient-specific treatments. Given that mutations in Ras differentially affect clinical outcomes, lengths of treatment, and treatment effectiveness, personalized medicine will be key in the fight against Ras (Asati et al. 2017). Studies have shown that a cancer with a G12D mutation responds to anti-Ras therapies, but other G12 mutations are unaffected (Mao et al. 2013). Additionally, screening for K-Ras biomarkers in patients to determine the most effective treatment has shown promise in improving individual clinical outcomes (Smith et al. 2010; Miller and Miller 2011). Given the hints from recent structural and biochemical studies, the hypothesis that each mutation can uniquely alter the balance of conformational states, especially in the most oncogenic isoform K-Ras, should be considered in future efforts to target these proteins. Overall, the idea that H-Ras can serve as an accurate model for the other isoforms should be reevaluated, with more focus on the details of the unique characteristics of each Ras isoform and how these can be exploited for design of potent and specific inhibitors.

REFERENCES

Abankwa D, Gorfe AA, Hancock JF. 2007. Ras nanoclusters: Molecular structure and assembly. *Semin Cell Dev Biol* **18:** 599–607.

Abankwa D, Gorfe AA, Hancock JF. 2008. Mechanisms of Ras membrane organization and signalling: Ras on a rocker. *Cell Cycle* **7:** 2667–2673.

Abankwa D, Gorfe AA, Inder K, Hancock JF. 2010. Ras membrane orientation and nanodomain localization generate isoform diversity. *Proc Natl Acad Sci* **107:** 1130–1135.

Abraham SJ, Nolet RP, Calvert RJ, Anderson LM, Gaponenko V. 2009. The hypervariable region of K-Ras4B is responsible for its specific interactions with calmodulin. *Biochemistry* **48:** 7575–7583.

Ahearn IM, Haigis K, Bar-Sagi D, Philips MR. 2012. Regulating the regulator: Post-translational modification of RAS. *Nat Rev Mol Cell Biol* **13:** 39–51.

Ahmadian MR, Stege P, Scheffzek K, Wittinghofer A. 1997. Confirmation of the arginine-finger hypothesis for the GAP-stimulated GTP-hydrolysis reaction of Ras. *Nature* **4:** 686–689.

Araki M, Shima F, Yoshikawa Y, Muraoka S, Ijiri Y, Nagahara Y, Shirono T, Kataoka T, Tamura A. 2011. Solution structure of the state 1 conformer of GTP-bound H-Ras protein and distinct dynamic properties between the state 1 and state 2 conformers. *J Biol Chem* **286:** 39644–39653.

Asati V, Mahapatra DK, Bharti SK. 2017. K-Ras and its inhibitors towards personalized cancer treatment: Pharmacological and structural perspectives. *Eur J Med Chem* **125:** 299–314.

Boriack-Sjodin PA, Margarit SM, Bar-Sagi D, Kuriyan J. 1998. The structural basis of the activation of Ras by Sos. *Nature* **394:** 337–343.

Bos JL, Rehmann H, Wittinghofer A. 2007. GEFs and GAPs: Critical elements in the control of small G proteins. *Cell* **129:** 865–877.

Bourne HR, Sanders DA, McCormick F. 1991. The GTPase superfamily: Conserved structure and molecular mechanism. *Nature* **349:** 117–127.

Buhrman G, Wink G, Mattos C. 2007. Transformation efficiency of RasQ61 mutants linked to structural features of the switch regions in the presence of Raf. *Structure* **15:** 1618–1629.

Buhrman G, Holzapfel G, Fetics S, Mattos C. 2010. Allosteric modulation of Ras positions Q61 for a direct role in catalysis. *Proc Natl Acad Sci* **107:** 4931–4936.

Buhrman G, O'Connor C, Zerbe B, Kearney BM, Napoleon R, Kovrigina EA, Vajda S, Kozakov D, Kovrigin EL, Mattos C. 2011a. Analysis of binding site hot spots on the surface of Ras GTPase. *J Mol Biol* **413:** 773–789.

Buhrman G, Kumar VS, Cirit M, Haugh JM, Mattos C. 2011b. Allosteric modulation of Ras-GTP is linked to signal transduction through RAF kinase. *J Biol Chem* **286:** 3323–3331.

Cammarata MB, Schardon CL, Mehaffey MR, Rosenberg J, Singleton J, Fast W, Brodbelt JS. 2016. Impact of G12 mutations on the structure of K-Ras probed by ultraviolet photodissociation mass spectrometry. *J Am Chem Soc* doi: 10.1021/jacs.6b04474.

Chen M, Peters A, Huang T, Nan X. 2016. Ras dimer formation as a new signaling mechanism and potential cancer therapeutic target. *Mini Rev Med Chem* **16:** 391–403.

Dharmaiah S, Bindu L, Tran TH, Gillette WK, Frank PH, Ghirlando R, Nissley DV, Esposito D, McCormick F, Stephen AG, et al. 2016. Structural basis of recognition of farnesylated and methylated KRas4b by PDEd. *Proc Natl Acad Sci* **13:** E6766–E6775.

Ehrhardt A, Ehrhardt GR, Guo X, Schrader JW. 2002. Ras and relatives—Job sharing and networking keep an old family together. *Exp Hematol* **30:** 1089–1106.

Fehrenbacher N, Bar-Sagi D, Philips M. 2009. Ras/MAPK signaling from endomembranes. *Mol Oncol* **3:** 297–307.

Fetics SK, Guterres H, Kearney BM, Buhrman G, Ma B, Nussinov R, Mattos C. 2015. Allosteric effects of the oncogenic RasQ61L mutant on Raf-RBD. *Structure* **23:** 505–516.

Franken SM, Scheidig AJ, Krengel U, Rensland H, Lautwein A, Geyer M, Scheffzek K, Goody RS, Kalbitzer HR, Pai EF, et al. 1993. Three-dimensional structures and properties of a transforming and a nontransforming glycine-12 mutant of *p21*$^{H\text{-}Ras}$. *Biochemistry* **32:** 8411–8420.

Gorfe AA, Grant BJ, McCammon JA. 2008. Mapping the nucleotide and isoform-dependent structural and dynamical features of Ras proteins. *Structure* **16:** 885–896.

Grant BJ, Gorfe AA, McCammon JA. 2009. Ras conformational switching: Simulating nucleotide-dependent conformational transitions with accelerated molecular dynamics. *PLoS Comput Biol* **5:** e1000325.

Hancock JF, Parton RG. 2005. Ras plasma membrane signalling platforms. *Biochem J* **389:** 1–11.

Hancock JF, Peterson H, Marshall CJ. 1990. A polybasic domain or palmitoylation is required in addition to the CAAX motif to localize p21Ras to the plasma membrane. *Cell* **63:** 133–139.

Harrison RA, Lu J, Carrasco M, Hunter J, Manandhar A, Gondi S, Westover KD, Engen JR. 2016. Structural dynamics in Ras and related proteins upon nucleotide switching. *J Mol Biol* **428:** 4723–4735.

Henis YI, Hancock JF, Prior IA. 2009. Ras acylation, compartmentalization and signaling nanoclusters. *Mol Membr Biol* **26:** 80–92.

Hobbs GA, Der CJ, Rossman KL. 2016. RAS isoforms and mutations in cancer at a glance. *J Cell Sci* **129:** 287–292.

Holderfield M, Morrison DK. 2017. RAS signaling: Divide and conquer. *Nat Chem Biol* **13:** 7–8.

Jang H, Muratcioglu S, Gursoy A, Keskin O, Nussinov R. 2016. Membrane-associated Ras dimers are isoform-specific: K-Ras dimers differ from H-Ras dimers. *Biochem J* **473:** 1719–1732.

John J, Sohmen R, Feuerstein J, Linke R, Wittinghofer A, Goody RS. 1990. Kinetics of interaction of nucleotides with nucleotide-free H-ras p21. *Biochemistry* **29:** 6058–6065.

Johnson CW, Reid D, Parker JA, Salter S, Knihtila R, Kuzmic P, Mattos C. 2017. The small GTPases K-Ras, N-Ras and H-Ras have distinct biochemical properties determined by allosteric effects. *J Biol Chem* **292:** 12981–12993.

Kapoor A, Travesset A. 2015. Differential dynamics of RAS isoforms in GDP- and GTP-bound states. *Proteins* **83:** 1091–1106.

Kauke MJ, Traxlmayr MW, Parker JA, Kiefer JK, Knihtila R, McGee J, Verdine G, Mattos C, Wittrup KD. 2017. An engineered protein antagonist of K-Ras/B-Raf interaction. *Sci Rep* **7:** 5831–5939.

Kearney BM, Johnson CW, Roberts DM, Swartz P, Mattos C. 2014. DRoP: A water analysis program identifies Ras-GTP-specific pathway of communication between membrane-interacting regions and the active site. *J Mol Biol* **426:** 611–629.

Keeton AB, Salter EA, Piazza GA. 2017. The RAS-effector interaction as a drug target. *Cancer Res* **77:** 221–226.

Knihtila R, Holzapfel G, Weiss K, Meilleur F, Mattos C. 2015. Neutron crystal structure of RAS GTPase puts in question the protonation state of the GTP γ-phosphate. *J Biol Chem* **290:** 31025–31036.

Leshchiner ES, Parkhitko A, Bird GH, Luccarelli J, Bellairs JA, Escudero S, Opoku-Nsiah K, Godes M, Perrimon N, Walensky LD. 2015. Direct inhibition of oncogenic KRAS by hydrocarbon-stapled SOS1 helices. *Proc Natl Acad Sci* **112:** 1761–1766.

Lim SM, Westover KD, Ficarro SB, Harrison RA, Choi HG, Pacold ME, Carrasco M, Hunter J, Kim ND, Xie T, et al. 2014. Therapeutic targeting of oncogenic K-Ras by a covalent catalytic site inhibitor. *Angew Chem Int Ed Engl* **53:** 199–204.

Lu S, Jang H, Nussinov R, Zhang J. 2016a. The structural basis of oncogenic mutations G12, G13 and Q61 in small GTPase K-Ras4B. *Sci Rep* **6:** 21949.

Lu S, Jang H, Gu S, Zhang J, Nussinov R. 2016b. Drugging Ras GTPase: A comprehensive mechanistic and signaling structural view. *Chem Soc Rev* **45:** 4929–4952.

Mao C, Huang YF, Yang ZY, Zheng DY, Chen JZ, Tang JL. 2013. KRAS p.G13D mutation and codon 12 mutations are not created equal in predicting clinical outcomes of cetuximab in metastatic colorectal cancer: A systematic review and meta-analysis. *Cancer* **119:** 714–721.

Marcus K, Mattos C. 2015. Direct attack on RAS: Intramolecular communication and mutation-specific effects. *Clin Cancer Res* **21:** 1810–1818.

Maurer T, Garrenton LS, Oh A, Pitts K, Anderson DJ, Skelton NJ, Fauber BP, Pan B, Malek S, Stokoe D, et al. 2012. Small-molecule ligands bind to a distinct pocket in Ras and inhibit SOS-mediated nucelotide exchange activity. *Proc Natl Acad Sci* **109:** 5299–5304.

McCubrey JA, Steelman LS, Chappell WH, Abrams SL, Wong EW, Chang F, Lehmann B, Terrian DM, Milella M, Tafuri A, et al. 2007. Roles of the Raf/MEK/ERK pathway in cell growth, malignant transformation and drug resistance. *Biochim Biophys Acta* **1773:** 1263–1284.

Miller MS, Miller LD. 2011. RAS mutations and oncogenesis: Not all RAS mutations are created equally. *Front Genet* **2:** 100.

Ostrem JM, Peters U, Sos ML, Wells JA, Shokat KM. 2013. K-Ras^{G12C} inhibitors allosterically control GTP affinity and effector interactions. *Nature* **503:** 548–551.

Pan JY, Wessling-Resnick M. 1998. GEF-mediated GDP/GTP exchange by monomeric GTPases: A regulatory role for Mg^{2+}? *BioEssays* **20:** 516–521.

Parker JA, Mattos C. 2015. The Ras-membrane interface: Isoform-specific differences in the catalytic domain. *Mol Cancer Res* **13:** 595–603.

Prakash P, Sayyed-Ahmad A, Cho KJ, Dolino DM, Chen W, Li H, Grant BJ, Hancock JF, Gorfe AA. 2017. Computational and biochemical characterization of two partially overlapping interfaces and multiple weak-affinity K-Ras dimers. *Sci Rep* **7:** 40109.

Prior IA, Lewis PD, Mattos C. 2012. A comprehensive survey of Ras mutations in cancer. *Cancer Res* **72:** 2457–2467.

Pylayeva-Gupta Y, Grabocka E, Bar-Sagi D. 2011. RAS oncogenes: Weaving a tumorigenic web. *Nat Rev Cancer* **11:** 761–774.

Raimondi F, Portella G, Orozco M, Fanelli F. 2011. Nucleotide binding switches the information flow in Ras GTPases. *PLoS Comput Biol* **7:** e1001098.

Rosnizeck IC, Graf T, Spoerner M, Tränkle J, Filchtinski D, Herrmann C, Gremer L, Vetter IR, Wittinghofer A, König B, et al. 2010. Stabilizing a weak binding state for effectors in the human Ras protein by cyclen complexes. *Angew Chem Int Ed Engl* **49:** 3830–3833.

Rudack T, Xia F, Schlitter J, Kötting C, Gerwert K. 2012. The role of magnesium for geometry and charge in GTP hydrolysis, revealed by quantum mechanics/molecular mechanics simulations. *Biophys J* **103:** 293–302.

Santos E. 2014. Dimerization opens new avenues into Ras signaling research. *Sci Signal* **7:** e12.

Scheffzek K, Ahmadian MR, Kabsch W, Wiesmüller L, Lautwein A, Schmitz F, Wittinghofer A. 1997. The Ras-RasGAP complex: Structural basis for GTPase activation and its loss in oncogenic Ras mutants. *Science* **277:** 333–338.

Shima F, Ijiri Y, Muraoka S, Liao J, Ye M, Araki M, Matsumoto K, Yamamoto N, Sugimoto T, Yoshikawa Y, et al. 2010. Structural basis for conformational dynamics of GTP-bound Ras protein. *J Biol Chem* **285:** 22696–22705.

Smith MJ, Ikura M. 2014. Integrated RAS signaling defined by parallel NMR detection of effectors and regulators. *Nature Chem Biol* **10:** 223–230.

Smith G, Bounds R, Wolf H, Steele RJ, Carey FA, Wolf CR. 2010. Activating K-Ras mutations outwith "hotspot" codons in sporadic colorectal tumours—Implications for personalised cancer medicine. *Br J Cancer* **102:** 693–703.

Spencer-Smith R, Koide A, Zhou Y, Eguchi RR, Sha F, Gajwani P, Santana D, Gupta A, Jacobs M, Herrero-Garcia E, et al. 2016. Inhibition of Ras function through targeting an allosteric regulatory site. *Nature Chem Biol* **13:** 62–68.

Spoerner M, Herrmann C, Vetter IR, Kalbitzer HR, Wittinghofer A. 2001. Dynamic properties of the Ras switch I region and its importance for binding to effectors. *Proc Natl Acad Sci* **98:** 4944–4949.

Spoerner M, Hozsa C, Poetzl A, Reiss K, Ganser P, Geyer M, Kalbitzer HR. 2010. Conformational states of human rat sarcoma (Ras) protein complexed with its natural ligand GTP and their role for effector interaction and GTP hydrolysis. *J Biol Chem* **285:** 39768–39778.

Sun Q, Burke JP, Phan J, Burns MC, Olejniczak ET, Waterson AG, Lee T, Rossanese OW, Fesik SW. 2012. Discovery of small molecules that bind to K-Ras and inhibit Sos-mediated activation. *Agnew Chem Int Ed* **51:** 6140–6143.

Tanaka T, Williams RL, Rabbitts TH. 2007. Tumor prevention by a single antibody domain targeting the interaction of signal transduction proteins with RAS. *EMBO J* **26:** 3250–3259.

Thompson H. 2013. US National Cancer Institute's new Ras project targets an old foe. *Nat Med* **19:** 949–950.

Tian T, Harding A, Inder K, Plowman S, Parton RG, Hancock JF. 2007. Plasma membrane nanoswitches generate high-fidelity Ras signal transduction. *Nat Cell Biol* **9:** 905–914.

Tsai FD, Lopes MS, Zhou M, Court H, Ponce O, Fiordalisi JJ, Gierut JJ, Cox AD, Haigis KM, Philips MR. 2015. K-Ras4A splice variant is widely expressed in cancer and uses a hybrid membrane-targeting motif. *Proc Natl Acad Sci* **112:** 779–784.

Upadhyaya P, Qian Z, Habir NA, Pei D. 2014. Direct Ras inhibitors identified from a structurally rigidified bicyclic peptide library. *Tetrahedron* **70:** 7714–7720.

Valencia A, Chardin P, Wittinghofer A, Sander C. 1991. The Ras protein family: Evolutionary tree and the role of conserved amino acids. *Biochemistry* **30:** 4637–4648.

Vetter IR. 2014. The structure of the G domain of the Ras superfamily. In *Ras superfamily small G proteins: Biology and mechanisms 1: General* (ed. Wittinghofer A), pp. 25–50. Springer, Vienna.

Whyte DB, Kirschmeier P, Hockenberry TN, Nunez-Oliva I, James L, Catino JJ, Bishop WR, Pai JK. 1997. K- and N-Ras are geranylgeranylated in cells treated with farnesyl protein transferase inhibitors. *J Biol Chem* **272:** 14459–14464.

Zhou Y, Liang H, Rodkey T, Ariotti N, Parton RG, Hancock JF. 2014. Signal integration by lipid-mediated spatial cross talk between Ras nanoclusters. *Mol Cell Biol* **34:** 862–876.

 Cite this article as *Cold Spring Harb Perspect Med* doi: 10.1101/cshperspect.a031427

KRAS: The Critical Driver and Therapeutic Target for Pancreatic Cancer

Andrew M. Waters and Channing J. Der

University of North Carolina at Chapel Hill, Lineberger Comprehensive Cancer Center, Chapel Hill, North Carolina 27599

Correspondence: cjder@med.unc.edu

RAS genes (*HRAS*, *KRAS*, and *NRAS*) comprise the most frequently mutated oncogene family in human cancer. With the highest *RAS* mutation frequencies seen with the top three causes of cancer deaths in the United States (lung, colorectal, and pancreatic cancer), the development of anti-RAS therapies is a major priority for cancer research. Despite more than three decades of intense effort, no effective RAS inhibitors have yet to reach the cancer patient. With bitter lessons learned from past failures and with new ideas and strategies, there is renewed hope that undruggable RAS may finally be conquered. With the KRAS isoform mutated in 84% of all *RAS*-mutant cancers, we focus on KRAS. With a near 100% *KRAS* mutation frequency, pancreatic ductal adenocarcinoma (PDAC) is considered the most RAS-addicted of all cancers. We review the role of KRAS as a driver and therapeutic target in PDAC.

Mutationally activated *RAS* genes (*HRAS*, *KRAS*, and *NRAS*) comprise the most frequently mutated gene family in cancer (27%; Catalogue of Somatic Mutations in Cancer [COSMIC] v80). KRAS is the predominant isoform mutated in cancer and the isoform mutated exclusively in pancreatic ductal adenocarcinoma (PDAC). With *KRAS* mutations found in nearly all PDAC, this cancer type is arguably the most RAS-addicted cancer. With substantial experimental evidence that mutant KRAS is essential for PDAC growth, the National Cancer Institute identified targeting KRAS as one of four major priorities for pancreatic cancer research. The current standards of care for PDAC consist of conventional cytotoxic drugs (Wolfgang et al. 2013). Although effective targeted therapies are now available for lung and colorectal cancer, no effective targeted therapies have been found for PDAC. With deaths caused by pancreatic cancer on the rise, the need for new therapies is now dire.

The KRAS small GTPase functions as a simple binary ON–OFF molecular switch, cycling between an active guanosine triphosphate (GTP)-bound and inactive guanosine diphosphate (GDP)-bound state (Vigil et al. 2010a). In normal quiescent cells, RAS is predominantly GDP-bound and inactive. Upon extracellular stimuli activation of receptor tyrosine kinases (RTKs) and other cell-surface receptors, there is rapid and transient formation of RAS-GTP, leading to engagement of effector proteins that then regulate a diversity of intracellular signaling networks (Cox and Der 2010) and thereby tightly control mitogenic processes. Cancer-associated *RAS* genes harbor missense mutations

Cite this article as *Cold Spring Harb Perspect Med* doi: 10.1101/cshperspect.a031435

that encode single amino acid substitutions primarily (98%) at one of three mutational hot spots: glycine-12 (G12), glycine-12 (G13), or glutamine-61 (Q61). These mutations render RAS persistently GTP-bound and constitutively active independent of extracellular stimuli, resulting in overstimulation of effector signaling pathways to drive cancer growth. Thus, by analogy to the successful development of clinically effective adenosine triphosphate (ATP)-competitive inhibitors for protein kinases, small molecule GTP antagonists should provide a straightforward strategy to target mutant RAS. However, with picomolar affinity for GTP, and with millimolar GTP cellular concentrations, this approach has not been feasible. Furthermore, when the structure of RAS was determined, it did not reveal a surface topology amenable to the design of high-affinity small-molecule antagonists, deterring efforts to develop direct RAS inhibitors. Consequently, much of the past and current efforts have centered on indirect strategies. However, recent success in the identification of direct RAS-binding small molecules has fueled excitement that perhaps RAS is druggable after all. In this review, we first provide an overview of the role of KRAS in PDAC. We provide a snapshot of past and ongoing efforts and direct and indirect strategies to develop the long elusive anti-RAS drug for cancer treatment. We then focus on the development of inhibitors of KRAS effector signaling.

KRAS MUTATION AND PANCREATIC CANCER TUMORIGENESIS

The three *RAS* genes encode four 188–189 amino acid proteins that share 82%–90% amino acid sequence identity and share near-identical structural and biochemical properties (Fig. 1A) However, they are differentially expressed and mutated with different frequencies in cancer (Prior et al. 2012; Cox et al. 2014). *KRAS* is the predominant mutated *RAS* gene in cancers (84% of all RAS missense mutations), followed by NRAS (12%), with HRAS rarely mutated (4%) (COSMIC v80) (Fig. 1B).

Significant cancer type preferences exist among the *RAS* genes (Cox et al. 2014). *KRAS* mutations predominate in lung, colorectal, and pancreatic cancer, whereas *NRAS* mutations predominate in cutaneous melanomas and acute myelogenous leukemia, and *HRAS* mutations are found in bladder and head and neck squamous cell carcinomas. Although already known in 1988 (Almoguera et al. 1988), subsequent comprehensive exome-wide deep sequencing verified that KRAS is mutationally activated in 94% of PDAC (Fig. 2A) (Jones et al. 2008; Biankin et al. 2012; Sausen et al. 2015; Waddell et al. 2015; Witkiewicz et al. 2015). These studies also verified an already well-established portrait of PDAC in which there are four major genetic alterations associated with the initiation and progression of PDAC, with the majority of gene alterations found in >10% of PDAC. In addition to *KRAS*, the *TP53* (64%), *SMAD4* (21%), and *CDKN2A* (17%) tumor suppressor genes are significantly mutated (Fig. 2A). Missense mutations and intragenic or homozygous deletion mutations and promoter methylation of *CDKN2A* (encoding p16INK4A and p19ARF), together with promoter silencing, result in a near universal loss of *CDKN2A* function in PDAC (Schutte et al. 1997).

KRAS mutation is the initiating genetic event for PDAC. The progression of normal pancreatic tissue to PDAC involves a stepwise genetic transition projected to span 12 years (Iacobuzio-Donahue et al. 2012). Most commonly, pancreatic duct epithelium transitions to advancing stages of noninvasive microscopic ductal lesions, or pancreatic intraepithelial neoplasms (PanINs). In early-stage PanIN formation, flat pancreatic epithelial cells take on a cuboidal appearance, increase mucin production, and acquire atypical cytological and morphological features (Hruban et al. 2000, 2004; Distler et al. 2014). High-grade PanINs are usually characterized by a papillary morphology (Distler et al. 2014). Greater than 90% frequency of *KRAS* mutations are found in PanIN-1 lesions (Kanda et al. 2012).

Genetically engineered mouse models also support the initiating role of *KRAS* mutation in PDAC and additionally support the cooperating role of subsequent tumor suppressor loss in facilitating KRAS-driven progression to the malignant disease (Gopinathan et al. 2015; Lee

 Cite this article as *Cold Spring Harb Perspect Med* doi: 10.1101/cshperspect.a031435

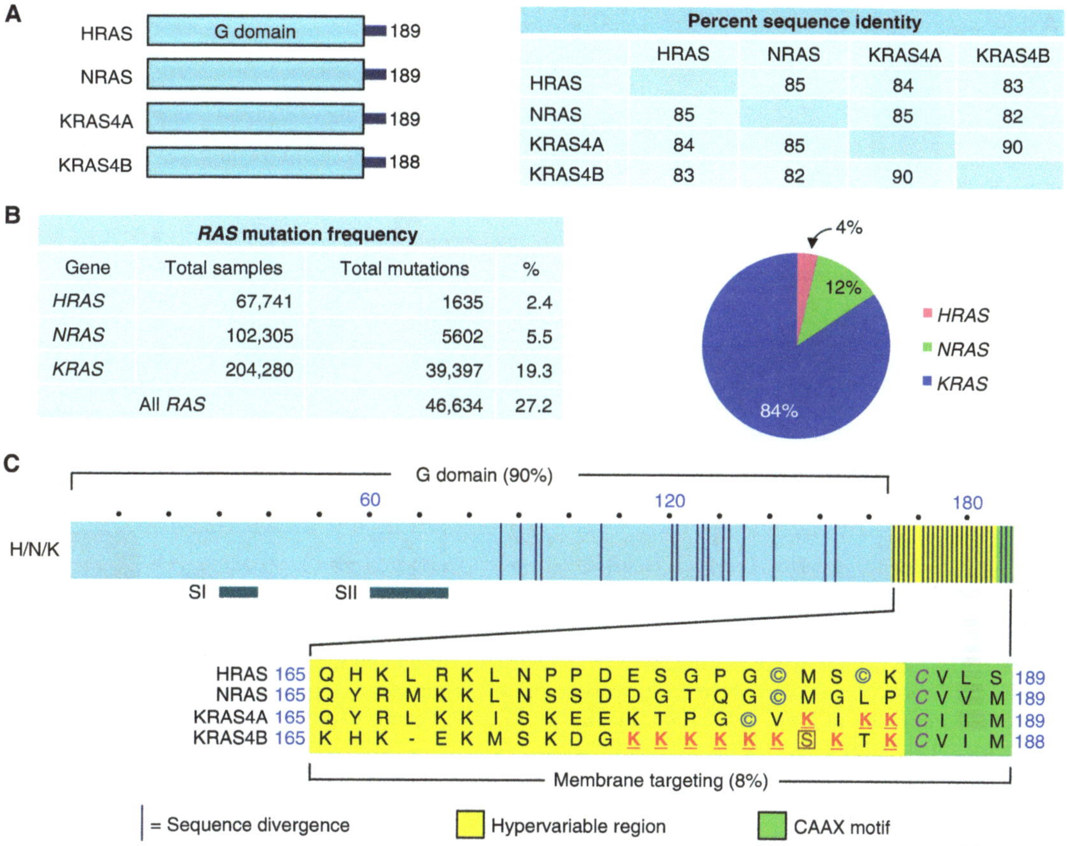

Figure 1. RAS proteins. (*A*) Sequence identity of human RAS proteins. Amino acid sequence identity was determined by CLUSTALW multiple sequence alignment. (*B*) RAS mutation frequencies. Data were compiled from Catalogue of Somatic Mutations in Cancer (COSMIC) v80. (*C*) RAS domains. The amino-terminal amino acids (1–164) comprise the G domain involved in guanosine triphosphate (GTP) binding and hydrolysis and interaction with guanine nucleotide exchange factors (GEFs), GTPase-activating proteins (GAPs), and effectors. RAS protein structure changes in the Switch I (SI; amino acids 30–38) and II (SII; amino acids 60–76) regions during guanosine diphosphate (GDP)–GTP cycling, with the GTP-bound form having higher affinity for effectors. Circled cysteine amino acids indicate sites of covalent modification by addition of a palmitate fatty acid. The boxed serine residue is phosphorylated by protein kinase C. The underlined lysine residues promote membrane targeting. The italicized cysteines in the CAAX motif are sites of covalent modification by addition of a farnesyl isoprenoid and carboxylmethylation.

et al. 2016). *Kras*^{G12D} mutation alone resulted in PanIN formation and infrequent and protracted onset of PDAC (Aguirre et al. 2003; Hingorani et al. 2003). When coupled with mutational inactivation of *Tp53* (*Tp53*^{R175H}) (Hingorani et al. 2005), *Cdkn2a* (Aguirre et al. 2003; Bardeesy et al. 2006), or *Smad4* (Kojima et al. 2007), PanIN formation was accelerated, progressing to rapid and high-frequency occurrence of metastatic PDAC.

It is currently not clear why PDAC is associated exclusively with *KRAS* mutations. One explanation for the isoform specificity of RAS-driven cancers could be that certain tissues are exposed to different carcinogens and environmental insults that lead to mutations in specific *RAS* genes. An alternative and intriguing possibility is that each isoform has specific functional properties that lead to these tissue preferences. One proposed basis for why KRAS is the pre-

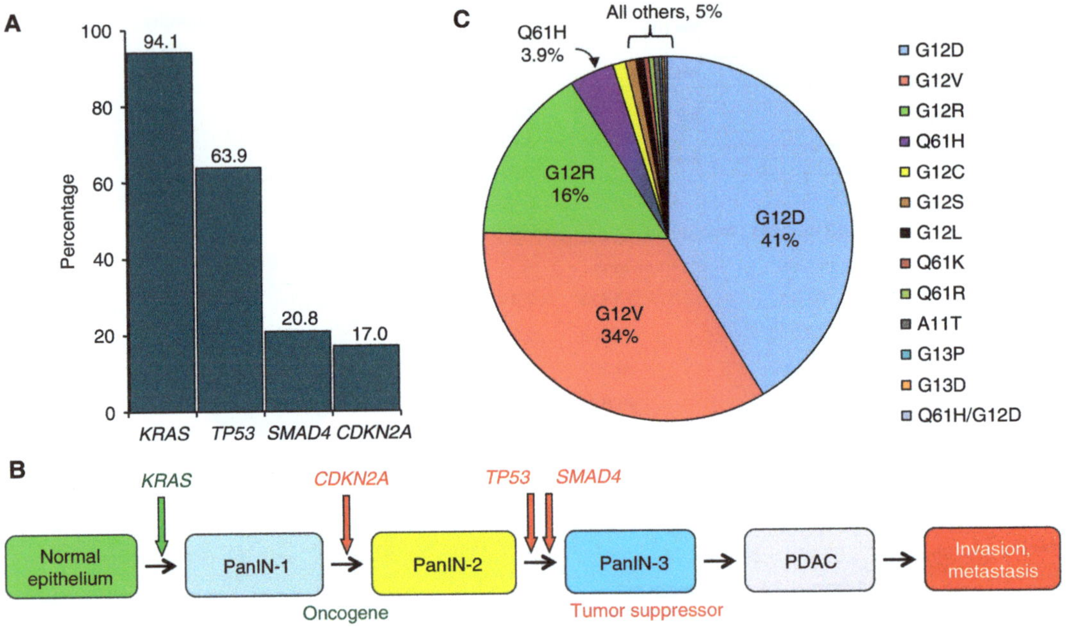

Figure 2. *KRAS* mutations in pancreatic cancer. (*A*) Frequency of mutations in the four major genes in pancreatic cancer. (*B*) Genetic progression of pancreatic cancer. (*C*) KRAS amino acid substitutions in pancreatic cancer. (Compiled from data in Jones et al. 2008, Biankin et al. 2012, Sausen et al. 2015, Waddell et al. 2015, and Witkiewicz et al. 2015.)

ferred cancer isoform is protein expression levels. Although nearly identical on the amino acid level, *KRAS*, *NRAS*, and *HRAS* differ in the frequency of rare codon utilization, with the high frequency in *KRAS* contributing to a reduced rate of translation and lower RAS protein expression (Lampson et al. 2013). Elevated ectopic expression of mutant RAS in normal cells induces senescence rather than cell proliferation. Concurrent suppression of p53 function prevents senescence, allowing RAS-driven growth enhancement. With KRAS mutations, protein expression levels are hypothesized to be sufficiently high for tumor initiation but low enough to evade apoptosis and senescence. In contrast, mutant HRAS protein expression levels are high enough to elicit an apoptotic or senescent response. In support of this hypothesis, in vivo studies that show enriching the *KRAS* codons with more frequently translated *HRAS* codons, even though the amino acids are the same, resulted in reduced *Kras*-induced tumor formation in mice (Pershing et al. 2015).

Another unresolved issue is the role of the remaining *KRAS* wild-type (WT) allele in *KRAS*-mutant PDAC. There is evidence that suggests that the WT allele serves a paradoxical role as a tumor suppressor, with loss of the WT allele facilitating PDAC progression (Qiu et al. 2011). When evaluated in a $Kras^{G12D}$-driven mouse model of PDAC, loss of the WT allele was associated with progression to metastatic disease. Loss of the *KRAS* WT allele is also seen in human PDAC. In contrast, a driver role for the remaining WT isoforms, HRAS and NRAS, has also been observed in *KRAS*-mutant pancreatic cell lines (Lim et al. 2008; Grabocka et al. 2014). However, whether their roles are simply quantitative or qualitatively distinct remains to be determined.

In addition to tissue-based isoform specificity, the frequency of different *KRAS* missense mutations can vary strikingly in different cancer types. For example, G12C mutations are rare in PDAC (1% of all *KRAS* mutations) but are the major *KRAS* mutation in lung adenocarcinoma

(LAC) (43% of all *KRAS* mutations) (Fig. 2C) (Cox et al. 2014; Stephen et al. 2014; Hobbs et al. 2016). In contrast, G12R mutations are infrequently seen overall in cancer (<3%), yet comprise 16% of all *KRAS* mutations in PDAC. It is likely these different frequencies in *KRAS* mutations reflect, in part, distinct roles in cancer development and growth. Until recently, it was assumed that the different mutations cause essentially identical consequences on RAS function; clearly this is not the case, as it was recently shown that there are altered nucleotide exchange kinetics and effector binding abilities of different mutant KRAS proteins (Smith et al. 2013; Hunter et al. 2015). Thus, a shift away from the long-standing "all RAS mutations are created equal" paradigm has emerged, in which the study of specific RAS mutants is being pursued in search of mutation-selective functional vulnerabilities that can be exploited by different therapeutic strategies. The impact of this shift is evidenced by the mutation-specific inhibitors that have been developed to target G12C-mutant KRAS in LAC (Ostrem et al. 2013; Lim et al. 2014). There is limited evidence that mutant specificity also plays an important role in pancreatic cancer prognosis. One study suggests that *KRAS* mutations are associated with more aggressive cancers and a shorter survival time (Bournet et al. 2016). Another found that *KRAS* Q61 mutations were associated with improved survival compared with *KRAS* G12 mutant patients (Witkiewicz et al. 2015).

PANCREATIC DUCTAL ADENOCARCINOMA CHEMOTHERAPY

In 2016, pancreatic cancer surpassed breast cancer and became the third leading cause of cancer deaths in the United States (Siegel et al. 2016). With a continued increase in incidence, pancreatic cancer is projected to become the second leading cause of cancer death by around 2020 (Rahib et al. 2014). With no biomarkers for early detection, the deadly nature of PDAC is largely the result of late onset of symptoms when the cancer has already reached the metastatic state. The 5-year survival rate of pancreatic cancer is 8% (Siegel et al. 2016), giving this cancer the dubious distinction of being the deadliest cancer in the United States.

Surgery, followed by adjuvant chemotherapy, is the only potentially curative treatment for PDAC, but only 15%–20% of patients are candidates for surgery. Chemotherapy for PDAC involves conventional cytotoxic drug combinations. The only targeted therapy approved for the treatment of PDAC, the epidermal growth factor receptor (EGFR) inhibitor erlotinib, when used in combination with gemcitabine, extended life by a statistically significant but clinically unimpressive 12 days more than gemcitabine alone (Moore et al. 2007). Since 1997, gemcitabine was the standard of care for PDAC. In 2011, the FOLFIRINOX (folinic acid, flurouracil, irinotecan, and oxaliplatin) chemotherapy regimen, an extremely toxic chemotherapy cocktail, was approved for PDAC. In 2013, the nab-paclitaxel (albumin-conjugated paclitaxel)-gemcitabine combination was found to increase life expectancy almost 2 months compared to gemcitabine and approved as a second standard of care for PDAC (Von Hoff et al. 2013). Therapeutic approaches as a whole have been largely unsuccessful in PDAC, with no treatment extending life longer than 1 year from diagnosis.

Although *KRAS* mutations are the initiating genetic step, continued mutant KRAS function is still required to maintain the growth of metastatic PDAC. RNA suppression studies in PDAC cell lines showed the strong requirement for tumorigenic growth (Brummelkamp et al. 2002; Lim and Counter 2005). *Kras* inactivation in $Kras^{G12D}$-driven PDAC, in the continued presence of a *Tp53* deficiency, showed rapid regression of primary (Collins et al. 2012a; Ying et al. 2012) and metastatic (Collins et al. 2012b) tumor growth. These studies support the significance of mutant KRAS as a therapeutic target in PDAC. Finally, although one study suggested that only a subset of *KRAS*-mutant PDAC cell lines were addicted to mutant KRAS (Singh et al. 2009), a subsequent study determined that all *KRAS*-mutant PDAC cell lines showed significant growth impairment on acute or sustained KRAS suppression (Hayes et al. 2016). The different conclusions from these two studies reflect the specific definition of "addiction" applied in each study.

Thus, whereas there will likely exist intertumor variability in the specific mutant KRAS-driven cellular phenotypes, all *KRAS*-mutant PDAC will likely display KRAS addiction.

Although clearly a validated therapeutic target, KRAS-targeted therapies will also be subject to the limitations seen with targeted and conventional chemotherapy: the development of cancer-cell resistance. One mechanism by which PDAC loses its addiction to KRAS is through activation and expression of the Hippo pathway component YAP1 (Kapoor et al. 2014; Shao et al. 2014).

TARGETING KRAS: DO EFFECTOR INHIBITORS HOLD THE GREATEST PROMISE?

The main past and current strategies for developing therapeutics to block mutant KRAS function have focused on indirect approaches: to target proteins that support KRAS function and promote KRAS-driven cancer growth (Papke and Der 2017). These approaches have centered on targeting (1) proteins that promote KRAS association with the plasma membrane, (2) KRAS effector signaling, (3) components that support KRAS-dependent metabolic processes, and (4) synthetic lethal interactors critical for the growth of mutant but not WT KRAS cancer cells. More recently, RAS-binding small molecules that disrupt KRAS function have been identified, raising hope that RAS can be targeted directly. An ambitious goal of pancreatic cancer research is to double the rate of survival by the year 2020. With this urgency at hand, of these approaches, arguably inhibitors of KRAS effector signaling hold the greatest promise for the most immediate transition to the PDAC patient within the next several years. Therefore, for this review, we have focused on the development of inhibitors of KRAS effector signaling.

KRAS-GTP binds preferentially to downstream effectors that regulate a complex diversity of cytoplasmic signaling networks. The majority of effectors are characterized by the presence of a RAS-binding domain (RBD) or RAS association (RA) domain (Fig. 3). There are at least 11 different effector families with distinct catalytic functions. It is likely that KRAS drives tumorigenesis by the integrated result of multiple effector signaling pathways, with multiple KRAS effector pathways contributing to the tumor etiology. Thus, the concept of targeting KRAS effector signaling is not straightforward. Which effector pathway(s) are the best to target? Will concurrent inhibition of multiple effectors be required?

There is substantial experimental evidence from cell culture and mouse model studies that validate driver roles for components of four KRAS effector signaling networks in cancer (Cox et al. 2014). However, some cautious interpretation is important when assessing the therapeutic value of effectors ablated by pharmacologic inhibitors. First, a majority of studies use genetic approaches to ablate effector expression and loss of protein expression, which may not accurately model target inhibition by pharmacologic approaches. Second, genetically engineered mouse model studies typically involve ablation of effector expression concurrently with induced expression of mutant RAS. Hence, they assess the role of effectors in tumor initiation and progression, and not maintenance of an already established metastatic PDAC. With these caveats in mind, we summarize the experimental evidence validating the therapeutic potential of targeting different effector signaling components, with a focus on the RAF effector signaling network.

The RACGEF-RAC1 Small GTPase Effector Signaling Network

The least validated and therapeutically tractable KRAS effector pathway implicated in oncogenesis is the RACGEF-RAC1 pathway (Fig. 3). Early studies in HRAS-transformed rodent fibroblasts provided the first indication of the role of RAC1 in RAS oncogenic functions (Bar-Sagi and Feramisco 1986; Khosravi-Far et al. 1995; Qiu et al. 1995). The subsequent identification of *RAC1* missense mutations (P29S), primarily in cutaneous melanomas, provided strong evidence for a driver role for RAC1 activation in cancer (Hodis et al. 2012; Krauthammer et al. 2012). Further, *Rac1*-deficient mice exhibited

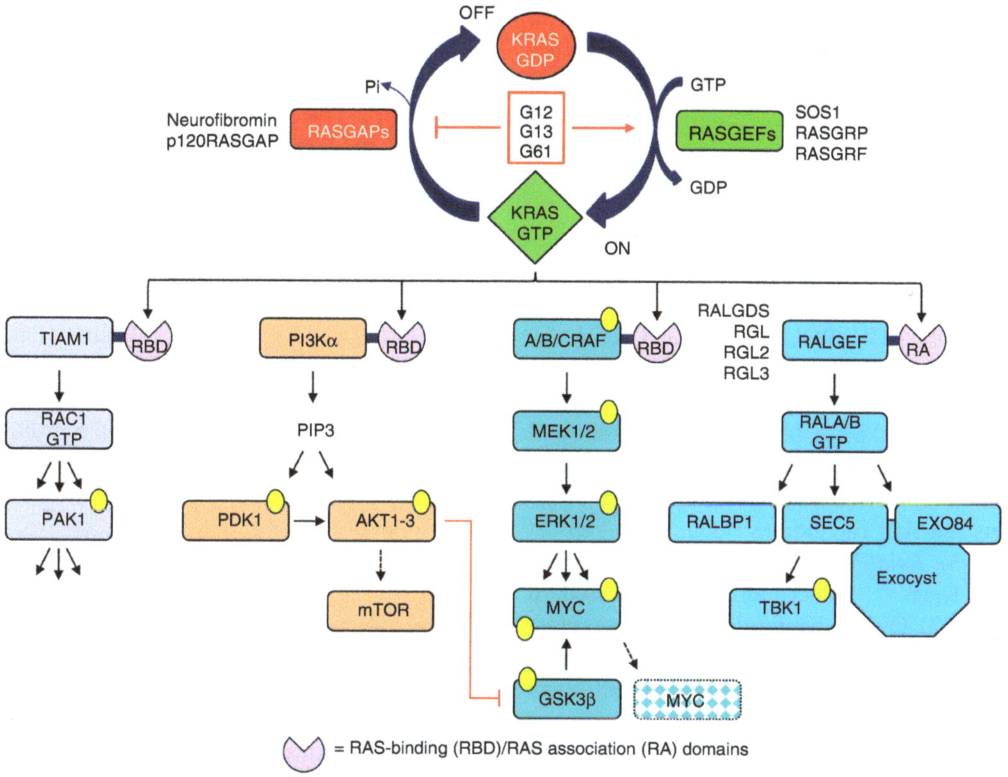

Figure 3. KRAS guanosine diphosphate (GDP)–guanosine triphosphate (GTP) regulation and effector signaling. RAS-selective guanine nucleotide exchange factors (GEFs) and GTP-activating proteins (GAPs) regulate GDP–GTP cycling. Cancer associated mutations, found primarily (>99%) at residues G12, G13, or Q61 disrupt GDP–GTP cycling by impairing intrinsic and GAP-stimulated GTP hydrolysis and/or accelerating GDP–GTP exchange. KRAS effectors are characterized by a RAS-binding domain (RBD) or RAS association (RA) domains. AKT1 phosphorylates and inactivates GSK3β, preventing MYC phosphorylation at T58 and FBW7-mediated polyubiquitination and degradation.

reduced Kras^{G12D}-driven pancreatic tumor formation and increased survival (Heid et al. 2011). However, constitutively active Rac1^{G12V} alone was not sufficient to induce PanIN and PDAC formation (Eser et al. 2013).

One RACGEF, TIAM1, has been identified as a RAS effector (Lambert et al. 2002). Like the majority of KRAS effectors, TIAM1 contains an RBD that facilitates association with RAS-GTP. The importance of TIAM1 in RAS-driven tumorigenesis was shown by the impaired onset of Hras-driven skin carcinoma formation in *Tiam1*-deficient mice (Malliri et al. 2002). Phosphoinositide 3-kinase (PI3K) activation can also activate RACGEFs (e.g., PREX1/2). Although tool compound inhibitors of RACGEF activation of RAC1 have been described (e.g., NSC 23766), it has been difficult to advance clinically effective and selective inhibitors for cancer therapy.

Like KRAS, activated RAC1-GTP can interact with a spectrum of catalytically distinct effectors. Although the effectors critical to facilitate RAC1 as a cancer driver remain unclear, there is experimental support for the PAK1 serine/threonine kinase as an important effector in cancer (Baker et al. 2014). A *Pak1* genetic deficiency or treatment with a pharmacologic inhibitor of Pak1 impaired Kras^{G12D}-driven lung tumor formation (Chow et al. 2012). Another effector that will likely contribute to RAC1-dependent cancer growth is the p100β catalytic subunit of PI3K (Fritsch et al. 2013). Inhibitors of PAK1 and PI3Kβ are under preclinical and clinical development.

The RALGEF-RAL Small GTPase Effector Signaling Network

The next-best-validated RAS effector family is a family of four RA domain-containing RALGEFs (RALGDS, RGL, RGL2, and RGL3) (Fig. 3). RALGEFs act as guanine nucleotide exchange factors and activate RALA and RALB RAS-like small GTPases. Mice deficient in *Ralgds* showed impaired incidence of chemical carcinogen-induced *Hras* mutation and formation of squamous cell skin carcinomas (Gonzalez-Garcia et al. 2005). *Rala*/*Ralb*-deficient mice showed impaired $Kras^{G12D}$-induced lung tumors (Peschard et al. 2012). Our studies found RGL2 overexpression in PDAC cell lines and patient tumors (Vigil et al. 2010b). Biochemical inhibition of RALGEF function as well as genetic suppression of RGL2 expression reduced PDAC cell line steady-state RAL activity, growth in soft agar, and Matrigel invasion. Interestingly, RGL2 exhibited both RAL-dependent and -independent roles in PDAC growth.

Like RAS, RAL proteins are activated by exchanging GDP in their active site for GTP. Although components of these effector pathways are not found mutated in cancer, sustained hyperactivation of this pathway and elevated levels of RAL-GTP were found more frequently in PDAC than the two more classical KRAS effector pathways, ERK mitogen-activated protein kinase (MAPK) and PI3K (Lim et al. 2006).

Interestingly, despite their strong sequence and biochemical identity, RALA and RALB have distinct roles in PDAC growth. RALA is necessary for PDAC anchorage-independent growth in vitro and tumorigenic growth (Lim et al. 2005). In contrast, RALB is dispensable during tumor initiation but important for PDAC invasion and metastases (Lim et al. 2006).

Active RAL-GTP can interact with a diversity of catalytically distinct effectors, several with demonstrated roles in supporting RAL-dependent cancer growth (Gentry et al. 2014). Of these, the best-characterized effectors are the EXO84 and SEC5 subunits of the octomeric exocyst complex that regulates vesicular trafficking. SEC5 can also act as a RALB effector independent of exocyst function, regulating the activity of the TBK1 serine/threonine kinase. Therefore, it was very intriguing when TBK1 was identified as a synthetic lethal interactor of mutant KRAS (Barbie et al. 2009). However, subsequent studies found that TBK1 function was not tightly linked with *RAS*-mutant cancer (Muvaffak et al. 2014). Another effector implicated in RALB-dependent PDAC invasion is RALBP1/RLIP76 (Neel et al. 2012), a multidomain protein that can act as a GTP-activating protein (GAP) for the Cdc42 and RAC1 RHO family small GTPases.

The PI3K-AKT-Mechanistic Target of Rapamycin (mTOR) Effector Signaling Network

There is substantial experimental evidence supporting the critical role of the catalytic subunits of class I PI3K (p110α/δ/γ) as critical effectors of mutant KRAS-driven oncogenesis (Fig. 3) (Castellano and Downward 2011). The p110 subunits contain amino-terminal RBDs and RAS-GTP binding promotes PI3K association with the cell membrane and additionally relieves autoinhibition. Mice harboring a *Pik3ca* locus that encoded a p110α variant with RBD mutations that impaired RAS-GTP binding showed strikingly impaired incidence of mutant $Kras^{G12D}$-driven lung tumor or mutant Hras-induced skin tumor initiation and maintenance (Gupta et al. 2007; Castellano et al. 2013).

The important driver role for PI3K is supported by frequent activating *PIK3CA* mutations in cancer (10% all cancers; COSMIC v80). However, two studies reached different conclusions regarding the sufficiency of PI3K activation in driving PDAC development. One study found that the cancer-associated constitutively activated $Pik3ca^{H1047R}$ mutant alone could not drive PanIN or PDAC development (Collisson et al. 2012). In contrast, a second study found that this mutant alone phenocopied $Kras^{G12D}$ and induced metastatic PDAC (Eser et al. 2013). A possible basis for these different findings may be the different pancreas cell types targeted.

PI3K phosphorylates phosphotidylinositol-4,5-bisphosphate (PIP2) and stimulates formation of plasma membrane–associated phosphotidylinositol-3,4,5-bisphosphate (PIP3) (Castellano and Downward 2011). PIP3 can then activate a multitude of proteins that include RACGEFs, PDK1, and the AKT1/2/3 serine–threonine kinases. Activated AKT can phosphorylate many other proteins that can then promote cell growth. The PTEN lipid phosphatase and tumor suppressor negatively regulates the pathway by dephosphorylating PIP3 to PIP2 and is an important tumor suppressor in PDAC (Ying et al. 2016) and *PTEN* loss is found commonly in cancer. However, *PIK3CA* and *PTEN* mutations are rare in PDAC. Further, whole-exome deep-sequencing data of pancreatic cancer patients shows that 93% of the rare *PIK3CA* mutations co-occur with a *KRAS* mutation, suggesting activated KRAS alone is not sufficient to effectively activate PI3K (Jones et al. 2008; Biankin et al. 2012; Sausen et al. 2015; Waddell et al. 2015; Witkiewicz et al. 2015). Supporting a limited linkage between *KRAS* and PI3K, KRAS suppression did not alter AKT activation levels in a majority of KRAS-mutant PDAC cell lines (Hayes et al. 2016).

The RAF-MEK-ERK Signaling Network

The three RAF serine-threonine kinases (ARAF, BRAF, and CRAF/RAF1) are the most significant effectors of KRAS-driven PDAC (Fig. 4). The importance of the extracellular regulated kinase (ERK) mitogen-activated protein kinase (MAPK) pathway in driving KRAS-dependent cancer growth is supported by the occurrence of *BRAF* mutations in cancer (18% all cancers; COSMIC v80). Although found infrequently in PDAC (<2%), *BRAF*-activating V600E mutations were mutually exclusive with *KRAS* mutations in PDAC (Witkiewicz et al. 2015). The widely studied BxPC-3 PDAC cell line, commonly used as a KRAS WT control in comparison with *KRAS*-mutant cell lines, harbors an in-frame genomic DNA deletion in *BRAF* that encodes a constitutively activated and transforming BRAF protein with a five-amino-acid deletion (Chen et al. 2016; Foster et al. 2016). Database analyses found that similar *BRAF* deletions were found in 4%–5% of *KRAS* WT PDAC and were mutually exclusive of *BRAF* missense mutations. Hence, BxPC-3 cells are not bona fide KRAS WT PDAC cells in that a major KRAS effector pathway is chronically activated in this cell line. Unlike $BRAF^{V600E}$, *BRAF* deletion mutants were insensitive to the BRAF-selective inhibitor vemurafenib but sensitive to the pan-RAF inhibitor LY3009120.

Importantly, the predominant cancer-associated constitutive-activated $Braf^{V600E}$ mutant phenocopied $Kras^{G12D}$ and initiated PanIN formation in the pancreas of *KRAS* WT mice, and, when combined with a *Tp53* mutation, resulted in lethal PDAC (Collisson et al. 2012). Moreover, a genetic deficiency in both *Mek1* and *Mek2* or both *Erk1* and *Erk2* blocked $Kras^{G12D}$-driven mouse lung development (Blasco et al. 2011). Interestingly, genetic ablation of *Raf1/Craf* but not *Braf* also impaired $Kras^{G12}$-driven mouse lung tumor development, suggesting that RAF1 is the critical RAF isoform involved in $Kras^{G12D}$-induced lung tumor formation (Blasco et al. 2011; Karreth et al. 2011). However, surprisingly, a *Raf1* deficiency did not impair $Kras^{G12D}$-induced PDAC formation and progression and did not improve mouse survival, indicating tissue-specific utilization of effectors in KRAS-driven PDAC initiation (Eser et al. 2013). These observations argue that the RAF-MEK-ERK cascade is the key effector in supporting KRAS-dependent tumor initiation, progression, and maintenance.

GTP-bound RAS interacts with the amino-terminal RBD of RAF, increasing accumulation at the plasma membrane and leading to activating phosphorylation events, conformational changes and relief of autoinhibition, and RAF homo-/heterodimerization, promoting RAF kinase activation (Morrison 2012; Freeman et al. 2013). RAF subsequently phosphorylates and activates the MEK1 and MEK2 dual specificity kinases, which then phosphorylate and activate the ERK1 and ERK2 serine/threonine kinases. Activated ERK1/2 then phosphorylate >200 cytoplasmic and nuclear substrates, many of

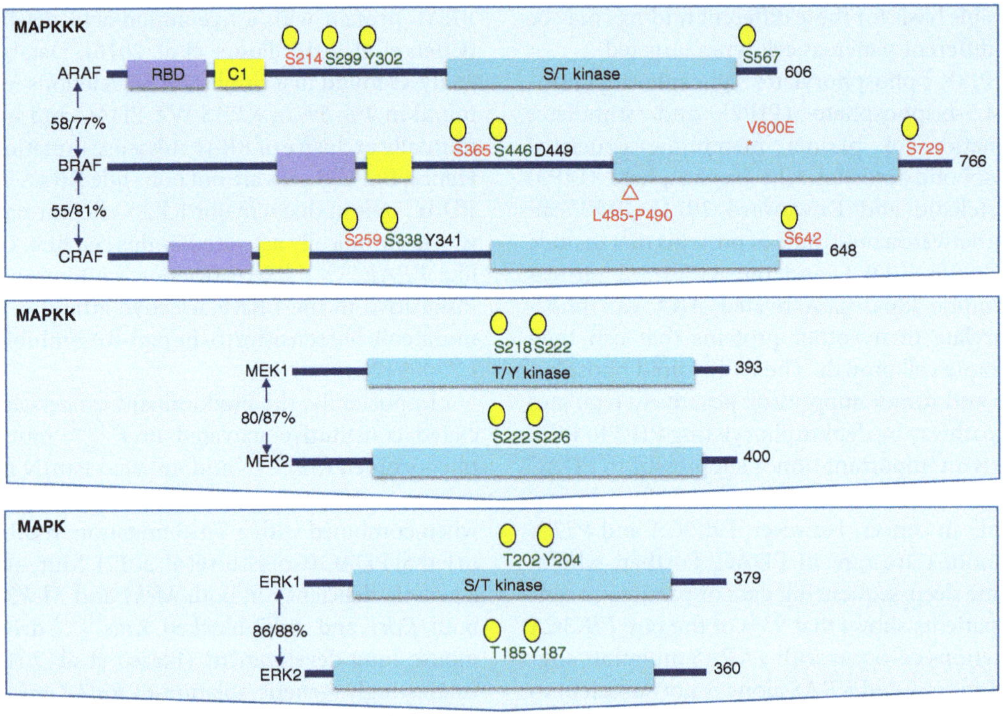

Figure 4. Components of the three-tiered extracellular regulated kinase (ERK) mitogen-activated protein kinase (MAPK) cascade. Each level of the ERK MAPK cascade is comprised of highly related isoforms. *BRAF* missense or deletion mutants (shown is the deletion in the BxPC-1 cell line) are found in pancreatic ductal adenocarcinoma (PDAC). RAF isoforms are regulated by positive (green) and negative (red) regulatory phosphorylation activities. Shown are the phosphorylation sites that activate MEK1/2 or ERK1/2. Domain topology was determined in simple modular architecture research tool (SMART) (smart.embl-heidelberg.de). Total protein and kinase domain sequence identities are indicated (%/%) as determined by CLUSTALW multiple sequence alignment. *, point mutation; Δ, deletion mutation.

which are transcription factors that regulate the expression of genes involved in mitogenic signaling (McKay and Morrison 2007; Shaul and Seger 2007). Because the only well-established substrates of RAF are MEK1/2, and ERK1/2 are the only well-defined substrates of MEK1/2, the RAF-MEK-ERK protein kinase cascade is commonly depicted as a simple linear, unidirectional signaling pathway. However, with inputs and outputs at each level, with feedback regulatory mechanisms, and with a spectrum of scaffolds that regulate signaling input and output, this three-tier protein kinase cascade is the core of a complex signaling network.

With a diversity of substrates, the key substrates important for ERK-dependent cancer growth remain poorly defined. An important growth-promoting ERK substrate is the transcription factor MYC (Farrell and Sears 2014), an oncoprotein required for maintenance of RAS-driven tumors (Soucek et al. 2008). MYC protein levels are elevated in mouse and human PanIN and tumor tissue (Lin et al. 2013). *Myc* overexpression alone has been shown to drive PanIN and metastatic PDAC development. MYC protein is normally unstable with a short half-life (~15 min). ERK stabilizes MYC by phosphorylation at residue S62, preventing FBXW7 E3 ligase-mediated protein degradation driven by GSK3β phosphorylation of MYC at T58 (Fig. 3) (Farrell and Sears 2014), and MYC protein half-life is prolonged in PDAC cell lines (Farrell et al. 2014; Hayes et al. 2016). Sensitivity to pharmacologic inhibitors of ERK was associated with MYC protein loss in growth-inhibited but not resistant PDAC cell lines (Hayes et al.

 Cite this article as *Cold Spring Harb Perspect Med* doi: 10.1101/cshperspect.a031435

2016). ERK MAPK-directed regulation of MYC also drives the expression of genes that support the elevated glucose metabolism in *Kras*-mutant mouse PDAC cells in vivo (Ying et al. 2012).

Excessive ERK activation can drive senescence and growth cessation. To maintain a mitogenic level of ERK activity, the ERK MAPK cascade is also dynamically regulated by negative feedback signaling mechanisms that modulate the level of ERK activity (Ryan et al. 2015). Multiple ERK substrates can negatively regulate upstream signaling components that dampen the signaling strength of the cascade at different levels. For example, ERK phosphorylates CRAF at multiple negative regulatory sites that decrease RAS-driven RAF dimerization (McKay and Morrison 2007).

Substantial cross talk between KRAS effector pathways exists. This is particularly evident with the RAF and PI3K effector signaling networks. Inhibition of ERK signaling can cause compensatory activation of PI3K-AKT-mTORC1 signaling (Collisson et al. 2012). ERK inhibitor treatment stimulated increased AKT activation in sensitive but not resistant PDAC cell lines, and concurrent inhibition of PI3K synergistically enhanced ERK inhibitor growth suppression. Conversely, inhibition of PI3K signaling increased ERK signaling (Soares et al. 2015). The compensatory interrelationship between RAF and PI3K signaling contributes to the synergistic growth suppression seen in Kras-driven cancer models with concurrent pharmacologic inhibition of these two effector pathways. However, largely as a result of toxicity issues, such combinations have not shown the same promising antitumor activities in the clinical evaluation of *KRAS*-mutant cancers.

THERAPEUTICALLY TARGETING THE RAF EFFECTOR PATHWAYS

Because the ERK MAPK pathway is hyperactivated through mutations in *RAS*, *BRAF*, *EGFR*, *NF1*, and other critical cancer drivers, there has been intensive effort by the pharmaceutical industry to develop inhibitors of ERK MAPK signaling. Currently, there are more than 30 inhibitors under clinical evaluation, with two BRAF-selective (vemurafenib and debrafenib) and two MEK1/2-selective inhibitors (trametinib and cobimetinib) approved for use in *BRAF*-mutant melanoma (Fig. 5). In this section, we summarize the current status of inhibitors of each level of the cascade for the treatment of *KRAS*-mutant PDAC.

RAF Inhibitors

The first-generation RAF inhibitors (vemurafenib and debrafenib) are BRAF-selective and have been effective in a subset of *BRAF*-mutant cancers, in particular, melanoma. They have not been effective with *BRAF*-mutant colorectal carcinoma due to up-regulation of EGFR signaling (Corcoran et al. 2012). Although there are cancer-type differences in the sensitivity to BRAF-selective inhibitors, the infrequent *BRAF*^{V600F} mutant PDACs were responsive in vitro (Witkiewicz et al. 2015). However, PDAC with the atypical BRAF deletion mutants were surprisingly not responsive to vemurafenib (Peng et al. 2015).

In *RAS*-mutant cells, the BRAF-selective inhibitors cause paradoxical activation of ERK signaling (Hatzivassiliou et al. 2010; Heidorn et al. 2010; Poulikakos et al. 2010). This occurs through a mechanism involving RAS-GTP-mediated RAF dimerization, with the predominant dimer BRAF-CRAF heterodimers. BRAF but not CRAF is inhibited in the dimer, with the inactivated BRAF causing allosteric activation of the associated CRAF protein, leading to MEK-ERK activation. This reactivation of ERK MAPK signaling is associated with increased proliferation in cell culture and tumor formation in patients (Oberholzer et al. 2012; Su et al. 2012). Because *KRAS* mutation frequencies are found in PanIN-1 lesions (Witkiewicz et al. 2015), an obvious concern is that treatment with BRAF-selective inhibitors may accelerate PDAC development in *BRAF*-mutant cancer patients.

More recently developed second-generation RAF inhibitors are not limited by paradoxical activation. Pan-RAF inhibitors can overcome this paradoxical activation by blocking WT A-, B-, and CRAF in addition to BRAF V600E. Although dimer formation is induced with pan-RAF inhibitors, MEK activation is abrogated

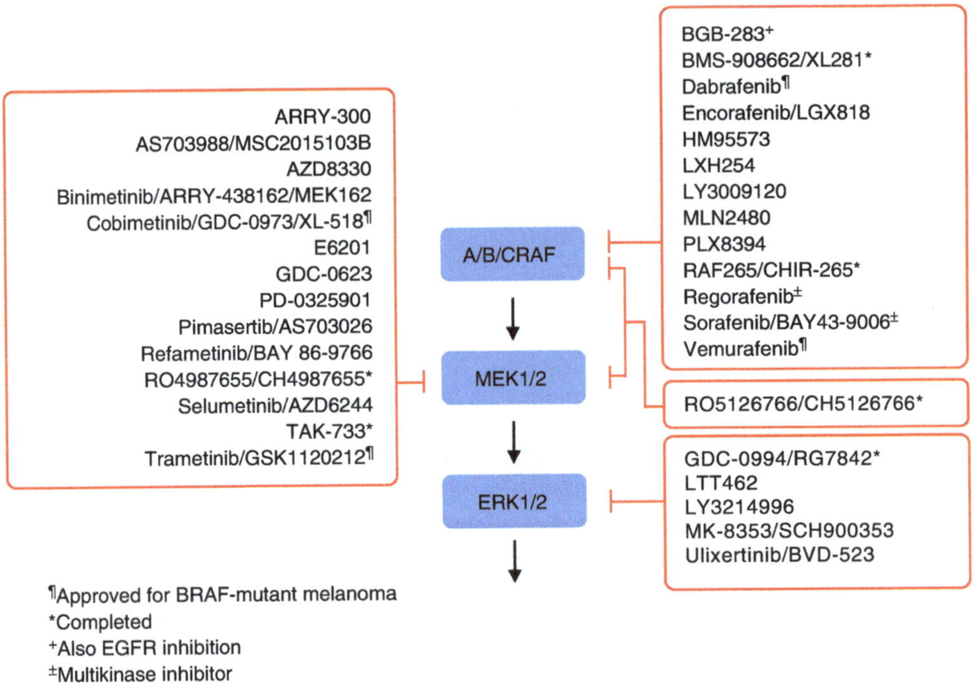

Figure 5. Clinical candidate inhibitors of the RAF-MEK-extracellular regulated kinase (ERK) mitogen-activated protein kinase (MAPK) cascade. (Compiled from data in www.clinicaltrials.gov.)

because each dimer molecule is inhibitor-bound and lacks kinase activity (Henry et al. 2015; Peng et al. 2015). The pan-RAF inhibitor LY3009120 was found to be active in *RAS*-mutant cancer cell lines, including *KRAS*-mutant PDAC. Several pan-RAF inhibitors are currently in phase I clinical trials. Another second-generation RAF inhibitor is the "paradox breaker" PLX8394 that also effectively inhibits CRAF and BRAF (Basile et al. 2014; Choi et al. 2014; Zhang et al. 2015). PLX8394 effectively blocked ERK activation and growth of *RAS*-mutant, vemurafenib-resistant melanoma cells.

MEK Inhibitors

MEK1/2-selective inhibitors have not shown potent activity in *RAS*-mutant cancers. This reduced potency is the result, in part, of vertical compensation mechanisms that lead to reactivation of ERK through up-regulation of upstream pathway components such as RTKs, BRAF, or KRAS (Little et al. 2011; Ryan et al. 2015). An additional basis of resistance involves MEK inhibitor treatment-induced activation of the PI3K-AKT-mTOR pathway. Further, reactivation of phospho-ERK on MEK inhibition with trametinib has been reported owing to compensatory mechanisms that involve kinome reprogramming of parallel pathways (Ryan et al. 2015), but these resistance mechanisms are still poorly understood (Samatar and Poulikakos 2014). Cell-culture analyses found that only a small subset of *KRAS*-mutant PDAC cell lines are sensitive to MEK inhibitor treatment (Witkiewicz et al. 2015; Hayes et al. 2016). Trametinib did not show efficacy in clinical evaluation with *KRAS*-mutant PDAC (Infante et al. 2014; Kasuga et al. 2015; Tolcher et al. 2015).

ERK Inhibitors

Because both RAF- and MEK-selective inhibitors are limited, in part, by mechanisms that cause reactivation of ERK downstream from the inhibitor block point, a logical approach to overcome this involve inhibitors that directly target ERK. Analyses with the ERK1/2-selective

 Cite this article as *Cold Spring Harb Perspect Med* doi: 10.1101/cshperspect.a031435

inhibitors SCH772984 or ulixertinib/BVD-523 found that ~50% of PDAC cell lines were sensitive (Morris et al. 2013; Ryan et al. 2015; Hayes et al. 2016). ERK inhibitor treatment also blocked tumorigenic growth in PDAC cell lines and PDX tumor mouse models. BVD-523 (NCT02608 229) and LY3214996 (NCT0285 7270) are currently being evaluated in phase I clinical trials in PDAC in combination with nab-paclitaxel and gemcitabine.

Treatment-induced loss of MYC protein was seen in sensitive but not resistant cell lines, suggesting that ERK-independent mechanisms that prevent MYC degradation may be a mechanism of de novo resistance.

Concurrent Inhibition of the RAF and PI3K Effector Signaling Networks

Pharmacological inhibition of the PI3K pathway by using AKT inhibitors with single-agent strategies has produced negative results in vitro and in vivo, and activated PI3K cannot phenocopy mutant KRAS in PDAC mouse models. However, MEK inhibition leads to increased AKT signaling in vivo and combined inhibition of MEK and AKT resulted in robust synergistic inhibition of PDAC growth in vitro and in vivo (Collisson et al. 2012). Similarly, ERK inhibitor treatment also increased AKT activity in sensitive but not resistant PDAC cell lines, and cotreatment with a PI3K inhibitor synergistically enhanced ERK inhibitor-mediated growth suppression and caused apoptotic cell death (Hayes et al. 2016). Concurrent treatment with a PI3K inhibitor did not, however, overcome de novo resistance. Several completed or ongoing clinical trials have evaluated or are evaluating combinations of inhibitors of specific components of the RAF and PI3K effector pathways (www.clinicaltrials.gov).

CONCLUSIONS AND FUTURE PERSPECTIVES

Despite the failures in developing anti-KRAS therapies, targeting KRAS remains one of the most promising directions in pancreatic cancer research. Although multiple directions continue to be pursued, with inhibitors already under clinical evaluation, targeting the key KRAS effector pathway—the RAF-MEK-ERK protein kinase cascade—arguably remains the most promising for short-term clinical success. However, the essential role of this pathway as a driver of KRAS-dependent PDAC growth has ensured that cancer cells have compensatory mechanisms that overcome the effectiveness of inhibitors when used as monotherapy. Instead, combination approaches will be needed to achieve a prolonged antitumor response that is less vulnerable to mechanisms of acquired resistance.

ACKNOWLEDGMENTS

We apologize to all colleagues whose work could not be cited because of space restrictions. Our research is funded by grants from the National Cancer Institute (CA042978, CA175747, CA199235, CA203657, and CA196510), the Department of Defense, the Pancreatic Cancer Action Network/American Association for Cancer Research, and the Lustgarten Foundation. The authors declare no financial interest.

REFERENCES

Aguirre AJ, Bardeesy N, Sinha M, Lopez L, Tuveson DA, Horner J, Redston MS, DePinho RA. 2003. Activated Kras and Ink4a/Arf deficiency cooperate to produce metastatic pancreatic ductal adenocarcinoma. *Genes Dev* **17:** 3112–3126.

Almoguera C, Shibata D, Forrester K, Martin J, Arnheim N, Perucho M. 1988. Most human carcinomas of the exocrine pancreas contain mutant c-K-*ras* genes. *Cell* **53:** 549–554.

Baker NM, Yee Chow H, Chernoff J, Der CJ. 2014. Molecular pathways: Targeting RAC-p21-activated serine-threonine kinase signaling in RAS-driven cancers. *Clin Cancer Res* **20:** 4740–4746.

Barbie DA, Tamayo P, Boehm JS, Kim SY, Moody SE, Dunn IF, Schinzel AC, Sandy P, Meylan E, Scholl C, et al. 2009. Systematic RNA interference reveals that oncogenic *KRAS*-driven cancers require TBK1. *Nature* **462:** 108–112.

Bardeesy N, Aguirre AJ, Chu GC, Cheng KH, Lopez LV, Hezel AF, Feng B, Brennan C, Weissleder R, Mahmood U, et al. 2006. Both $p16^{Ink4a}$ and the $p19^{Arf}$-p53 pathway constrain progression of pancreatic adenocarcinoma in the mouse. *Proc Natl Acad Sci* **103:** 5947–5952.

Bar-Sagi D, Feramisco JR. 1986. Induction of membrane ruffling and fluid-phase pinocytosis in quiescent fibroblasts by ras proteins. *Science* **233:** 1061–1068.

Basile KJ, Le K, Hartsough EJ, Aplin AE. 2014. Inhibition of mutant BRAF splice variant signaling by next-generation, selective RAF inhibitors. *Pigment Cell Melanoma Res* **27:** 479–484.

Biankin AV, Waddell N, Kassahn KS, Gingras MC, Muthuswamy LB, Johns AL, Miller DK, Wilson PJ, Patch AM Wu J, et al. 2012. Pancreatic cancer genomes reveal aberrations in axon guidance pathway genes. *Nature* **491:** 399–405.

Blasco RB, Francoz S, Santamaria D, Canamero M, Dubus P, Charron J, Baccarini M, Barbacid M. 2011. c-Raf, but not B-Raf, is essential for development of *K-Ras* oncogene-driven non-small cell lung carcinoma. *Cancer Cell* **19:** 652–663.

Bournet B, Muscari F, Buscail C, Assenat E, Barthet M, Hammel P, Selves J, Guimbaud R, Cordelier P, Buscail L. 2016. *KRAS* G12D mutation subtype is a prognostic factor for advanced pancreatic adenocarcinoma. *Clin Transl Gastroenterol* **7:** e157.

Brummelkamp TR, Bernards R, Agami R. 2002. Stable suppression of tumorigenicity by virus-mediated RNA interference. *Cancer Cell* **2:** 243–247.

Castellano E, Downward J. 2011. RAS interaction with PI3K: More than just another effector pathway. *Genes Cancer* **2:** 261–274.

Castellano E, Sheridan C, Thin MZ, Nye E, Spencer-Dene B, Diefenbacher ME, Moore C, Kumar MS, Murillo MM, Gronroos E, et al. 2013. Requirement for interaction of PI3-kinase p110α with RAS in lung tumor maintenance. *Cancer Cell* **24:** 617–630.

Chen SH, Zhang Y, Van Horn RD, Yin T, Buchanan S, Yadav V, Mochalkin I, Wong SS, Yue YG, Huber L, et al. 2016. Oncogenic *BRAF* deletions that function as homodimers and are sensitive to inhibition by RAF dimer inhibitor LY3009120. *Cancer Discov* **6:** 300–315.

Choi J, Landrette SF, Wang T, Evans P, Bacchiocchi A, Bjornson R, Cheng E, Stiegler AL, Gathiaka S, Acevedo O, et al. 2014. Identification of PLX4032-resistance mechanisms and implications for novel RAF inhibitors. *Pigment Cell Melanoma Res* **27:** 253–262.

Chow HY, Jubb AM, Koch JN, Jaffer ZM, Stepanova D, Campbell DA, Duron SG, O'Farrell M, Cai KQ, Klein-Szanto AJ, et al. 2012. p21-Activated kinase 1 is required for efficient tumor formation and progression in a Ras-mediated skin cancer model. *Cancer Res* **72:** 5966–5975.

Collins MA, Bednar F, Zhang Y, Brisset JC, Galban S, Galban CJ, Rakshit S, Flannagan KS, Adsay NV, Pasca di Magliano M. 2012a. Oncogenic Kras is required for both the initiation and maintenance of pancreatic cancer in mice. *J Clin Invest* **122:** 639–653.

Collins MA, Brisset JC, Zhang Y, Bednar F, Pierre J, Heist KA, Galban CJ, Galban S, di Magliano MP. 2012b. Metastatic pancreatic cancer is dependent on oncogenic Kras in mice. *PloS ONE* **7:** e49707.

Collisson EA, Trejo CL, Silva JM, Gu S, Korkola JE, Heiser LM, Charles RP, Rabinovich BA, Hann B, Dankort D, et al. 2012. A central role for RAF→MEK→ERK signaling in the genesis of pancreatic ductal adenocarcinoma. *Cancer Discov* **2:** 685–693.

Corcoran RB, Ebi H, Turke AB, Coffee EM, Nishino M, Cogdill AP, Brown RD, Della Pelle P, Dias-Santagata D, Hung KE, et al. 2012. EGFR-mediated re-activation of MAPK signaling contributes to insensitivity of *BRAF* mutant colorectal cancers to RAF inhibition with vemurafenib. *Cancer Discov* **2:** 227–235.

Cox AD, Der CJ. 2010. Ras history: The saga continues. *Small GTPases* **1:** 2–27.

Cox AD, Fesik SW, Kimmelman AC, Luo J, Der CJ. 2014. Drugging the undruggable RAS: Mission possible? *Nat Rev Drug Discov* **13:** 828–851.

Distler M, Aust D, Weitz J, Pilarsky C, Grutzmann R. 2014. Precursor lesions for sporadic pancreatic cancer: PanIN, IPMN, and MCN. *Biomed Res Int* **2014:** 474905.

Eser S, Reiff N, Messer M, Seidler B, Gottschalk K, Dobler M, Hieber M, Arbeiter A, Klein S, Kong B, et al. 2013. Selective requirement of PI3K/PDK1 signaling for Kras oncogene-driven pancreatic cell plasticity and cancer. *Cancer Cell* **23:** 406–420.

Farrell AS, Sears RC. 2014. MYC degradation. *Cold Spring Harb Perspect Med* **4:** a014365.

Farrell AS, Allen-Petersen B, Daniel CJ, Wang X, Wang Z, Rodriguez S, Impey S, Oddo J, Vitek MP, Lopez C, et al. 2014. Targeting inhibitors of the tumor suppressor PP2A for the treatment of pancreatic cancer. *Mol Cancer Res* **12:** 924–939.

Foster SA, Whalen DM, Ozen A, Wongchenko MJ, Yin J, Yen I, Schaefer G, Mayfield JD, Chmielecki J, Stephens PJ, et al. 2016. Activation mechanism of oncogenic deletion mutations in BRAF, EGFR, and HER2. *Cancer Cell* **29:** 477–493.

Freeman AK, Ritt DA, Morrison DK. 2013. The importance of Raf dimerization in cell signaling. *Small GTPases* **4:** 180–185.

Fritsch R, de Krijger I, Fritsch K, George R, Reason B, Kumar MS, Diefenbacher M, Stamp G, Downward J. 2013. RAS and RHO families of GTPases directly regulate distinct phosphoinositide 3-kinase isoforms. *Cell* **153:** 1050–1063.

Gentry LR, Martin TD, Reiner DJ, Der CJ. 2014. Ral small GTPase signaling and oncogenesis: More than just 15 minutes of fame. *Biochim Biophys Acta* **1843:** 2976–2988.

Gonzalez-Garcia A, Pritchard CA, Paterson HF, Mavria G, Stamp G, Marshall CJ. 2005. RalGDS is required for tumor formation in a model of skin carcinogenesis. *Cancer Cell* **7:** 219–226.

Gopinathan A, Morton JP, Jodrell DI, Sansom OJ. 2015. GEMMs as preclinical models for testing pancreatic cancer therapies. *Dis Model Mech* **8:** 1185–1200.

Grabocka E, Pylayeva-Gupta Y, Jones MJ, Lubkov V, Yemanaberhan E, Taylor L, Jeng HH, Bar-Sagi D. 2014. Wild-type H- and N-Ras promote mutant K-Ras-driven tumorigenesis by modulating the DNA damage response. *Cancer Cell* **25:** 243–256.

Gupta S, Ramjaun AR, Haiko P, Wang Y, Warne PH, Nicke B, Nye E, Stamp G, Alitalo K, Downward J. 2007. Binding of ras to phosphoinositide 3-kinase p110α is required for ras-driven tumorigenesis in mice. *Cell* **129:** 957–968.

Hatzivassiliou G, Song K, Yen I, Brandhuber BJ, Anderson DJ, Alvarado R, Ludlam MJ, Stokoe D, Gloor SL, Vigers G, et al. 2010. RAF inhibitors prime wild-type RAF to activate the MAPK pathway and enhance growth. *Nature* **464:** 431–435.

Hayes TK, Neel NF, Hu C, Gautam P, Chenard M, Long B, Aziz M, Kassner M, Bryant KL, Pierobon M, et al. 2016. Long-term ERK inhibition in KRAS-mutant pancreatic cancer is associated with MYC degradation and senescence-like growth suppression. *Cancer Cell* **29:** 75–89.

Heid I, Lubeseder-Martellato C, Sipos B, Mazur PK, Lesina M, Schmid RM, Siveke JT. 2011. Early requirement of Rac1 in a mouse model of pancreatic cancer. *Gastroenterology* **141:** 719–730, 730.e711–717.

Heidorn SJ, Milagre C, Whittaker S, Nourry A, Niculescu-Duvas I, Dhomen N, Hussain J, Reis-Filho JS, Springer CJ, Pritchard C, et al. 2010. Kinase-dead BRAF and oncogenic RAS cooperate to drive tumor progression through CRAF. *Cell* **140:** 209–221.

Henry JR, Kaufman MD, Peng SB, Ahn YM, Caldwell TM, Vogeti L, Telikepalli H, Lu WP, Hood MM, Rutkoski TJ, et al. 2015. Discovery of 1-(3,3-dimethylbutyl)-3-(2-fluoro-4-methyl-5-(7-methyl-2-(methylamino)pyrido[2,3-d]pyrimidin-6-yl)phenyl)urea (LY3009120) as a pan-RAF inhibitor with minimal paradoxical activation and activity against BRAF or RAS mutant tumor cells. *J Med Chem* **58:** 4165–4179.

Hingorani SR, Petricoin EF, Maitra A, Rajapakse V, King C, Jacobetz MA, Ross S, Conrads TP, Veenstra TD, Hitt BA, et al. 2003. Preinvasive and invasive ductal pancreatic cancer and its early detection in the mouse. *Cancer Cell* **4:** 437–450.

Hingorani SR, Wang L, Multani AS, Combs C, Deramaudt TB, Hruban RH, Rustgi AK, Chang S, Tuveson DA. 2005. $Trp53^{R172H}$ and $Kras^{G12D}$ cooperate to promote chromosomal instability and widely metastatic pancreatic ductal adenocarcinoma in mice. *Cancer Cell* **7:** 469–483.

Hobbs GA, Der CJ, Rossman KL. 2016. RAS isoforms and mutations in cancer at a glance. *J Cell Sci* **129:** 1287–1292.

Hodis E, Watson IR, Kryukov GV, Arold ST, Imielinski M, Theurillat JP, Nickerson E, Auclair D, Li L, Place C, et al. 2012. A landscape of driver mutations in melanoma. *Cell* **150:** 251–263.

Hruban RH, Goggins M, Parsons J, Kern SE. 2000. Progression model for pancreatic cancer. *Clin Cancer Res* **6:** 2969–2972.

Hruban RH, Takaori K, Klimstra DS, Adsay NV, Albores-Saavedra J, Biankin AV, Biankin SA, Compton C, Fukushima N, Furukawa T, et al. 2004. An illustrated consensus on the classification of pancreatic intraepithelial neoplasia and intraductal papillary mucinous neoplasms. *Am J Surg Pathol* **28:** 977–987.

Hunter JC, Manandhar A, Carrasco MA, Gurbani D, Gondi S, Westover KD. 2015. Biochemical and structural analysis of common cancer-associated KRAS mutations. *Mol Cancer Res* **13:** 1325–1335.

Iacobuzio-Donahue CA, Velculescu VE, Wolfgang CL, Hruban RH. 2012. Genetic basis of pancreas cancer development and progression: Insights from whole-exome and whole-genome sequencing. *Clin Cancer Res* **18:** 4257–4265.

Infante JR, Somer BG, Park JO, Li CP, Scheulen ME, Kasubhai SM, Oh DY, Liu Y, Redhu S, Steplewski K, et al. 2014. A randomised, double-blind, placebo-controlled trial of trametinib, an oral MEK inhibitor, in combination with gemcitabine for patients with untreated metastatic adenocarcinoma of the pancreas. *Eur J Cancer* **50:** 2072–2081.

Jones S, Zhang X, Parsons DW, Lin JC, Leary RJ, Angenendt P, Mankoo P, Carter H, Kamiyama H, Jimeno A, et al. 2008. Core signaling pathways in human pancreatic cancers revealed by global genomic analyses. *Science* **321:** 1801–1806.

Kanda M, Matthaei H, Wu J, Hong SM, Yu J, Borges M, Hruban RH, Maitra A, Kinzler K, Vogelstein B, et al. 2012. Presence of somatic mutations in most early-stage pancreatic intraepithelial neoplasia. *Gastroenterology* **142:** 730–733.e739.

Kapoor A, Yao W, Ying H, Hua S, Liewen A, Wang Q, Zhong Y, Wu CJ, Sadanandam A, Hu B, et al. 2014. Yap1 activation enables bypass of oncogenic Kras addiction in pancreatic cancer. *Cell* **158:** 185–197.

Karreth FA, Frese KK, DeNicola GM, Baccarini M, Tuveson DA. 2011. C-Raf is required for the initiation of lung cancer by $K\text{-}Ras^{G12D}$. *Cancer Discov* **1:** 128–136.

Kasuga A, Nakagawa K, Nagashima F, Shimizu T, Naruge D, Nishina S, Kitamura H, Kurata T, Takasu A, Fujisaka Y, et al. 2015. A phase I/Ib study of trametinib (GSK1120212) alone and in combination with gemcitabine in Japanese patients with advanced solid tumors. *Invest New Drugs* **33:** 1058–1067.

Khosravi-Far R, Solski PA, Clark GJ, Kinch MS, Der CJ. 1995. Activation of Rac1, RhoA, and mitogen-activated protein kinases is required for Ras transformation. *Mol Cell Biol* **15:** 6443–6453.

Kojima K, Vickers SM, Adsay NV, Jhala NC, Kim HG, Schoeb TR, Grizzle WE, Klug CA. 2007. Inactivation of Smad4 accelerates $Kras^{G12D}$-mediated pancreatic neoplasia. *Cancer Res* **67:** 8121–8130.

Krauthammer M, Kong Y, Ha BH, Evans P, Bacchiocchi A, McCusker JP, Cheng E, Davis MJ, Goh G, Choi M, et al. 2012. Exome sequencing identifies recurrent somatic RAC1 mutations in melanoma. *Nat Genet* **44:** 1006–1014.

Lambert JM, Lambert QT, Reuther GW, Malliri A, Siderovski DP, Sondek J, Collard JG, Der CJ. 2002. Tiam1 mediates Ras activation of Rac by a PI(3)K-independent mechanism. *Nat Cell Biol* **4:** 621–625.

Lampson BL, Pershing NL, Prinz JA, Lacsina JR, Marzluff WF, Nicchitta CV, MacAlpine DM, Counter CM. 2013. Rare codons regulate KRas oncogenesis. *Curr Biol* **23:** 70–75.

Lee JW, Komar CA, Bengsch F, Graham K, Beatty GL. 2016. Genetically engineered mouse models of pancreatic cancer: The KPC model ($LSL\text{-}Kras^{G12D/+};LSL\text{-}Trp53^{R172H/+}$; *Pdx-1-Cre*), its variants, and their application in immuno-oncology drug discovery. *Curr Protoc Pharmacol* **73:** 14.39.1–14.39.20.

Lim KH, Counter CM. 2005. Reduction in the requirement of oncogenic Ras signaling to activation of PI3K/AKT pathway during tumor maintenance. *Cancer Cell* **8:** 381–392.

Lim KH, Baines AT, Fiordalisi JJ, Shipitsin M, Feig LA, Cox AD, Der CJ, Counter CM. 2005. Activation of RalA is critical for Ras-induced tumorigenesis of human cells. *Cancer Cell* **7:** 533–545.

Lim KH, O'Hayer K, Adam SJ, Kendall SD, Campbell PM, Der CJ, Counter CM. 2006. Divergent roles for RalA and

RalB in malignant growth of human pancreatic carcinoma cells. *Curr Biol* **16:** 2385–2394.

Lim KH, Ancrile BB, Kashatus DF, Counter CM. 2008. Tumour maintenance is mediated by eNOS. *Nature* **452:** 646–649.

Lim SM, Westover KD, Ficarro SB, Harrison RA, Choi HG, Pacold ME, Carrasco M, Hunter J, Kim ND, Xie T, et al. 2014. Therapeutic targeting of oncogenic K-Ras by a covalent catalytic site inhibitor. *Angew Chem Int Ed Engl* **53:** 199–204.

Lin WC, Rajbhandari N, Liu C, Sakamoto K, Zhang Q, Triplett AA, Batra SK, Opavsky R, Felsher DW, DiMaio DJ, et al. 2013. Dormant cancer cells contribute to residual disease in a model of reversible pancreatic cancer. *Cancer Res* **73:** 1821–1830.

Little AS, Balmanno K, Sale MJ, Newman S, Dry JR, Hampson M, Edwards PA, Smith PD, Cook SJ. 2011. A correction to the research article titled: "Amplification of the driving oncogene, *KRAS* or *BRAF*, underpins acquired resistance to MEK1/2 inhibitors in colorectal cancer cells" by A. S. Little, K. Balmanno, M. J. Sale, S. Newman, J. R. Dry, M. Hampson, P. A. W. Edwards, P. D. Smith, S. J. Cook. *Sci Signal* **4:** er2.

Malliri A, van der Kammen RA, Clark K, van der Valk M, Michiels F, Collard JG. 2002. Mice deficient in the Rac activator Tiam1 are resistant to Ras-induced skin tumours. *Nature* **417:** 867–871.

McKay MM, Morrison DK. 2007. Integrating signals from RTKs to ERK/MAPK. *Oncogene* **26:** 3113–3121.

Moore MJ, Goldstein D, Hamm J, Figer A, Hecht JR, Gallinger S, Au HJ, Murawa P, Walde D, Wolff RA, et al. 2007. Erlotinib plus gemcitabine compared with gemcitabine alone in patients with advanced pancreatic cancer: A phase III trial of the National Cancer Institute of Canada Clinical Trials Group. *J Clin Oncol* **25:** 1960–1966.

Morris EJ, Jha S, Restaino CR, Dayananth P, Zhu H, Cooper A, Carr D, Deng Y, Jin W, Black S, et al. 2013. Discovery of a novel ERK inhibitor with activity in models of acquired resistance to BRAF and MEK inhibitors. *Cancer Discov* **3:** 742–750.

Morrison DK. 2012. MAP kinase pathways. *Cold Spring Harb Perspect Biol* **4:** a011254.

Muvaffak A, Pan Q, Yan H, Fernandez R, Lim J, Dolinski B, Nguyen TT, Strack P, Wu S, Chung R, et al. 2014. Evaluating TBK1 as a therapeutic target in cancers with activated IRF3. *Mol Cancer Res* **12:** 1055–1066.

Neel NF, Rossman KL, Martin TD, Hayes TK, Yeh JJ, Der CJ. 2012. The RalB small GTPase mediates formation of invadopodia through a GTPase-activating protein-independent function of the RalBP1/RLIP76 effector. *Mol Cell Biol* **32:** 1374–1386.

Oberholzer PA, Kee D, Dziunycz P, Sucker A, Kamsukom N, Jones R, Roden C, Chalk CJ, Ardlie K, Palescandolo E, et al. 2012. *RAS* mutations are associated with the development of cutaneous squamous cell tumors in patients treated with RAF inhibitors. *J Clin Oncol* **30:** 316–321.

Ostrem JM, Peters U, Sos ML, Wells JA, Shokat KM. 2013. K-Ras^{G12C} inhibitors allosterically control GTP affinity and effector interactions. *Nature* **503:** 548–551.

Papke B, Der CJ. 2017. Drugging RAS: Know the enemy. *Science* **355:** 1158–1163.

Peng SB, Henry JR, Kaufman MD, Lu WP, Smith BD, Vogeti S, Rutkoski TJ, Wise S, Chun L, Zhang Y, et al. 2015. Inhibition of RAF isoforms and active dimers by LY3009120 leads to anti-tumor activities in *RAS* or *BRAF* mutant cancers. *Cancer Cell* **28:** 384–398.

Pershing NL, Lampson BL, Belsky JA, Kaltenbrun E, MacAlpine DM, Counter CM. 2015. Rare codons capacitate Kras-driven de novo tumorigenesis. *J Clin Invest* **125:** 222–233.

Peschard P, McCarthy A, Leblanc-Dominguez V, Yeo M, Guichard S, Stamp G, Marshall CJ. 2012. Genetic deletion of RALA and RALB small GTPases reveals redundant functions in development and tumorigenesis. *Curr Biol* **22:** 2063–2068.

Poulikakos PI, Zhang C, Bollag G, Shokat KM, Rosen N. 2010. RAF inhibitors transactivate RAF dimers and ERK signalling in cells with wild-type BRAF. *Nature* **464:** 427–430.

Prior IA, Lewis PD, Mattos C. 2012. A comprehensive survey of Ras mutations in cancer. *Cancer Res* **72:** 2457–2467.

Qiu RG, Chen J, Kirn D, McCormick F, Symons M. 1995. An essential role for Rac in Ras transformation. *Nature* **374:** 457–459.

Qiu W, Sahin F, Iacobuzio-Donahue CA, Garcia-Carracedo D, Wang WM, Kuo CY, Chen D, Arking DE, Lowy AM, Hruban RH, et al. 2011. Disruption of p16 and activation of Kras in pancreas increase ductal adenocarcinoma formation and metastasis in vivo. *Oncotarget* **2:** 862–873.

Rahib L, Smith BD, Aizenberg R, Rosenzweig AB, Fleshman JM, Matrisian LM. 2014. Projecting cancer incidence and deaths to 2030: The unexpected burden of thyroid, liver, and pancreas cancers in the United States. *Cancer Res* **74:** 2913–2921.

Ryan MB, Der CJ, Wang-Gillam A, Cox AD. 2015. Targeting *RAS*-mutant cancers: Is ERK the key? *Trends Cancer* **1:** 183–198.

Samatar AA, Poulikakos PI. 2014. Targeting RAS-ERK signalling in cancer: Promises and challenges. *Nat Rev Drug Discov* **13:** 928–942.

Sausen M, Phallen J, Adleff V, Jones S, Leary RJ, Barrett MT, Anagnostou V, Parpart-Li S, Murphy D, Kay Li Q, et al. 2015. Clinical implications of genomic alterations in the tumour and circulation of pancreatic cancer patients. *Nat Commun* **6:** 7686.

Schutte M, Hruban RH, Geradts J, Maynard R, Hilgers W, Rabindran SK, Moskaluk CA, Hahn SA, Schwarte-Waldhoff I, Schmiegel W, et al. 1997. Abrogation of the Rb/p16 tumor-suppressive pathway in virtually all pancreatic carcinomas. *Cancer Res* **57:** 3126–3130.

Shao DD, Xue W, Krall EB, Bhutkar A, Piccioni F, Wang X, Schinzel AC, Sood S, Rosenbluh J, Kim JW, et al. 2014. KRAS and YAP1 converge to regulate EMT and tumor survival. *Cell* **158:** 171–184.

Shaul YD, Seger R. 2007. The MEK/ERK cascade: From signaling specificity to diverse functions. *Biochim Biophys Acta* **1773:** 1213–1226.

Siegel RL, Miller KD, Jemal A. 2016. Cancer statistics, 2016. *CA Cancer J Clin* **66:** 7–30.

Singh A, Greninger P, Rhodes D, Koopman L, Violette S, Bardeesy N, Settleman J. 2009. A gene expression signature associated with "K-Ras addiction" reveals regulators of EMT and tumor cell survival. *Cancer Cell* **15:** 489–500.

Smith MJ, Neel BG, Ikura M. 2013. NMR-based functional profiling of RASopathies and oncogenic RAS mutations. *Proc Natl Acad Sci* **110:** 4574–4579.

Soares HP, Ming M, Mellon M, Young SH, Han L, Sinnet-Smith J, Rozengurt E. 2015. Dual PI3K/mTOR inhibitors induce rapid overactivation of the MEK/ERK pathway in human pancreatic cancer cells through suppression of mTORC2. *Mol Cancer Ther* **14:** 1014–1023.

Soucek L, Whitfield J, Martins CP, Finch AJ, Murphy DJ, Sodir NM, Karnezis AN, Swigart LB, Nasi S, Evan GI. 2008. Modelling Myc inhibition as a cancer therapy. *Nature* **455:** 679–683.

Stephen AG, Esposito D, Bagni RK, McCormick F. 2014. Dragging ras back in the ring. *Cancer Cell* **25:** 272–281.

Su F, Viros A, Milagre C, Trunzer K, Bollag G, Spleiss O, Reis-Filho JS, Kong X, Koya RC, Flaherty KT, et al. 2012. *RAS* mutations in cutaneous squamous-cell carcinomas in patients treated with BRAF inhibitors. *N Engl J Med* **366:** 207–215.

Tolcher AW, Bendell JC, Papadopoulos KP, Burris HA III, Patnaik A, Jones SF, Rasco D, Cox DS, Durante M, Bellew KM, et al. 2015. A phase IB trial of the oral MEK inhibitor trametinib (GSK1120212) in combination with everolimus in patients with advanced solid tumors. *Ann Oncol* **26:** 58–64.

Vigil D, Cherfils J, Rossman KL, Der CJ. 2010a. Ras superfamily GEFs and GAPs: Validated and tractable targets for cancer therapy? *Nat Rev Cancer* **10:** 842–857.

Vigil D, Martin TD, Williams F, Yeh JJ, Campbell SL, Der CJ. 2010b. Aberrant overexpression of the Rgl2 Ral small GTPase-specific guanine nucleotide exchange factor promotes pancreatic cancer growth through Ral-dependent and Ral-independent mechanisms. *J Biol Chem* **285:** 34729–34740.

Von Hoff DD, Ervin T, Arena FP, Chiorean EG, Infante J, Moore M, Seay T, Tjulandin SA, Ma WW, Saleh MN, et al. 2013. Increased survival in pancreatic cancer with nab-paclitaxel plus gemcitabine. *N Engl J Med* **369:** 1691–1703.

Waddell N, Pajic M, Patch AM, Chang DK, Kassahn KS, Bailey P, Johns AL, Miller D, Nones K, Quek K, et al. 2015. Whole genomes redefine the mutational landscape of pancreatic cancer. *Nature* **518:** 495–501.

Witkiewicz AK, McMillan EA, Balaji U, Baek G, Lin WC, Mansour J, Mollaee M, Wagner KU, Koduru P, Yopp A, et al. 2015. Whole-exome sequencing of pancreatic cancer defines genetic diversity and therapeutic targets. *Nat Commun* **6:** 6744.

Wolfgang CL, Herman JM, Laheru DA, Klein AP, Erdek MA, Fishman EK, Hruban RH. 2013. Recent progress in pancreatic cancer. *CA Cancer J Clin* **63:** 318–348.

Ying H, Kimmelman AC, Lyssiotis CA, Hua S, Chu GC, Fletcher-Sananikone E, Locasale JW, Son J, Zhang H, Coloff JL, et al. 2012. Oncogenic Kras maintains pancreatic tumors through regulation of anabolic glucose metabolism. *Cell* **149:** 656–670.

Ying H, Dey P, Yao W, Kimmelman AC, Draetta GF, Maitra A, DePinho RA. 2016. Genetics and biology of pancreatic ductal adenocarcinoma. *Genes Dev* **30:** 355–385.

Zhang C, Spevak W, Zhang Y, Burton EA, Ma Y, Habets G, Zhang J, Lin J, Ewing T, Matusow B, et al. 2015. RAF inhibitors that evade paradoxical MAPK pathway activation. *Nature* **526:** 583–586.

Mechanistic and Preclinical Insights from Mouse Models of Hematologic Cancer Characterized by Hyperactive Ras

Anica Wandler[1] and Kevin Shannon[1,2]

[1]Department of Pediatrics, Helen Diller Family Cancer Research Building, University of California, San Francisco, San Francisco, California 94158-9001

[2]Comprehensive Cancer Center, Helen Diller Family Cancer Research Building, University of California, San Francisco, San Francisco, California 94158-9001

Correspondence: shannonk@peds.ucsf.edu

RAS genes are mutated in 5%–40% of a spectrum of myeloid and lymphoid cancers with *NRAS* affected 2–3 times more often than *KRAS*. Genomic analysis indicates that *RAS* mutations generally occur as secondary events in leukemogenesis, but are integral to the disease phenotype. The tractable nature of the hematopoietic system has facilitated generating accurate mouse models of hematologic malignancies characterized by hyperactive Ras signaling. These strains provide robust platforms for addressing how oncogenic *Ras* expression perturbs proliferation, differentiation, and self-renewal programs in stem and progenitor cell populations, for testing potential therapies, and for investigating mechanisms of drug response and resistance. This review summarizes recent insights from key studies in mouse models of hematologic cancer that are broadly relevant for understanding Ras biology and for ongoing efforts to implement rational therapeutic strategies for cancers with oncogenic *RAS* mutations.

Somatic *RAS* gene mutations were first identified in leukemia in the 1980s (reviewed in Bos 1989). Subsequent studies encompassing thousands of hematologic cancers have defined the overall incidence of *NRAS*, *KRAS*, and *HRAS* mutations in different blood cancers, the frequency of individual amino acid substitutions, the likely order in which these mutations are acquired during leukemogenesis, and the status of *RAS* mutations detected at diagnosis, during disease remission, and after relapse (reviewed in Ward et al. 2012). The *NF1* tumor suppressor gene, which encodes a GTPase activating protein (GAP) that is a core component of the Ras/GAP molecular switch (reviewed in Cichowski and Jacks 2001), is also recurrently mutated in a number of hematologic cancers (Ward et al. 2012). We first summarize key findings from studies of human patients that have informed our current understanding of the role of these mutations in blood cancers and refer readers to a recent review for additional background information.

The overall prevalence of *RAS*/*NF1* mutations in hematologic malignancies ranges from ~5% in myelodysplastic syndrome (MDS) to

Cite this article as *Cold Spring Harb Perspect Med* doi: 10.1101/cshperspect.a031526

~20% in acute myeloid leukemia (AML) to ~40% in juvenile myelomonocytic leukemia (JMML). In contrast to most solid cancers, *NRAS* mutations predominate over *KRAS* mutations by a ratio of ~3:1 in hematologic malignancies (Ward et al. 2012). Deep sequencing of human tumors indicates that *RAS* and *NF1* mutations do not initiate most hematologic cancers, but typically comprise secondary events that cooperate with antecedent mutations in genes encoding transcription factors and proteins that regulate epigenetic programs in hematopoietic stem and progenitor cells (HSPCs) (Jan et al. 2012; Shlush et al. 2014; Lindsley et al. 2015). JMML, an aggressive cancer of young children, is a notable exception to this general rule (Jankowska et al. 2011; Sakaguchi et al. 2013; Stieglitz et al. 2015). Interestingly, *RAS* (and *FLT3*) mutations identified at diagnosis are invariably undetectable in blood and bone marrow samples from patients with AML and acute lymphoblastic leukemia (ALL) analyzed during remission (Lindsley et al. 2015). This contrasts with mutations in genes such as *DMNT3A*, *TET2*, and *IDH1/2*, which frequently persist (Chou et al. 2012; Corces-Zimmerman et al. 2014; Klco et al. 2015). Furthermore, *RAS* mutations that are present at diagnosis do not invariably reappear at disease relapse, underscoring the impact of treatment-induced selective pressure on the dominant clone and highlighting the subsequent genetic evolution that occurs in response to it. Finally, and importantly, when relapse follows a period of remission in patients with acute leukemia, molecular analysis frequently reveals outgrowth of a rare clone with intrinsic drug resistance that was already present at diagnosis (reviewed in Jan and Majeti 2013).

The broad subject of mouse cancer models driven by oncogenic *Ras* genes is reviewed elsewhere in the literature (Drosten et al. 2017). Here we focus on mouse models of hematologic malignancies characterized by *RAS* and *NF1* mutations, summarize key insights from these systems, and discuss the advantages and potential liabilities of different experimental approaches. Importantly, as no mechanism-based treatments exist for the ~25% of human cancers with *KRAS*, *NRAS*, or *NF1* mutations, we emphasize the use of these models as platforms for investigating mechanisms of drug response and resistance that may inform efforts to target oncogenic Ras signaling in both hematologic and nonhematologic cancers.

TRANSGENIC AND RETROVIRAL TRANSDUCTION/TRANSPLANTATION MODELS

Oncogenic *KRAS*, *NRAS*, and *HRAS* mutations are distributed nonrandomly across different cancers. The prevalence of *NRAS* and, to a lesser extent, *KRAS* mutations in human myeloid malignancies prompted efforts to develop mouse models by expressing different mutant *Ras* alleles under the control of various promoters. Transgenic animals in which the mouse mammary tumor virus (MMTV) long terminal repeat (LTR) was used to drive oncogenic *Nras* expression developed T- and B-cell lymphoblastic lymphomas and mammary carcinomas (Mangues et al. 1996). Using this same promoter to drive transgenic expression of *Hras* caused B-cell lymphoblastic lymphomas at low frequency. While disappointing at the time because *RAS* gene mutations are more frequent in myeloid malignances and were not thought to contribute to lymphoid cancers, *RAS* and *NF1* mutations were later identified in certain aggressive types of ALL such as early T-cell precursor ALL (ETP-ALL) and hypodiploid B lineage ALL (Zhang et al. 2012; Holmfeldt et al. 2013).

A seminal study showed that retroviral transduction of murine bone marrow with the *BCR-ABL* transgene, a potent upstream activator of Ras and of the Raf/MEK/ERK effector pathway (Sawyers 1999), followed by transplantation into irradiated recipient mice, caused myeloproliferative neoplasms (MPNs) that closely resembled human chronic myeloid leukemia (CML) (Daley et al. 1990). Following this paradigm, a number of groups pursued a similar strategy with *RAS* oncogenes (Fig. 1), which was facilitated by development of the murine stem cell virus (MSCV) vector. In one study, ~60% of irradiated recipient mice injected

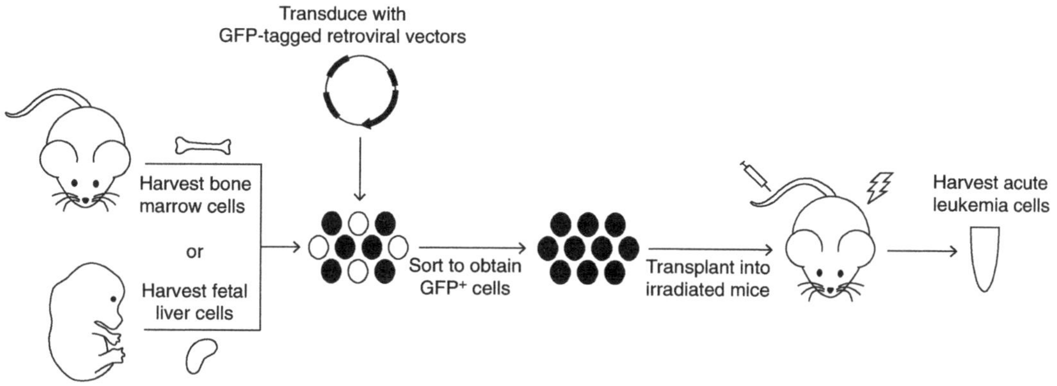

Figure 1. Retroviral transduction/transplantation to generate acute leukemias. Bone marrow or fetal liver cells are isolated from donor mice, transduced with retroviral vectors, sorted to obtain a pure population, and then transplanted into irradiated recipient mice. These animals develop acute leukemias that can be harvested and analyzed. GFP, Green fluorescent protein.

with bone marrow cells that had been transduced with a retrovirus expressing an oncogenic *Nras* allele developed a spectrum of myeloid malignancies (MacKenzie et al. 1999). However, the phenotypic variability, incomplete penetrance, and prolonged latency combined with both impaired in vitro proliferation and a high rate of apoptosis in *Nras*-infected cells provided an early indication that levels of oncogenic *Ras* expression strongly modulate cellular phenotypes. Consistent with this idea, positioning the *Nras* oncogene downstream of an internal ribosomal entry site (IRES) in the MSCV retroviral backbone enhanced induction of MPNs and AML, which was likely caused by a lower level of expression (Parikh et al. 2007). A direct comparison of the tumorigenic capacity of oncogenic *Hras*, *Nras*, or *Kras* was performed by expressing each oncogene downstream of an IRES in the MSCV vector that also contained a green fluorescent protein (GFP) selectable marker (MSCV-IRES-GFP [MIG]). In this study, transduced (GFP-positive) bone marrow HSPCs expressing oncogenic *Nras* caused myeloid malignancies that resembled human chronic myelomonocytic leukemia (CMML) at low viral titers and AML at higher titers. HSPCs expressing oncogenic *Kras* uniformly caused a CMML-like disease in recipient mice, the latency of which correlated with viral titer. Finally, expressing oncogenic *Hras* exclusively produced an invasive AML-like disease and no effect of viral titer was observed (Parikh et al. 2007). The robust ability of oncogenic *Hras* to induce myeloid malignancies despite the very low frequency of *HRAS* mutations in patient samples (Ward et al. 2012), coupled with the observed effects of viral titer, suggests that the superphysiological levels of Ras expressed in these systems strongly influence disease phenotypes. Importantly, the hematologic cancers that emerge from retroviral transduction/transplantation experiments invariably exhibit clonal retroviral integrations (MacKenzie et al. 1999; Parikh et al. 2006, 2007), arguing that a specific level of oncogenic Ras protein expression is strongly selected for, or that misregulation of genes near the integration site or other cooperating somatic mutations are required to induce leukemia in vivo.

Several recent studies have deployed modified retroviral transduction/transplantation or transgenic models to examine Ras membrane trafficking, the roles of different effector pathways in leukemogenesis, the ability of oncogenic *Ras* expression to cooperate with other mutations to generate myeloid malignancies, and to perform preclinical testing of various therapeutic strategies. As described in detail in Ahearn et al. (2018), Ras processing was initially pursued as a therapeutic target in the 1980s and is an area of renewed interest. Retroviral transduction of a palmitoylation-defective form of oncogenic *Nras* followed by transplantation of sorted GFP-positive HSPCs abrogated myeloid disease and biochemical activation of downstream effectors in recipient mice, suggesting a requirement for *Nras* palmitoylation and subsequent plasma membrane localization to induce myeloid malignancies (Cuiffo and Ren 2010). However, a subsequent study showed that expressing $Nras^{G12D}$ from the MSCV promoter has dominant negative effects on myeloid progenitor colony growth and Ras-regulated activation of effector pathways ex vivo (Xu et al. 2012). These data, in turn, raise the possibility that nonphysiologic levels of oncogenic N-Ras^{G12D} expression impaired the growth of transduced HSPCs before and after transplantation. Another recent application of the MIG vector transduction/transplantation system assessed the ability of "second-site" amino acid substitutions that impair the ability of K-Ras^{G12D} to bind Raf or PI3 kinase (PI3K) to initiate leukemia in vivo (Shieh et al. 2013). HSPCs transduced with these mutant $Kras^{G12D}$ oncogenes efficiently generated aggressive T-ALLs in recipient mice, and these leukemias restored downstream effector activation by either acquiring "third-site" mutations within the *Kras* transgenes or by silencing phosphatase and tensin homolog (PTEN) expression (Shieh et al. 2013). Inducible transgenic expression of oncogenic *Nras* under control of the *Vav* promoter caused mast cell leukemia that was reversed by administering doxycycline to repress expression from a tetracycline response element upstream of *Nras*. This inducible allele also cooperated with an *Mll-Af9* fusion gene to produce AMLs that required continuous oncogenic *Nras* expression to promote self-renewal of the leukemic population (Kim et al. 2009; Sachs et al. 2014). Similarly, retroviral transduction was used to coexpress oncogenic *Nras* with *AML1/ETO* or *MLL* fusion genes under control of the MSCV promoter in fetal liver cells, which were then transplanted into irradiated recipient mice to generate AMLs. These leukemias were subsequently used in preclinical studies to uncover differential effects of standard induction chemotherapy regimens resulting from a difference in the ability of each fusion protein to activate a p53 response (Zuber et al. 2009). This transduction/transplantation model was used more recently in combination with a tetracycline-regulated RNA interference (RNAi) system (Zuber et al. 2011a) to identify BRD4 as a therapeutic target in AML (Zuber et al. 2011b), and to uncover potential mechanisms of resistance to small molecule inhibitors of this epigenetic modifier (Rathert et al. 2015).

In summary, retroviral transduction/transplantation and transgenic models have helped elucidate important aspects of Ras biology and have also been used to assess drug response and resistance. As described below, the development of "knockin" mouse strains has provided investigators in the field with more robust and reliable experimental models that overcome the potential confounding effects of nonphysiologic levels of oncogenic Ras expression that also vary across different leukemias because of differences in the genomic location of the integrated transgene.

IN VIVO EFFECTS OF *NF1* INACTIVATION AND OF ENDOGENOUS $KRAS^{G12D}$/$NRAS^{G12D}$ EXPRESSION IN THE HEMATOPOIETIC COMPARTMENT

The World Health Organization classifies JMML and CMML as MPN/MDS "overlap" disorders. These aggressive hematologic cancers share overlapping clinical and biologic features, including excessive proliferation of cells in the monocytic lineage, anemia, splenomegaly, progression to AML in a subset of patients, and resistance to cytotoxic chemotherapy. One important difference between CMML and JMML is that mutations in genes that broadly regulate epigenetic programs appear to initiate CMML, but not JMML (Jankowska et al. 2011; Sakaguchi et al. 2013; Stieglitz et al. 2015). This observation, in turn, suggests that fetal HSPCs are uniquely susceptible to *RAS*-induced transformation in vivo. Indeed, JMML likely represents the clearest example of a human cancer driven by aberrant Ras signaling as ~90% of JMML genomes harbor driver mutations in one of five genes involved in Ras signaling

 Cite this article as *Cold Spring Harb Perspect Med* doi: 10.1101/cshperspect.a031526

(*NF1*, *NRAS*, *KRAS*, *PTPN11*, *CBL*), but very few additional somatic alterations (Sakaguchi et al. 2013; Stieglitz et al. 2015; reviewed in Chang et al. 2014).

Given these human data, it was exciting when investigators showed that expressing Cre recombinase in the blood and bone marrow of mice harboring conditional mutant alleles of *Kras*, *Nras*, or *Nf1* induced myeloid malignancies that recapitulated many aspects of JMML and CMML. A key advance in the field was the development of a conditional *Lox-Stop-Lox (LSL) Kras*^{G12D} "knockin" allele that allowed investigators to express oncogenic *Kras* from its endogenous locus following Cre-mediated excision of the inhibitory *LSL* cassette (Jackson et al. 2001). Although embryonic expression of oncogenic *Kras* with ubiquitous or germline promoters was lethal, instilling an *Adeno-Cre* vector into the lungs of *LSL-Kras*^{G12D} mice generated multifocal adenocarcinomas that correlated with the viral titer (Jackson et al. 2001). To examine the effects of oncogenic *Kras* expression in the hematopoietic compartment, two groups intercrossed *LSL-Kras*^{G12D} and *Mx1-Cre* mice (Braun et al. 2004; Chan et al. 2004). In this system, the interferon-inducible *Mx1* promoter drives Cre recombinase expression following in vivo administration of polyI-polyC (pI-pC), a synthetic double-stranded RNA that induces endogenous interferon expression (Kühn et al. 1995). Mice expressing *Kras*^{G12D} under control of the *Mx1* promoter developed fully penetrant and aggressive MPNs characterized by leukocytosis, splenomegaly, and bone marrow myeloid hyperplasia. Myeloid progenitor cells from these mice form colonies in methylcellulose media in the absence of cytokine growth factors and also display hypersensitivity to granulocyte-macrophage colony-stimulating factor (GM-CSF), a cellular hallmark of JMML (Emanuel et al. 1991). Taken together, these studies established a method for conditional expression of oncogenic *Ras* alleles from their endogenous loci in HSPCs, showed the ability of endogenous *Kras*^{G12D} to perturb hematopoietic growth at the single-cell level, and generated a new and robust model of JMML and CMML. An unexpected finding was that endogenous *Kras*^{G12D} expression in primary hematopoietic cells resulted in a modest increase in the levels of activated Ras-GTP, but low basal ERK and Akt phosphorylation (Braun et al. 2004). This observation provided one of the first hints that hyperactive Ras signaling induces potent negative biochemical feedback in primary cells, which was also observed in a subsequent analysis of defined bone marrow cell populations isolated from *Mx1-Cre; LSL-Kras*$^{G12D/+}$ mice (Van Meter et al. 2007).

To compare the effects of endogenous *Kras*^{G12D} versus *Nras*^{G12D} expression in colonic epithelium, a conditional *LSL-Nras*^{G12D} allele was expressed using the *Fabpl-Cre* promoter. Whereas mice expressing *Kras*^{G12D} developed colonic hyperplasia, *Nras*^{G12D} expression failed to induce proliferation of colon cells but instead conferred resistance to apoptosis (Haigis et al. 2008). Similarly, inducing expression of *Nras*^{G12D} in the hematopoietic compartment of *Mx1-Cre; LSL-Nras*$^{G12D/+}$ mice had a different outcome than expressing *Kras*^{G12D} in the same strain background. Although *Mx1-Cre; LSL-Kras*$^{G12D/+}$ mice invariably died from progressive MPNs by 4 months of age, *Mx1-Cre; LSL-Nras*$^{G12D/+}$ mice exhibited indolent myeloid disease that developed with greatly increased latency. Approximately 80% of these mice ultimately succumbed to a variety of hematologic malignancies by 1 year of age, which included MPNs, an MDS-like disorder, lymphoproliferative disease, and histiocytic sarcoma. Although bone marrow cells from *Mx1-Cre; LSL-Nras*$^{G12D/+}$ mice showed cytokine-independent growth and GM-CSF hypersensitivity in methylcellulose cultures, these phenotypes were much less pronounced than in *Kras* mutant progenitors. Total Ras protein expression and Ras-GTP levels were also significantly higher in myeloid lineage cells expressing *Kras*^{G12D} as compared to those expressing *Nras*^{G12D} (Li et al. 2011). These differences in K-Ras^{G12D} and N-Ras^{G12D} protein expression further suggested that *Kras*/*Nras* oncogene dosage might contribute to hematologic phenotypes. This hypothesis was tested directly by generating *Mx1-Cre* mice expressing one or two *Nras*^{G12D} alleles and showing that the homozygous *Nras*$^{G12D/G12D}$ genotype caused an acute MPN that more closely resembled the aggressive myeloid disease induced by *Kras*^{G12D} expression. In addition, transplanting bone marrow cells isolated from heterozygous *Nras*^{G12D} mutant mice into lethally irradiated recipients resulted in predominantly CMML and rarely T-ALL, whereas transplantation of *Nras*$^{G12D/G12D}$ cells exclusively caused T-ALL (Wang et al. 2010). Analyzing hemizygous *Mx1-Cre; Nras*$^{G12D/-}$ mice confirmed that gene dosage, and not tumor suppressor activity of wild-type (WT) *Nras*, was responsible for the more aggressive biologic behavior of homozygous mutant *Nras*$^{G12D/G12D}$ hematopoietic cells (Xu et al. 2013). Finally, heterozygous *Nras*^{Q61L} expression driven by *Mx1-Cre* caused an intermediate phenotype of MPNs in mice that was more aggressive and penetrant than heterozygous *Nras*^{G12D} expression but less severe than homozygous *Nras*^{G12D} expression (Kong et al. 2016). Because N-Ras^{Q61L} is more activated biochemically than N-Ras^{G12D}, these studies provide additional evidence that the degree and duration of oncogenic Ras output modulates cellular phenotypes. In this context, it is striking that *Nras*^{G12D}, which has less potent effects on the proliferation and differentiation of primary hematopoietic cells as compared to either *Nras*^{Q61L} or *Kras*^{G12D}, is also the most common mutation found in human blood cancers. Together, these data support the existence of substantial positive and negative pressure that selects for the outgrowth of HSPCs with specific levels of aberrant Ras pathway activation in vivo.

The consequences of disrupting *Nf1* were first examined using gene targeting to generate mice expressing a null allele (Brannan et al. 1994; Jacks et al. 1994). Homozygous *Nf1* mutant embryos die in utero from cardiac defects. Interestingly, however, *Nf1*-deficient fetal liver cells displayed GM-CSF hypersensitivity in methylcellulose cultures and caused JMML-like MPNs upon transplantation into lethally irradiated recipients (Bollag et al. 1996; Largaespada et al. 1996). Heterozygous *Nf1* mutant mice are phenotypically normal over the first year of life and subsequently develop pheochromocytomas and JMML-like MPNs with incomplete penetrance (Jacks et al. 1994). These data are consistent with the tumor spectrum observed in persons with neurofibromatosis type 1 (NF1), a familial cancer syndrome caused by germline *NF1* mutations. The tumor predispositions and other phenotypic features of NF1 disease, such as the role of germline and somatic *NF1* mutations in tumorigenesis, and the role of NF1 GAP activity in regulating Ras signaling are discussed in detail in Ratner and Miller (2015). Heterozygous *Nf1* inactivation also cooperated with whole-body and focal radiation and the alkylating agent cyclophosphamide to induce breast cancers, myeloid malignancies, sarcomas, and pheochromocytomas (Chao et al. 2005; Nakamura et al. 2011).

To circumvent the embryonic lethality resulting from homozygous *Nf1* inactivation and to create a mouse model of human NF1 disease, a conditional mutant *Nf1* allele was generated (Zhu et al. 2001). Driving Cre recombinase expression from the Synapsin I promoter showed that tissue-restricted *Nf1* inactivation caused abnormal development of the cerebral cortex and increased astrocyte proliferation (Zhu et al. 2001). *Mx1-Cre; Nf1*$^{flox/flox}$ mice developed fully penetrant MPNs with a median survival of 7.5 months—an intermediate hematologic phenotype between those observed in *Mx1-Cre; LSL-Nras*$^{G12D/+}$ and *Mx1-Cre; LSL-Kras*$^{G12D/+}$ mice. This myeloid disease models aspects of JMML, including splenomegaly, bone marrow infiltration by myeloid cells, and GM-CSF hypersensitivity. Transplanting *Nf1*-deficient bone marrow caused MPNs in lethally but not sublethally irradiated recipient mice (Le et al. 2004). Mouse models exhibiting conditional *Nf1* inactivation in the hematopoietic compartment have facilitated comparing the consequences of direct versus indirect activation of Ras signaling, and have also provided an additional model of MPNs for performing preclinical studies.

BIOLOGIC AND PRECLINICAL STUDIES IN *KRAS*, *NRAS*, AND *NF1* MUTANT MICE

In addition to generating models that recapitulate the characteristics of hematologic malignancies driven by hyperactive Ras, conditional expression of *Kras*^{G12D} or *Nras*^{G12D} or ablation of *Nf1* has elucidated fundamental aspects of Ras biology and tumorigenesis. The relative fitness of *Nf1* mutant and WT HSPCs was assessed by performing repopulation experiments (Fig. 2) in which defined numbers of sorted *Nf1*-deficient bone marrow cells and WT "competitors" were injected into irradiated recipient mice (Zhang et al. 2001). In these studies, *Nf1* mutant HSPCs outcompeted WT cells, but failed to induce MPNs unless a high ratio of mutant-to-WT cells was injected (Zhang et al. 2001). Similarly, *Nras*^{G12D} expression caused increased proliferation of highly purified hematopoietic stem cells (HSCs) and enhanced their competitive fitness but did not induce MPNs in recipients when coinjected with competitor cells. These data have important implications for modeling leukemogenesis, as they indicate that WT cells can suppress the tumorigenic potential of primary cells with oncogenic alterations that activate Ras signaling. In addition, endogenous *Nras*^{G12D} expression unexpectedly increased the frequency of cell divisions in a subpopulation of HSCs, but reduced this parameter in others. Enhanced self-renewal potential and competitive fitness was restricted to the infrequently dividing HSC population and was shown to be STAT5-dependent (Li et al. 2013). Oncogenic *Kras* expression increased HSC proliferation and conferred a competitive advantage over WT cells under repopulating conditions (Sabnis et al. 2009). In contrast to *Nf1*$^{-/-}$ and *Nras*^{G12D}-expressing HSPC, *Kras*^{G12D}-expressing HSCs efficiently induced T-ALL under competitive repopulation conditions. These aggressive cancers were characterized by secondary mutations in *Notch1*, demonstrating the acquisition of cooperating mutations during progression to acute leukemia (Kindler et al. 2008; Sabnis et al. 2009). These studies show how genetically accurate mouse models are robust platforms for identifying cancer-initiating populations and for addressing fundamental questions regarding how endogenous oncogenic Ras expression perturbs stem cell fates. They also underscore the relative potency of oncogenic *Kras* versus oncogenic *Nras* or mutant *Nf1* in leukemia initiation, albeit under the somewhat artificial experimental condition of competitive repopulation in irradiated congenic recipient mice.

The observed GM-CSF hypersensitivity of *Nf1*-deficient progenitors in methylcellulose cultures raised the question of whether abnormal responses to exogenous stimuli are integral to the development of MPNs in vivo. To address this, fetal liver cells homozygous for mutations in both *Nf1* and *Gmcsf* were transplanted into WT or *Gmcsf*-deficient recipients (Birnbaum et al. 2000). Remarkably, GM-CSF expression in either donor cells or the host bone marrow microenvironment was sufficient to induce MPNs, but *Gmcsf* inactivation in both contexts severely prolonged disease latency. Further-

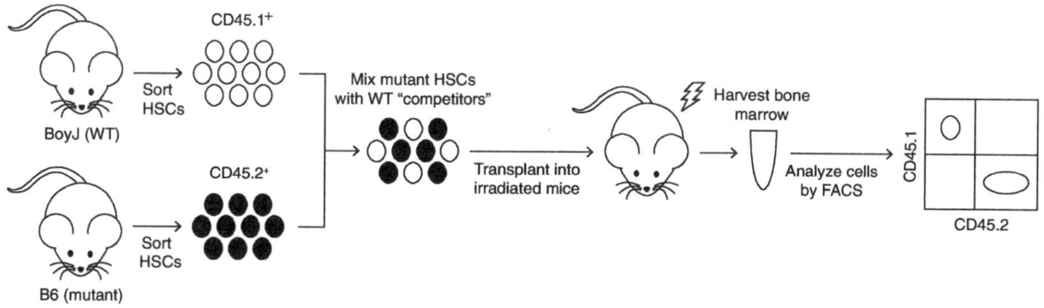

Figure 2. Competitive repopulation to assess fitness of hematopoietic stem cells (HSCs). Bone marrow cells are isolated from wild-type (WT) or mutant donor mice of different genetic backgrounds, then HSCs are sorted and mixed at a fixed ratio. These competitor cells are transplanted into irradiated recipient mice, then bone marrow is harvested and analyzed using fluorescence-activated cell sorting (FACS) to determine the percentage of mutant versus WT donor cells in the resulting population.

more, transplanting doubly mutant bone marrow cells from WT recipients with established MPNs into secondary *Gmcsf* mutant recipients caused disease regression, which was reversed by exogenous GM-CSF administration (Birnbaum et al. 2000). In a related study, a mutation in the β common subunit of the GM-CSF receptor also attenuated the development of MPNs in *Nf1* mutant mice (Kim et al. 2009). These data indicate that *Nf1* inactivation cooperates with upstream inputs from activated growth factor receptors to drive aberrant proliferation in vivo, and are consistent with biochemical studies of primary *Kras* and *Nf1* mutant cells, showing that Ras effector pathways are not fully activated under basal conditions but respond robustly to growth factor stimulation (Diaz-Flores et al. 2013).

Two recent studies examined additional requirements for the development of MPNs in *Nf1* and *Nras* mutant mice. In the first, genetic disruption of *Erk1/2* abrogated myeloid disease in *Mx1-Cre; Nf1*$^{flox/flox}$ mice, supporting an essential role of aberrant Raf/MEK/ERK signaling in induction of MPNs (Staser et al. 2013). More recently, genetic ablation of *Zdhhc9*, which encodes a palmitoyltransferase, impaired the ability of *Nras*^{G12D} to induce MPNs and T-ALL (Liu et al. 2016). In addition to uncovering genes required for the development of Ras-driven diseases in hematopoietic cells, these studies credentialed the Raf/MEK/ERK pathway as a therapeutic target in MPNs characterized by hyperactive Ras signaling and support inhibition of the palmitoylation/depalmitoylation cycle as a potentially effective therapeutic strategy for *NRAS* mutant hematologic cancers.

Nf1, *Nras*, and *Kras* mutant mice with MPNs were also harnessed to directly test small molecules with the potential to inhibit hyperactive Ras signaling. The initial preclinical trial administered L744,832, a potent and selective farnesyltransferase inhibitor (Kohl et al. 1995), to mice transplanted with homozygous *Nf1*-deficient fetal liver cells (Mahgoub et al. 1999). Although this compound reduced Raf/MEK/ERK pathway activation and abrogated GM-CSF colony growth ex vivo, it had no effect on MPNs in vivo. Consistent with these data, L744,832 inhibited H-Ras prenylation in recipient mice but a substantial fraction of total Ras was processed normally (Mahgoub et al. 1999), which was almost certainly caused by "bypass" N-Ras and K-Ras prenylation by gerenylgerenyl transferase. This study accurately predicted the disappointing efficacy of farnesyltransferase inhibitors in the clinic.

Directly inhibiting the oncogenic Ras/GAP switch continues to pose great challenges, and this has stimulated the development of numerous small molecule inhibitors of various Ras effectors. MEK is a particularly promising therapeutic target in many cancers, and the Food and Drug Administration has approved two allosteric inhibitors (trametinib and cobimetinib) for advanced melanoma. Preclinical studies evaluating efficacy of the "first-generation" MEK inhibitor CI-1040 in *Mx1-Cre; Nf1*$^{flox/flox}$ mice with MPNs showed no beneficial effects (Lauchle et al. 2009). By contrast, *Mx1-Cre; LSL-Kras*^{G12D} mice that received PD0325901 (PD901), a "second-generation" MEK inhibitor with optimized pharmacologic properties (Brown et al. 2007), had remarkable hematologic improvement and greatly enhanced survival (Lyubynska et al. 2011). This study also uncovered an ability of PD901 to block GM-CSF hypersensitivity in oncogenic *Kras* mutant bone marrow cells in vitro. However, *Kras*^{G12D} hematopoietic cells persisted after treatment, indicating that MEK inhibition rebalanced growth and differentiation in vivo. Consistent with this observation, some *Kras* mutant mice progressed to T-ALL despite continuous treatment with PD901, which supports differential dependence on the Raf/MEK/ERK pathway in myeloid versus lymphoid cells. To determine whether the discordant results of these preclinical trials reflected differential sensitivity of *Nf1* and *Kras* mutant hematopoietic cells to MEK inhibition or indicated that sustained Raf/MEK/ERK pathway inhibition is essential for therapeutic efficacy, PD901 was administered to *Mx1-Cre; Nf1*$^{flox/flox}$ mice with MPNs and induced profound reductions in leukocytosis and splenomegaly (Chang et al. 2013). As in mice with *Kras* mutant MPNs, MEK inhibition failed to eradicate *Nf1* mutant HSPCs but rather exerted antiproliferative and prodifferentiation effects in the myeloid and erythroid lineages.

Preclinical trials in mouse models of MPNs and other NF1-associated tumors (Jessen et al. 2013) informed a recent phase I clinical trial of the MEK inhibitor selumetinib, which showed tumor regression in most children with plexiform neurofibromas (Dombi et al. 2016). A national clinical trial of MEK inhibition in patients with relapsed/refractory JMML is also expected to open later this year. While data from *Kras* and *Nf1* mutant mice support the idea that MEK inhibition will induce hematologic improvement without eradicating mutant HSPCs, certain clinical observations suggest that this approach has curative potential. Specifically, infants with developmental disorders of the Noonan syndrome (NS) spectrum sometimes develop a transient MPN that is clinically indistinguishable from JMML, but resolves spontaneously (Kratz et al. 2005; Lauchle et al. 2006; Loh 2011). The causative germline *PTPN11* and *KRAS* mutations in these patients typically encode mutant proteins that are less activated biochemically than the strong somatic gain-of-function alleles detected in JMML (Keilhack et al. 2005; Schubbert et al. 2006, 2007). In these cases, modestly activated SHP-2 and K-Ras mutant proteins likely fail to confer a durable growth advantage. By contrast, stronger gain-of-function somatic mutations such as K-Ras^{G12D} and SHP-2^{E76K} appear to be capable of fully transforming fetal hematopoietic cells, leading to JMML. If this idea is correct, it raises the provocative possibility that pharmacologic inhibition could reduce Raf/MEK/ERK signaling below a critical threshold level and thereby cure some JMML patients by allowing the normal developmental program to reassert itself (Fig. 3). The design of the forthcoming clinical trial includes correlative pharmacokinetic and molecular studies to directly address any underlying mechanisms of response.

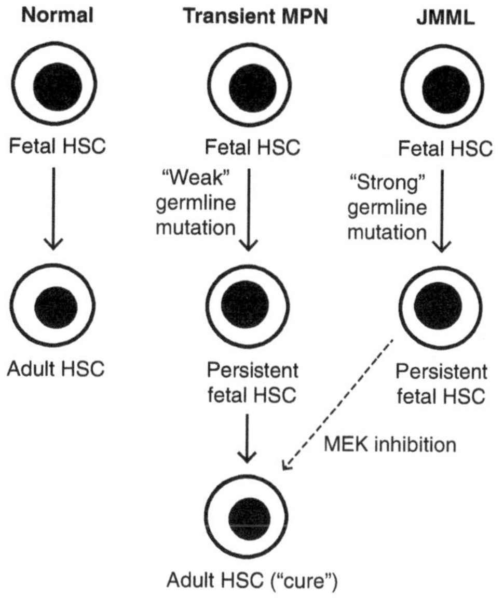

Figure 3. Potential therapeutic responses to MEK inhibition in juvenile myelomonocytic leukemia (JMML). Normal hematopoiesis is dynamically regulated during development with maturation from a fetal-like stem/progenitor to adult populations. The observation of a self-limited JMML-like myeloproliferative neoplasm (MPN) in infants with disorders of the Noonan syndrome spectrum suggests that a threshold of aberrant Ras/Raf/MEK/ERK signaling is required to fully transform fetal stem/progenitor cells. Based on this idea and preclinical data from *Kras* and *Nf1* mutant mice with MPNs, possible beneficial outcomes of sustained pharmacologic MEK inhibition in JMML include (1) clinical improvement (reduction in leukocyte counts, improvement of anemia, less splenomegaly) with persistence of the malignant clone; or (2) eradication of the JMML clone caused by restoration of normal developmental processes regulating hematopoietic cell fates (dotted line). Clinical trials with molecular assessment of mutant allele burden in patient specimens are required to investigate these potential mechanisms of response. HSC, Hematopoietic stem cell.

Mx1-Cre; Kras^{G12D} mice with MPNs were also used to test the efficacy of GDC-0941, a pan-PI3K inhibitor (Akutagawa et al. 2016). Treatment resulted in disease remission similar to that observed with PD901. However, biochemical studies of bone marrow cells showed reduced Raf/MEK/ERK pathway activation in response to GDC-0941, an observation that is consistent with a previous study showing that PI3K functions both upstream and downstream of K-Ras^{G12D} in myeloid lineage cells stimulated with cytokine growth factors (Diaz-Flores et al. 2013). To more specifically inhibit PI3K/Akt signaling, cohorts of *Kras* and *Nf1* mutant mice were treated with the Akt inhibitor MK-2206 and preclinical efficacy was observed in both models (Akutagawa et al. 2016). These data highlight the utility of preclinical studies in mouse models of oncongenic Ras-driven MPNs to elucidate potential cross talk between different Ras effector pathways.

MODELS OF ACUTE LEUKEMIA CHARACTERIZED BY *NF1* INACTIVATION OR ENDOGENOUS *KRAS*^{G12D}/*NRAS*^{G12D} EXPRESSION

Whereas *Nf1* inactivation and endogenous *Kras*^{G12D} or *Nras*^{G12D} expression in hematopoietic cells generated robust and tractable models of CMML and JMML, these mice do not develop acute leukemia. This observation and genomic analysis of human leukemias infer a requirement of cooperating mutations for progression to AML or ALL. Insertional mutagenesis is an unbiased forward genetic strategy for generating diverse primary leukemias. The first successful example of this method utilized the BXH-2 mouse strain that expresses a B-ecotropic murine leukemia virus (MuLV) to induce AML with long latency (Jenkins et al. 1982; Bedigian et al. 1984). Heterozygous *Nf1* inactivation both increased the incidence of myeloid leukemia and accelerated disease onset in BXH-2 mice (Largaespada et al. 1996). Interestingly, these *Nf1* mutant AMLs frequently exhibited loss of heterozygosity at the *Nf1* locus, consistent with data from JMML patients with NF1 showing loss of the normal allele (Shannon et al. 1994). Over a decade later, retroviral insertional mutagenesis (RIM) was used to generate AMLs with unique cooperating genetic lesions in *Mx1-Cre; Nf1*$^{flox/flox}$ mice (Lauchle et al. 2009). This experimental approach involved simultaneously injecting newborn mice with pI-pC to induce *Nf1* inactivation and with the MOL4070LTR retrovirus to generate integrations throughout the genome (Fig. 4A). In this system, *Nf1* cooperated with MOL4070LTR to induce AMLs that were heterogeneous with respect to morphology, myeloid surface markers, and retroviral insertions. *Myb* was identified as a common insertion site in multiple independent *Nf1* mutant AMLs, providing evidence for cooperation between these two genetic loci in myeloid leukemogenesis (Lauchle et al. 2009).

RIM was also performed in *Mx1-Cre; Nras*$^{G12D/+}$ mice using a modified approach (Fig. 4B) in which mice were injected with MOL4070LTR shortly after birth and pI-pC was administered at weaning (Li et al. 2011). This experimental system more closely recapitulates the pathogenesis of human AML in which *RAS* mutations generally comprise secondary events that cooperate with antecedent oncogenic translocations or with mutations in genes encoding hematopoietic transcription factors or proteins that broadly regulate epigenetic programs (Jan et al. 2012; Shlush et al. 2014; Lindsley et al. 2015). The *Nras* mutant AMLs generated in this study shared phenotypic features with the M4/M5 subtype of human AML, frequently exhibited loss of the WT *Nras* allele, and showed heterogeneous activation of Ras effector pathways. These AMLs were transplantable into sublethally irradiated recipient mice, and showed a clonal pattern of retroviral insertions that was stable upon secondary transplantation. The most frequent common insertion site was upstream of a gene encoding the zinc-finger transcription factor Evi-1 and resulted in a substantial increase in transcript expression specifically in leukemias harboring this integration (Li et al. 2011). Interestingly, *NRAS* mutations in human AML are frequently associated with both M4/M5 morphology and with *EVI1* translocations (Bowen et al. 2005; Bacher et al. 2006).

In contrast to *Nf1* and *Nras* mutant mice, *Mx1-Cre; Kras*$^{G12D/+}$ mice that were injected with MOL4070LTR uniformly died from MPNs. However, transplanting their bone marrow into sublethally irradiated recipients unexpectedly generated T-ALLs in ~80% and AMLs in the remaining ~20% of mice (Fig. 4C), which emerged with reduced latency compared to virus-injected WT animals. Primary T-ALLs were characterized by thymic masses, leukocytosis, extensive tissue infiltration, and an immature double-positive stage of lymphocyte development. Secondary transplant recipients also

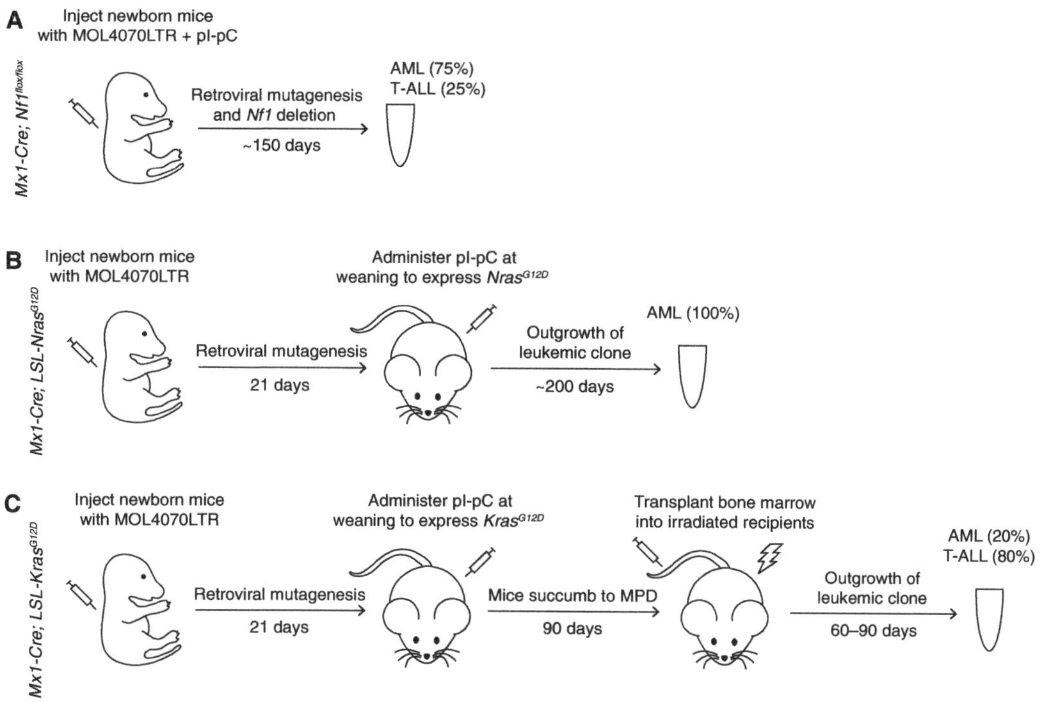

Figure 4. Conditional Ras activation in the hematopoietic compartment. (*A*) Newborn mice are injected with MOL4070LTR to generate diverse genomic integrations and with polyI-polyC (pI-pC) to inactivate *Nf1* in hematopoietic cells, resulting in predominantly acute myeloid leukemia (AML) with less frequent emergence of T-cell acute lymphoblastic leukemia (T-ALL). (*B*) Retroviral mutagenesis is performed in newborn mice, then pI-pC is administered at weaning to express *Nras*^{G12D} in hematopoietic cells and generate AML. (*C*) Mice are injected with MOL4070LTR shortly after birth, then with pI-pC at weaning to activate *Kras*^{G12D}. When these animals develop a lethal myeloproliferative disorder (MPD), bone marrow is transplanted into irradiated recipient mice to generate both AML and T-ALL.

developed T-ALL with more profound bone marrow infiltration and variable thymic involvement (Dail et al. 2010). A common insertion site in *Ikzf1*, which encodes the lymphoid cell-specific transcription factor and known ALL tumor suppressor Ikaros (Rebollo and Schmitt 2003; Mullighan et al. 2007), was identified in *Kras*-driven T-ALLs and integrations in this gene appeared to result in proteins with dominant negative activity. Downstream Ras effector pathway activation was highly variable in *Kras* mutant and WT leukemias. Strikingly, somatic *Notch1* mutations were identified in all *Mx1-Cre; Kras*$^{G12D/+}$ T-ALLs examined (Dail et al. 2010). This unanticipated observation demonstrates that cooperating events can arise from somatic alterations that are independent of retroviral insertions in RIM models of acute leukemia. AMLs generated by transplanting *Mx1-Cre; Kras*$^{G12D/+}$ bone marrow into sublethally irradiated recipients also contained clonal retroviral integrations within genes such as *Myb* and *Evi5* (Burgess et al. 2017).

Taken together, data from RIM screens performed with the same virus in mice expressing mutant *Nf1*, *Kras*, and *Nras* in hematopoietic cells highlight the extent to which specific Ras pathway alterations can modulate the spectrum of acute leukemias that develop. This finding is consistent with the variable latency and penetrance of MPNs observed across the three genotypes described above. Importantly, these panels of genetically diverse transplantable leukemias were subsequently deployed to perform controlled preclinical trials of signal transduction inhibitors.

PRECLINICAL TRIALS AND STUDIES OF DRUG RESISTANCE IN MOUSE MODELS OF ACUTE LEUKEMIA

Acute leukemias generated by RIM recapitulate the genetic heterogeneity seen in advanced human cancers, and transplantation into recipient mice is an attractive preclinical platform (Fig. 5) for assessing therapeutic responses and elucidating mechanisms of drug resistance because (1) primary cancers are treated in immunocompetent mice; (2) retroviral integration patterns can be used to track the emergence of drug resistant clones; and (3) relapsed leukemia cells can be retransplanted to verify intrinsic resistance, test alternative therapies, and validate candidate resistance mechanisms. To examine the sensitivity of myeloid malignancies characterized by *Nf1* inactivation to MEK inhibition, studies were initiated to assess the growth of bone marrow cells from WT mice, *Nf1* mutant mice with MPNs, and *Nf1*-deficient AMLs generated by RIM in methylcellulose medium containing GM-CSF and various doses of CI-1040 (Lauchle et al. 2009). In this in vitro system, AML cells unexpectedly exhibited markedly enhanced sensitivity to MEK inhibition compared to either WT cells or those harvested from mice with MPNs. Transplanting *Nf1*-deficient AMLs into recipient mice and treating them in vivo revealed impressive efficacy of this compound and of PD901, which was characterized by a rapid clearance of circulating leukemic blasts and a greater than threefold increase in median survival compared to vehicle-treated animals. However, all of these mice eventually relapsed despite sustained treatment. AMLs that relapsed after a prolonged response showed intrinsic drug resistance upon retransplantation and retreatment and also exhibited clonal evolution by the criterion of one or more novel retroviral integrations detected on Southern blots. Cloning retroviral insertions from resistant AMLs identified *RasGrp1* overexpression and disruption of one allele of *Mapk14* as candidate resistance mechanisms that were validated functionally (Lauchle et al. 2009).

Recipient mice transplanted with $Nras^{G12D}$ AMLs showed a modest but significant survival benefit in response to treatment with PD901 or trametinib. In contrast, these AMLs did not respond to the pan-PI3K inhibitor GDC-0941 as a single agent, and combining this drug with PD901 also had limited effects. As was observed in *Nf1*-deficient AMLs, all of the recipient mice given MEK and/or PI3K inhibitors relapsed on treatment. However, none of these relapsed AMLs harbored novel retroviral integrations and all showed a similar response to MEK inhibition upon retransplantation. Further analysis revealed that MEK inhibition induces neither differentiation nor apoptosis of leukemia cells, but rather exerts its antileukemic effect by reducing proliferation of $Nras^{G12D}$ AMLs (Bur-

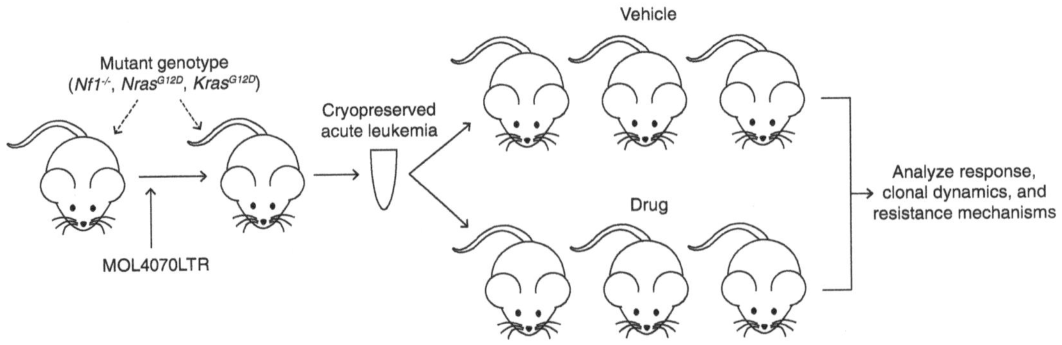

Figure 5. Preclinical studies using acute leukemias generated by retroviral insertional mutagenesis (RIM). Mice expressing mutant *Nf1*, *Nras*, or *Kras* under the control of *Mx1-Cre* are subjected to RIM to generate large panels of cryopreserved acute leukemias. These cells can be serially transplanted into recipient mice, exposed to various drugs or combinations in vivo, and then harvested at relapse and analyzed.

Cite this article as *Cold Spring Harb Perspect Med* doi: 10.1101/cshperspect.a031526

gess et al. 2014). Although this study showed modest efficacy of potent small-molecule MEK inhibitors, the lack of clonal evolution or acquired resistance in relapsed AMLs was an important limitation. Recently, an unbiased screen identified the creatine kinase pathway as a novel therapeutic vulnerability in AML cell lines (Fenouille et al. 2017). This finding was validated by transplanting primary *Nras* mutant AMLs generated by RIM, treating them with a chemical inhibitor of creatine biosynthesis, and showing that leukemia cells with an *Evi1* insertion were much more sensitive to the drug (Fenouille et al. 2017).

Treatment with PD901 also extended the survival of mice transplanted with *Kras*^{G12D} AMLs with one leukemia showing an exceptional response, harboring a novel integration at relapse, and demonstrating both clonal evolution and intrinsic drug resistance upon retransplantation/retreatment. Extensive analysis of the parental AML and resistant subclone revealed a uniparental disomy event at the *Kras* locus that was followed by loss of the WT allele in the parental leukemia. This genetic configuration conferred a competitive advantage upon these primary AML cells, but also rendered them hypersensitive to MEK inhibition. Interestingly, the resistant AML exhibited trisomy 6 and a *Kras* genotype of two mutant and one WT allele. These data showed a role for serial genetic chances at the *Kras* locus in modulating Raf/MEK/ERK pathway dependence, which was subsequently also observed in a panel of human colorectal cancer cell lines. Furthermore, allelic imbalance at the *KRAS* locus was identified in over half of a large panel of advanced human cancers with diverse histologies harboring oncogenic *KRAS* mutations (Burgess et al. 2017).

Preclinical studies in *Kras*^{G12D} T-ALLs uncovered a pattern of dependencies on the downstream Ras effector pathways that was distinct from AML. Whereas these leukemias showed a modest response to PD901 and were refractory to GDC-0941, treatment with both inhibitors markedly extended survival (Dail et al. 2010, 2014). As in the AML models, all recipient mice ultimately died from drug-resistant leukemia, which exhibited remarkable clonal heterogeneity with numerous relapsed samples harboring different novel retroviral integrations. T-ALLs that relapsed after a prolonged response to treatment unexpectedly showed loss of retroviral integrations and somatic mutations that activated *Notch1* and a corresponding elevation in PI3K pathway activation (Dail et al. 2014). These data have important implications for the treatment of T-ALL, as they suggest that the rational strategy of combining PI3K and NOTCH1 inhibitors might drive clonal outgrowth of drug-resistant cells.

CONCLUDING REMARKS

The tractable nature of the hematopoietic compartment and the availability of robust methodologies for isolating and functionally analyzing HSPCs in vivo have facilitated examining the effects of Ras pathway activation on HSPC biology and behavior. Extensive characterization of mouse models of MPNs driven by mutant *Nf1*, *Kras*, or *Nras* has revealed important differences at the level of individual HSPCs and, more globally, with respect to in vivo disease phenotypes and the extent to which downstream Ras effector pathways are activated following oncogene expression. Taken together, these studies indicate that heterozygous *Kras*^{G12D} is a more activating gain-of-function mutation than homozygous *Nf1* inactivation, which is, in turn, more activating than a heterozygous *Nras*^{G12D} mutation. It will be interesting to perform similar comparisons in other tissue lineages. In addition to demonstrating differential effects of individual mutant alleles, these studies also underscore the importance of posttranslational modifications of Ras in tumorigenesis. Consistent with these findings, a recent genome-wide CRISPR-mediated screen to identify synthetic lethal interactions found that genes encoding proteins involved in Ras processing and components of the Raf/MEK/ERK pathway were essential specifically in human AML cell lines harboring oncogenic *RAS* mutations (Wang et al. 2017). Another article in this collection focuses on the topic of uncovering and validating candidate synthetic lethal interactions with oncogenic *RAS* (Aguirre and Hahn 2017).

Developing predictive preclinical models for testing new agents and drug combinations remains a fundamental challenge for advancing cancer medicine. In this regard, the ability to transplant human and mouse leukemias into cohorts of congenic recipient mice is a major advantage. Collections of patient-derived xenografts (PDXs) are also valuable for preclinical evaluation because they are derived from human cancers and can reflect the heterogeneity of a given patient population. However, the use of immunocompromised mice for engraftment precludes examination of any potential effects of the microenvironment on response and resistance to therapeutic agents, and the reduced genetic complexity of PDX models that results from the barrier of xenotransplantation (Klco et al. 2014) has potential limitations for characterizing relapse mechanisms. Whereas acute leukemias generated by RIM also recapitulate the genetic heterogeneity found in human tumors, they can be transplanted into congenic immunocompetent mice and interact with host cells in hematopoietic tissues. Although these models are dependent upon genetic and biochemical similarities between mouse and human cells, recent work has shown that insights from preclinical studies in RIM-induced AMLs can generate mechanistic hypotheses that are relevant to human cancers (Burgess et al. 2017). Mouse models of hematologic cancers generated by RIM can also be used to uncover mechanisms of resistance to conventional chemotherapeutic agents and to identify novel compounds that might overcome resistance to these drugs. Ideally, performing comprehensive in vivo preclinical drug testing that utilizes the RIM-driven leukemias described in this review in combination with banks of well-characterized human PDX models (Townsend et al. 2016) will identify promising therapeutic approaches for hematologic and solid cancers with oncogenic *RAS* mutations. The tractable experimental nature of the hematopoietic system makes it a particularly appealing model for addressing this fundamental problem.

ACKNOWLEDGMENTS

We are indebted to our long-term collaborators Gideon Bollag, Kevin Haigis, Tyler Jacks, Michelle Le Beau, Scott Lowe, Luis Parada, David Tuveson, and Linda Wolff for reagents, advice, and assistance with much of the work from our laboratory described in this review, which is supported by the National Institutes of Health (NIH) Grants R01 CA180037 and R37 CA72614, by the Leukemia and Lymphoma Society Specialized Center of Research Grant LLS 7019-04, by the Department of Defense (DOD) NF1 Research Program, and by the Rally Foundation for Childhood Cancer Research. A.W. is supported by a Postdoctoral Fellowship, PF-14-070-01-TBG, from the American Cancer Society, including a supplement from the Hillcrest Committee. K.S. is an American Cancer Society Research Professor.

REFERENCES

**Reference is also in this collection.*

* Aguirre AJ, Hahn WC. 2017. Synthetic lethal vulnerabilities in *KRAS*-mutant cancers. *Cold Spring Harb Perspect Med* doi: 10.1101/cshperspect.a031518.

* Ahearn I, Zhou M, Philips MR. 2018. Posttranslational modifications of RAS proteins. *Cold Spring Harb Perspect Med* doi: 10.1101/cshperspect.a031484.

Akutagawa J, Huang TQ, Epstein I, Chang T, Quirindongo-Crespo M, Cottonham CL, Dail M, Slusher BS, Friedman LS, Sampath D, et al. 2016. Targeting the PI3K/Akt pathway in murine MDS/MPN driven by hyperactive Ras. *Leukemia* **30**: 1335–1343.

Bacher U, Haferlach T, Schoch C, Kern W, Schnittger S. 2006. Implications of NRAS mutations in AML: A study of 2502 patients. *Blood* **107**: 3847–3853.

Bedigian HG, Johnson DA, Jenkins NA, Copeland NG, Evans R. 1984. Spontaneous and induced leukemias of myeloid origin in recombinant inbred BXH mice. *J Virol* **51**: 586–594.

Birnbaum RA, O'Marcaigh A, Wardak Z, Zhang YY, Dranoff G, Jacks T, Clapp DW, Shannon KM. 2000. *Nf1* and *Gmcsf* interact in myeloid leukemogenesis. *Mol Cell* **5**: 189–195.

Bollag G, Clapp DW, Shih S, Adler F, Zhang YY, Thompson P, Lange BJ, Freedman MH, McCormick F, Jacks T, et al. 1996. Loss of NF1 results in activation of the Ras signaling pathway and leads to aberrant growth in haematopoietic cells. *Nat Genet* **12**: 144–148.

Bos JL. 1989. Ras oncogenes in human cancer: A review. *Cancer Res* **49**: 4682–4689.

Bowen DT, Frew ME, Hills R, Gale RE, Wheatley K, Groves MJ, Langabeer SE, Kottaridis PD, Moorman AV, Burnett AK, et al. 2005. RAS mutation in acute myeloid leukemia is associated with distinct cytogenetic subgroups but does not influence outcome in patients younger than 60 years. *Blood* **106**: 2113–2119.

Brannan CI, Perkins AS, Vogel KS, Ratner N, Nordlund ML, Reid SW, Buchberg AM, Jenkins NA, Parada LF, Copeland NG. 1994. Targeted disruption of the neurofibromatosis type-1 gene leads to developmental abnormalities in heart and various neural crest-derived tissues. *Genes Dev* **8**: 1019–1029.

Braun BS, Tuveson DA, Kong N, Le DT, Kogan SC, Rozmus J, Le Beau MM, Jacks TE, Shannon KM. 2004. Somatic activation of oncogenic Kras in hematopoietic cells initiates a rapidly fatal myeloproliferative disorder. *Proc Natl Acad Sci* **101**: 597–602.

Brown AP, Carlson TCG, Loi C-M, Graziano MJ. 2007. Pharmacodynamic and toxicokinetic evaluation of the novel MEK inhibitor, PD0325901, in the rat following oral and intravenous administration. *Cancer Chemother Pharmacol* **59**: 671–679.

Burgess MR, Hwang E, Firestone AJ, Huang T, Xu J, Zuber J, Bohin N, Wen T, Kogan SC, Haigis KM, et al. 2014. Preclinical efficacy of MEK inhibition in Nras-mutant AML. *Blood* **124**: 3947–3955.

Burgess MR, Hwang E, Mroue R, Bielski CM, Wandler AM, Huang BJ, Firestone AJ, Young A, Lacap JA, Crocker L, et al. 2017. KRAS allelic imbalance enhances fitness and modulates MAP kinase dependence in cancer. *Cell* **168**: 817–829.e15.

Chan IT, Kutok JL, Williams IR, Cohen S, Kelly L, Shigematsu H, Johnson L, Akashi K, Tuveson DA, Jacks T, et al. 2004. Conditional expression of oncogenic K-ras from its endogenous promoter induces a myeloproliferative disease. *J Clin Invest* **113**: 528–538.

Chang T, Krisman K, Theobald EH, Xu J, Akutagawa J, Lauchle JO, Kogan S, Braun BS, Shannon K. 2013. Sustained MEK inhibition abrogates myeloproliferative disease in Nf1 mutant mice. *J Clin Invest* **123**: 335–339.

Chang TY, Dvorak CC, Loh ML. 2014. Bedside to bench in juvenile myelomonocytic leukemia: Insights into leukemogenesis from a rare pediatric leukemia. *Blood* **124**: 2487–2497.

Chao RC, Pyzel U, Fridlyand J, Kuo Y-M, Teel L, Haaga J, Borowsky A, Horvai A, Kogan SC, Bonifas J, et al. 2005. Therapy-induced malignant neoplasms in *Nf1* mutant mice. *Cancer Cell* **8**: 337–348.

Chou WC, Peng KY, Lei WC, Ko BS, Tsay W, Kuo CH, Tien HF. 2012. Persistence of mutant isocitrate dehydrogenase in patients with acute myeloid leukemia in remission. *Leukemia* **26**: 527–529.

Cichowski K, Jacks T. 2001. NF1 tumor suppressor gene function: Narrowing the GAP. *Cell* **104**: 593–604.

Corces-Zimmerman MR, Hong W-J, Weissman IL, Medeiros BC, Majeti R. 2014. Preleukemic mutations in human acute myeloid leukemia affect epigenetic regulators and persist in remission. *Proc Natl Acad Sci* **111**: 2548–2553.

Cuiffo B, Ren R. 2010. Palmitoylation of oncogenic NRAS is essential for leukemogenesis. *Blood* **115**: 3598–3605.

Dail M, Li Q, McDaniel A, Wong J, Akagi K, Huang B, Kang HC, Kogan SC, Shokat K, Wolff L, et al. 2010. Mutant Ikzf1, $Kras^{G12D}$, and Notch1 cooperate in T lineage leukemogenesis and modulate responses to targeted agents. *Proc Natl Acad Sci* **107**: 5106–5111.

Dail M, Wong J, Lawrence J, O'Connor D, Nakitandwe J, Chen S-C, Xu J, Lee LB, Akagi K, Li Q, et al. 2014. Loss of oncogenic Notch1 with resistance to a PI3K inhibitor in T-cell leukaemia. *Nature* **513**: 512–516.

Daley GQ, Van Etten RA, Baltimore D. 1990. Induction of chronic myelogenous leukemia in mice by the $P210^{bcr/abl}$ gene of the Philadelphia chromosome. *Science* **247**: 824–830.

Diaz-Flores E, Goldschmidt H, Depeille P, Ng V, Akutagawa J, Krisman K, Crone M, Burgess MR, Williams O, Houseman B, et al. 2013. PLC-γ and PI3K link cytokines to ERK activation in hematopoietic cells with normal and oncogenic Kras. *Sci Signal* **6**: ra105.

Dombi E, Baldwin A, Marcus LJ, Fisher MJ, Weiss B, Kim A, Whitcomb P, Martin S, Aschbacher-Smith LE, Rizvi TA, et al. 2016. Activity of selumetinib in neurofibromatosis type 1-related plexiform neurofibromas. *N Engl J Med* **375**: 2550–2560.

* Drosten M, Guerra C, Barbacid M. 2017. Genetically engineered mouse models of K-Ras-driven lung and pancreatic tumors: Validation of therapeutic targets. *Cold Spring Harb Perspect Med* doi: 10.1101/cshperspect.a031542.

Emanuel PD, Bates LJ, Castleberry RP, Gualtieri RJ, Zuckerman KS. 1991. Selective hypersensitivity to granulocyte-macrophage colony-stimulating factor by juvenile chronic myeloid leukemia hematopoietic progenitors. *Blood* **77**: 925–929.

Fenouille N, Bassil CF, Ben-Sahra I, Benajiba L, Alexe G, Ramos A, Pikman Y, Conway AS, Burgess MR, Li Q, et al. 2017. The creatine kinase pathway is a metabolic vulnerability in EVI1-positive acute myeloid leukemia. *Nat Med* **23**: 301–313.

Haigis KM, Kendall KR, Wang Y, Cheung A, Haigis MC, Glickman JN, Niwa-Kawakita M, Sweet-Cordero A, Sebolt-Leopold J, Shannon KM, et al. 2008. Differential effects of oncogenic K-Ras and N-Ras on proliferation, differentiation and tumor progression in the colon. *Nat Genet* **40**: 600–608.

Holmfeldt L, Wei L, Diaz-Flores E, Walsh M, Zhang J, Ding L, Payne-Turner D, Churchman M, Andersson A, Chen SC, et al. 2013. The genomic landscape of hypodiploid acute lymphoblastic leukemia. *Nat Genet* **45**: 242–252.

Jacks T, Shih TS, Schmitt EM, Bronson RT, Bernards A, Weinberg RA. 1994. Tumour predisposition in mice heterozygous for a targeted mutation in Nf1. *Nat Genet* **7**: 353–361.

Jackson EL, Willis N, Mercer K, Bronson RT, Crowley D, Montoya R, Jacks T, Tuveson DA. 2001. Analysis of lung tumor initiation and progression using conditional expression of oncogenic K-ras. *Genes Dev* **15**: 3243–3248.

Jan M, Majeti R. 2013. Clonal evolution of acute leukemia genomes. *Oncogene* **32**: 135–140.

Jan M, Snyder TM, Corces-Zimmerman MR, Vyas P, Weissman IL, Quake SR, Majeti R. 2012. Clonal evolution of preleukemic hematopoietic stem cells precedes human acute myeloid leukemia. *Sci Transl Med* **4**: 149ra118.

Jankowska AM, Makishima H, Tiu RV, Szpurka H, Huang Y, Traina F, Visconte V, Sugimoto Y, Prince C, O'Keefe C, et al. 2011. Mutational spectrum analysis of chronic myelomonocytic leukemia includes genes associated with epigenetic regulation: UTX, EZH2, and DNMT3A. *Blood* **118**: 3932–3941.

Jenkins NA, Copeland NG, Taylor BA, Bedigian HG, Lee BK. 1982. Ecotropic murine leukemia virus DNA content of normal and lymphomatous tissues of BXH-2 recombinant inbred mice. *J Virol* **42**: 379–388.

Jessen WJ, Miller SJ, Jousma E, Wu J, Rizvi TA, Brundage ME, Eaves D, Widemann B, Kim M-O, Dombi E, et al. 2013. MEK inhibition exhibits efficacy in human and mouse neurofibromatosis tumors. *J Clin Invest* **123**: 340–347.

Keilhack H, David FS, McGregor M, Cantley LC, Neel BG. 2005. Diverse biochemical properties of Shp2 mutants. Implications for disease phenotypes. *J Biol Chem* **280**: 30984–30993.

Kim WI, Matise I, Diers MD, Largaespada DA. 2009. RAS oncogene suppression induces apoptosis followed by more differentiated and less myelosuppressive disease upon relapse of acute myeloid leukemia. *Blood* **113**: 1086–1096.

Kindler T, Cornejo MG, Scholl C, Liu J, Leeman DS, Haydu JE, Fröhling S, Lee BH, Gilliland DG. 2008. *K-Ras^{G12D}*-induced T-cell lymphoblastic lymphoma/leukemias harbor *Notch1* mutations and are sensitive to γ-secretase inhibitors. *Blood* **112**: 3373–3382.

Klco JM, Spencer DH, Miller CA, Griffith M, Lamprecht TL, O'Laughlin M, Fronick C, Magrini V, Demeter RT, Fulton RS, et al. 2014. Functional heterogeneity of genetically defined subclones in acute myeloid leukemia. *Cancer Cell* **25**: 379–392.

Klco JM, Miller CA, Griffith M, Petti A, Spencer DH, Ketkar-Kulkarni S, Wartman LD, Christopher M, Lamprecht TL, Helton NM, et al. 2015. Association between mutation clearance after induction therapy and outcomes in acute myeloid leukemia. *JAMA* **314**: 811–822.

Kohl NE, Omer CA, Conner MW, Anthony NJ, Davide JP, deSolms SJ, Giuliani EA, Gomez RP, Graham SL, Hamilton K. 1995. Inhibition of farnesyltransferase induces regression of mammary and salivary carcinomas in ras transgenic mice. *Nat Med* **1**: 792–797.

Kong G, Chang YI, You X, Ranheim EA, Zhou Y, Burd CE, Zhang J. 2016. The ability of endogenous *Nras* oncogenes to initiate leukemia is codon-dependent. *Leukemia* **30**: 1935–1938.

Kratz CP, Niemeyer CM, Castleberry RP, Cetin M, Bergsträsser E, Emanuel PD, Hasle H, Kardos G, Klein C, Kojima S, et al. 2005. The mutational spectrum of *PTPN11* in juvenile myelomonocytic leukemia and Noonan syndrome/myeloproliferative disease. *Blood* **106**: 2183–2185.

Kühn R, Schwenk F, Aguet M, Rajewsky K. 1995. Inducible gene targeting in mice. *Science* **269**: 1427–1429.

Largaespada DA, Brannan CI, Jenkins NA, Copeland NG. 1996. Nf1 deficiency causes Ras-mediated granulocyte/macrophage colony stimulating factor hypersensitivity and chronic myeloid leukaemia. *Nat Genet* **12**: 137–143.

Lauchle JO, Braun BS, Loh ML, Shannon K. 2006. Inherited predispositions and hyperactive Ras in myeloid leukemogenesis. *Pediatr Blood Cancer* **46**: 579–585.

Lauchle JO, Kim D, Le DT, Akagi K, Crone M, Krisman K, Warner K, Bonifas JM, Li Q, Coakley KM, et al. 2009. Response and resistance to MEK inhibition in leukaemias initiated by hyperactive Ras. *Nature* **461**: 411–414.

Le DT, Kong N, Zhu Y, Lauchle JO, Aiyigari A, Braun BS, Wang E, Kogan SC, Le Beau MM, Parada L, et al. 2004. Somatic inactivation of Nf1 in hematopoietic cells results in a progressive myeloproliferative disorder. *Blood* **103**: 4243–4250.

Li Q, Haigis KM, McDaniel A, Harding-Theobald E, Kogan SC, Akagi K, Wong JCY, Braun BS, Wolff L, Jacks T, et al. 2011. Hematopoiesis and leukemogenesis in mice expressing oncogenic *Nras^{G12D}* from the endogenous locus. *Blood* **117**: 2022–2032.

Li Q, Bohin N, Wen T, Ng V, Magee J, Chen SC, Shannon K, Morrison SJ. 2013. Oncogenic *Nras* has bimodal effects on stem cells that sustainably increase competitiveness. *Nature* **504**: 143–147.

Lindsley RC, Mar BG, Mazzola E, Grauman PV, Shareef S, Allen SL, Pigneux A, Wetzler M, Stuart RK, Erba HP, et al. 2015. Acute myeloid leukemia ontogeny is defined by distinct somatic mutations. *Blood* **125**: 1367–1376.

Liu P, Jiao B, Zhang R, Zhao H, Zhang C, Wu M, Li D, Zhao X, Qiu Q, Li J, et al. 2016. Palmitoylacyltransferase Zdhhc9 inactivation mitigates leukemogenic potential of oncogenic Nras. *Leukemia* **30**: 1225–1228.

Loh ML. 2011. Recent advances in the pathogenesis and treatment of juvenile myelomonocytic leukaemia. *Br J Haematol* **152**: 677–687.

Lyubynska N, Gorman MF, Lauchle JO, Hong WX, Akutagawa JK, Shannon K, Braun BS. 2011. A MEK inhibitor abrogates myeloproliferative disease in *Kras* mutant mice. *Sci Transl Med* **3**: 76ra27.

MacKenzie KL, Dolnikov A, Millington M, Shounan Y, Symonds G. 1999. Mutant N-*ras* induces myeloproliferative disorders and apoptosis in bone marrow repopulated mice. *Blood* **93**: 2043–2056.

Mahgoub N, Taylor BR, Gratiot M, Kohl NE, Gibbs JB, Jacks T, Shannon KM. 1999. In vitro and in vivo effects of a farnesyltransferase inhibitor on Nf1-deficient hematopoietic cells. *Blood* **94**: 2469–2476.

Mangues R, Symmans WF, Lu S, Schwartz S, Pellicer A. 1996. Activated N-ras oncogene and N-ras proto-oncogene act through the same pathway for in vivo tumorigenesis. *Oncogene* **13**: 1053–1063.

Mullighan CG, Goorha S, Radtke I, Miller CB, Coustan-Smith E, Dalton JD, Girtman K, Mathew S, Ma J, Pounds SB, et al. 2007. Genome-wide analysis of genetic alterations in acute lymphoblastic leukaemia. *Nature* **446**: 758–764.

Nakamura JL, Phong C, Pinarbasi E, Kogan SC, Vandenberg S, Horvai AE, Faddegon BA, Fiedler D, Shokat K, Houseman BT, et al. 2011. Dose-dependent effects of focal fractionated irradiation on secondary malignant neoplasms in Nf1 mutant mice. *Cancer Res* **71**: 106–115.

Parikh C, Subrahmanyam R, Ren R. 2006. Oncogenic NRAS rapidly and efficiently induces CMML- and AML-like diseases in mice. *Blood* **108**: 2349–2357.

Parikh C, Subrahmanyam R, Ren R. 2007. Oncogenic NRAS, KRAS, and HRAS exhibit different leukemogenic potentials in mice. *Cancer Res* **67**: 7139–7146.

Rathert P, Roth M, Neumann T, Muerdter F, Roe JS, Muhar M, Deswal S, Cerny-Reiterer S, Peter B, Jude J, et al. 2015. Transcriptional plasticity promotes primary and acquired resistance to BET inhibition. *Nature* **525**: 543–547.

Ratner N, Miller SJ. 2015. A RASopathy gene commonly mutated in cancer: The neurofibromatosis type 1 tumour suppressor. *Nat Rev Cancer* **15**: 290–301.

Rebollo A, Schmitt C. 2003. Ikaros, Aiolos and Helios: Transcription regulators and lymphoid malignancies. *Immunol Cell Biol* **81**: 171–175.

Sabnis AJ, Cheung LS, Dail M, Kang HC, Santaguida M, Hermiston ML, Passegué E, Shannon K, Braun BS. 2009. Oncogenic Kras initiates leukemia in hematopoietic stem cells. *PLoS Biol* **7**: e59.

Sachs Z, LaRue RS, Nguyen HT, Sachs K, Noble KE, Mohd Hassan NA, Diaz-Flores E, Rathe SK, Sarver AL, Bendall SC, et al. 2014. *NRAS*^{G12V} oncogene facilitates self-renewal in a murine model of acute myelogenous leukemia. *Blood* **124**: 3274–3283.

Sakaguchi H, Okuno Y, Muramatsu H, Yoshida K, Shiraishi Y, Takahashi M, Kon A, Sanada M, Chiba K, Tanaka H, et al. 2013. Exome sequencing identifies secondary mutations of SETBP1 and JAK3 in juvenile myelomonocytic leukemia. *Nat Genet* **45**: 937–941.

Sawyers CL. 1999. Chronic myeloid leukemia. *N Engl J Med* **340**: 1330–1340.

Schubbert S, Zenker M, Rowe SL, Böll S, Klein C, Bollag G, van der Burgt I, Musante L, Kalscheuer V, Wehner LE, et al. 2006. Germline *KRAS* mutations cause Noonan syndrome. *Nat Genet* **38**: 331–336.

Schubbert S, Shannon K, Bollag G. 2007. Hyperactive Ras in developmental disorders and cancer. *Nat Rev Cancer* **7**: 295–308.

Shannon KM, O'Connell P, Martin GA, Paderanga D, Olson K, Dinndorf P, McCormick F. 1994. Loss of the normal NF1 allele from the bone marrow of children with type 1 neurofibromatosis and malignant myeloid disorders. *N Engl J Med* **330**: 597–601.

Shieh A, Ward AF, Donlan KL, Harding-Theobald ER, Xu J, Mullighan CG, Zhang C, Chen SC, Su X, Downing JR, et al. 2013. Defective K-Ras oncoproteins overcome impaired effector activation to initiate leukemia in vivo. *Blood* **121**: 4884–4893.

Shlush LI, Zandi S, Mitchell A, Chen WC, Brandwein JM, Gupta V, Kennedy JA, Schimmer AD, Schuh AC, Yee KW, et al. 2014. Identification of pre-leukaemic haematopoietic stem cells in acute leukaemia. *Nature* **506**: 328–333.

Staser K, Park SJ, Rhodes SD, Zeng Y, He YZ, Shew MA, Gehlhausen JR, Cerabona D, Menon K, Chen S, et al. 2013. Normal hematopoiesis and neurofibromin-deficient myeloproliferative disease require Erk. *J Clin Invest* **123**: 329–334.

Stieglitz E, Taylor-Weiner AN, Chang TY, Gelston LC, Wang YD, Mazor T, Esquivel E, Yu A, Seepo S, Olsen SR, et al. 2015. The genomic landscape of juvenile myelomonocytic leukemia. *Nat Genet* **47**: 1326–1333.

Townsend EC, Murakami MA, Christodoulou A, Christie AL, Köster J, DeSouza TA, Morgan EA, Kallgren SP, Liu H, Wu SC, et al. 2016. The public repository of xenografts enables discovery and randomized phase II-like trials in mice. *Cancer Cell* **29**: 574–586.

Van Meter MEM, Díaz-Flores E, Archard JA, Passegué E, Irish JM, Kotecha N, Nolan GP, Shannon K, Braun BS. 2007. K-Ras^{G12D} expression induces hyperproliferation and aberrant signaling in primary hematopoietic stem/progenitor cells. *Blood* **109**: 3945–3952.

Wang J, Liu Y, Li Z, Du J, Ryu MJ, Taylor PR, Fleming MD, Young KH, Pitot H, Zhang J. 2010. Endogenous oncogenic *Nras* mutation promotes aberrant GM-CSF signaling in granulocytic/monocytic precursors in a murine model of chronic myelomonocytic leukemia. *Blood* **116**: 5991–6002.

Wang T, Yu H, Hughes NW, Liu B, Kendirli A, Klein K, Chen WW, Lander ES, Sabatini DM. 2017. Gene essentiality profiling reveals gene networks and synthetic lethal interactions with oncogenic Ras. *Cell* **168**: 890–903.e15.

Ward AF, Braun BS, Shannon KM. 2012. Targeting oncogenic Ras signaling in hematologic malignancies. *Blood* **120**: 3397–3406.

Xu J, Hedberg C, Dekker FJ, Li Q, Haigis KM, Hwang E, Waldmann H, Shannon K. 2012. Inhibiting the palmitoylation/depalmitoylation cycle selectively reduces the growth of hematopoietic cells expressing oncogenic Nras. *Blood* **119**: 1032–1035.

Xu J, Haigis KM, Firestone AJ, McNerney ME, Li Q, Davis E, Chen S-C, Nakitandwe J, Downing J, Jacks T, et al. 2013. Dominant role of oncogene dosage and absence of tumor suppressor activity in Nras-driven hematopoietic transformation. *Cancer Discov* **3**: 993–1001.

Zhang Y, Taylor BR, Shannon K, Clapp DW. 2001. Quantitative effects of Nf1 inactivation on in vivo hematopoiesis. *J Clin Invest* **108**: 709–715.

Zhang J, Ding L, Holmfeldt L, Wu G, Heatley SL, Payne-Turner D, Easton J, Chen X, Wang J, Rusch M, et al. 2012. The genetic basis of early T-cell precursor acute lymphoblastic leukaemia. *Nature* **481**: 157–163.

Zhu Y, Romero MI, Ghosh P, Ye Z, Charnay P, Rushing EJ, Marth JD, Parada LF. 2001. Ablation of NF1 function in neurons induces abnormal development of cerebral cortex and reactive gliosis in the brain. *Genes Dev* **15**: 859–876.

Zuber J, Radtke I, Pardee TS, Zhao Z, Rappaport AR, Luo W, McCurrach ME, Yang M-M, Dolan ME, Kogan SC, et al. 2009. Mouse models of human AML accurately predict chemotherapy response. *Genes Dev* **23**: 877–889.

Zuber J, McJunkin K, Fellmann C, Dow LE, Taylor MJ, Hannon GJ, Lowe SW. 2011a. Toolkit for evaluating genes required for proliferation and survival using tetracycline-regulated RNAi. *Nat Biotechnol* **29**: 79–83.

Zuber J, Shi J, Wang E, Rappaport AR, Herrmann H, Sison EA, Magoon D, Qi J, Blatt K, Wunderlich M, et al. 2011b. RNAi screen identifies Brd4 as a therapeutic target in acute myeloid leukaemia. *Nature* **478**: 524–528.

Genetically Engineered Mouse Models of K-Ras-Driven Lung and Pancreatic Tumors: Validation of Therapeutic Targets

Matthias Drosten, Carmen Guerra, and Mariano Barbacid

Molecular Oncology Programme, Centro Nacional de Investigaciones Oncológicas (CNIO), E-28029 Madrid, Spain

Correspondence: mdrosten@cnio.es; mcguerra@cnio.es; mbarbacid@cnio.es

K-RAS signaling has been intensely studied for over 40 years. Yet, as of today, no drugs have been approved to treat *K-RAS* mutant cancers. Since the turn of the century, scientists have used genetically engineered mouse (GEM) models to reproduce *K-RAS* mutant cancers in a laboratory setting to elucidate those molecular events responsible for the onset and progression of these tumors and to identify suitable therapies. In this review, we outline a brief description of available GEM models for two tumor types known to be driven by *K-RAS* mutations: lung adenocarcinoma and pancreatic ductal adenocarcinoma. In addition, we summarize a series of studies that have used these GEM tumor models to validate, either by genetic or pharmacological approaches, the therapeutic potential of a variety of targets, with the ultimate goal of translating these results to the clinical setting.

Saturation analysis of more than 4700 human tumors has identified more than 200 bona fide cancer genes (Lawrence et al. 2014). Among the five cancer genes most frequently mutated appears to be *K-RAS*, an oncogene mutated in more than a quarter of all human tumors. *K-RAS* mutations are not equally distributed. They appear in pancreatic ductal adenocarcinoma (PDAC) (>90%), colorectal carcinoma (45%), lung adenocarcinoma (LAC) (25%), and endometrial carcinoma (15%) with lower incidence rates in several other tumor types (Pylayeva-Gupta et al. 2011; Eser et al. 2014). U.S. estimates for 2016 alone indicate that there will be more than 330,000 new cases and close to 120,000 deaths caused by *K-RAS* mutant pancreatic and lung tumors, two tumor types in which experimental evidence indicates that these mutations are the initiating events of the neoplastic process.

K-RAS signaling has been intensely studied for over 40 years. These studies have revealed that the main signaling pathways that convey *K-RAS* oncogenic activity are kinase cascades that could be amenable to interference with selective inhibitors. Yet, as of today, not a single targeted drug has been approved to treat *K-RAS* mutant cancers. Thus, it is essential to get a better understanding about K-RAS signaling and to develop experimental systems in which scientists can test novel therapeutic strategies that can be eventually translated to the clinic.

Since the turn of the century, scientists have focused on the generation of genetically engi-

Cite this article as *Cold Spring Harb Perspect Med* doi: 10.1101/cshperspect.a031542

neered mouse (GEM) models of *K-RAS* mutant cancer that faithfully reproduce the natural history of the human disease. Most of these models carry endogenous mutations in those loci known to be the key drivers in human cancer such as *K-Ras*, *p53*, *Ink4a/Arf*, *Lkb1*, *Pten*, etc. In this review, we summarize current information regarding available GEM models for two tumor types known to be driven by *K-RAS* mutations, LAC, and PDAC. In addition, we summarize those studies that have used these GEM tumor models to validate, either by genetic or pharmacological approaches, the therapeutic potential of selected targets with the ultimate goal of translating these results to the clinical setting.

MOUSE MODELS OF *K-RAS* MUTANT LUNG ADENOCARCINOMA

Several laboratories have developed GEM models that faithfully reproduce human LAC by introducing oncogenic mutations into resident *K-Ras* alleles. These GEM models allow systematic and reproducible interrogation of the therapeutic properties of suspected targets using sophisticated genetic strategies. However, although these GEM models often recapitulate the natural history and histologic properties of human tumors, in general, they display a more limited mutational complexity than those human tumors they are intended to model (McFadden et al. 2016).

The first GEM model of *K-RAS* mutant LAC was developed by Jacks, Tuveson, and colleagues by generating mutant *K-Ras* alleles, designated as *K-Ras*^{LA1} and *K-Ras*^{LA2}, which after spontaneous somatic recombination encoded a functional *K-Ras*^{G12D} oncoprotein (Johnson et al. 2001). These mice developed LACs with 100% incidence as well as thymic lymphomas and papillomas with reduced prevalence. Soon thereafter, these investigators developed a more sophisticated model that expressed an endogenous *K-Ras*^{G12D} allele (*K-Ras*^{LSLG12D}) in a temporally and spatially controllable manner by Cre-mediated recombination (Jackson et al. 2001). Selective activation of K-Ras^{G12D} expression in lung tissue can be easily accomplished by intranasal instillation or intratracheal infection with adenoviral particles expressing the Cre recombinase. A similar strain carrying a G12V mutation was generated in our laboratory (Guerra et al. 2003). This strain, *K-Ras*$^{+/LSLG12Vgeo}$, also carries sequences that encode a bacterial protein (geo) that provides β-galactosidase activity that serves as a color marker (Guerra et al. 2003). Intratracheal infection with adenoviral particles yields results very similar to those reported by Tuveson, Jacks, and coworkers. More recently, Jacks and colleagues have generated a variant of their *K-Ras*$^{+/LSLG12D}$ strain in which the loxP sites flanking the transcriptional termination element were replaced by frt sequences (Young et al. 2011). This strain, *K-Ras*$^{+/FSFG12D}$, allows uncoupling of oncogene expression (mediated by the yeast Flp recombinase) from target inactivation or any other genetic manipulation that relies on the use of the Cre recombinase. It is worth mentioning that Varmus and colleagues developed another transgenic model by generating a strain that expressed the K-Ras4b^{G12D} isoform in alveolar type II cells under the control of a doxycyline response element (*Spc-rtTA-tetO-K-RAS4b*^{G12D}) (Fisher at al. 2001). This model gives rise to LACs similar to the GEM strains described above, and has also been used to show that these tumors require continuous expression of the K-Ras4b^{G12D} oncoprotein for maintenance.

Although all strains described above develop LACs that are highly reminiscent of human tumors, human LACs carry additional oncogenic insults that result in more aggressive neoplasias that are likely to alter therapeutic responses (Skoulidis et al. 2015). Some of these cooperating oncogenic insults have also been modeled in GEM models of *K-Ras* LAC by concomitantly ablating tumor suppressor genes such as *p53* (Jackson et al. 2005), *Ink4a/Arf* (Ji et al. 2007), or *Lkb1* (Ji et al. 2007). Loss of *p53* or expression of mutant *p53* alleles causes lung tumors to progress much more rapidly and eventually yield metastatic lesions (Jackson et al. 2005). Similar results were obtained by deletion of the *Ink4a/Arf* locus. Ablation of *Lkb1* caused tumors to progress even faster and to metastasize with higher incidence (Ji et al. 2007).

 Cite this article as *Cold Spring Harb Perspect Med* doi: 10.1101/cshperspect.a031542

TARGET VALIDATION: TUMOR INITIATION MODELS

GEM tumor models were first used to genetically validate targets of potential therapeutic interest by knocking-out candidate genes in the germline (providing they did not significantly affect viability) or by using conditional floxed alleles that were ablated concomitantly with the induction of *K-Ras* oncogene expression (Kwon and Berns 2013; Drosten and Barbacid 2016). Thus, the observed therapeutic responses may have limited clinical applicability, because the requisites to block tumor initiation are likely to be far less stringent than those necessary to induce tumor regression (Drosten and Barbacid 2016). The majority of these studies have focused on targets that are thought to play a role as K-Ras downstream effector molecules such as the members of the mitogen-activated protein kinase (MAPK) pathway (Raf, Mek, and Erk kinases, Etv5, Jnk, Cdk4, Smyd3), other Ras-like molecules (Ral, Rac) or metabolic enzymes (Acsl, Siva, Hk2, Cox2). Other studies have validated targets known to be involved in the Notch (Notch1, Psen1, Psen2, Rbpj), NF-κB (IκBα, p65/RelA, Ikkβ), or PI3K (p110α, p85α, p85β, mTRAIL-R) pathways. Genes encoding targets involved in other cellular processes such as ribosomal RNA (rRNA) synthesis (Ect2), telomere elongation (Trf1), senescence prevention (Wt1), and control of cell proliferation (Fadd, Csn1k1, Foxm1) have also been validated (Table 1). Altogether, these studies have helped to unravel mechanisms implicated in the early steps of tumor development. Because of space limitations, they will not be discussed in this review but they are listed in Table 1.

TARGET VALIDATION: THERAPEUTIC MODELS

K-Ras^{LA2} Model

This model develops spontaneous tumors. Hence, target validation can be performed using floxed alleles that can be inactivated on activation of a Cre recombinase in tumor-bearing animals (Table 2). Downward and colleagues used a conditional *Pik3ca*^{RBDlox} allele that on Cre-mediated recombination expresses a p110α PI3K subunit defective in Ras binding (Castellano et al. 2013). Acute expression of this p110α mutant isoform in established *K-Ras*^{G12D} lung tumors caused long-term tumor stasis and, in some cases, partial regression (Castellano et al. 2013). The same laboratory also identified a requirement for the Gata2 transcriptional network. Conditional elimination of Gata2 from established lung tumors led to near-complete tumor regression rates (Kumar et al. 2012). The same model was also used to ablate Hexokinase 2 from established tumors causing reduced tumor progression rates (Patra et al. 2013).

K-Ras^{LSLG12D} and K-Ras^{LSLG12Vgeo} Models

Since these models rely on the Cre recombinase for tumor induction, they are not particularly well suited to validate therapeutic targets in preexisting tumors because target ablation often relies on the Cre/loxP system. However, there are some examples in which target inactivation was performed by different experimental approaches. For instance, the *K-Ras*$^{+/LSLG12D}$ strain was used to validate Myc by expressing a dominant-negative peptide (Omomyc) using a doxycycline-dependent expression system (Soucek et al. 2008). Induction of Omomyc expression led to overt tumor regression even in aggressive tumors driven by mutant K-Ras expression and p53 ablation without significant toxicity under conditions of controlled Omomyc expression (Soucek et al. 2013). This model was also used to block the NF-κB pathway by infecting *K-Ras*$^{+/LSLG12D}$*;p53*$^{lox/lox}$ mice with a lentivirus that expressed a dominant negative IκBα super repressor under the control of a doxycycline-inducible promoter. Tumors still grew but at a lower rate than in control mice (Meylan et al. 2009). Finally, Tran et al. (2012) observed tumor stasis when they reduced Twist1 expression in *K-Ras*$^{+/LSLG12D}$ mice carrying advanced lung tumors using a doxycycline-dependent transgene. Finally, the related *K-Ras*$^{+/LSLG12Vgeo}$ model has been used to validate the therapeutic effect of Cdk4 by using con-

Table 1. Target validation in tumor initiation models driven by K-Ras oncogenes

Model	Target(s)	Tumor phenotype	References
$K\text{-}Ras^{+/LA2}$	$p110\alpha^{RBDmut/RBDmut}$	No tumor development	Castellano et al. 2013
$K\text{-}Ras^{+/LSLG12D}$	$c\text{-}Raf^{lox/lox}$	No tumor development	Karreth et al. 2011
$K\text{-}Ras^{+/LSLG12D}$	$RalA^{lox/lox}$;$RalB^{-/-}$	No tumor development	Peschard et al. 2012
$K\text{-}Ras^{+/LSLG12D}$	$Rac1^{lox/lox}$	No tumor development	Kissil et al. 2007
$K\text{-}Ras^{+/LSLG12D}$	$Jnk1^{lox/lox}$;$Jnk2^{-/-}$	No tumor development	Cellurale et al. 2011
$K\text{-}Ras^{+/LSLG12D}$	$Etv5^{lox/lox}$	No tumor development	Zhang et al. 2017
$K\text{-}Ras^{+/LSLG12D}$	$Pik3r1^{lox/lox}$;$Pik3r2^{-/-}$	Decreased tumor development	Engelman et al. 2008
$K\text{-}Ras^{+/LSLG12D}$	$Notch1^{lox/lox}$	Decreased tumor development	Licciulli et al. 2013
$K\text{-}Ras^{+/LSLG12D}$	$Wt1^{lox/lox}$	Decreased tumor development	Vicent et al. 2010
$K\text{-}Ras^{+/LSLG12D}$	$Fadd^{lox/lox}$	Decreased tumor development	Bowman et al. 2015
$K\text{-}Ras^{+/LSLG12D}$	$Csn1k1^{lox/lox}$	Decreased tumor development	Bowman et al. 2015
$K\text{-}Ras^{+/LSLG12D}$	$Siva^{lox/lox}$	Decreased tumor development	Van Nostrand et al. 2015
$K\text{-}Ras^{+/LSLG12D}$	$Smyd3^{-/-}$	Decreased tumor development	Mazur et al. 2014
$K\text{-}Ras^{+/LSLG12D}$	$Acsl3^{-/-}$	Decreased tumor development	Padanad et al. 2016
$K\text{-}Ras^{+/LSLG12D}$	$Cox2^{-/-}$	Decreased tumor development	Pan et al. 2015
$K\text{-}Ras^{+/LSLG12D}$	$Hk2^{lox/lox}$	Decreased tumor development	Patra et al. 2013
$K\text{-}Ras^{+/LSLG12D}$;$p53^{lox/lox}$	DN-IκBα (Tet-On)	Decreased tumor development	Meylan et al. 2009
$K\text{-}Ras^{+/LSLG12D}$;$p53^{lox/lox}$	$p65^{lox/lox}$	Decreased tumor development	Basséres et al. 2010
$K\text{-}Ras^{+/LSLG12D}$;$p53^{lox/lox}$	$Ect2^{lox/lox}$	Decreased tumor development	Justilien et al. 2017
$K\text{-}Ras^{+/LSLG12D}$;$p53^{lox/lox}$	$Trkb^{lox/lox}$	Reduced metastasis	Sinkevicius et al. 2014
$K\text{-}Ras^{+/LSLG12D}$; $p53^{+/LSLR172H}$	$mTRAIL\text{-}R^{lox/lox}$	Decreased tumor development	von Karstedt et al. 2015
$K\text{-}Ras^{+/LSLG12D}$;$sh\text{-}p53$	$Ikk\beta^{lox/lox}$	Decreased tumor development	Xia et al. 2012
$K\text{-}Ras^{+/LSLG12Vgeo}$	$c\text{-}Raf^{lox/lox}$	No tumor development	Blasco et al. 2011
$K\text{-}Ras^{+/LSLG12Vgeo}$	$Mek1^{lox/lox}$;$Mek2^{-/-}$	No tumor development	Blasco et al. 2011
$K\text{-}Ras^{+/LSLG12Vgeo}$	$Erk1^{-/-}$;$Erk2^{lox/lox}$	No tumor development	Blasco et al. 2011
$K\text{-}Ras^{+/LSLG12Vgeo}$	$Cdk4^{-/-}$	No tumor development	Puyol et al. 2010
$K\text{-}Ras^{+/LSLG12Vgeo}$	$Psen1^{lox/lox}$;$Psen2^{-/-}$	Decreased tumor development	Maraver et al. 2012
$K\text{-}Ras^{+/LSLG12Vgeo}$	$Rbpj^{lox/lox}$	Decreased tumor development	Maraver et al. 2012
$K\text{-}Ras^{+/LSLG12Vgeo}$;$p53^{-/-}$	$Trf1^{lox/lox}$	Decreased tumor development	García-Beccaria et al. 2015
$Tet\text{-}op\text{-}K\text{-}Ras4b^{G12D}$	$Foxm1^{lox/lox}$	Decreased tumor development	Wang et al. 2014

ditional $Cdk4^{frt}$ alleles (Puyol et al. 2010). However, this approach is not recommended because the Flp recombinase is less efficient than the bacterial Cre recombinase, resulting in limited target ablation.

Spc-rtTA-tetO-K-RAS4b^{G12D} Model

In this model, the K-Ras4b protein is selectively expressed in lung tissue (mainly SP-C^{+} AT2 cells) on treatment with doxycycline (Fisher et al. 2001). Once the tumors are formed, targets to be validated can be ablated by incorporating their corresponding floxed alleles along with an inducible CreERT2 recombinase followed by exposure to tamoxifen (Tam). This model has been used to show that ablation of the metabolic enzyme lactate dehydrogenase A caused limited regression of K-Ras^{G12D}-induced tumors (Xie et al. 2014).

K-Ras$^{+/FSFG12D}$ Model

This tumor model allows tumor induction on expression of the Flp recombinase in lung tissue, most frequently via intranasal instillation or intratracheal infection with adenoviral particles. Once tumors are formed, targets can be

Table 2. Target validation in established *K-Ras*-driven lung adenocarcinomas

Model	Target	Tumor phenotype	Systemic toxicity	References
K-Ras$^{+/LA2}$	*p110α*$^{RBDlox/-}$	Tumor stasis and partial regression	No significant toxicity	Castellano et al. 2013
K-Ras$^{+/LA2}$	*Gata2*$^{lox/lox}$	Regression of established tumors	No significant toxicity	Kumar et al. 2012
K-Ras$^{+/LSLG12D}$	Omomyc (c-Myc)	Long-term tumor eradication	Reversible toxicity	Soucek et al. 2013
K-Ras$^{+/LSLG12D}$	Twist1 (Tet-On)	Tumor stasis, senescence	ND[a]	Tran et al. 2012
K-Ras$^{+/LSLG12D}$; *p53*$^{ER/ER}$	Omomyc (c-Myc)	Long-term tumor eradication	Reversible toxicity	Soucek et al. 2013
K-Ras$^{+/LSLG12D}$; *p53*$^{lox/lox}$	DN-IκBα (Tet-On)	Inhibition of tumor progression, apoptosis	ND	Meylan et al. 2009
K-Ras$^{+/LSLG12Vgeo}$	*Cdk4*$^{frt/frt}$	Inhibition of tumor progression	No significant toxicity	Puyol et al. 2010
Tet-op-K-Ras4b^{G12D}	*Ldha*$^{lox/lox}$	Inhibition of tumor progression	Nonlethal anemia	Xie et al. 2014
K-Ras$^{+/FSFG12D}$; *p53*$^{frt/frt}$	*Atg7*$^{lox/lox}$	Inhibition of tumor progression	Long-term toxicity	Karsli-Uzunbas et al. 2014

[a]ND, Not determined.

validated by using conditional floxed alleles along with an inducible CreERT2 recombinase activated by exposure to Tam or by using inducible Cre transgenes. White and coworkers have used this model, including conditional *p53*^{frt} alleles (Lee et al. 2012), to show that the autophagy-related protein Atg7 was required for progression of *K-Ras/p53*-driven LAC tumors (Karsli-Uzunbas et al. 2014).

PHARMACOLOGICAL STUDIES

GEM tumor models, if properly designed, closely recapitulate the natural history of those human tumors they try to model. The fact that GEM tumors are initiated by the same endogenous mutations responsible for their human counterparts also serve as an excellent tool to validate pharmacological approaches before they are translated to the clinic. Indeed, GEM tumor models have proven their value in mouse coclinical trial platforms, where mice are subjected to experimental protocols that mirror clinical trials in human patients (Clohessy and Pandolfi 2015).

Johnson and colleagues have treated *K-Ras*$^{+/LSLG12D}$;*p53*$^{frt/frt}$ mice infected with adenoviruses expressing Flp and Cre recombinases to generate aggressive LACs, which were subsequently treated with an antibody against vascular endothelial growth factor (VEGF) in combination with standard chemotherapy under experimental conditions that mirrored prior clinical trials (Singh et al. 2010). The similarity between the results obtained in the mouse tumor models with those previously observed in the clinic proved the value of these models to predict clinical outcomes (Singh et al. 2010). Another coclinical trial showed that treatment with the MEK inhibitor Selumetinib in combination with chemotherapy was beneficial only in *K-Ras*$^{+/LSLG12D}$ or in compound *K-Ras*$^{+/LSLG12D}$;*p53*$^{lox/lox}$ mice, but not in *K-Ras*$^{+/LSLG12D}$;*Lkb1*$^{lox/lox}$ animals, thus indicating that loss of Lkb1 modified the therapeutic response (Chen et al. 2012). The same group has further shown that blocking two of the most relevant K-Ras downstream pathways with MEK and PI3K inhibitors resulted in synergistic shrinkage of lung tumors in *K-Ras*$^{+/LSLG12D}$ mice (Engelman et al. 2008). Another study has shown that combining the MEK inhibitor Trametinib with inhibition of FGFR1 was also effective in *K-Ras*$^{+/LSLG12D}$;*p53*$^{-/-}$ tumors (Manchado et al. 2016). Unfortunately, the differential tolerance of rodents and humans to MEK inhibitors have limited the clinical applicability of these studies.

Other investigators have reported that metastases from *K-Ras*$^{+/LSLG12D}$;*Lkb1*$^{lox/lox}$ mice showed increased Src signaling and that combination of MEK, PI3K, and SRC inhibitors was able to cause regression of these metastasis (Carretero et al. 2010). In another study, Scaglioni and colleagues observed that loss of *Ink4a/Arf* caused aberrant signaling through RhoA and FAK in *K-Ras*$^{+/LSLG12D}$ mice (Konstantinidou et al. 2013). Treatment with the FAK inhibitor PF562271 caused regression of lung tumors in *K-Ras*$^{+/LSLG12D}$;*Ink4a/Arf*$^{-/-}$ mice, but not of those initiated in *K-Ras*$^{+/LSLG12D}$ mice. Work from our laboratory has identified DDR1 as a promising target by focusing on genes induced during the earliest stages of tumor development (Ambrogio et al. 2016). Pharmacological treatment of established *K-Ras/p53*-driven lung tumors with a DDR1 inhibitor in combination with a Notch inhibitor, a pathway activated by DDR1, resulted in significant tumor regression with less toxic effects that standard chemotherapy.

More recently, the particular metabolic requirements of *K-Ras* mutant LAC were also exploited pharmacologically. Shaw and colleagues have shown that blocking fatty acid oxidation by pharmacologically blocking the enzyme Acetyl-CoA carboxylase with an allosteric inhibitor suppressed growth of tumors derived from both *K-Ras*$^{+/LSLG12D}$;*p53*$^{lox/lox}$ and *K-Ras*$^{+/LSLG12D}$;*Lkb1*$^{lox/lox}$ mice (Svennson et al. 2016). Similarly, Vander Heiden and coworkers have revealed increased glucose contribution to the tricarboxylic acid (TCA) cycle in *K-Ras* mutant lung tumors (Davidson et al. 2016). This property made inhibition of glutaminase ineffective, but suggested that inhibition of mitochondrial metabolism of pyruvate could be a beneficial strategy.

MOUSE MODELS OF PDAC

PDAC is one of the most lethal types of cancer for which there are no effective therapies available (Hidalgo 2010). Current experimental and pathological evidence suggests that PDAC stems from *K-RAS* mutations in acinar cells. When these cells undergo acinar-ductal cell metaplasia, a physiological and reversible process, they become blocked in a ductal-like phenotype that eventually progresses to form preneoplastic lesions known as pancreatic intraepithelial neoplasias (PanINs). Indeed, these benign lesions are present in many healthy pancreata (Maitra and Hruban 2008). Unfortunately, low-grade PanINs, such as PanIN1A, can progress to more aggressive lesions such as PanIN1B and PanIN2, possibly caused by additional mutations such as loss or inactivation of the *INK4a* tumor suppressor (Kanda et al. 2012). These lesions further progress to yield high-grade PanIN3, considered already as a form of in situ carcinoma, as well as invasive PDAC on accumulation of additional genetic insults, mainly inactivation of other tumor suppressors such as *P53*, *SMAD4*, or *BRCA2* (Maitra and Hruban 2008). During the last decade or so, several experimental models of human PDAC have been generated, most of them by engineering within the mouse genome those mutations responsible for the development of the human tumors. A brief description of the models most frequently used is described below and listed in Table 3.

K-Ras^{LSLG12D};Pdx1-Cre/P48-Cre Model

The first experimental model that faithfully reproduced the natural history of human PDAC in mice involved expression of an endogenous *K-Ras*^{G12D} oncogene in all epithelial pancreatic lineages during early embryonic development by crossing the conditional *K-Ras*$^{+/LSLG12D}$ strain described above with transgenic mice expressing the Cre recombinase under the control of the *Pdx1*- or the *Ptf1a/P48*-specific pancreatic promoters (Hingorani et al. 2003). Both compound strains, *K-Ras*$^{+/LSLG12D}$;*Pdx1-Cre* or *K-Ras*$^{+/LSLG12D}$;*Ptf1a/P48-Cre*, generically known as KC mice, develop the full spectrum of PanIN lesions and PDAC tumors histologically indistinguishable from those present in human patients (Hingorani et al. 2003). A variant of the KC mice in which the Cre recombinase is driven by a *Nestin-Cre* transgene has also

Table 3. Genetically engineered mouse pancreatic ductal adenocarcinoma (PDAC) models

Driver genes(s)	Recombinase	References
K-Ras$^{+/LSLG12D}$	*Pdx1-Cre*	Hingorani et al. 2003
K-Ras$^{+/LSLG12D}$	*Ptf1a/P48-Cre*	Hingorani et al. 2003
K-Ras$^{+/LSLG12D}$	*Nestin-Cre*	Carriere et al. 2007
K-Ras$^{+/LSLG12D}$	*Pdx1-CreERT*	Gidekel-Friedlander et al. 2009
K-Ras$^{+/LSLG12D}$	*Ela-CreER*	De La O et al. 2008
K-Ras$^{+/LSLG12D}$	*Ela-CreERT2*	Habbe et al. 2008
K-Ras$^{+/LSLG12D}$	*Mist1-CreERT2*	Habbe et al. 2008
K-Ras$^{+/LSLG12D}$	*proCPA1*-CreERT2	Gidekel-Friedlander et al. 2009
K-Ras$^{+/LSLG12D}$*;p53*$^{+/LSLR172H}$	*Pdx1-Cre* and *Ptf1a/P48-Cre*	Hingorani et al. 2005
K-Ras$^{+/LSLG12D}$*;p53*$^{lox/lox}$	*Pdx1-Cre* and *Ptf1a/P48-Cre*	Bardeesy et al. 2006
K-Ras$^{+/LSLG12D}$*;Ink4a/Arf*$^{lox/lox}$	*Pdx1-Cre*	Aguirre et al. 2003
K-Ras$^{+/LSLG12D}$*;Brca2*$^{lox/lox}$	*Pdx1-Cre*	Skoulidis et al. 2010
K-Ras$^{+/LSLG12D}$*;Lkb1*$^{+/lox}$	*Pdx1-Cre*	Morton et al. 2010
K-Ras$^{+/LSLG12D}$*;p21Cip1*$^{+/-}$	*Pdx1-Cre*	Morton et al. 2010
K-Ras$^{+/LSLG12D}$*;Smad4*$^{lox/lox}$	*Pdx1-Cre* and *Ptf1a/P48-Cre*	Bardeesy et al. 2006
K-Ras$^{+/LSLG12D}$*;Smad4*$^{lox/lox}$*;p53*$^{+/LSLR172H}$	*Ptf1a/P48-Cre*	Whittle et al. 2015
K-Ras$^{+/LSLG12D}$*;Smad4*$^{lox/lox}$*;Ink4a/Arf*$^{lox/lox}$	*Pdx1-Cre* and *Ptf1a/P48-Cre*	Bardeesy et al. 2006
K-Ras$^{+/LSLG12D}$*;TGFβRII*$^{lox/lox}$	*Ptf1a/P48-Cre*	Ijichi et al. 2006
K-Ras$^{+/LSLG12D}$*;Pten*$^{lox/lox}$	*Pdx1-Cre*	Ying et al. 2011
K-Ras$^{+/LSLG12D}$*;p53*$^{lox/lox}$	*Pdx1-CreERT*	Gidekel-Friedlander et al. 2009
K-Ras$^{+/LSLG12D}$*;p53*$^{lox/lox}$	*proCPA1-CreERT2*	Gidekel-Friedlander et al. 2009
K-Ras$^{+/LSLG12D}$*;Ink4a/Arf*$^{lox/lox}$	*Pdx1-CreERT*	Gidekel-Friedlander et al. 2009
K-Ras$^{+/LSLG12D}$*;Ink4a/Arf*$^{lox/lox}$	*proCPA1*-CreERT2	Gidekel-Friedlander et al. 2009
K-Ras$^{+/LSLG12Vgeo}$	*Elas-tTA;Tet-O-Cre*	Guerra et al. 2007
K-Ras$^{+/LSLG12Vgeo}$*;p53*$^{lox/lox}$	*Elas-tTA;Tet-O-Cre*	Guerra et al. 2011
K-Ras$^{+/LSLG12Vgeo}$*;Ink4a/Arf*$^{lox/lox}$	*Elas-tTA;Tet-O-Cre*	Guerra et al. 2011
TetO-K-Ras4b^{G12D}	*Ptf1a/P48-Cre;R26-LSLrtTA-IRES-EGFP*	Collins et al. 2012
TetO-LSL-K-Ras4b^{G12D}	*Ptf1a/P48-Cre;R26-LSLrtTA-IRES-EGFP*	Ying et al. 2012
K-Ras$^{+/FSFG12D}$	*Pdx1-Flp*	Schönhuber et al. 2014
K-Ras$^{+/FSFG12D}$	*Pdx1-Flp;R26-CAG-FSF-CreERT2*	Schönhuber et al. 2014
K-Ras$^{+/FSFG12D}$*;p53*$^{lox/lox}$	*Pdx1-Flp;R26-CAG-FSF-CreERT2*	Schönhuber et al. 2014
K-Ras$^{+/FSFG12V}$	*Elas-tTA;Tet-O-Flp*	MT Blasco, C Guerra, and M Barbacid, Unpubl.
K-Ras$^{+/FSFG12V}$	*Elas-tTA;Tet-O-Flp;hUBC-CreERT2*	MT Blasco, C Guerra, and M Barbacid, Unpubl.
K-Ras$^{+/FSFG12V}$*;p53*$^{frt/frt}$	*Elas-tTA;Tet-O-Flp*	MT Blasco, C Guerra, and M Barbacid, Unpubl.
K-Ras$^{+/FSFG12V}$*;p53*$^{frt/frt}$	*Elas-tTA;Tet-O-Flp;hUBC-CreERT2*	MT Blasco, C Guerra, and M Barbacid, Unpubl.

been generated (Carriere et al. 2007). Yet, the incidence of PDAC tumors in these strains is rather low, at least at 1 year of age. Additional mutations present in human tumors, including inactivation or deletion of tumor suppressors such as *p53* (KPC strain), *Ink4a/Arf*, *Brca2*, *Lkb1*, *p21Cip1*, or *Smad4* significantly accelerate progression of PanIN lesions to invasive PDAC and induce metastasis in a significant percentage of cases (Table 3). Similar results have been obtained on elimination of the genes encoding the transforming growth factor β (TGF-β) receptor II (*Tgfbr2*) and the Pten phosphatase (*Pten*) (Ijichi et al. 2006; Ying et al. 2011) (Table 3).

K-Ras^{LSLG12Vgeo};Elas-tTA;TetO-Cre Model

An independent GEM tumor model that also recapitulates human PDAC was generated by expressing an endogenous *K-Ras*^{G12V} oncogene in pancreatic acinar cells during late embryonic development (Guerra et al. 2007). This model, *K-Ras*$^{+/LSLG12Vgeo}$*;Elas-tTA;TetO-Cre,* uses an inducible Tet-Off strategy to express the Cre recombinase under the control of the elastase promoter (Guerra et al. 2007). This strain develops PanIN lesions as well as full-blown PDAC tumors with similar latencies and penetrance as the KC mice (Hingorani et al. 2003; Guerra et al. 2007). *K-Ras*$^{+/LSLG12Vgeo}$*;Elas-tTA;TetO-Cre* mice also show complete tumor penetrance and reduced survival on inactivation of the *p53* or *Ink4a/Arf* tumor suppressors (Guerra et al. 2011). This model has the advantage that Cre expression can be inhibited by doxycycline. Hence, continuous exposure of pregnant mothers and their offspring to this antibiotic prevents activation of these genetic insults. Doxycycline withdrawal at any time during postnatal development or in adult mice allows activation of *K-Ras* oncogene expression as well as ablation of the *p53* and/or *Ink4a/Arf* tumor suppressors. This strategy has been used to show that adult pancreatic acinar cells are completely resistant to transformation by the *K-Ras* oncogene even in the absence of *p53* (Guerra et al. 2011). These studies have also served to show the importance of pancreatic injury in the development of PDACs in adult mice (Guerra et al. 2007, 2011).

Inducible CreERT2-Based Models

Other investigators have developed GEM models of PDAC that allow expression of *K-Ras*^{G12D} oncogenes in adult mice by using the inducible CreERT2 recombinase driven by specific promoters. They include *Ela-CreER* (De La O et al. 2008), *Ela-CreERT2* (Habbe et al. 2008), *Mist1-CreERT2* (Habbe et al. 2008), *Pdx1-CreERT* (Gidekel-Friedlander et al. 2009), and *proCPA1-CreERT2* (Gidekel-Friedlander et al. 2009). Although these strains display differences in the incidence of PanIN development, it is clear that the susceptibility of acinar cells to *K-Ras* oncogene-induced transformation is progressively lost during postnatal development. The "age limit" for adult acinar cells to become resistant to *K-Ras*-induced transformation may depend on the genetic background, the expression pattern of the Cre recombinase, and the different conditional strategies.

Inducible K-Ras4b^{G12D} Transgenic Models

Some PDAC models allow to turn on and off the *K-Ras* driver oncogene at will (Collins et al. 2012). These models consist of a transgene that encodes the K-Ras4B^{G12D} oncoprotein by using a *TetO-K-Ras4b*^{G12D} complementary DNA (cDNA) (Fisher et al. 2001) or the variant construct *TetO-LSL-K-Ras4b*^{G12D} (Ying et al. 2012). Expression of K-Ras4B^{G12D} requires a reverse tetracycline-controlled transactivator (rtTA), which in these models was knocked in within the *Rosa26* locus (*R26-LSL-rtTA-IRES-EGFP* allele) (Belteki et al. 2005). Expression of the rtTA is restricted to pancreatic cells by means of the *Ptf1a/P48-Cre* transgene. The resulting strains, *TetO-K-Ras4b*^{G12D}*;R26-LSL-rtTA-IRES-EGFP;Ptf1a/P48-Cre* and *TetO-LSL-K-Ras4b*^{G12D}*;R26-LSL-rtTA-IRES-EGFP;Ptf1a/P48-Cre* express the K-Ras4B^{G12D} oncoprotein on exposure to doxycycline leading to the frequent generation of PanIN lesions but not PDAC tumors. Addition of a *p53* null (Collins et al. 2012) or floxed (Ying et al. 2012) alleles to these mice resulted in the induction of PDAC with full penetrance. These models have the disadvantage that the levels of *K-Ras* oncogene expression are dictated by the integration site of the *TetO-K-Ras4b*^{G12D} and *TetO-LSL-K-Ras4b*^{G12D} transgenes rather than by the normal *K-Ras* regulatory sequences. However, they allow the possibility to turn on and off the expression of the K-Rasb^{G12D} oncoprotein at will.

 Cite this article as *Cold Spring Harb Perspect Med* doi: 10.1101/cshperspect.a031542

Flp/frt-Based Models

More recently, a new generation of PDAC models has been developed by shifting their dependency from the bacterial Cre/loxP to the yeast Flp/frt recombinase system. As in the case of the lung tumor models, the use of two independent recombinase systems allows the spatial and temporal separation of tumor development from target ablation/inactivation, an experimental approach essential to evaluate the therapeutic potential of selected targets in preexisting tumors. Saur and colleagues used this dual recombinant system by expressing the Flp recombinase under the control of the *Pdx1* promoter (Schönhuber et al. 2014). The *Pdx1-Flp* transgene was incorporated to the *K-Ras*$^{+/FSFG12D}$ and *K-Ras*$^{+/FSFG12D}$*;p53*$^{lox/lox}$ strains to recapitulate the KC and KPC mice, respectively, originally developed by Tuveson and colleagues (Hingorani et al. 2003, 2005). These strains can be combined with any conditional Cre-based strains such as those that make use of the Tam-dependent CreERT2 or those that express Cre under an inducible promoter to ablate/inactivate those targets whose therapeutic activity needs to be evaluated. In the original model described by Saur and colleagues, the CreERT2 recombinase was part of a *CAG-FSF-CreERT2* transgene inserted within the *Rosa26* locus (*Rosa26*^{CAGFSFCreERT2} allele). A second *Rosa26* allele carrying a *CAG-LSL-Tomato* transgene was added to the model to mark those cells that express an active CreERT2 recombinase on exposure to Tam. Saur and colleagues also developed a variant model in which they incorporate *p53*^{lox} alleles to increase the tumorigenic properties of the K-Ras^{G12D} expressing cells. These models, however, have a limitation because the CreERT2 recombinase can only be expressed in those cells derived from the pancreatic precursors that expressed the *Pdx1-Flp* transgene. Thus, preventing the analysis of potential toxicities caused by the ablation/inactivation of the target in nonpancreatic tissues. This limitation could be easily overcome by using a *Rosa26* allele that allows the systemic expression of the Cre-ERT2 recombinase.

We have also developed a variant PDAC model that uses a two-recombinase strategy by replacing the *TetO-Cre* transgene present in the *K-Ras*$^{+/LSLG12Vgeo}$*;Elas-tTA;TetO-Cre* mice by a *TetO-Flp* insert. Likewise, the *K-Ras*^{LSLG12Vgeo} allele was replaced with a *K-Ras*^{FSFG12V} allele (MT Blasco, C Guerra, and M Barbacid, unpubl.). The resulting strain, *K-Ras*$^{+/FSFG12V}$*; Elas-tTA;TetO-Flp*, develops PanIN lesions and PDAC with similar incidence and latencies to mice carrying the Cre recombinase. Exposure of pregnant mothers and their offspring to doxycycline can also prevent expression of the oncogene. Finally, addition of *p53*^{frt} alleles to these mice (Lee et al. 2012) allows the rapid development of PDAC tumors with complete penetrance and kinetics similar to those observed in *K-Ras*$^{+/LSLG12Vgeo}$*;Elas-tTA;TetO-Cre;p53*$^{lox/lox}$ animals. Finally, a ubiquitously expressed hUBC-CreERT2 transgene was added to these mice to allow the ablation/inactivation of targets encoded by their corresponding floxed alleles on Tam exposure.

PDAC TUMORS: TARGET VALIDATION

The above tumor models have been instrumental in identifying targets with therapeutic properties in PDAC tumors (reviewed in Guerra and Barbacid 2013 and Gopinathan et al. 2015). Unfortunately, most of these studies have used strategies in which the target was ablated or inactivated before or at the time of tumor initiation, hence limiting the translational value of these results to a clinical scenario. Yet, targets whose inhibition can prevent or delay tumor development should be taken into consideration for testing in second generation models that allow their inactivation in advanced tumors. A summary of targets known to block or inhibit PanIN/PDAC development when ablated or inactivated at the time of *K-Ras* oncogene expression is shown in Table 4. They include molecules involved in epidermal growth factor receptor (EGFR) signaling (EGFR, ADAM17, Stat3), Notch signaling (Notch2), NF-κB pathway (Ikkβ), PI3K/mTOR pathway (p110α, Pdk1, Rictor, mTRAIL-R), actin reorganization (Rac1), lysosomal proteases

Table 4. Target validation in tumor initiation models of pancreatic ductal adenocarcinoma (PDAC) driven by K-Ras oncogenes

Model	Target	Tumor phenotype	References
K-Ras$^{+/LSLG12D}$;*Pdx1-Cre*	*Bmi1*$^{lox/lox}$	No PanIN/PDAC	Bednar et al. 2015
K-Ras$^{+/LSLG12D}$;*Pdx1-Cre*	*CathepsinB*$^{lox/lox}$	Reduced PanIN burden	Gopinathan et al. 2012
K-Ras$^{+/LSLG12D}$;*Pdx1-Cre*	*eNOS*$^{lox/lox}$	Reduced PanIN/PDAC formation	Lampson et al. 2012
K-Ras$^{+/LSLG12D}$;*Pdx1-Cre*	*Ikkβ*$^{lox/lox}$	Reduced PanIN/PDAC formation	Maniati et al. 2011
K-Ras$^{+/LSLG12D}$;*Pdx1-Cre*	*Ikkβ*$^{lox/lox}$	Reduced PanIN/PDAC formation	Ling et al. 2012
K-Ras$^{+/LSLG12D}$;*Pdx1-Cre*	*Stat3*$^{lox/lox}$	Reduced PanIN/PDAC formation	Corcoran et al. 2011
K-Ras$^{+/LSLG12D}$;*Pdx1-Cre*	*mTRAIL-R*$^{lox/lox}$	Reduced PanIN/PDAC formation	von Karstedt et al. 2015
K-Ras$^{+/LSLG12D}$;*Ptf1a/P48-Cre*	*EGFR*$^{lox/lox}$	No PanIN/PDAC	Ardito et al. 2012
K-Ras$^{+/LSLG12D}$;*Ptf1a/P48-Cre*	*ADAM17*$^{lox/lox}$	No PanIN/PDAC	Ardito et al. 2012
K-Ras$^{+/LSLG12D}$;*Ptf1a/P48-Cre*	*Notch2*$^{lox/lox}$	No PanIN/PDAC	Mazur et al. 2010
K-Ras$^{+/LSLG12D}$;*Ptf1a/P48-Cre*	*Stat3*$^{lox/lox}$	Reduced PanIN/PDAC formation	Fukuda et al. 2011
K-Ras$^{+/LSLG12D}$;*Ptf1a/P48-Cre*	*Stat3*$^{lox/lox}$	Reduced PanIN/PDAC formation	Lesina et al. 2011
K-Ras$^{+/LSLG12D}$;*Ptf1a/P48-Cre*	*Rictor*$^{lox/lox}$	Reduced PanIN/PDAC formation	Driscoll et al. 2016
K-Ras$^{+/LSLG12D}$;*Ptf1a/P48-Cre*	*Pdk1*$^{lox/lox}$	No PanIN/PDAC	Eser et al. 2013
K-Ras$^{+/LSLG12D}$;*Elas-CreERT*	*Ikkβ*$^{lox/lox}$	Reduced PanIN/PDAC formation	Daniluk et al. 2012
K-Ras$^{+/LSLG12D}$;*p53*$^{+/LSLR172H}$;*Pdx1-Cre*	*Rictor*$^{lox/lox}$	Increased survival	Driscoll et al. 2016
K-Ras$^{+/LSLG12D}$;*p53*$^{+/LSLR172H}$;*Pdx1-Cre*	*Pik3ca*$^{loxKD/loxKD}$	No tumor development	Baer et al. 2014
K-Ras$^{+/LSLG12D}$;*p53*$^{+/LSLR172H}$;*Pdx1-Cre*	*CathepsinB*$^{lox/lox}$	Increased survival	Gopinathan et al. 2012
K-Ras$^{+/LSLG12D}$;*p53*$^{+/LSLR172H}$;*Pdx1-Cre*	*eNOS*$^{lox/lox}$	Increased survival	Lampson et al. 2012
K-Ras$^{+/LSLG12D}$;*p53*$^{+/LSLR172H}$;*Pdx1-Cre*	*mTRAIL-R*$^{lox/lox}$	Increased survival	von Karstedt et al. 2015
K-Ras$^{+/LSLG12D}$;*p53*$^{+/LSLR172H}$;*Ptf1a/P48-Cre*	*Rac1*$^{lox/lox}$	No tumor development	Wu et al. 2014
Ras$^{+/LSLG12D}$;*p53*$^{+/LSLR172H}$;*Ptf1a/P48-Cre*	*Pik3ca*$^{lox/lox}$	No tumor development	Wu et al. 2014
K-Ras$^{+/LSLG12D}$;*p53*$^{lox/lox}$;*Ptf1a/P48-Cre*	*EGFR*$^{lox/lox}$	Delayed tumor development	Ardito et al. 2012
K-Ras$^{+/LSLG12D}$;*Ink4a/Arf*$^{-/-}$;*Pdx1-Cre*	*Bmi1*$^{lox/lox}$	Lack of PDAC rescue	Bednar et al. 2015
K-Ras$^{+/LSLG12D}$;*Ink4a/Arf*$^{lox/lox}$;*Pdx1-Cre*	*Ikkβ*$^{lox/lox}$	Reduced PanIN/PDAC formation	Ling et al. 2014
K-Ras$^{+/LSLG12Vgeo}$;*Elas-tTA;TetO-Cre*	*EGFR*$^{lox/lox}$	No PanIN or PDAC	Navas et al. 2012
K-Ras$^{+/LSLG12Vgeo}$;*p53*$^{lox/lox}$;*Elas-tTA;TetO-Cre*	*EGFR*$^{lox/lox}$	Delayed tumor development	Navas et al. 2012

(CathepsinB), ROS-related molecules (eNOS), and epigenetic regulators (Bmi1) (Table 4). So far, there are no studies in which targets have been validated in advanced established tumors possibly because of the recent generation of the two-recombinase models described above. Schönhuber et al. (2014) ablated expression of the Pdk1 kinase in PanIN containing young mice. Examinations of these animals 6 months later revealed the almost complete disappearance of these lesions. Whether Pdk1 ablation or inactivation will have a similar therapeutic effect in advanced PDAC remains to be determined.

K-RAS is considered to be a "nondruggable target" with the possible exception of the K-RAS^{G12C} isoform (Ostrem et al. 2013; Patricelli et al. 2016). Yet, it is of significant interest to determine whether direct targeting of this oncoprotein will result in complete tumor regression. Removal of doxycycline exposure results in significant tumor regression in both *TetO-K-Ras4b*^{G12D}*;R26-LSL-rtTA-IRES-EGFP;Ptf1a/P48-Cre* and *TetO-LSL-K-Ras4b*^{G12D}*;R26-LSL-rtTA-IRES-EGFP; Ptf1a/P48-Cre* mice (Collins et al. 2012; Ying et al. 2012). However, prolonged observation of these tumors revealed that as many as 70% of them relapsed between 2 and 10 months. Whereas half of these relapsed tumors reexpressed the *K-Ras4b*^{G12D} oncogene, the other half harbors other oncogenic insults, including *Yap1* amplification (Kapoor et al. 2014). Thus, indicating that even effective inhibition of K-RAS, the initial event in PDAC development, may not be sufficient to induce the complete and permanent regression of these tumors.

PHARMACOLOGICAL STUDIES

GEM PDAC models have also been instrumental in pharmacological studies. The fact that these tumors display the same histological features as human tumors, including the formation of an abundant desmoplastic stroma, makes them more informative than classical xenograft or even PDX models. Some of these studies include those involving the γ-secretase inhibitor, MRK-003, which blocked PanIN formation in KPC mice treated at very early stages (Plentz et al. 2009). In PDAC-bearing mice, this inhibitor, in combination with gemcitabine, induced increased survival (nine vs. 26 days). In a more recent study, Tuveson and colleagues (Chio et al. 2016) treated KPC mice with an AKT inhibitor (MK2206) and an inhibitor of glutathione biosynthesis (BSO), a combination that mimics Nfr2 down-regulation. In a week-long treatment, the combination slowed down tumor growth but did not cause tumor regression. Whether longer treatments will have a more profound therapeutic effect remains to be determined. Treatment of advanced PDAC-bearing KPC mice with JQ1, a BET inhibitor, in combination with the HDAC inhibitor SAHA, led to a significant increase in survival (15–47 days) (Mazur et al. 2015). More importantly, none of the treated mice died of PDAC. Indeed, histological analysis only revealed residual tumor clusters. Similar results were obtained in a related PDAC tumor model in which the *p53* conditional alleles were replaced by those of the *Ink4a/Arf* locus (Mazur et al. 2015).

TARGETING THE STROMA

GEM tumor models have also been used to ascertain the role of the desmoplastic stroma in PDAC. Although it has been generally accepted that the stroma enhances PDAC growth, two recent studies performed in KPC mice suggest that inhibiting the stroma may result in more aggressive pancreatic cancer (Ozdemir et al. 2014; Rhim et al. 2014). As expected, ablation of *Shh* in pancreatic cells (*K-Ras*$^{+/LSLG12D}$*; p53*$^{+/lox}$*;Pdx1-Cre* mice) resulted in a significant decrease in desmoplastic stroma. Yet, the PDAC tumors were more aggressive and significantly decreased survival of these mice. These results have been interpreted as evidence that inhibition of the desmoplastic stroma contributes rather than inhibits tumor development. However, it is also possible that loss of Shh may alter the tumorigenic properties of the tumor cells in a cell-autonomous manner. In the second study, Ozdemir et al. (2014) use the *K-Ras*$^{+/LSLG12D}$*;Tgfbr2*$^{lox/lox}$*;Ptf1a-Cre* model to ablate cancer-associated fibroblasts (CAFs), the main cellular component of the desmoplastic stroma, using a TK^+/ganciclovir strategy. Ablation of CAFs resulted in the death of the mice caused by pulmonary emphysema, thus precluding from establishing a direct correlation between the reduction in CAFs and PDAC-dependent survival. Yet, analysis of the tumors appeared to indicate a more aggressive phenotype. In support of these studies, pharmacological inhibition of stromal cells using the smoothened inhibitor IPI-926 favored tumor progression and reduced survival (Rhim et al. 2014). These results, however, are at variance with a previous study in which treatment of KPC mice with the same compound allowed improved gemcitabine delivery and increased therapeutic efficacy (Olive et al. 2009).

In a different set of studies, exposure of KPC mice to a new paclitaxel formulation (nanoparticle albumin-bound [nab]-paclitaxel), in combination with gemcitabine, slowed tumor growth and in some cases induced tumor regression (Frese et al. 2012; Alvarez et al. 2013). Interestingly, the advantage of this formulation appears to be primarily caused by a decrease in the main gemcitabine-metabolizing enzyme, cytidine deaminase, rather than an effect on the stroma (Frese et al. 2012).

The desmoplastic stroma of PDAC tumors is rich in glycosaminoglycan hyaluronan (HA), a property that may prevent drugs from reaching the tumor cells (Toole et al. 2008). Treatment of KPC mice with PEGPH20, a PEGylated human recombinant PH20 hyaluronidase, in combination with gemcitabine, led to inhibition of PDAC tumor growth and prolonged survival (Provenzano et al. 2012; Jacobetz et al. 2013). Based on these studies, several clinical trials combining PEGPH20 with gemcitabine have already been initiated.

Regardless of whether the desmoplastic stroma has pro- or antitumoral effects, this tissue clearly serves as a barrier for T cells, thus preventing the potential use of checkpoint inhibitors, such as anti-CTLA4 or anti-PD1 antibodies, to treat PDAC patients. In a recent study, Feig and colleagues (2013) observed that CAF depletion allowed checkpoint antagonists to control tumor growth. The inhibitory activity of CAFs was mediated by the chemokine CXCL12, because treatment of KPC mice with a CXCL12 receptor inhibitor, AMD3100, induced the rapid accumulation of T cells among the tumor cells and acted synergistically with an antibody against the PD1 ligand, PD1-L1 (Feig et al. 2013). Whether these strategies can be successfully translated to the clinic remains to be determined.

CONCLUDING REMARKS

Several challenges need to be resolved before the results obtained with GEM tumor models, such as those summarized here, can smoothly and steadily reach the clinic. First of all, target or drug validation studies have to be routinely performed in tumor-bearing mice, ideally harboring advanced or metastatic tumors. In addition, target ablation/inactivation should take place systemically and not only in the tumor cells or in selected tissues to mimic future studies with selected inhibitors. In pharmacological studies, it is essential to take into consideration the differential susceptibility to secondary toxic effects of humans versus rodents (e.g., MEK inhibitors) as well as their distinct pharmacodynamic and pharmacokinetic properties.

Another challenge involves the need to carry out combination studies. The mutational load of most human tumors strongly suggests that the successful use of precision medicine will require interference with several targets, either at the same time or sequentially. Use of single targeted therapies (erlotinib, vemurafenib, etc.) has already highlighted their weaknesses: mainly incomplete responses and appearance of resistance mechanisms. Combination studies in GEM tumor models will represent an important challenge for genetic strategies caused by the already significant complexity of the GEM strains. Along the same lines, future target validation experiments should make use of conditional knocked-in alleles in which targets are inactivated (e.g., by expressing a kinase dead isoform) rather than ablated as with standard conditional floxed alleles, because, as of today, protein ablation cannot be achieved by standard medicinal chemistry. It will also be important that future target validations studies, either ge-

 Cite this article as *Cold Spring Harb Perspect Med* doi: 10.1101/cshperspect.a031542

netic or pharmacological, give more attention to the ultimate outcome of their results. A 50% or even an 80% decrease in tumor size in a GEM tumor model may have little value in the clinic, where these responses can be often obtained with classical chemotherapy treatments. Likewise, results that yield increased survival by a few days or weeks may not represent significant benefits if translated to human patients.

Finally, we should always keep in mind that no matter how similar GEM tumor models are to the human disease, mouse tumors are considerably smaller, they take less time to develop, and have a more limited mutational load. In any case, we hope that studies using GEM tumor models such as those summarized here will become an indispensable tool in preclinical studies and serve to streamline the translation of novel therapeutic strategies into future clinical trials.

ACKNOWLEDGMENTS

This work is supported by grants from the European Research Council (ERC-ADG/695566-THERACAN), the Spanish Ministry of Economy and Competitiveness (SAF2014-59864-R), and the Spanish Association against Cancer (AECC, GC16173694BARB).

REFERENCES

Aguirre AJ, Bardeesy N, Sinha M, Lopez L, Tuveson DA, Horner J, Redston MS, DePinho RA. 2003. Activated Kras and Ink4a/Arf deficiency cooperate to produce metastatic pancreatic ductal adenocarcinoma. *Genes Dev* **17:** 3112–3126.

Alvarez R, Musteanu M, Garcia-Garcia E, Lopez-Casas PP, Megias D, Guerra C, Muñoz M, Quijano Y, Cubillo A, Rodriguez-Pascual J, et al. 2013. Stromal disrupting effects of nab-paclitaxel in pancreatic cancer. *Br J Cancer* **109:** 926–933.

Ambrogio C, Gómez-López G, Falcone M, Vidal A, Nadal E, Crosetto N, Blasco RB, Fernández-Marcos PJ, Sánchez-Céspedes M, Ren X, et al. 2016. Combined inhibition of DDR1 and Notch signaling is a therapeutic strategy for KRAS-driven lung adenocarcinoma. *Nat Med* **22:** 270–277.

Ardito CM, Gruner BM, Takeuchi KK, Lubeseder-Martellato C, Teichmann N, Mazur PK, DelGiorno KE, Carpenter ES, Halbrook CJ, Hall JC, et al. 2012. EGF receptor is required for KRAS-induced pancreatic tumorigenesis. *Cancer Cell* **22:** 304–317.

Baer R, Cintas C, Dufresne M, Cassant-Sourdy S, Schönhuber N, Planque L, Lulka H, Couderc B, Bousquet C, Garmy-Susini B, et al. 2014. Pancreatic cell plasticity and cancer initiation induced by oncogenic Kras is completely dependent on wild-type PI 3-kinase p110α. *Genes Dev* **28:** 2621–2635.

Bardeesy N, Aguirre AJ, Chu GC, Cheng KH, Lopez LV, Hezel AF, Feng B, Brennan C, Weissleder R, Mahmood U, et al. 2006. Both p16(Ink4a) and the p19(Arf)-p53 pathway constrain progression of pancreatic adenocarcinoma in the mouse. *Proc Natl Acad Sci* **103:** 5947–5952.

Basséres DS, Ebbs A, Levantini E, Baldwin AS. 2010. Requirement of the NK-κB subunit p65/RelA for K-Ras-induced lung tumorigenesis. *Cancer Res* **70:** 3537–3546.

Bednar F, Schofield HK, Collins MA, Yan W, Zhang Y, Shyam N, Eberle JA, Almada LL, Olive KP, et al. 2015. Bmi1 is required for the initiation of pancreatic cancer through an Ink4a-independent mechanism. *Carcinogenesis* **36:** 730–738.

Belteki G, Haigh J, Kabacs N, Haigh K, Sison K, Costantini F, Whitsett J, Quaggin SE, Nagy A. 2005. Conditional and inducible transgene expression in mice through the combinatorial use of Cre-mediated recombination and tetracycline induction. *Nucleic Acids Res* **33:** e51.

Blasco RB, Francoz S, Santamaría D, Cañamero M, Dubus P, Charron J, Baccarini M, Barbacid M. 2011. C-Raf, but not B-Raf, is essential for development of K-Ras oncogene-driven non-small-cell lung carcinoma. *Cancer Cell* **19:** 1–12.

Bowman BM, Sebolt KA; Hoff BA, Boes JL, Daniels DL, Heist KA, Galbán CJ, Patel RM, Zhang J, Beer DG, et al. 2015. Phosphorylation of FADD by the kinase CK1α promotes KRASG12D-induced lung cancer. *Sci Signal* **8:** ra9.

Carretero J, Shimamura T, Rikova K, Jackson AL, Wilkerson MD, Borgman CL, Buttarazzi MS, Sanofsky BA, McNamara KL, Brandstetter KA, et al. 2010. Integrative genomic and proteomic analyses identify targets for Lkb1-deficient metastatic lung tumors. *Cancer Cell* **17:** 547–559.

Carriere C, Seeley ES, Goetze T, Longnecker DS, Korc M. 2007. The Nestin progenitor lineage is the compartment of origin for pancreatic intraepithelial neoplasia. *Proc Natl Acad Sci* **104:** 4437–4442.

Castellano E, Sheridan C, Thin MZ, Nye E, Spencer-Dene B, Diefenbacher ME, Moore C, Kumar MS, Murillo MM, Grönroos E, et al. 2013. Requirement for interaction of PI3-kinase p110α with Ras in lung tumor maintenance. *Cancer Cell* **24:** 617–630.

Cellurale C, Sabio G, Kennedy NJ, Das M, Barlow M, Sandy P, Jacks T, Davis RJ. 2011. Requirement of c-Jun NH_2-terminal kinase for Ras-initiated tumor formation. *Mol Cell Biol* **31:** 1565–1576.

Chen Z, Cheng K, Walton Z, Wang Y, Ebi H, Shimamura T, Liu Y, Tupper T, Ouyang J, Li J, et al. 2012. A murine lung cancer co-clinical trial identifies genetic modifiers of therapeutic response. *Nature* **483:** 613–617.

Chio II, Jafarnejad SM, Ponz-Sarvise M, Park Y, Rivera K, Palm W, Wilson J, Sangar V, Hao Y, Ohlund D, et al. 2016. NRF2 promotes tumor maintenance by modulating mRNA translation in pancreatic cancer. *Cell* **166:** 963–976.

Clohessy JG, Pandolfi PP. 2015. Mouse hospital and co-clinical trial project-from bench to bedside. *Nat Rev Clin Oncol* **12:** 491–498.

Collins MA, Bednar F, Zhang Y, Brisset JC, Galban S, Galban CJ, Rakshit S, Flannagan KS, Adsay NV, di Magliano MP. 2012. Oncogenic Kras is required for both the initiation and maintenance of pancreatic cancer in mice. *J Clin Invest* **122:** 639–653.

Corcoran RB, Contino G, Deshpande V, Tzatsos A, Conrad C, Benes CH, Levy DE, Settleman J, Engelman JA, Bardeesy N. 2011. STAT3 plays a critical role in KRAS-induced pancreatic tumorigenesis. *Cancer Res* **71:** 5020–5029.

Daniluk J, Liu Y, Deng D, Chu J, Huang H, Gaiser S, Cruz-Monserrate Z, Wang H, Ji B, Logsdon CG. 2012. An NF-κB pathway-mediated positive feedback loop amplifies Ras activity to pathological levels in mice. *J Clin Invest* **122:** 1519–1528.

Davidson SM, Papgiannakopoulos T, Olenchock BA, Heyman JE, Keibler MA, Luengo A, Bauer MR, Jha AK, O'Brian JP, Pierce KA, et al. 2016. Environment impacts the metabolic dependencies of Ras-driven non-small cell lung cancer. *Cell Metab* **23:** 517–528.

De La O J, Emerson LL, Goodman JL, Froebe SC, Illum BE, Curtis AB, Murtaugh LC. 2008. Notch and Kras reprogram pancreatic acinar cells to ductal intraepithelial neoplasia. *Proc Natl Acad Sci* **105:** 18907–18912.

Driscoll DR, Karim SA, Sano M, Gay DM, Jacob W, Yu J, Mizukami Y, Gopinathan A, Jodrell DI, Evans TR, et al. 2016. mTORC2 signaling drives the development and progression of pancreatic cancer. *Cancer Res* **76:** 6911–6923.

Drosten M, Barbacid M. 2016. Modeling K-Ras-driven lung adenocarcinoma in mice: Preclinical validation of therapeutic targets. *J Mol Med* **94:** 121–135.

Engelman JA, Chen L, Tan X, Crosby K, Guimaraes AR, Upadhyay R, Maira M, McNamara K, Perera SA, Song Y, et al. 2008. Effective use of PI3K and MEK inhibitors to treat Kras G12D and PIK3CA H1047R murine lung cancers. *Nat Med* **14:** 1351–1359.

Eser S, Reiff N, Messer M, Seidler B, Gottschalk K, Dobler M, Hieber M, Arbeiter A, Klein S, Kong B, et al. 2013. Selective requirement of PI3K/PDK1 signaling for Kras oncogene-driven pancreatic cell plasticity and cancer. *Cancer Cell* **23:** 406–420.

Eser S, Schnieke A, Schneider G, Saur D. 2014. Oncogenic KRAS signaling in pancreatic cancer. *Br J Cancer* **111:** 817–822.

Feig C, Jones JO, Kraman M, Wells RJ, Deonarine A, Chan DS, Connell CM, Roberts EW, Zhao Q, Caballero OL, et al. 2013. Targeting CXCL12 from FAP-expressing carcinoma-associated fibroblasts synergizes with anti-PD-L1 immunotherapy in pancreatic cancer. *Proc Natl Acad Sci* **110:** 20212–20217.

Fisher GH, Wellen SL, Klimstra D, Lenczowski JM, Tichelaar JW, Lizak MJ, Whitsett JA, Koretsky A, Varmus HE. 2001. Induction and apoptotic regression of lung adenocarcinomas by regulation of a K-Ras transgene in the presence and absence of tumor suppressor genes. *Genes Dev* **15:** 3249–3262.

Frese KK, Neesse A, Cook N, Bapiro TE, Lolkema MP, Jodrell DI, Tuveson DA. 2012. nab-Paclitaxel potentiates gemcitabine activity by reducing cytidine deaminase levels in a mouse model of pancreatic cancer. *Cancer Discov* **2:** 260–269.

Fukuda A, Wang SC, Morris JP IV, Folias AE, Liou A, Kim GE, Akira S, Boucher KM, Firpo MA, Mulvihill SJ, et al. 2011. Stat3 and MMP7 contribute to pancreatic ductal adenocarcinoma initiation and progression. *Cancer Cell* **19:** 441–455.

García-Beccaria M, Martínez P, Méndez-Pertuz M, Martínez S, Blanco-Aparicio C, Cañamero M, Mulero F, Ambrogio C, Flores JM, Megias D, et al. 2015. Therapeutic inhibition of TRF1 impairs the growth of p53–deficient K-RasG12V-induced lung cancer by indiction of telomeric DNA damage. *EMBO Mol Med* **7:** 930–949.

Gidekel-Friedlander SY, Chu GC, Snyder EL, Girnius N, Dibelius G, Crowley D, Vasile E, DePinho RA, Jacks T. 2009. Context-dependent transformation of adult pancreatic cells by oncogenic K-Ras. *Cancer Cell.* **16:** 379–389.

Gopinathan A, Denicola GM, Frese KK, Cook N, Karreth FA, Mayerle J, Lerch MM, Reinheckel T, Tuveson DA. 2012. Cathepsin B promotes the progression of pancreatic ductal adenocarcinoma in mice. *Gut* **61:** 877–884.

Gopinathan A, Morton JP, Jodrell DI, Sansom OJ. 2015. GEMMs as preclinical models for testing pancreatic cancer. *Dis Model Mech* **8:** 1185–1200.

Guerra C, Barbacid M. 2013. Genetically engineered mouse models of pancreatic adenocarcinoma. *Mol Oncol* **7:** 232–247.

Guerra C, Mijimolle N, Dhawahir A, Dubus P, Barradas M, Serrano M, Campuzano V, Barbacid M. 2003. Tumor induction by an endogenous K-ras oncogene is highly dependent on cellular context. *Cancer Cell* **4:** 111–120.

Guerra C, Schuhmacher AJ, Cañamero M, Grippo PJ, Verdaguer L, Perez-Gallego L, Dubus P, Sandgren EP, Barbacid M. 2007. Chronic pancreatitis is essential for induction of pancreatic ductal adenocarcinoma by K-Ras oncogenes in adult mice. *Cancer Cell* **11:** 291–302.

Guerra C, Collado M, Navas C, Schuhmacher AJ, Hernandez-Porras I, Canamero M, Rodriguez-Justo M, Serrano M, Barbacid M. 2011. Pancreatitis-induced inflammation contributes to pancreatic cancer by inhibiting oncogene-induced senescence. *Cancer Cell* **19:** 728–739.

Habbe N, Shi G, Meguid RA, Fendrich V, Esni F, Chen H, Feldmann G, Stoffers DA, Konieczny SF, Leach SD, et al. 2008. Spontaneous induction of murine pancreatic intraepithelial neoplasia (mPanIN) by acinar cell targeting of oncogenic Kras in adult mice. *Proc Natl Acad Sci* **105:** 18913–18918.

Hidalgo M. 2010. Pancreatic cancer. *N Engl J Med* **362:** 1605–1617.

Hingorani SR, Petricoin EF, Maitra A, Rajapakse V, King C, Jacobetz MA, Ross S, Conrads TP, Veenstra TD, Hitt BA, et al. 2003. Preinvasive and invasive ductal pancreatic cancer and its early detection in the mouse. *Cancer Cell* **4:** 437–450.

Hingorani SR, Wang L, Multani AS, Combs C, Deramaudt TB, Hruban RH, Rustgi AK, Chang S, Tuveson DA. 2005. $Trp53^{R172H}$ and $Kras^{G12D}$ cooperate to promote chromosomal instability and widely metastatic pancreatic ductal adenocarcinoma in mice. *Cancer Cell* **7:** 469–483.

Ijichi H, Chytil A, Gorska AE, Aakre ME, Fujitani Y, Fujitani S, Wright CV, Moses HL. 2006. Aggressive pancreatic ductal adenocarcinoma in mice caused by pancreas-specific blockade of transforming growth factor-β signaling in cooperation with active Kras expression. *Genes Dev* **20:** 3147–3160.

Jackson EL, Willis N, Mercer K, Bronson RT, Crowley D, Montoya R, Jacks T, Tuveson DA. 2001. Analysis of lung tumor initiation and progression using conditional expression of oncogenic K-ras. *Genes Dev* **15:** 3243–3248.

Jackson EL, Olive KP, Tuveson DA, Bronson R, Crowley D, Brown M, Jacks T. 2005. The differential effects of mutant p53 alleles on advanced murine lung cancer. *Cancer Res* **65:** 10280–10288.

Jacobetz MA, Chan DS, Neesse A, Bapiro TE, Cook N, Frese KK, Feig C, Nakagawa T, Caldwell ME, Zecchini HI, et al. 2013. Hyaluronan impairs vascular function and drug delivery in a mouse model of pancreatic cancer. *Gut* **62:** 112–120.

Ji H, Ramsey MR, Hayes DN, Fan C, McNamara K, Kozlowski P, Torrice C, Wu MC, Shimamura T, Perera SA, et al. 2007. LKB1 modulates lung cancer differentiation and metastasis. *Nature* **448:** 807–810.

Johnson L, Mercer K, Greenbaum D, Bronson RT, Crowley D, Tuveson DA, Jacks T. 2001. Somatic activation of the K-ras oncogene causes early onset lung cancer in mice. *Nature* **410:** 1111–1116.

Justilien V, Ali SA, Jamieson L, Yin N, Cox AD, Der CJ, Murray NR, Fields AP. 2017. Ect2-dependent rRNA synthesis is required for KRAS-TRP53-driven lung adenocarcinoma. *Cancer Cell* **31:** 256–269.

Kanda M, Matthaei H, Wu J, Hong SM, Yu J, Borges M, Hruban RH, Maitra A, Kinzler K, Vogelstein B, et al. 2012. Presence of somatic mutations in most early-stage pancreatic intraepithelial neoplasia. *Gastroenterology* **142:** 730–733.

Kapoor A, Yao W, Ying H, Hua S, Liewen A, Wang Q, Zhong Y, Wu CJ, Sadanandam A, Hu B, et al. 2014. Yap1 activation enables bypass of oncogenic Kras addiction in pancreatic cancer. *Cell* **158:** 185–197.

Karreth FA, Frese KK, DeNicola GM, Baccarini M, Tuveson DA. 2011. C-Raf is required for the initiation of lung cancer by K-Ras^{G12D}. *Cancer Discov* **1:** 128–136.

Karsli-Uzunbas G, Huo JY, Price S, Teng X, Laddha SV, Khor S, Kalaany NY, Jacks T, Chan CS, Rabinowitz JD, et al. 2014. Autophagy is required for glucose homeostasis and lung tumor maintenance. *Cancer Discov* **4:** 914–927.

Kissil JL, Walmsley MJ, Hanlon L, Haigis KM, Bender Kim CF, Sweet-Cordero A, Eckman MS, Tuveson DA, Capobianco AJ, Tybulewicz VL, et al. 2007. Requirement for Rac1 in a K-ras induced lung cancer in the mouse. *Cancer Res* **67:** 8089–8094.

Konstantinidou G, Ramadori G, Torti F, Kangasniemi K, Ramirez RE, Cai Y, Behrens C, Dellinger MT, Brekken RA, Wistuba II, et al. 2013. RHOA-FAK is a required signaling axis for the maintenance of KRAS-driven lung adenocarcinomas. *Cancer Discov* **3:** 444–457.

Kumar MS, Hancock DC, Molina-Arcas M, Steckel M, East P, Diefenbacher M, Armenteros-Monterroso E, Lassailly F, Matthews N, Nye E, et al. 2012. The GATA2 transcriptional network is requisite for *RAS* oncogene-driven non-small-cell lung cancer. *Cell* **149:** 642–655.

Kwon MC, Berns A. 2013. Mouse models for lung cancer. *Mol Oncol* **7:** 165–177.

Lampson BL, Kendall SD, Ancrile BB, Morrison MM, Shealy MJ, Barrientos KS, Crowe MS, Kashatus DF, White RR, Gurley SB, et al. 2012. Targeting eNOS in pancreatic cancer. *Cancer Res* **72:** 4472–4482.

Lawrence MS, Stojanov P, Mermel CH, Robinson JT, Garraway LA, Golub TR, Meyerson M, Gabriel SB, Lander ES, Getz G. 2014. Discovery and saturation analysis of cancer genes across 21 tumour types. *Nature* **505:** 495–501.

Lee CL, Moding EJ, Huang X, Li Y, Woodlief LZ, Rodrigues RC, Ma Y, Kirsch DG. 2012. Generation of primary tumors with Flp recombinase in FRT-flanked p53 mice. *Dis Model Mech* **5:** 397–402.

Lesina M, Kurkowski MU, Ludes K, Rose-John S, Treiber M, Klöppel G, Yoshimura A, Reindl W, Sipos B, Akira S, et al. 2011. Stat3/Socs3 activation by IL-6 transsignaling promotes progression of pancreatic intraepithelial neoplasia and development of pancreatic cancer. *Cancer Cell* **19:** 456–469.

Licciulli S, Avila JL, Hanlon L, Troutman S, Cesaroni M, Kota S, Keith B, Simon MC, Puré E, Radtke F, et al. 2013. Notch1 is required for Kras-induced lung adenocarcinoma and controls tumor cell survival via p53. *Cancer Res* **73:** 5974–5984.

Ling J, Kang Y, Zhao R, Xia Q, Lee DF, Chang Z, Li J, Peng B, Fleming JB, Wang H, et al. 2012. Kras^{G12D}-induced IKK2/β/NF-κB activation by IL-1α and p62 feedforward loops is required for development of pancreatic ductal adenocarcinoma. *Cancer Cell* **21:** 105–120.

Maitra A, Hruban RH. 2008. Pancreatic cancer. *Annu Rev Pathol* **3:** 157–188.

Manchado E, Weissmueller S, Morris IV JP, Chen CC, Wullenkord R, Lujambio A, de Stanchina E, Poirier JT, Gainor JF, Corcoran RB, et al. 2016. A combinatorial strategy for treating KRAS-mutant lung cancer. *Nature* **534:** 647–651.

Maniati E, Bossard M, Cook N, Candido JB, Emami-Shahri N, Nedospasov SA, Balkwill FR, Tuveson DA, Hagemann T. 2011. Crosstalk between the canonical NF-κB and Notch signaling pathways inhibits Pparγ expression and promotes pancreatic cancer progression in mice. *J Clin Invest* **121:** 4685–4699.

Maraver A, Fernandez-Marcos PJ, Herranz D, Cañamero M, Muñoz-Martin M, Gómez-Lopez G, Mulero F, Megías D, Sanchez-Carbayo M, Shen J, et al. 2012. Therapeutic effect of γ-secretase inhibition in *Kras*^{G12V}-driven non-small cell lung carcinoma by derepression of DUSP1 and inhibition of ERK. *Cancer Cell* **22:** 222–234.

Mazur PK, Einwachter H, Lee M, Sipos B, Nakhai H, Rad R, Zimber-Strobl U, Strobl LJ, Radtke F, Klöppel G, et al. 2010. Notch2 is required for progression of pancreatic intraepithelial neoplasia and development of pancreatic ductal adenocarcinoma. *Proc Natl Acad Sci* **107:** 13438–13443.

Mazur PK, Reynoird N, Khatri P, Jansen PW, Wilkinson AW, Liu S, Barbash O, Van Aller GS, Huddleston M, Dhanak D, et al. 2014. SMYD3 links lysine methylation of MAP3K2 to Ras-driven cancer. *Nature* **510:** 283–287.

Mazur PK, Herner A, Mello SS, Wirth M, Hausmann S, Sánchez-Rivera FJ, Lofgren SM, Kuschma T, Hahn SA,

Vangala D, et al. 2015. Combined inhibition of BET family proteins and histone deacetylases as a potential epigenetics-based therapy for pancreatic ductal adenocarcinoma. *Nat Med* **21:** 1163–1171.

McFadden DG, Politi K, Bhutkar A, Chen FK, Song X, Pirun M, Santiago PM, Kim-Kiselak C, Platt JT, Lee E, et al. 2016. Mutational landscape of EGFR-, MYC-, and Kras-driven genetically engineered mouse models of lung adenocarcinoma. *Proc Natl Acad Sci* **113:** E6409–E6417.

Meylan E, Dooley AL, Feldser DM, Shen L, Turk E, Ouyang C, Jacks T. 2009. Requirement for NK-κB signalling in a mouse model of lung adenocarcinoma. *Nature* **462:** 104–107.

Morton JP, Jamieson NB, Karim SA, Athineos D, Ridgway RA, Nixon C, McKay CJ, Carter R, Brunton VG, Frame MC, et al. 2010. LKB1 haploinsufficiency cooperates with Kras to promote pancreatic cancer through suppression of p21-dependent growth arrest. *Gastroenterology* **139:** 586–597.

Navas C, Hernandez-Porras I, Schumacher AJ, Sibilia M, Guerra C, Barbacid M. 2012. EGF receptor signaling is essential for K-*Ras* oncogene-driven pancreatic ductal adenocarcinoma. *Cancer Cell* **22:** 318–330.

Olive KP, Jacobetz MA, Davidson CJ, Gopinathan A, McIntyre D, Honess D, Madhu B, Goldgraben MA, Caldwell ME, Allard D, et al. 2009. Inhibition of Hedgehog signaling enhances delivery of chemotherapy in a mouse model of pancreatic cancer. *Science* **324:** 1457–1461.

Ostrem JM, Peters U, Sos ML, Wells JA, Shokat KM. 2013. K-Ras^{G12C} inhibitors allosterically control GTP affinity and effector interactions. *Nature* **503:** 548–551.

Ozdemir BC, Pentcheva-Hoang T, Carstens JL, Zheng X, Wu CC, Simpson TR, Laklai H, Sugimoto H, Kahlert C, Novitskiy SV, et al. 2014. Depletion of carcinoma-associated fibroblasts and fibrosis induces immunosuppression and accelerates pancreas cancer with reduced survival. *Cancer Cell* **25:** 719–734.

Padanad MS, Konstantinidou G, Venkateswaran N, Melegari M, Rindhe S, Mitsche M, Yang C, Batten K, Huffman KE, Liu J, et al. 2016. Fatty acid oxidation mediated by Acyl-CoA synthetase long chain 3 is required for mutant *KRAS* lung tumorigenesis. *Cell Rep* **16:** 1614–1628.

Pan Y, Jiang Y, Tan L, Ravoori MK, Gagea M, Kundra V, Fischer SM, Yang P. 2015. Deletion of cyclooxygenase-2 inhibits K-ras-induced lung carcinogenesis. *Oncotarget* **6:** 38816–38826.

Patra KC, Wang Q, Bhaskar PT, Miller L, Wang Z, Wheaton W, Chandel N, Laakso M, Muller WJ, Allen EL, et al. 2013. Hexokinase 2 is required for tumor initiation and maintenance and its systemic deletion is therapeutic in mouse models of cancer. *Cancer Cell* **24:** 213–228.

Patricelli MP, Janes MR, Li LS, Hansen R, Peters U, Kessler LV, Chen Y, Kucharski JM, Feng J, et al. 2016. Selective inhibition of oncogenic KRAS output with small molecules targeting the inactive state. *Cancer Discov* **6:** 316–329.

Peschard P, McCarthy A, Leblanc-Dominguez V, Yeo M, Guichard S, Stamp G, Marshall CJ. 2012. Genetic deletion of RALA and RALB small GTPases reveals redundant functions in development and tumorigenesis. *Curr Biol* **22:** 2063–2068.

Plentz R, Park JS, Rhim AD, Abravanel D, Hezel AF, Sharma SV, Gurumurthy S, Deshpande V, Kenific C, Settleman J, et al. 2009. Inhibition of γ-secretase activity inhibits tumor progression in a mouse model of pancreatic ductal adenocarcinoma. *Gastroenterology* **136:** 1741–1749.

Provenzano PP, Cuevas C, Chang AE, Goel VK, Von Hoff DD, Hingorani SR. 2012. Enzymatic targeting of the stroma ablates physical barriers to treatment of pancreatic ductal adenocarcinoma. *Cancer Cell* **21:** 418–429.

Puyol M, Martín A, Dubus P, Mulero F, Pizcueta P, Khan G, Guerra C, Santamaría D, Barbacid M. 2010. A synthetic lethal interaction between *K-Ras* oncogenes and Cdk4 unveils a therapeutic strategy for non-small-cell lung carcinoma. *Cancer Cell* **18:** 63–73.

Pylayeva-Gupta Y, Grabocka E, Bar-Sagi D. 2011. RAS oncogenes: Weaving a tumorigenic web. *Nat Rev Cancer* **11:** 761–774.

Rhim AD, Oberstein PE, Thomas DH, Mirek ET, Palermo CF, Stephen A, Sastra SA, Dekleva EN, Saunders T, Becerra CP, et al. 2014. Stromal elements act to restrain, rather than support, pancreatic ductal adenocarcinoma. *Cancer Cell* **25:** 735–747.

Schönhuber N, Seidler B, Schuck K, Veltkamp C, Schachtler C, Zukowska M, Eser S, Feyerabend TB, Paul MC, Eser P, et al. 2014. A next-generation dual-recombinase system for time- and host-specific targeting of pancreatic cancer. *Nat Med* **20:** 1340–1337.

Singh M, Lima A, Molina R, Hamilton P, Clermont AC, Devasthali V, Thompson JD, Cheng JH, Reslan HB, Ho CC, et al. 2010. Assessing therapeutic responses in *Kras* mutant cancers using genetically engineered mouse models. *Nat Biotechnol* **28:** 585–593.

Sinkevicius KW, Kriegel C, Bellaria KJ, Lee J, Lau AN, Leeman KT, Zhou P, Beede AM, Fillmore CM, Caswell D, et al. 2014. Neurotrophin receptor TrkB promotes lung adenocarcinoma metastasis. *Proc Natl Acad Sci* **111:** 10299–10304.

Skoulidis F, Cassidy LD, Pisupati V, Jonasson JG, Bjarnason H, Eyfjord JE, Karreth FA, Lim M, Barber LM, et al. 2010. Germline *Brca2* heterozygosity promotes *Kras*^{G12D}-driven carcinogenesis in a murine model of familial pancreatic cancer. *Cancer Cell* **18:** 499–509.

Skoulidis F, Byers LA, Diao L, Papadimitrakopoulou VA, Tong P, Izzo J, Behrens C, Kadara H, Parra ER, Canales JR, et al. 2015. Co-occurring genomic alterations define major subsets of *KRAS*-mutant lung adenocarcinoma with distinct biology, immune profiles, and therapeutic vulnerabilities. *Cancer Discov* **5:** 860–877.

Soucek L, Whitfield J, Martins CP, Finch AJ, Murphy DJ, Sodir NM, Karnezis AN, Swigart LB, Nasi S, Evan GI. 2008. Modelling Myc inhibition as a cancer therapy. *Nature* **455:** 679–683.

Soucek L, Whitfield JR, Sodir NM, Massó-Vallés D, Serrano E, Karnezis AN, Swigart LB, Evan GI. 2013. Inhibition of Myc family proteins eradicates KRas-driven lung cancer in mice. *Genes Dev* **27:** 504–513.

Svennson RU, Parker SJ, Eichner LJ, Kolar MJ, Wallace M, Brun SN, Lombardo PS, Van Nostrand JL, Hutchins A, Vera L, et al. 2016. Inhibition of acetyl-CoA carboxylase suppresses fatty acid synthesis and tumor growth of non-small-cell lung cancer in preclinical models. *Nat Med* **22:** 1108–1119.

Toole BP, Slomiany MG. 2008. Hyaluronan: A constitutive regulator of chemoresistance and malignancy in cancer cells. *Semin Cancer Biol* **18:** 244–250.

Tran PT, Shroff EH, Burns TF, Thiyagarajan S, Das ST, Zabuawala T, Chen J, Cho YJ, Luong R, Tamayo P, et al. 2012. Twist1 suppresses senescence programs and thereby accelerates and maintains mutant Kras-induced lung tumorigenesis. *PLoS Genet* **8:** e1002650.

Van Nostrand JL, Brisac A, Mello SS, Jacobs SB, Luong R, Attardi LD. 2015. The p53 target gene SIVA enables non-small-cell lung cancer development. *Cancer Discov* **5:** 622–635.

Vicent S, Chen R, Sayles LC, Lin C, Walker RG, Gillespie AK, Subramanian A, Hinkle G, Yang X, Saif S, et al. 2010. Wilms tumor 1 (WT1) regulates KRAS-driven oncogenesis and senescence in mouse and human models. *J Clin Invest* **120:** 3940–3952.

von Karstedt S, Conti A, Nobis M, Montinaro A, Hartwig T, Lemke J, Legler K, Annewanter F, Campbell AD, Taraborrelli L, et al. 2015. Cancer cell-autonomous TRAIL-R signaling promotes KRAS-driven cancer progression, invasion, and metastasis. *Cancer Cell* **27:** 561–573.

Wang IC, Ustiyan V, Zhang Y, Cai Y, Kalin TV, Kalinichenko VV. 2014. Foxm1 transcription factor is required for the initiation of lung tumorigenesis by oncogenic $Kras^{G12D}$. *Oncogene* **33:** 5391–5396.

Whittle MC, Izeradjene K, Rani PG, Feng L, Carlson MA, DelGiorno KE, Wood LD, Goggins M, Hruban RH, Chang AE, et al. 2015. RUNX3 controls a metastatic switch in pancreatic ductal adenocarcinoma. *Cell.* **161:** 1345–1360.

Wu CY, Carpenter ES, Takeuchi KK, Halbrook CJ, Peverley LV, Bien H, Hall JC, DelGiorno KE, Pal D, Song Y, et al. 2014. PI3K regulation of RAC1 is required for KRAS-induced pancreatic tumorigenesis in mice. *Gastroenterology* **147:** 1405–1416.e7.

Xia Y, Yeddula N, Leblanc M, Ke E, Zhang Y, Oldfield E, Shaw RJ, Verma IM. 2012. Reduced cell proliferation by IKK2 depletion in a mouse lung-cancer model. *Nat Cell Biol* **14:** 257–265.

Xie H, Hanai JI, Ren JG, Kats L, Burgess K, Bhargava P, Signoretti S, Billiard J, Duffy KJ, Grant A, et al. 2014. Targeting lactate dehydrogenase-A inhibits tumorigenesis and tumor progression in mouse models of lung cancer and impacts tumor initiating cells. *Cell Metab* **19:** 795–809.

Ying H, Elpek KG, Vinjamoori A, Zimmerman SM, Chu GC, Yan H, Fletcher-Sananikone E, Zhang H, Liu Y, Wang W, et al. 2011. PTEN is a major tumor suppressor in pancreatic ductal adenocarcinoma and regulates an NF-κB-cytokine network. *Cancer Discov.* **1:** 158–169.

Ying H, Kimmelman AC, Lyssiotis CA, Hua S, Chu GC, Fletcher-Sananikone E, Locasale JW, Son J, Zhang H, Coloff JL, et al. 2012. Oncogenic Kras maintains pancreatic tumors through regulation of anabolic glucose metabolism. *Cell* **149:** 656–670.

Young NP, Crowley D, Jacks T. 2011. Uncoupling cancer mutations reveals critical timing of p53 loss in sarcomagenesis. *Cancer Res* **71:** 4040–4047.

Zhang Z, Newton K, Kummerfeld SK, Webster J, Kirkpatrick DS, Phu L, Eastham-Anderson J, Liu J, Lee WP, et al. 2017. Transcription factor Etv5 is essential for the maintenance of alveolar type II cells. *Proc Natl Acad Sci* **114:** 3903–3908.

Kras in Organoids

Derek Cheng and David Tuveson

CSHL Cancer Center, Cold Spring Harbor Laboratory, Cold Spring Harbor, New York 11743

Correspondence: dtuveson@cshl.edu

Oncogenic Kras are genetic dependencies for the majority of pancreatic and colorectal adenocarcinomas; however, much remains to be understood regarding its tropism to these carcinomas. Recently developed organoid technology presents a more representative model culture system for pancreatic and colon epithelial tissues as well as better fostering the culture of nonimmortalized cells than two-dimensional culture. These advantages enable cancer researchers to directly compare tumor and normal tissue models to better study tumor initiation as well as therapeutic efficacy. Although in vivo models better model the complexity of multiple cell types, the organoid system allows for easier genetic manipulations and isolation of specific cell types. Furthermore, syngeneic orthotopically transplanted organoids recapitulate tumor histologically and gene expression of the tumors from which they were derived. Thus, organoids may extend the use of genetically engineered mouse models. These advantages of organoid cultures allow for many questions, including but not limited to studying the interaction between different cell types within a tumor and elucidating dependencies of Kras-driven tumors.

The role of oncogenic Kras is especially of interest in the context of pancreatic ductal adenocarcinoma (PDAC) and colorectal (CRC) adenocarcinoma, as it is respectively mutated in ~90% and 50% of cases (Gao et al. 2013). As a well-established oncogenic driver and dependency for tumor maintenance, Kras is a prime target for tumor-specific therapy (Tuveson et al. 2004; Jancík et al. 2010; Collins et al. 2012); however, many questions and challenges remain regarding Kras biology and Kras targeting. For example, despite being similar to the other Ras family members, Nras and Hras, only Kras has a higher prevalence of being mutated in the tumors just mentioned. Perhaps there are protein interactors and effectors that are specific to Kras that have yet to be discovered and can account for these differences. It also remains to be understood why the prevalence of Kras mutations in these specific cancer types is higher than in other cancers. This raises the question of whether Kras must be studied in a physiologically relevant model system that recapitulates traits of epithelial tumors and their corresponding normal cells of origin.

ORGANOID CULTURE

Organoid culture, which highly supports the growth of epithelial cells and preserves their functions, has been recently used to study Kras-related diseases (Sato et al. 2011; Li et al. 2014; Boj et al. 2015). Organoid culture was first

Cite this article as *Cold Spring Harb Perspect Med* doi: 10.1101/cshperspect.a031575

developed for intestinal epithelial cell growth with $Lgr5^+$ intestinal stem cells, and has now been adapted for other tissue applications (Dutta et al. 2017). Because of the various culture conditions and methods established from different laboratories, a standard definition of "organoid" has not been set. However, to be called an organoid, the culture must show organ-like properties such as being able to self-organize into a functional structure mimicking the architecture of the organ from which they were derived (Sato et al. 2009; Boj et al. 2015; Baker et al. 2016). To achieve a defined architecture, organoid cells are embedded in or grown on top of semisolid matrices such as agarose, collagen, or Matrigel. It has been suggested that organoids be reserved for three-dimensional (3D) cultures that are generated directly from dissociated primary tissue, embryonic stem, or iPS cells differentiated in vitro and propagated in a matrix rather than cells initially cultured as monolayers and assayed in 3D culture (Petersen et al. 1992). Organoid culture conditions have been established from both mouse and human gastrointestinal tissues, including intestinal (Sato et al. 2009), gastric (Barker et al. 2010), colon (Sato et al. 2011), pancreatic (Huch et al. 2013a; Boj et al. 2015), hepatic (Huch et al. 2013b), prostate (Gao et al. 2014), and breast (Rowald et al. 2016).

Organoid culture has some advantages over monolayer in vitro culture methods as well as in vivo models. In contrast to two-dimensional (2D) culture, cells in organoid culture maintain tissue structure by self-organizing and retaining features of polarity, with the basal side in contact with the Matrigel and an apical side toward the lumen of the organoid sphere (Sato et al. 2009; Huch et al. 2013b; Boj et al. 2015). Additionally, the organoid culture can be established with higher efficiency than 2D cell lines (Boj et al. 2015). More importantly, they allow for growth of nontransformed normal cells, whereas the 2D culture of these cells remains difficult. Organoid cultures can serve as a complementary tool, as well as an extension to genetically engineered mouse models. As an in vitro model system, organoids can be easily propagated and genetically manipulated to study specific targets compared with in vivo models. Syngeneic orthotopically transplanted organoids show architecture that resembles the tissue from which they were derived and therefore offer an expedited in vivo model system (Boj et al. 2015). These advantages make organoid culture a flexible tool to answer important questions, including how oncogenic Kras affects pancreatic ductal and colon epithelial cells.

Kras IN MURINE PANCREATIC ORGANOIDS

Since Kras mutations occur in >90% of PDACs, the most common type of pancreatic cancer, these mutations are believed to be the driver of most PDAC cases (Almoguera et al. 1988). Therefore, the study of oncogenic Kras in pancreatic cells is ideal and, fortunately, organoids can increase the use of existing genetically engineered Kras-driven PDAC mouse models. Historically, a major challenge of studying Kras-driven cancer has been the difficulty in comparing in vitro cells with and without activated Kras in a disease-relevant context. Normal pancreatic ductal cells require immortalization for growth in 2D and would need wild-type (WT) and mutant Kras exogenously expressed (Furukawa et al. 1996). To address this hurdle, LSL-Kras mouse embryonic fibroblasts (MEFs) may be used to study the effect of the presence of oncogenic Kras. However, experiments should ideally be performed in a more relevant cell type such as pancreatic ductal cells. To model PDAC progression, $Kras^{+/LSL\text{-}G12D}$; Pdx1-Cre (KC) mice were designed to express mutant $Kras^{G12D}$ endogenously in the pancreas. These mice develop preinvasive ductal lesions that resemble human PanINs and, upon aging, can stochastically develop primary and metastatic PDAC (Hingorani et al. 2003). Although KC mice serve as a reliable source for murine PanIN tissue, $Kras^{+}/LSL^{G12D}$; $Trp53^{+}$/LSL-R172H; Pdx1-Cre (KPC) mice further expedite PDAC formation, providing a source for murine PDAC tumors and metastases (Hingorani et al. 2005). Together, WT, KC, and KPC mice cover the spectrum of PDAC progression from normal to PanIN, PDAC, and metastasis. Although some comparisons can be made between the

 Cite this article as *Cold Spring Harb Perspect Med* doi: 10.1101/cshperspect.a031575

tissues of normal, KC, and KPC mice, there is still a demand for in vitro systems to more efficiently study the effects of genetic manipulation and drug response in the form of screens, and organoids have begun to meet these needs.

Kras-Driven Mouse PanIN and Tumor Organoids Compared with Normal

Boj et al. (2015) have shown that normal mouse and human pancreatic ductal cells as well as PanIN and tumor counterparts can be maintained in organoid culture for >30 passages. The ability to culture normal ductal cells enables the comparison of Kras WT cells with Kras mutant cells derived from tumor and PanIN lesions. It also enables the generation of isogenic Kras lines for delineating the biological consequences of acute Kras activation.

Organoids generated from WT, KC, and KPC mice serve as in vitro tools that complement their respective in vivo models. Genetic manipulation, large-scale screens, and characterization is often more feasible in in vitro systems because of the ability to propagate cells in a well-controlled setting. Organoids generated from normal ducts from WT mice, PanIN from KC mice, and tumor from KPC mice designated as mN, mP, and mT, respectively, were established for this purpose. These organoids were verified to express genes of the ductal lineage such as CK19 and Sox9 (Cleveland et al. 2012). mP organoids also up-regulated genes that are associated with a PanIN disease state, such as Muc5ac, Muc6, and Tff1. As expected from expressing $Kras^{G12D}$, mT organoids had increased levels of Kras-GTP but, interestingly, there was also elevated expression of the Kras protein (Boj et al. 2015). Although there is evidence from genetically engineered mouse models that mitogen-activated protein kinase (MAPK)/ERK and PI3K/AKT signaling may drive PDAC progression (Collisson et al. 2012; Eser et al. 2013), mT and metastatic (mM) organoids revealed Erk and Akt phosphorylation similar to that of normal organoids but with relatively increased downstream S6 phosphorylation. The lack of phosphorylation of ERK in organoids is also consistent with what is seen in PanIN lesions of KC mice and LSL-Kras MEFs and may be caused by negative feedback mechanisms (Karreth and Tuveson 2009).

Culture conditions may not only affect cell signaling but also more dramatic changes such as karyotypic stability. Loss of heterozygosity (LOH) for Trp53 has been commonly seen in mouse PDAC based on studies of 2D cell lines (Hingorani et al. 2005). Interestingly, murine pancreatic tumor cells cultured as organoids maintain a stable diploid karyotype but the same cells plated as 2D cell lines spontaneously become aneuploid and lose their WT copy of p53 (Boj et al. 2015). This LOH after plating mouse tumor cells in 2D suggests that culture conditions can greatly affect the genomic makeup of the cells, as well as the expression of genes. 2D culture may select for p53 LOH cells and, therefore, may not accurately reflect tumor heterogeneity. Because the number of copies of WT and mutant Kras has been shown to affect MAPK dependence, organoid culture may provide a way to study tumor cells without chromosomal imbalance (Burgess et al. 2017).

Organoids are also useful for transplantation experiments. Mouse organoids can be orthotopically transplanted into syngeneic C57Bl/6 or Nu/Nu mice. Although transplanted mN organoids form ductal structures resembling normal ductal architecture comprised of simple cuboidal cells, mP organoids formed small cysts lined with simple cuboidal cells as well as patches of mucin-containing columnar epithelial cells resembling PanIN morphology. Like PanIN tissue, these transplanted dysplastic cells had increased proliferation and stromal response. When mT organoids are orthotopically transplanted, they initially form early and late PanIN-like lesions and eventually develop into invasive or metastatic PDAC. Although mT organoids take several months to progress to invasive PDAC, organoids derived from mM quickly engraft as invasive PDAC. In addition to displaying dysplasia, transplanted tumor organoids also induce a stroma response leading to lowly vascularized tumors. The organoid transplantation model retains many of the characteristics of autochthonous PanIN and PDAC tissue and offers an alternative PDAC progression model.

Kras Activation in Mouse Pancreatic Organoids

The ability to culture normal organoids permits the generation of isogenic cultures allowing for the study of Kras activation. To generate these lines, organoids derived from pancreatic ducts harboring the conditional $Kras^{LSL\text{-}G12D}$ allele were infected with adenoviral-Cre recombinase and were compared with the control-infected organoids (Boj et al. 2015). $Kras^{G12D}$-expressing organoids maintained expression of genes specific to ductal cells rather than acinar and endocrine lineages but showed an increase in Kras expression compared with control-infected organoids. $Kras^{G12D}$-expressing organoids also showed a greater proliferation rate compared with control-infected organoids. Additionally, transplanting $Kras^{G12D}$-expressing organoids gave rise to PanIN-like lesions, whereas control-infected organoids formed normal-appearing ductal architecture. These PanIN-like lesions showed tall columnar morphology and produced mucin. These findings are consistent with the observation that endogenous $Kras^{G12D}$ expression is sufficient in giving rise to PanIN lesions in mice (Hingorani et al. 2003). Because the organoid cultures described above support ductal cells and no other lineages, these studies show that ductal cells, when expressing $Kras^{G12D}$, are competent in forming PanIN lesions. Similar to mP organoids, which also express $Kras^{G12D}$, these $Kras^{G12D}$ organoids also up-regulate genes associated with early PanIN lesions (Muc5ac, Muc6, Tff1) when compared with control-infected organoids. RNA sequencing characterization of mN, mP, and mT revealed that several genes were up-regulated in mP and mT organoids. Of these, Gcnt1, Gcnt3, Acsm3, Agr2, Syt16, Nt5e, and Ugdh, were also up-regulated in $Kras^{G12D}$ organoids compared to infected control organoids, suggesting that up-regulation of these genes is downstream from Kras activation. In short, activation of Kras in vitro can yield PanIN-like organoids, so it follows that PanIN-associated cellular changes may be Kras-driven behaviors. Gene set enrichment analysis (GSEA) on the RNA-seq and proteomic data of mouse organoids revealed that mP organoids up-regulated genes are involved in glutathione metabolism, which is involved with redox regulation (Boj et al. 2015).

The study of $Kras^{G12D}$ using organoids has particularly uncovered the dependencies of Kras cells on the oxidative state of pancreatic tumor cells. Kras activation has been shown to induce Nrf2 protein expression and/or its target gene, Nqo1, in KPC mouse tumors and their primary cultures (DeNicola et al. 2011). With regard to its relevance in human disease, Nrf2 protein is increased in human tumor organoids compared with normal counterparts (Chio et al. 2016). As a master regulator of an antioxidant response, the transcription factor Nrf2 keeps reactive oxygen species (ROS) levels down and is a dependency of Kras-driven PanIN formation (DeNicola et al. 2011). To investigate the biochemical effects and function of Nrf2 in $Kras^{G12D}$-driven tumor maintenance, Chio et al. (2016) performed Nrf2 knockdown and ablation experiments in both human and mouse organoid models. As seen in the KC and KPC mouse organoid models, $Kras^{G12D}$ induces the Nrf2 antioxidant program, leading to an increase in intracellular glutathione (GSH) and decreased levels of ROS; features reverted on NRF2 ablation (Chio et al. 2016). Nrf2 knockdown in human organoids inhibits the growth of some human tumor organoids (KRAS mutant) although being tolerated by human normal organoids (KRAS WT) (Chio et al. 2016). Interestingly, the surviving tumor organoids showed elevated ROS and a decrease in cell proliferation, which was alleviated by the addition of antioxidant NAC. These Nrf2-depletion experiments suggest that Kras mutant tumor organoids depend more on Nrf2 for cell proliferation than Kras WT normal organoids, likely the result of elevation of ROS in tumor cells. Dependencies of Kras cells such as Nrf2 suppression of ROS as seen in organoids may elucidate novel therapeutic approaches in pancreatic cancer.

Coculture Studies with Mouse Kras Organoids

Organoids derived from KPC tumors are also a relevant model for studying cell-to-cell interactions. Organoid cultures allow us to identify distinct populations of cancer-associated fibroblasts (CAFs) and confirm their presence in vivo. Although Kras is mutated in the epithelial cells that drive PDAC, up to 90% of the total volume of the tumor is made of desmoplasia, consisting of extracellular matrix, immune cells, vasculature, and cancer-associated fibroblasts (Moir et al. 2015). The organoid system allows for the study of interactions between carcinoma cells and pancreatic stellate cells (PSCs), which support such desmoplasia. Because the organoid system maintains spatial organization of cells, the phenomenon of heterogeneity in fibroblasts was discovered (Öhlund et al. 2017). PSCs and tumor organoids cocultured in Matrigel can proliferate in the presence of reduced media. Interestingly, the PSCs and tumor organoids promote the proliferation of each other. Although PSCs are relatively quiescent when plated alone in Matrigel, the PSCs show morphological changes and deposit collagen I when in close contact with tumor organoids. This PSC activation in vitro recapitulates the desmoplasia characteristic of PDAC (Öhlund et al. 2017). Although tumor organoids are capable of triggering PSC activation, it remains to be tested whether Kras mutant cells induce specific changes in PSCs that Kras WT cells do not.

In addition to PSCs, peritoneal mesothelial cells and adipocytes have been shown to affect the behavior of pancreatic tumor cells. In another coculture system, Abe et al. (2017) showed that 3D coculture and cotransplantation of peritoneal mesothelial cells with pancreatic cancer cells derived from KPC tumors lead to increased invasion of the cancer cells. Another cell–cell interaction relevant to cancer may be that of the cancer cell to neighboring adipocytes. Okumura et al. (2017) observed an increase in metastasis in KPC mice fed on a high-fat diet. This adipose tissue–derived conditioned media increased the migration, invasiveness, and gemcitabine resistance of KPC tumor cells. These changes were attributed to direct uptake of released fatty acids from adipocytes as seen with 3D culture experiments with conditioned media from adipose tissue embedded in collagen I (Okumura et al. 2017). It remains to be shown whether the interaction of KPC tumor organoids with PSC, peritoneal mesothelial cells, and adipocytes are specific to Kras mutant cells, to tumor cells, or to proliferating epithelial cells in general.

LESSONS FROM HUMAN PANCREATIC ORGANOIDS

Human disease may be more complex than our currently used mouse models. Thus, human organoids are of interest to model PDAC as well as other cancers. Boj et al. (2015) have established human pancreatic organoid cultures from normal tissue (hN), PDAC biopsies (hT), and fine needle aspirations (hFNA) and metastases (hM) with >75% efficiency. Unsurprisingly, Kras was mutated in seven of eight organoids derived from tumor and both organoids derived from metastases; however, human PDAC is a complex disease. There are some notable differences between human organoids and mouse organoids. hN organoids can only be propagated for ~20 passages, causing some limitation to experiments that can be performed on them. hT organoids often grow as dysplastic tall columnar cells and the organoids can proliferate as filled spheres rather than the simple cuboidal lined cysts of hN. Because organoids are grown in vitro, the culturing conditions can be controlled leading to identification of specific dependencies for growth factors and nutrients. In the case of human pancreatic ductal organoids, tumor organoids can be grown in the absence of epithelial growth factor (EGF), whereas normal organoids cannot (Boj et al. 2015). Likely, the expression of mutant Kras allows these organoids to become EGF-independent, whereas normal organoids remain EGF-dependent; however, this concept has yet to be validated in isogenic lines.

Although representative of the tumor from which they are derived, human tumor organoids vary in other mutations such as p53, SMAD4, and CDKN2A (Boj et al. 2015). Because of the heterogeneity of human tumors, isogenic experiments in a human context would be beneficial. Introducing oncogenic Kras ectopically via lentiviral infection and knocking out p53, CDKN2, and SMAD4 simultaneously by CRISPR in

CD133$^+$ human pancreatic ductal cells and orthotopically transplanting these cells into the splenic lobe of pancreas of immunocompromised NOD SCID IL2rg null (NSG) mice was sufficient in PanIN-1 formation. The further introduction of ERBB2 overexpression into these organoids followed by transplantation causes formation of PanIN-2-like formation. Although this model system only recreated PanIN lesions, it is a transplantation model that uses human cells. Although human cells seem to require more mutations to drive transformation compared with the Kras and p53 in mouse organoids, it is consistent with how human organoids have also been shown to require more components in organoid media for propagation. These observations may suggest that the human disease may be more complex than the mouse model, and more genetic changes are required to supplement the driving Kras mutation in human disease.

Kras IN COLON ORGANOIDS

Kras is also a known CRC driver gene. Li et al. (2014) used the single air–liquid interface method to model malignancies of the pancreas, stomach, and colon. This organoid method cultures primary mouse epithelial and mesenchymal cells, yielding detailed in situ histologic end points for oncogenic transformation in vitro, and tumorigenicity after transplantation in vivo. Consistent with the findings from the KPC organoid system by Boj et al. (2015), K, P, and KP organoids also showed increased tumorigenicity when Cre expression activated Kras^{G12D} expression in various primary gastrointestinal epithelial cells (Li et al. 2014).

Because organoids recapitulate many Ras-driven phenomenon, organoids may provide a more informative model for screening Ras pathway inhibitors. Because nontransformed cells with WT Kras can be cultured alongside Kras mutant tumor cells, the organoid platform can better aid us in the search for targeted therapy with an effective therapeutic index. Verissimo et al. (2016) tested clinically used epidermal growth factor receptor (EGFR)-RAS-ERK pathway inhibitors in organoids derived from normal colon tissue and colon cancer. They found features that are seen in in vivo data and clinical trials. Clinically, treatment of CRC tumors is dependent on Kras mutation status, namely, the effectiveness of EGFR inhibition against WT Kras tumors but not mutant Kras tumors. Interestingly, the drug effects were similar when comparing normal organoids to patient-derived CRC organoids with WT Kras. Drug resistance against all tested EGFR inhibitors correlated with the mutational status of KRAS, which is consistent with the EGF withdrawal experiments done in human pancreatic organoids mentioned earlier (Boj et al. 2015). MEK (selumetinib) and ERK (SCH772984) inhibitors were effective in all organoid lines tested. The combination of EGFR and MEK inhibition is additive and leads to more consistent inhibitory effects on the CRC organoids. Combination therapies such as dual pan-HER and MEK inhibition as well as dual MEK and ERK inhibition, are in clinical trials. Although the CRC organoid model show synergistic effects of combination therapy suppressing the growth of Ras mutant organoids, this leads to cell-cycle arrest rather than cell death but sensitizes cells to Bcl2. Although the tested drugs did not selectively inhibit tumor organoids over normal organoids, this study shows that the organoid platform is a preclinical drug-screening tool that is representative of patient tumor response. As new anti-Ras therapies such as direct allosteric inhibitors and drugs that block Kras from interacting with other proteins are developed, organoids may be the ideal model system in which to test their efficacy preclinically (Maurer et al. 2012; Ostrem et al. 2013; Shima et al. 2013).

ACHIEVING A BETTER MODEL

Ras continues to be a highly important topic because Kras is a driver of many tumors and remains without effective therapies that target it directly or indirectly in the clinic. Genetically engineered LSL-Kras^{G12D} mouse models have enabled the study of endogenously expressed oncogenic Kras, and now organoid models have extended the use of this model and enabled the study of Kras^{G12D} in epithelial cells. Organ-

 Cite this article as *Cold Spring Harb Perspect Med* doi: 10.1101/cshperspect.a031575

oids provide an in vitro cancer progression model from normal cells to PanIN, tumor, and metastatic cells. Progression models are essential to studying cancer as the time to diagnosis of cancer often limits what physicians can do for their patients. Being able to dissect the biology of early tumors and high-risk PanIN lesions, researchers may be able to identify early biomarkers leading to improved diagnostics or targets for cancer prevention, both of which could lead to better patient outcomes. Furthermore, organoids provide a way to study the interaction of cancer cells with neighboring cells such as cancer-associated fibroblasts. As the model becomes even better characterized, it will be interesting to see other coculture systems to better model the tumors, which are complex in the cell types present. With the ability of propagating Kras WT and Kras^{G12D} organoids under the same conditions, screening for mutant Kras-specific therapies can be facilitated and the confounding factors inherent when comparing across cell lines are minimized. The LSL-Kras^{G12D} model still has the caveat of expressing one allele of WT Kras when Cre is not present and one allele each of WT and mutant when Cre is present. Therefore, the next advance in building a better Kras model to study pancreatic disease may be a system where comparisons between Kras WT-expressing cells and Kras^{G12D}-expressing cells will have both alleles functional. Theoretically, this could be accomplished with the advances in CRISPR technology in the organoid model system. In summary, organoids open the door to ask many Kras-related questions that, when answered, could impact how we treat Kras-driven cancers.

REFERENCES

Abe T, Ohuchida K, Koikawa K, Endo S, Okumura T, Sada M, Horioka K, Zheng B, Moriyama T, Nakata K, et al. 2017. Cancer-associated peritoneal mesothelial cells lead the formation of pancreatic cancer peritoneal dissemination. *Int J Oncol* **50:** 457–467.

Almoguera C, Shibata D, Forrester K, Martin J, Arnheim N, Perucho M. 1988. Most human carcinomas of the exocrine pancreas contain mutant c-K-*ras* genes. *Cell* **53:** 549–554.

Baker LA, Tiriac H, Clevers H, Tuveson DA. 2016. Modeling pancreatic cancer with organoids. *Trends Cancer* **2:** 176–190.

Barker N, Huch M, Kujala P, van de Wetering M, Snippert HJ, van Es JH, Sato T, Stange DE, Begthel H, van den Born M, et al. 2010. Lgr5^{+ve} stem cells drive self-renewal in the stomach and build long-lived gastric units in vitro. *Cell Stem Cell* **6:** 25–36.

Boj SF, Hwang CI, Baker LA, Chio II, Engle DD, Corbo V, Jager M, Ponz-Sarvise M, Tiriac H, Spector MS, et al. 2015. Organoid models of human and mouse ductal pancreatic cancer. *Cell* **160:** 324–338.

Burgess MR, Hwang E, Mroue R, Bielski CM, Wandler AM, Huang BJ, Firestone AJ, Young A, Lacap JA, Crocker L, et al. 2017. KRAS allelic imbalance enhances fitness and modulates MAP kinase dependence in cancer. *Cell* **168:** 817–829.e15.

Chio IIC, Jafarnejad SM, Ponz-Sarvise M, Park Y, Rivera K, Palm W, Wilson J, Sangar V, Hao Y, Öhlund D, et al. 2016. NRF2 promotes tumor maintenance by modulating mRNA translation in pancreatic cancer. *Cell* **166:** 963–976.

Cleveland MH, Sawyer JM, Afelik S, Jensen J, Leach SD. 2012. Exocrine ontogenies: On the development of pancreatic acinar, ductal and centroacinar cells. *Semin Cell Dev Biol* **23:** 711–719.

Collins MA, Brisset JC, Zhang Y, Bednar F, Pierre J, Heist KA, Galbán CJ, Galbán S, di Magliano MP. 2012. Metastatic pancreatic cancer is dependent on oncogenic Kras in mice. *PLoS ONE* **7:** e49707.

Collisson EA, Trejo CL, Silva JM, Gu S, Korkola JE, Heiser LM, Charles RP, Rabinovich BA, Hann B, Dankort D, et al. 2012. A central role for RAF→MEK→ERK signaling in the genesis of pancreatic ductal adenocarcinoma. *Cancer Discov* **2:** 685–693.

DeNicola GM, Karreth FA, Humpton TJ, Gopinathan A, Wei C, Frese K, Mangal D, Yu KH, Yeo CJ, Calhoun ES, et al. 2011. Oncogene-induced Nrf2 transcription promotes ROS detoxification and tumorigenesis. *Nature* **475:** 106–109109.

Dutta D, Heo I, Clevers H. 2017. Disease modeling in stem cell-derived 3D organoid systems. *Trends Mol Med* **23:** 393–410.

Eser S, Reiff N, Messer M, Seidler B, Gottschalk K, Dobler M, Hieber M, Arbeiter A, Klein S, Kong B, et al. 2013. Selective requirement of PI3K/PDK1 signaling for Kras oncogene-driven pancreatic cell plasticity and cancer. *Cancer Cell* **23:** 406–420.

Furukawa T, Duguid WP, Rosenberg L, Viallet J, Galloway DA, Tsao MS. 1996. Long-term culture and immortalization of epithelial cells from normal adult human pancreatic ducts transfected by the E6E7 gene of human papilloma virus 16. *Am J Pathol* **148:** 1763–1770.

Gao J, Aksoy BA, Dogrusoz U, Dresdner G, Gross B, Sumer SO, Sun Y, Jacobsen A, Sinha R, Larsson E, et al. 2013. Integrative analysis of complex cancer genomics and clinical profiles using the cBioPortal. *Sci Signal* **6:** 1.

Gao D, Vela I, Sboner A, Iaquinta PJ, Karthaus WR, Gopalan A, Dowling C, Wanjala JN, Undvall EA, Arora VK, et al. 2014. Organoid cultures derived from patients with advanced prostate cancer. *Cell* **159:** 176–187.

Hingorani SR, Petricoin EF, Maitra A, Rajapakse V, King C, Jacobetz MA, Ross S, Conrads TP, Veenstra TD, Hitt BA, et al. 2003. Preinvasive and invasive ductal pancreatic cancer and its early detection in the mouse. *Cancer Cell* **4:** 437–450.

Hingorani SR, Wang L, Multani AS, Combs C, Deramaudt TB, Hruban RH, Rustgi AK, Chang S, Tuveson DA. 2005. $Trp53^{R172H}$ and $Kras^{G12D}$ cooperate to promote chromosomal instability and widely metastatic pancreatic ductal adenocarcinoma in mice. *Cancer Cell* **7:** 469–483.

Huch M, Bonfanti P, Boj SF, Sato T, Loomans CJ, van de Wetering M, Sojoodi M, Li VS, Schuijers J, Gracanin A, et al. 2013a. Unlimited in vitro expansion of adult bipotent pancreas progenitors through the Lgr5/R-spondin axis. *EMBO J* **32:** 2708–2721.

Huch M, Dorrell C, Boj SF, van Es JH, Li VS, van de Wetering M, Sato T, Hamer K, Sasaki N, Finegold MJ, et al. 2013b. In vitro expansion of single $Lgr5^{+}$ liver stem cells induced by Wnt-driven regeneration. *Nature* **494:** 247–250.

Jancík S, Drábek J, Radzioch D, Hajdúch M. 2010. Clinical relevance of KRAS in human cancers. *J Biomed Biotechnol* doi: 10.1155/2010/150960.

Karreth FA, Tuveson DA. 2009. Modelling oncogenic Ras/Raf signalling in the mouse. *Curr Opin Genet Dev* **19:** 4–11.

Li XN, Nadauld L, Ootani A, Corney DC, Pai RK, Gevaert O, Cantrell MA, Rack PG, Neal JT, Chan CW, et al. 2014. Oncogenic transformation of diverse gastrointestinal tissues in primary organoid culture. *Nat Med* **20:** 769–777.

Maurer T, Garrenton LS, Oh A, Pitts K, Anderson DJ, Skelton NJ, Fauber BP, Pan B, Malek S, Stokoe D, et al. 2012. Small-molecule ligands bind to a distinct pocket in Ras and inhibit SOS-mediated nucleotide exchange activity. *Proc Natl Acad Sci* **109:** 5299–5304.

Moir JA, Mann J, White SA. 2015. The role of pancreatic stellate cells in pancreatic cancer. *Surg Oncol* **24:** 232–238.

Öhlund D, Handly-Santana A, Biffi G, Elyada E, Almeida AS, Ponz-Sarvise M, Corbo V, Oni TE, Hearn SA, Lee EJ, et al. 2017. Distinct populations of inflammatory fibroblasts and myofibroblasts in pancreatic cancer. *J Exp Med* **214:** 579–596.

Okumura T, Ohuchida K, Sada M, Abe T, Endo S, Koikawa K, Iwamoto C, Miura D, Mizuuchi Y, Moriyama T, et al. 2017. Extra-pancreatic invasion induces lipolytic and fibrotic changes in the adipose microenvironment, with released fatty acids enhancing the invasiveness of pancreatic cancer cells. *Oncotarget* **8:** 18280–18295.

Ostrem JM, Peters U, Sos ML, Wells JA, Shokat KM. 2013. $K\text{-}Ras^{G12C}$ inhibitors allosterically control GTP affinity and effector interactions. *Nature* **503:** 548–551.

Petersen OW, Rønnov-Jessen L, Howlett AR, Bissell MJ. 1992. Interaction with basement membrane serves to rapidly distinguish growth and differentiation pattern of normal and malignant human breast spithelial cells. *Proc Natl Acad Sci* **89:** 9064–9068.

Rowald K, Mantovan M, Passos J, Buccitelli C, Mardin BR, Korbel JO, Jechlinger M, Sotillo R. 2016. Negative selection and chromosome instability induced by Mad2 overexpression delay breast cancer but facilitate oncogene-independent outgrowth. *Cell Rep* **15:** 2679–2691.

Sato T, Vries RG, Snippert HJ, van de Wetering M, Barker N, Stange DE, van Es JH, Abo A, Kujala P, Peters PJ, et al. 2009. Single Lgr5 stem cells build crypt-villus structures in vitro without a mesenchymal niche. *Nature* **459:** 262–265.

Sato T, Stange DE, Ferrante M, Vries RG, Van Es JH, Van den Brink S, Van Houdt WJ, Pronk A, Van Gorp J, Siersema PD, et al. 2011. Long-term expansion of epithelial organoids from human colon, adenoma, adenocarcinoma, and Barrett's epithelium. *Gastroenterology* **141:** 1762–1772.

Shima F, Yoshikawa Y, Ye M, Araki M, Matsumoto S, Liao J, Hu L, Sugimoto T, Ijiri Y, Takeda A, et al. 2013. In silico discovery of small-molecule Ras inhibitors that display antitumor activity by blocking the Ras-effector interaction. *Proc Natl Acad Sci* **110:** 8182–8187.

Tuveson DA, Shaw AT, Willis NA, Silver DP, Jackson EL, Chang S, Mercer KL, Grochow R, Hock H, Crowley D, et al. 2004. Endogenous oncogenic $K\text{-}ras^{G12D}$ stimulates proliferation and widespread neoplastic and developmental defects. *Cancer Cell* **5:** 375–387.

Verissimo CS, Overmeer RM, Ponsioen B, Drost J, Mertens S, Verlaan-Klink I, Gerwen BV, van der Ven M, Wetering MV, Egan DA, et al. 2016. Targeting mutant RAS in patient-derived colorectal cancer organoids by combinatorial drug screening. *eLife* **5:** e18489.

Kras and Tumor Immunity: Friend or Foe?

Jane Cullis,[1] Shipra Das,[1] and Dafna Bar-Sagi

Department of Biochemistry and Molecular Pharmacology, New York University School of Medicine, New York, New York 10016

Correspondence: dafna.bar-sagi@nyumc.org

With the recent breakthroughs in immunotherapy as curative treatments in certain tumor types, there has been renewed interest in the relationship between immunity and tumor growth. Although we are gaining a greater understanding of the complex interplay of immune modulating components in the tumor microenvironment, the specific role that tumor cells play in shaping the immune milieu is still not well characterized. In this review, we focus on how mutant Kras tumor cells contribute to tumor immunity, with a specific focus on processes induced directly or indirectly by the oncogene.

It is well recognized that the growth of mutant Kras tumors is associated with an immunosuppressed state that is established via the dynamic interplay between components of both the innate and adaptive immune response (Fig. 1). For example, immunosuppressive cells commonly associated with mutant Kras tumors include myeloid cells, such as alternatively activated (M2) macrophages and myeloid-derived suppressor cells (MDSCs), as well as lymphoid cells, such as interleukin (IL)-17-producing T helper (Th)17 cells, $CD4^+FoxP3^+$ T regulatory (Treg) cells, and $CD19^+IL\text{-}10^+$ B regulatory (Breg) cells (Almand et al. 2001; Gabrilovich et al. 2001; Kusmartsev and Gabrilovich 2006). These cell types contribute to the suppression of tumoricidal cells such as $CD4^+$Th1 T cells, natural killer (NK) cells, and $CD8^+$ T cytotoxic (Tc) cells (Drake et al. 2006). The relative balance of these two antagonistic immune subpopulations profoundly impacts not only disease establishment and progression but also sensitivity to immunotherapy (Topalian et al. 2012; Pauken et al. 2015). In the sections that follow, we will elaborate on how mutant Kras-regulated signaling pathways affect the presence and function of these immune cell types. Moreover, we will describe how this contributes to the tumorigenic potential of Kras-mutant cancers with specific focus on pancreatic ductal adenocarcinoma (PDAC) and non-small-cell lung cancer (NSCLC), tumor types that harbor Kras mutations in more than 95% and 35% of cases, respectively (Seo et al. 2012; Rishi et al. 2015).

KRAS' IMMUNOLOGISTICS

The discovery that oncogenic Kras could induce nuclear factor (NF)-κB activation in fibroblasts and epithelial cells provided the first direct evidence of its capacity to drive proinflammatory signaling in transformed cells (Finco et al. 1997;

[1]These authors contributed equally to this work.

Cite this article as *Cold Spring Harb Perspect Med* doi: 10.1101/cshperspect.a031849

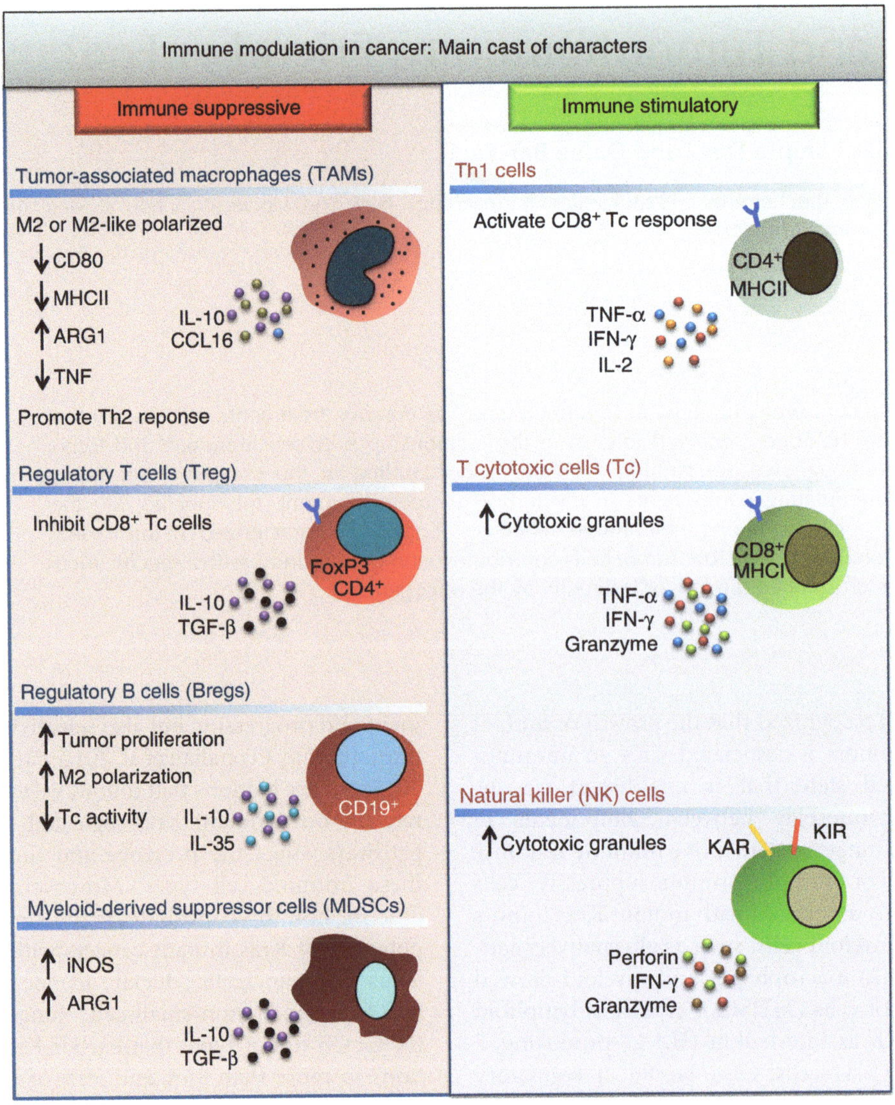

Figure 1. Main mediators of immune modulation in the tumor microenvironment (TME). Tumor-associated macrophages (TAMs), regulatory T (Treg) cells, regulatory B (Breg) cells, and myeloid-derived suppressor cells (MDSCs) induce a tumor-tolerant microenvironment through production of immune suppressive cytokines like interleukin (IL)-10, IL-35, and transforming growth factor β (TGF-β). These factors antagonize the tumoricidal activity of T helper (Th)1 cells, T cytotoxic (Tc) cells, and natural killer (NK) cells that produce immune stimulatory cytokines and cytolytic factors. MHC, Major histocompatibility complex; iNOS, inducible nitric oxide synthase; ARG1, arginase 1; TNF-α, tumor necrosis factor α; IFN-γ, interferon γ.

Kim et al. 2002). Finco and colleagues showed that NF-κB was a transcriptional target downstream from the Raf/mitogen-activated protein kinase (MAPK) pathway that was required to maintain the transformed phenotype of $Hras^{G12V}$-transformed cells, a finding that was later confirmed in the context of mutant Kras (Finco et al. 1997; Kim et al. 2002). Although it is well established that NF-κB engages cell-intrinsic signaling pathways that drive cellular transformation, it is also appreciated to play a critical role in shaping the immune microenvironment

 Cite this article as *Cold Spring Harb Perspect Med* doi: 10.1101/cshperspect.a031849

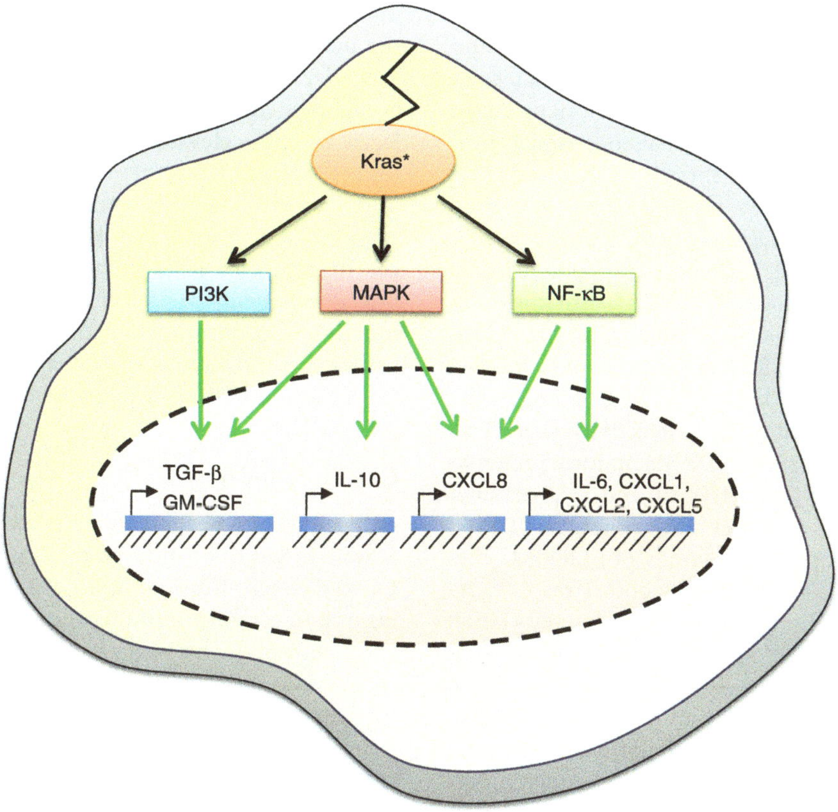

Figure 2. Secreted immunomodulatory factors transcriptionally induced by oncogenic Kras signaling. Transforming growth factor β (TGF-β) and granulocyte macrophage colony-stimulating factor (GM-CSF) are regulated via the concerted action of mitogen-activated protein kinases (MAPKs) and PI3K pathways, interleukin (IL)-10 is regulated via the MAPK pathway, CXCL8 is induced by both MAPK and canonical nuclear factor (NF)-κB pathways, IL-6 is regulated by the noncanonical RalB/TBK1/IKKE/NF-κB pathway, and CXCL1, CXCL2, and CXCL5 are induced via the classical NF-κB signaling pathway.

through the transcriptional induction of a plethora of cytokines and chemokines, including tumor necrosis factor α (TNF-α), IL-1α/β, IL-6, CXCL1, 2, 5, and 8, COX2, monocyte chemoattractant protein 1 (MCP-1), inducible nitric oxide synthase (iNOS), intracellular adhesion molecule 1 (ICAM1), and ELAM1 (Fig. 2) (Baud and Karin 2009). Mutant Kras can also induce the expression of cytokines via the classical Raf/MAPK and PI3K signaling pathways independently of NF-κB, such as in the case of IL-10, transforming growth factor β (TGF-β), and granulocyte macrophage colony-stimulating factor (GM-CSF) (Fig. 2). Below, we highlight those cytokine and growth factor families regulated directly by oncogenic Kras, the immune cells they affect, and how this modifies the tumorigenic potential of Kras-mutant tumors.

ELR^+ CXC Chemokines

The ELR^+ CXC family of chemokines perhaps best exemplifies the expanse of mutant Kras-dependent immunomodulation in human cancers, comprising CXCL1 (GRO-a/KC), CXCL2 (GRO-b/MIP2), CXCL3 (GRO-c), CXCL4, (PF-4), CXCL5 (ENA-78/LIX), CXCL6 (GCP-2), CXCL7 (NAP-2), CXCL8 (IL-8), CXCL9 (MIG), CXCL10 (IP-10), CXCL11 (I-TAC),

CXCL12 (stromal cell-derived factor 1 [SDF-1]), CXCL13 (BCA-1), CXCL14 (BRAK), and CXCL16. These chemokines are characterized by a canonical Cys-X-Cys (CXC) motif preceded by a Glu-Leu-Arg (ELR) sequence, which promotes their engagement with CXC receptors (CXCR1-5) that are predominantly expressed on myeloid cell types, including macrophages, neutrophils, and MDSCs (Rossi and Zlotnik 2000; Allen et al. 2007). CXCR1, also known as IL-8RA, binds to CXCL6 and 8 with high affinity, whereas CXCR2 has been shown to interact with CXCL1, 2, 3, 5, 6, 7, and 8, acting as the common mediator of their activity (Lee et al. 1992; Murphy 1994). CXCR3, in turn, preferentially binds to CXCL4, 9, 10, and 11 (Loetscher et al. 2001). O'Hayer and colleagues (2009) showed that the entire family of ELR CXC chemokines can be induced by both $Hras^{G12V}$ and $Kras^{G12D}$ in human embryonic kidney (HEK) cells. Up-regulation of all CXC chemokines was similarly observed in Kras-mutant human pancreatic cell lines (Purohit et al. 2016) and in the DMBA/TPA model of skin tumorigenesis, showing that CXC chemokine up-regulation occurs in physiologically relevant settings of Ras-induced tumorigenesis (O'Hayer et al. 2009).

CXCL8

Studies performed by Sparmann and Bar-Sagi (2004) showed that CXCL8 was transcriptionally induced by oncogenic Hras via the concerted action of AP-1 and NF-κB transcription factors (Fig. 2). Tumors generated from $Hras^{G12V}$-transformed cells showed a robust infiltration of macrophages and granulocytes that facilitated tumor growth and whose recruitment was prevented by treating mice with a neutralizing antibody to CXCL8. Critically, they showed that the tumor cells lacked expression of CXCR1 and 2, supporting the model that CXCL8 acted cell extrinsically. Similar findings were observed in the setting of endogenous mutant Kras expression in the $Kras^{G12D}$LA1 lung adenocarcinoma (LAC) mouse model, in which lung tumorigenesis correlated with increased CXCL8 levels in the lung tissue and neutrophil recruitment (Wislez et al. 2006). Moreover, CXCR2 inhibition with a neutralizing antibody reduced the incidence and progression of lung lesions. The relevance of these studies to human disease is highlighted by the observation that CXCL8 levels are often elevated in both the serum and tumor tissues of patients with pancreatic and lung cancers (Wigmore et al. 2002; Tas et al. 2006; Frick et al. 2008). In the case of lung cancer, high CXCL8 in the serum was specifically linked to patients whose tumors harbored Kras mutations, confirming the direct relationship between Kras mutation status and CXCL8 expression in human cancer (Tas et al. 2006).

CXCL1, 2, and 5

Other ELR^+ CXC chemokines, CXCL1, 2, and 5, were found to be induced by $Kras^{G12D}$ in the CC10-Cre/LSL-$Kras^{G12D}$ mouse model of LAC, leading to a robust infiltration of macrophages and neutrophils that was amplified with tumor progression (Ji et al. 2006). CXCL1 induction by Kras-mutant pancreatic cancer cells was required for angiogenesis and xenograft growth in studies performed by O'Hayer and colleagues (2009), as was similarly found for CXCL5 (Li et al. 2011). Importantly, Steele and colleagues (2016) associated increased CXCL1, 2, and 5 expression by Kras-mutant tumor cells in situ to elevated CXCR2 signaling in the myeloid compartment of $Kras^{G12D/+}p53^{R172H/+}$ transgenic (KPC) mice. Global deletion of CXCR2 or depletion of neutrophils and MDSCs in these mice resulted in decreased metastasis and improved T-cell entry into the primary tumor. The therapeutic significance of these findings was shown by the sensitization of KPC mice to PD-1 checkpoint therapy when treated with a small molecule inhibitor of CXCR2. Together, these studies support an important role for mutant Kras in establishing an immunosuppressive microenvironment that affects proper cytotoxic T-cell entry and activation via the CXC-dependent recruitment of neutrophils and MDSCs. As with CXCL8, the therapeutic potential of targeting other CXC chemokines in Kras-mutant tumors is exciting given the number of studies showing the elevated expression of CXCL2 and CXCL5 in pancreatic cancers and their association with poorer tumor differentiation, advanced clinical stage, and/or shorter overall survival (Frick et al. 2008; Li et al. 2011; Steele et al. 2016).

CYTOKINES

A number of Kras-regulated cytokines play dominant roles in shaping the immune microenvironment. Although the signaling pathway(s) leading to the mutant Kras-dependent activation of some cytokines, such as IL-6, IL-10, TGF-β, and GM-CSF, has been delineated, such mechanistic links are lacking for many other cytokines highly expressed in Kras-mutant tumors. Below, we will focus on those cytokines that are direct transcriptional targets of oncogenic Kras.

IL-6

IL-6 is a pleiotropic inflammatory cytokine that plays an essential role in immune and cancer cell cross talk in many tumor types, primarily via the activation of the Janus kinase (JAK)/signal transducers and activators of transcription (STAT) family of transcription factors (Holmer et al. 2014). Ancrile and colleagues showed that acute induction of oncogenic $Hras^{G12V}$ resulted in IL-6 gene and protein expression in multiple cell types (Ancrile et al. 2007). Zhu and colleagues (2014b) subsequently showed that IL-6 up-regulation was mediated via noncanonical NF-κB signaling via the Kras/RalB/TBK1/IKKE/NF-κB pathway (Fig. 2). Intriguingly, McAllister and colleagues (2014) found that $Kras^{G12D}$ could promote IL-17 receptor expression in preneoplastic pancreatic cells, which in turn induced IL-6 expression. These data suggest that multiple signaling pathways downstream from mutant Kras may contribute and/or synergize to sustain high IL-6 levels. In support of these findings, in Kras-mutant pancreatic and NSCLC cell lines, IL-6 is consistently and highly up-regulated (Wigmore et al. 2002; Yamaji et al. 2004; Bellone et al. 2006; Feurino et al. 2007). A striking number of studies have found that IL-6 levels are elevated in the serum and tumor tissue of pancreatic and NSCLC patients and correlate with tumor progression, the presence of metastases, and/or poorer overall survival (Okada et al. 1998; Wenger et al. 1999; Carpagnano et al. 2002; Ebrahimi et al. 2004; Huang et al. 2005; Bellone et al. 2006; Talar-Wojnarowska et al. 2009; Mroczko et al. 2010; Chang et al. 2013). Despite the established negative prognostic role of IL-6 in many cancers that commonly harbor Kras mutations, relatively little is known about how tumor-derived IL-6 directly modifies the immune microenvironment. This is attributed to the fact that tumor cells can express IL-6 receptors, allowing for the possibility that IL-6 promotes tumor growth in a cell-autonomous manner. Moreover, IL-6 is secreted by many cell types in the tumor microenvironment, making the determination of the relative functionally active sources of IL-6 challenging (Lesina et al. 2011; Zhang et al. 2013). Nevertheless, some studies have convincingly shown that mutant Kras-induced IL-6 secretion plays a critical role in shaping the immune microenvironment. Fukuda and colleagues showed that oncogenic Kras expression in the pancreas was required for sustained IL-6 and STAT3 signaling in the epithelial cell compartment (Fukuda et al. 2011). This was accompanied by an infiltration of myeloid cells, including MDSCs and immature myeloid cells, as well as B and T lymphocytes, whose recruitment was abrogated in STAT3-deficient $Kras^{G12D}$ transgenic mice and was paralleled with an inhibition of pancreatic intraepithelial neoplasia (PanIN) formation. Although this study did not further elaborate on the functional properties of the immune cell types recruited by IL-6/STAT3 signaling, they clearly show the capacity of tumor-derived IL-6 to substantially alter the immune cell composition to one favoring tumor progression. In line with these studies, IL-6-deficient $iKras^{G12D}$ mice showed decreased MDSC, macrophage, T-regulatory cell, and mast cell recruitment to the pancreas concomitant with abrogated PanIN progression (Zhang et al. 2013). Although IL-6 production was observed in Kras-mutant tumor cells in this model, it could not be causally linked to the immunosuppressive phenotype of the PanINs given their model of tissue-wide IL-6 deletion. Interestingly, in a $Kras^{G12D}$ model of lung cancer, IL-6 deletion resulted in a dramatic increase of iNOS-expressing macrophages in the lung, thereby promoting tissue damage and accelerating the onset of tumorigenesis (Qu et al. 2015). These studies support the role of IL-6 in promoting an immunosuppressive microenvironment while highlighting the context specificity of the tumorigenic outcome of immunostimulatory and immunosuppressive signaling. Of note, it is also likely that IL-6-producing immune cells synergize with Kras-mutant tumor cells to amplify IL-6 levels in the tumor, as studies have shown a positive feedback loop between IL-6 and Kras oncogenic activation (Fukuda et al. 2011).

IL-10

IL-10 is an anti-inflammatory cytokine that, like IL-6, primarily acts via the induction of the JAK/STAT signal activation. The effect of IL-10 secretion in the tumor microenvironment has primarily been studied in the context of Treg-cell differentiation and macrophage polarization. Zdanov and colleagues (2016) showed that Kras-mutant colorectal cells transcriptionally up-regulated IL-10 via the MEK/ERK/AP-1 pathway (Fig. 2). Critically, they showed that mutant Kras-dependent IL-10 secretion was required for the conversion of $CD4^+$ T cells to $CD4^+FoxP3^+$ Treg cells. Moreover, Treg-cell recruitment to lung tumors was abrogated with a small molecule inhibitor of Kras activity. In PDAC and NSCLC, increased IL-10 levels have long been observed in the serum and tumor tissue of patients (Fortis et al. 1996; Hatanaka et al. 2000; von Bernstorff et al. 2001; Bellone et al. 2006), with elevated serum levels of IL-10 correlating with significantly worse survival (Hatanaka et al. 2000; Ebrahimi et al. 2004; Hsu et al. 2015). Bellone and colleagues (1999) showed that IL-10 present in pancreatic tumor-cell-conditioned medium was required to abrogate peripheral blood mononuclear cell (PBMC) proliferation and Th1 cytokine production, confirming the important role of IL-10 in promoting a Th2 phenotype. In a mutant Kras-driven model of lung tumorigenesis, elevated IL-10 levels were observed in serum and tumor tissues and were required for the recruitment of M2 macrophages and Treg cells to the tumor (Hsu et al. 2015). Disruption of Treg-cell recruitment to Kras-mutant tumors has been shown to activate cytotoxic T-cell-dependent immune clearance and dramatically attenuate tumor growth in several mouse models, although in many studies the mechanistic link to tumor-cell-derived IL-10 was not validated (Granville et al. 2009; Ali et al. 2014). Given the abundant IL-10 secretion in human cancers harboring Kras mutations and its established effect on Treg-cell differentiation, however, the potential of targeting IL-10 to restore immune recognition is an exciting prospect to consider.

TGF-β

TGF-β is a pleiotropic cytokine that plays diverse cellular roles, including the regulation of cell proliferation, differentiation, migration, invasion, and survival. The development of TGF-β knockout mice in the 1990s revealed a central role for the cytokine in the regulation of the immune system, with a particularly strong effect on the inhibition of inflammation and autoimmune diseases (Shull et al. 1992; Kulkarni and Karlsson 1993). Subsequent studies focusing on cell-type-specific deletion of TGF-β in mice revealed a dominant role for TGF-β in the regulation of T lymphocyte proliferation, differentiation, and survival (Gorelik and Flavell 2002), although its immunosuppressive effects also extend across B-cell, NK-cell, dendritic-cell (DC), mast-cell, and granulocyte-cell populations (Li et al. 2006). TGF-β was originally identified as a secreted protein induced by oncogenic Hras expression in breast cancer cells by Dickson and colleagues (1987). Tsubaki and colleagues (Tsubaki et al. 2011) found that inhibition of either MAPK or PI3K pathways downstream of active Ras could suppress TGF-β induction, suggesting that both signaling axes contribute to TGF-β production (Fig. 2). Early studies by Bellone and colleagues (1999) showed that TGF-β was elevated in the medium of pancreatic cancer cell lines and in serum from pancreatic cancer patients. Using conditioned medium from pancreatic carcinoma cells, they found that TGF-β contributed to the inhibition of Th1 responses in PBMCs, as evidenced by the reversion of these effects using TGF-β-neutralizing antibodies. Similarly, Zdanov and colleagues (2016) showed that Kras^{G12V}-induced Treg-cell differentiation required TGF-β secretion from colorectal cancer cells. In this study, TGF-β secretion was induced only in cell lines harboring mutant Kras and was mediated via the MEK/ERK/AP-1 pathway. Moreover, small molecular inhibition of Kras in mice harboring lung-specific Kras mutations prevented the infiltration of Treg cells. Perhaps the most convincing evidence for the critical role of Kras-induced TGF-β secretion in dictating intratumoral Treg-cell differentiation was presented by Moo-Young and colleagues (2009). Adoptive transfer of $CD4^+FoxP3^-$ T cells into mice implanted with Kras-mutant, TGF-β-secreting tumor cells induced their differentiation into $CD4^+FoxP3^+$ Treg cells, which was reversed by systemic injection of a blocking antibody to TGF-β or T-cell-specific inhibition of the TGF-β receptor (Moo-Young et al. 2009). Moreover, Treg-cell differentiation was not observed in mice with Kras wild-type tumors that lacked TGF-β secretion. Intriguingly, another study showed that although global TGF-β receptor deficiency in EL-Kras^{G12D} mice restored antitumor immune function, epithelial-specific deletion of TGF-β receptor facilitated tumorigenesis, supporting the notion that tumor-cell-derived TGF-β acts non-cell autonomously to contribute to widespread immunosuppression (Principe et al. 2016). TGF-β levels are also elevated in NSCLC patients (Tateishi et al. 1991) and are secreted by NSCLC cells, leading to suppression of infiltrating T-cell cytotoxic responses (Ortegel et al. 2002). In NSCLC, tumor-cell-derived TGF-β has been further shown to suppress the activation of B cells, the maturation of DCs, and activation of NK and lymphokine-activated killer cells (Yoon 2014). These observations have led to the development of Belagenpumatucel-L, an allogeneic tumor-cell vaccine composed of irradiated NSCLC cell lines, to block TGF-β secretion, which is currently in phase III clinical trials (Nemunaitis et al. 2006, 2009).

Given the plethora of studies that have found elevated TGF-β levels in the serum and tumor tissue of pancreatic and NSCLC patients (Hasegawa et al. 2001; von Bernstoff et al. 2001; Bellone et al. 2006), the role of TGF-β in mediating immunosuppression in Kras-mutant tumors is likely of great relevance to human disease. That being said, it is not clear whether TGF-β secreted by Kras-mutant cells constitutes the major determinant of TGF-β-dependent immunosuppression in these tumor types. Many other cell types in the tumor microenvironment are rich sources of TGF-β, including fibroblasts, pancreatic stellate cells, and immune cells themselves. Mutant Kras may indirectly promote TGF-β-mediated immunosuppression via the activation of other cell types in the tumor microenvironment, including M2 macrophages and pancreatic stellate cells, which we will touch on in our discussions of these cell types below.

GM-CSF

GM-CSF is a hematopoietic growth factor that stimulates the development of neutrophils, DCs, granulocytes, and macrophages in the bone marrow. Pylayeva-Gupta and colleagues (2012) showed Kras^{G12D}-dependent transcriptional up-regulation of GM-CSF in pancreatic ductal epithelial cells (PDECs) that was mediated by the concerted activation of MAPK and PI3K pathways (Fig. 2). Human PDAC lesions showed selective up-regulation of GM-CSF relative to tissues from patients with other diseases of the pancreas, showing the specificity of GM-CSF up-regulation in mutant Kras-induced neoplasia. Importantly, GM-CSF production by Kras^{G12D}-expressing pancreatic tumor cells promoted the recruitment of MDSCs to the tumor microenvironment, which in turn fostered tumor growth by suppressing immune clearance by $CD8^+$ T cells (Bayne et al. 2012; Pylayeva-Gupta et al. 2012). Other groups have similarly found that PDAC cell lines secrete GM-CSF, resulting in MDSC differentiation and the inhibition of T-cell proliferation (Takeuchi et al. 2015; Kenkel et al. 2017). Kenkel and colleagues found that pancreatic tumor-cell-secreted GM-CSF induced DC differentiation and Treg-cell expansion at early sites of metastasis in KPC mice. Strikingly, targeted depletion of this DC population resulted in a reduction of Treg cells, activation of cytotoxic lymphocytes, and prevented metastasis formation. In line with the above findings, GM-CSF is present at high levels in pancreatic cancer patients and is correlated with reduced overall survival (Vasiliades et al. 2012; Delitto et al. 2015; Takeuchi et al. 2015). GM-CSF was also found to be elevated in NSCLC patients and to correlate with worse outcome (Baldwin et al. 1989). These data are somewhat paradoxical given that GM-CSF classically functions as an immune stimulant that in many contexts enhances antitumor immune responses. Indeed, the prognostic value of GM-CSF secretion varies widely depending on tumor type, and likely is caused by thresholds of GM-CSF levels that shift the balance from suppression to stimulation of specific immune cell types. In addition, the pleiotropic role of GM-CSF may reflect the differing roles of immune activation at different stages of tumorigenesis. In the context of pancreatic cancer, however, GM-CSF appears to drive immunosuppression via MDSC differentiation and cytotoxic T-cell suppression. Thus, inhibition of GM-CSF in this setting has the potential to be an effective therapeutic avenue to restore immune surveillance.

KRAS' COIMMUNIST SOCIETY

As described above, mutant Kras engages a pronounced number of immune modulating cytokines with clear and direct links to immune cell recruitment and/or differentiation. However, there are a number of cell types that play a critical role in the immune phenotype of mutant Kras tumors where the mechanistic link to Kras' oncogenic activity has not been fully elucidated. Below, we highlight some cell types that are prevalent in mutant Kras tumors and significantly impact mutant Kras-associated immunity.

Macrophages

Macrophage infiltration occurs early in NSCLC and PDAC tumorigenesis and is an established promoter of tumor progression. Depending on signals that prevail within their microenvironment, macrophages can adopt a variety of functional states. In response to bacterial products such as lipopolysaccharide (LPS) and Th1 cytokines, macrophages become immunostimulatory (M1). M1 macrophages are characterized by the expression of high levels of iNOS, major histocompatibility complex II (MHCII), cluster of differentiation 80 and 86 (CD80, CD86), TNF-α, contributing to their tumoricidal effect (Mantovani et al. 2002). In contrast, in the presence of Th2 cytokine macrophages acquire an alternatively activated state (M2) that is immunosuppressive, tumor promoting, and is characterized by the expression of arginase 1, cluster of differentiation 206 (CD206), and low levels of MHCII. In pancreatic cancer, tumor-associated macrophages (TAMs) show both M1 and M2 phenotypes, although higher M2:M1 ratios correlate with disease progression, metastasis, and shorter survival in patients (Kurahara et al. 2011; Ino et al. 2013). Other studies have argued that both M1 and M2 macrophages are tumor promoting but play distinct roles in the initiation (M1) and progression (M2) of tumorigenesis (Liou et al. 2015). Although cytokines discussed previously in this review such as IL-10 and TGF-β can drive M2 phenotypes, a direct link between mutant Kras-induced secretion of these cytokines and macrophage polarization in PDAC has not been stringently tested. Interestingly, Liou and Storz (2015) showed that $Kras^{G12D}$ expression in pancreatic acinar cells drove the expression of ICAM1, which served as a chemoattractant for M1 macrophages. Although the precise mechanism of ICAM1-dependent macrophage recruitment was not investigated, the authors speculated that proteases secreted by acinar cells processed ICAM1 into a soluble form that could bind to unidentified receptors present on macrophages. Liou and colleagues (2013; Liou and Storz 2015) further show that ICAM1-dependent macrophage recruitment promoted acinar-to-ductal metaplasia (ADM) via TNF-α and CCL5 production, thereby contributing to neoplastic transformation and providing functional evidence of a temporally dependent, tumor-promoting role of M1 macrophages in early pancreatic tumorigenesis. In advanced stages of pancreatic tumorigenesis, macrophages play a primarily immunosuppressive role and emerge as one of the most abundant immune cell types present in the tumor microenvironment (Mielgo et al. 2013). M2 macrophages have been shown to contribute to angiogenesis, invasion (via MIP-3α), metastasis (via vascular endothelial growth factor [VEGF]), and the promotion of cancer stem-cell-like properties (via aldehyde dehydrogenase [ADLH]). Zhu and colleagues (2014a), in turn, showed that CSF1 was produced by human PDAC cells and that CSF1 receptor blockade in an orthotopic model of PDAC abrogated the infiltration of M2 macrophages, with modest increases in cytotoxic T-cell activation.

In NSCLC, macrophages constitute a dominant immune cell type whose correlation with patient prognosis remains controversial (Dai et al. 2010; Remark et al. 2015; Busch et al. 2016). For example, Chen and colleagues found that TAMs were associated with negative prognosis in NSCLC patients, whereas Toomey and colleagues found no association between the presence of TAMs and overall survival (Chen et al. 2003; Toomey et al. 2003). In contrast, Dai and colleagues (2010) discovered that TAMs in the stroma correlated with worse survival, whereas TAMs in the tumor islets predicted better overall survival of NSCLC patients. In the CCSP-Kras^{G12D} murine model of lung cancer, Redente and colleagues (2010) showed that macrophages in and surrounding lung tumors showed an M2 phenotype early in tumor progression that was reversed to an unpolarized state on extinction of the mutant Kras transgene expression. Intriguingly, comparison of the immune profile of epidermal growth factor receptor (EGFR), Kras^{G12D}, and Kras$^{G12D/+}$p53$^{R172H/+}$ transgenic mice revealed that macrophage numbers increased over time specifically in the background of an oncogenic Kras allele, supporting the notion that mutant Kras is sufficient to direct macrophage infiltration (Busch et al. 2016). Although the precise mechanism of macrophage recruitment to mutant Kras-driven lung tumors has not been established, several groups have found that macrophage CXCR2 signaling was required for TAM recruitment in both transgenic and human xenograft models of Kras-mutant lung cancer, suggesting that the mutant Kras-dependent secretion of ELR-CXC chemokines may constitute one mechanism of macrophage attraction (Cortez-Retamozo et al. 2012; Schmall et al. 2015).

B Cells

Only recently has the significance of tumor-infiltrating B cells in Kras-mutant tumors been unveiled, although it has been largely restricted to studies performed in pancreatic cancer models. B cells constitute a significant proportion of lymphocytes in both early and late stages of pancreatic cancer (Gunderson et al. 2016; Pylayeva-Gupta et al. 2016), with a somewhat contradictory role in tumorigenesis. Although pan-B-cell numbers correlated with improved disease prognosis, elevated levels of B-cell-activating factors in tumors correlated with increased metastatic potential (Koizumi et al. 2013; Tewari et al. 2013). However, subsequent studies by three independent groups using distinct oncogenic Kras-driven mouse models showed a protumorigenic role for tumor-infiltrating B cells (Gunderson et al. 2016; Lee et al. 2016; Pylayeva-Gupta et al. 2016). Interestingly, the tumor-promoting functions of the B-cell population were mediated through distinct B-cell subpopulations at different stages of disease progression. Pylayeva-Gupta and colleagues (2016) identified a subset of regulatory B cells that promoted PanIN initiation and early progression through the secretion of IL-35. In parallel to these findings, Gunderson and colleagues (2016) identified a population of antibody-producing B cells that promoted macrophage M2 polarization and tumor growth in advanced PDAC. Furthermore, a tumor-protective role was attributed to the B1β subset of B lymphocytes in early pancreatic neoplasms by Lee and colleagues (2016). Collectively, these studies provide evidence that the role of B cells in promoting pancreatic cancer is complex and highly contextual. Although the importance of B cells in lung cancer is limited, regulatory B cells were found to be elevated in lung cancer patients (Zhou et al. 2014). In addition, Busch and colleagues (2016) observed a substantial recruitment of B cells in early and late stages of tumor progression in the CCSP-Kras^{G12D}, but not CCSP-EGFR^{L858R} transgenic LAC mouse model, providing further support that cells expressing mutant Kras secretes B-cell-attracting factors in multiple contexts. Ultimately, although these studies elucidate an important role for B cells in oncogenic Kras-driven tumors, the mechanism of B-cell recruitment remains an area of active research. In our own study, we have identified the role of fibroblast-derived chemoattractant CXCL13 in promoting B-cell infiltration in pancreatic neoplasia, although the connection to oncogenic Kras remains elusive and merits further investigation (Pylayeva-Gupta et al. 2016).

Th17 Cells

Th17 cells are a recently discovered lineage of Th cells that are potent immunostimulators through the production of the cytokine IL-17A. Murine models of Kras-mutant lung and pancreatic tumorigenesis showed a significant enrichment in Th17 cells (Chang et al. 2014; McAllister et al. 2014; Marshall et al. 2016). Similar findings were reported by Zhang and colleagues (2014) in the iKras^{G12D} model of pancreatic cancer, although in this case the induction of pancreatitis was required to detect Th17 cells in the tumor microenvironment. Furthermore, infiltration of Th17 cells and high plasma IL-17A levels correlated with poor prognosis in NSCLC and pancreatic cancer patients (McAllister et al. 2014; Duan et al. 2015; Marshall et al. 2016). Importantly, Busch and colleagues (2016) showed that Th17 cells were specifically recruited in mutant Kras transgenic mice compared to mutant EGFR-transgenic mice in lung cancer models. Although the mechanistic link between mutant Kras and Th17-cell recruitment and/or differentiation has not been specifically addressed, IL-6 and TGF-β, transcriptional targets of mutant Kras described in previous sections, are potent inducers of Th17 differentiation. Given that these cytokines are also produced by several additional tumor-associated cell types, including fibroblasts and immune cells (Löhr et al. 2001; Lesina et al. 2011), the recruitment of Th17 cells is likely facilitated by multiple cell types in the tumor microenvironment. In addition, in pancreatic cancer, mutant Kras-dependent up-regulation of the IL-17 receptor on tumor cells may also promote their interaction with Th17 cells (McAllister et al. 2014). Functionally, ablation of Th17 cells in oncogenic Kras models of lung and pancreatic cancer significantly abrogated tumor initiation and progression, with loss of IL-17A resulting in reduced tumor-cell proliferation, angiogenesis, and MDSC recruitment (Chang et al. 2014; McAllister et al. 2014). Zhang and colleagues (2014) further showed that genetic depletion of CD4^{+} T cells, including the Th17 population, resulted in the derepression of stromal CD8^{+} cytotoxic T cells and subsequent attenuation of tumor progression. Collectively, these data point to an important role for Th17 cells in mediating immunosuppression in Kras-mutant tumors and warrants a more thorough investigation of the mechanism of Th17 recruitment to the tumor microenvironment.

γδ T Cells

An additional source of IL-17 in the Kras-mutant tumor microenvironment is derived from γδ T cells, a distinct subclass of T cells that are characterized by the expression of the γδ T-cell receptor (TCR). γδ T cells constitute up to three-quarters of tumor-infiltrating T cells in human PDAC samples. Interestingly, however, Daley and colleagues (2016) described a tumor-protective role for γδ T cells in pancreatic cancer. Genetic depletion of γδ T cells in KC mice severely decreased ADM and subsequent PanIN progression and was accompanied by increased Th1 response and elevated cytotoxic T-cell activity. Surprisingly, deletion of γδ T cells in Kras-mutant LAC did not impact tumor burden, emphasizing the complexity of differential immune modulation depending on the tumor type (Busch et al. 2016). In the Daley and colleagues' (2016) study, γδ T-cell recruitment to the tumor was contingent on ligation of the CC chemokine receptor family members CCR2, CCR5, and CCR6, leading to the tempting speculation that mutant Kras may contribute to CC chemokine production (Daley et al. 2016). In line with the surprising tumor-protective role of abundantly expressed γδ T cells in pancreatic cancer, the functional significance of this cell population in mutant Kras-driven tumor progression remains controversial. This may be because of the selective recruitment of two distinct γδ T-cell subpopulations in different tumors: the antitumorigenic interferon (IFN)-γ^{+}CD27^{+} γδ T cells and the protumorigenic IL-17A^{+} CD27^{-} γδ T cells. Undoubtedly, further investigation is needed to untangle the specific and context-dependent roles of this T-cell population in Kras-mutant tumors, as well as their mechanism of recruitment to the tumor microenvironment.

Stellate Cells

A principal theme reinforced throughout the previous sections has been that the functional output of mutant Kras-induced immune signaling pathways is modified by tumor-cell interactions with neighboring cells and cytokines in their microenvironment. With this theme in mind, it is critical that we not only consider the immune cells directly stimulated by mutant Kras signaling, but other cell types that are regulated by Kras-mutant cells that may also modulate immune cell recruitment. This concept is best explored in PDAC because of its extensive stromal compartment composed of fibroblasts, pancreatic stellate cells (PSCs), extracellular matrix (ECM) proteins, endothelial cells, nerve cells, and adipocytes in addition to immune cells. PSCs are the most prevalent cell type in the pancreatic tumor stroma and are established contributors to disease progression (Omary et al. 2007). PSCs are myofibroblast-like cells that are normally present in a quiescent state in areas of the exocrine pancreas. In chronic pancreatitis and the early stages of pancreatic cancer, PSCs become activated (aPSCs), inducing their proliferation, migration, and secretion of ECM components and promotion of wound repair. Importantly, PSCs are activated in response to factors produced by Kras-mutant tumor cells, and in turn secrete cytokines that direct immune-cell recruitment. A major factor in the pancreatic tumor microenvironment (TME) responsible for the activation of PSCs is tumor-cell-derived TGF-β, which induces the proliferation and secretion of type I collagen and fibronectin in fibroblasts (Fig. 3) (Löhr et al. 2001). A number of other growth factors, including fibroblast growth factor (FGF), heparin-binding epidermal growth factor (HB-EGF), platelet-derived growth factor (PDGF), and sonic hedgehog (SHH) are also secreted by Kras-mutant PDAC cells and contribute to PSC activation (Means et al. 2003; Kordes et al. 2005; Bailey et al. 2008; Tian et al. 2012). In turn, aPSCs produce a plethora of chemokines, cytokines, and growth factors, including several ELR^{+} CXC chemokines, TGF-β, TNF-α, connective tissue growth factor (CTGF), MCP-1, regulated and normal T-cell regulated and secrete (RANTES), SDF-1, and IL-1β, 4, 6, 8, 13, and 15 (Andoh et al. 2000; Apte et al. 2012; Mace et al. 2013a,b; Xue et al. 2015). Given the expanse of the aPSC immunomodulatory secretome and the significant overlap with factors induced by mutant Kras, it comes as no surprise that aPSCs strongly amplify the immune phenotype of PDAC.

The cytokine profile of aPSCs offers a pertinent example of how parallel secretion of immunosuppressive factors by stellate cells and Kras-mutant cells can amplify immune cell recruitment to the tumor. For example, isolation of a mesenchymal stem cell (MSC) subtype of stellate cells from KC mice revealed the increased expression of macrophage colony-stimulating factor (M-CSF) and GM-CSF, in addition to IL-6, IL-10, and TGF-β, relative to wild-type mice (Fig. 3) (Mathew et al. 2016; Waghray et al. 2016). Coinjection of KC-derived MSCs with tumor cells induced a significant increase in macrophage number in the tumor and promoted their polarization to an M2 phenotype via elevated IL-6 and IL-10 secretion (Mathew et al. 2016). Critically, macrophage recruitment by the MSCs was required for the increase in tumor size observed following coimplantation of MSC and Kras$^{G12D/+}$p53$^{R172H/+}$ cells. These findings were paralleled in an orthotopic model of human PDAC, in which coinjec-

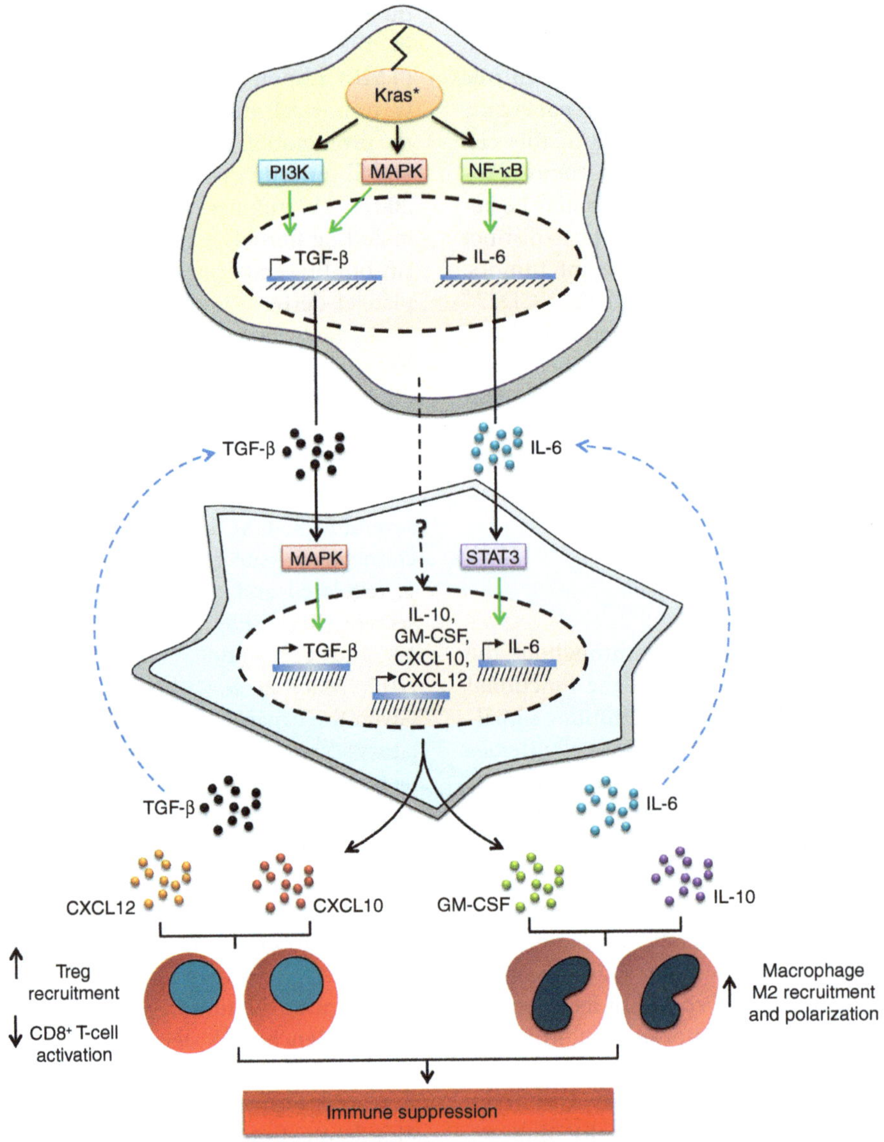

Figure 3. Cooperation of Kras-mutant tumor cells with pancreatic stellate cells in tumor microenvironment (TME) immunomodulation. Kras-mutant tumor-cell-derived transforming growth factor β (TGF-β) induces the activation of pancreatic stellate cells (PSCs) in the pancreatic TME, resulting in mitogen-activated protein kinase (MAPK) pathway activation and induction of TGF-β transcription and secretion by PSCs. Increased TGF-β levels can both promote T regulatory (Treg) recruitment and macrophage M2 polarization, thereby contributing to immunosuppression in the TME. Similarly, interleukin (IL)-6 produced by Kras-mutant tumor cells binds to IL-6 receptors on PSCs, resulting in intracellular signal transducers and activators of transcription 3 (STAT3) activation and IL-6 transcription and secretion into the TME. High IL-6 levels amplify the recruitment and M2 polarization of macrophages, thereby promoting tumor growth. Although the mechanism of transcriptional induction of IL-10, granulocyte macrophage colony-stimulating factor (GM-CSF), CXCL10, and CXCL12 by PSCs has not been directly addressed, their secretion by PSCs cooperates with IL-10, GM-CSF, CXCL10, and CXCL12 derived from Kras-mutant tumor cells to amplify Treg cell recruitment (CXCL10), dampen $CD8^+$ T-cell responses (CXCL12), or promote macrophage M2 polarization (IL-10, GM-CSF) in the pancreatic TME. NF-κB, Nuclear factor κB.

 Cite this article as *Cold Spring Harb Perspect Med* doi: 10.1101/cshperspect.a031849

tion of tumor cells with tumor-derived MSCs harboring stable GM-CSF knockdown resulted in a reduction in number and proportion of M2 macrophages in the tumor (Waghray et al. 2016). By contrast, Xue and colleagues found that aPSC-induced M2 macrophage polarization was dependent on IL-4 and IL-13 secretion in an experimental model of chronic pancreatitis, suggesting that increased levels of cytokines derived from both Kras-mutant cells and aPSCs may alter the cytokine dependence for M2 polarization to one favoring IL-6 and IL-10 (Xue et al. 2015). aPSCs were also reported to promote the differentiation of PBMCs to CD11b-expressing MDSCs via IL-6 production, which resulted in the inhibition of T-cell proliferation (Mace et al. 2013a,b).

The profoundly dampened cytotoxic T-cell responses observed in PDAC may also be attributed to synergistic effects of aPSCs and Kras-mutant tumor cells, but likely mediated through their concomitant secretion of CXC chemokines. Indeed, many reports have established that aPSCs can inhibit $CD8^+$ T-cell recruitment and activation (Watt and Kocher 2013). Ene-Obong and colleagues (2013) showed that aPSCs sequestered $CD8^+$ T cells and prevented their infiltration around tumor cells, resulting in reduced antitumor immunity. Similarly, fibroblast activation protein-positive aPSCs (FAP^+ aPSCs), an additional subtype of aPSCs, were reported to disrupt antitumor immunity via the repulsion of T cells from tumor cells via CXCL12 production in KPC mice (Feig et al. 2013). Intriguingly, although FAP^+ aPSCs were deemed the principal source of CXCL12, tumor cells also colocalized with the chemokine, suggesting that both tumor-cell- and aPSC-derived CXCL12 may cooperate to regulate T-cell responses (Fig. 3). Critically, on small molecule inhibition of the CXCL12 receptor, T-cell accumulation was restored and the checkpoint agonist PD-L1 was effective in inducing tumor regression. Lunardi and colleagues, in turn, found that human PDAC cells induced CXCL10 expression in aPSCs, which induced the recruitment of $CXCR3^+$ effector T cells and Treg cells and correlated with their presence in human PDAC samples (Fig. 3) (Lunardi et al. 2014). These data highlight the interdependence of chemokine secretion between aPSCs and PDAC cells and strongly argue in favor of the concept that mutant Kras synergizes with aPSCs to dictate the immune phenotype of PDAC.

In addition to the plethora of cytokines and chemokines that are common to aPSCs and Kras-mutant tumor cells, aPSCs secrete a number of unique factors that independently contribute to the immune contexture of PDAC. For example, galectin 1 was found to be a principal mediator of T-cell suppression in PanINs and PDAC and is uniquely expressed by aPSCs (Tang et al. 2012, 2015). Zambirinis and colleagues (2015) showed that CCL3 secretion by aPSCs was required for the recruitment of Treg cells in KPC mice. Ultimately, the immune effects of aPSCs are dictated by mutant Kras on multiple levels, starting with the initial requirement of oncogenic Kras-induced factors for their activation, and followed by their synergistic and distinct cytokine and chemokine profiles that together help shape the immune phenotype of PDAC.

MUTANT KRAS-DEPENDENT IMMUNE MODULATION: CONTEXT MATTERS

Inflammation: Making the Most of It

The potent effects of mutant Kras on the immunogenicity of the tumor microenvironment brings to question how it may cooperate with or regulate external inflammatory insults associated with tumor development. Kras is most frequently mutated in PDAC (95%), NSCLC (35%), and colorectal cancer (CRC) (40%) tumor types that are highly associated with the inflammatory disorders pancreatitis, chronic obstructive pulmonary disorder (COPD) and smoking, and ulcerative colitis (UC), respectively (Lowenfels et al. 1993; Ahrendt et al. 2001; Malka et al. 2002; Young et al. 2009; Jess et al. 2012). Below, we describe how mutant Kras' immunomodulatory outputs interact with pancreatitis and COPD to modulate the development of PDAC and NSCLC.

Using pancreatic acinar-cell-specific expression of mutant Kras, Guerra and colleagues (2011) described a highly refractory mouse model of PanIN that acquired tumorigenic properties following the induction of chronic pancreatitis. Mutation of Kras concomitant with loss of tumor suppressors, such as Trp53 or p16Ink4a/p19Arf, and caerulein-induced pancreatitis resulted in malignant PDAC disease. The authors proposed that pancreatitis contributed to PanIN progression through the induction of its own inflammatory signature, abrogating oncogenic Kras-induced senescence in mutant precursor cells and delaying tissue repair, thereby providing the mouse PanINs with the potential to expand and progress. Further mechanistic insight into the synergistic effects of pancreatitis and mutant Kras was provided by Fukuda and colleagues, who showed that the transient expression of IL-6 and activation of STAT3 signaling induced by acute pancreatitis was amplified and prolonged in the presence of a mutated Kras allele (Fukuda et al. 2011). This was also observed by McAllister and colleagues (2014), who further discovered that IL-17A levels and the recruitment of Th17 and IL-17^{+}/γδ T cells were synergistically increased by Kras^{G12D} and pancreatitis.

Similar observations were made in an LSL-Kras^{G12D}/CCSP^{Cre} model of lung cancer, wherein *Haemophilus influenzae* infection-mediated induction of COPD resulted in increased lung tumor burden, accompanied by a robust infiltration of myeloid cells (Moghaddam et al. 2009). Consistent with the pancreatitis studies, oncogenic Kras cooperation with COPD in lung cancer was also determined to be a consequence of amplified Th17 recruitment, relative to mutant Kras alone, leading to enhanced IL-17 production and MDSC infiltration (Chang et al. 2014) and resulting in an increase in tumor burden and disease progression. Together, these studies indicate that exogenous inflammatory conditions amplify Kras' immunodulatory effects to further promote oncogenic transformation.

The significance of mutant Kras' cooperation with additional inflammatory diseases in tumorigenesis is emphasized by the observation that oncogenic mutations in Kras have been detected in disease-free pancreas of healthy individuals (Terhune et al. 1998; Lüttges et al. 1999; Löhr et al. 2005). Furthermore, the reciprocity of this relationship is evident by the lack of oncogenic transformation observed in wild-type mice subjected to induction of pancreatitis or COPD. Hence, although oncogenic mutation of Kras or inflammation of the pancreas and the lung individually create a strong predisposition for transformation, synchronously they create an optimal immune microenvironment that accelerates tumor progression.

Different Folks for Different Strokes

As described previously, a significant proportion of mutant Kras-mediated immune modulation mechanisms are a function of its paracrine signaling to components of the tumor microenvironment. The inherent tissue-specific variations of the tumor stroma thus profoundly impact the immunological characteristics of different mutant Kras-expressing tumor types, which can in turn have significant implications for disease progression and therapeutic intervention. This is perhaps best exemplified by the variations observed in the CD8^{+} Tc cell compartment of Kras^{G12D}-bearing PDAC and NSCLC tumors.

Extensive characterization of Kras^{G12D}-driven PDACs, both in mouse models and human samples, has revealed the PDAC TME to be largely immunosuppressive, with high levels of myeloid cell infiltration, especially the M2 polarized TAMs and MDSCs (Mantovani et al. 2006; Clark et al. 2009; Markowitz et al. 2015). MDSC infiltration in the PanIN and PDAC microenvironment strongly correlates with paucity of infiltrating T cells, more specifically the tumoricidal CD8^{+} Tc cells (Ino et al. 2013; Vonderheide and Bayne 2013). Furthermore, the few CD8^{+} T cells that are present exist in a state of clonal anergy with an absence of markers of activation.

In remarkable contrast to this, the TME of Kras^{G12D} lung cancer mouse models, although also rich in macrophages and MDSCs, display significant expansion of CD8^{+} Tc cells (Busch et al. 2016). This observation is also corroborated in human NSCLC samples in which mu-

 Cite this article as *Cold Spring Harb Perspect Med* doi: 10.1101/cshperspect.a031849

tant Kras tumors associate with higher $CD8^+$ T-cell infiltration in the stroma.

Mechanistically, the scarcity of tumor-infiltrating $CD8^+$ T cells in $Kras^{G12D}$-driven PDAC has been attributed to the unique architecture of the PDAC stroma, characterized by intense desmoplasia, fibrosis, and presence of mutant Ras-activated PSCs. Early on, von Bernstorff and colleagues (2001) argued that tumor-infiltrating lymphocytes (TILs) become "trapped" within the peritumoral, fibrous stroma, thereby preventing them from reaching their malignant cellular targets. Ene-Obong and colleagues (2013) elaborated on these findings using the KPC PDAC mouse model and showed that $Kras^{G12D}$-induced aPSCs inhibited juxtatumoral $CD8^+$ T-cell infiltration through the expression of cytokines, chemokines, and adhesion molecules that regulate T-cell migration.

The difference in $CD8^+$ T-cell infiltration in PDAC versus NSCLC significantly impacts the design of therapeutic strategies for the two diseases. Owing to the lack of infiltrating $CD8^+$ Tc cells, PDAC has proved to be a poor candidate for immune checkpoint monotherapy (Sharma et al. 2015; Topalian et al. 2015). Conversely, $CD8^+$ Tc infiltration in $Kras^{G12D}$-driven lung cancer has prompted the exploration of checkpoint inhibitors as a viable therapeutic means. Consequently, the co-occurrence of Kras and p53 mutations in NSCLC is emerging as a predictive biomarker for immune checkpoint therapy, with those patients harboring both mutations showing significantly longer progression-free survival on PD-1 blockade therapy (Ji et al. 2006; Herter-Sprie et al. 2016; Dong et al. 2017).

In light of these outcomes, continued comparative analysis of oncogenic Kras' molecular and cellular modes of action in different tumor types is clearly imperative for understanding and devising organ-specific therapeutic strategies.

IS KRAS IMMUNE?

An Inflamed Achilles Heel

The disappointing outcome of the attempts to directly target oncogenic Ras (Cox et al. 2014) has generated an urgent need to develop alternative therapeutic strategies to treat mutant Ras-driven cancers. An alternative approach consists of disrupting the biological outputs influenced by oncogenic Ras in lieu of its specific molecular functions. The current chemotherapeutic treatment for pancreatic and lung cancer consists of cytotoxic drugs like gemcitabine and paclitaxel, which primarily target the high proliferative index of cancer cells and are not specific to the unique properties of Kras-mutant tumor cells. Of note, we have recently discovered an immunostimulatory effect of nab-paclitaxel on macrophages in PDAC that may prove useful in combination therapies with other immunomodulatory agents (Cullis et al. 2017). However, a single regimen with either drug so far provides only modest benefits, especially in advanced stages, because of a high degree of acquired or inherent resistance. Combinational therapy-targeting multiple facets of the tumor are therefore a need of the day.

Given the significant role of immunosuppression in promoting growth of Ras-mutant tumors, immunotherapy is an attractive therapeutic approach. Immunotherapy uses strategies aimed at activating an antitumor immune response, predominantly mediated through $CD8^+$ cytotoxic T cells, in addition to NK cells and macrophages with tumoricidal activity (Stambrook et al. 2017). These strategies can either be aimed directly at the $CD8^+$ Tc cells or indirectly lead to their activation by inhibiting the immunosuppressive cells and factors in the TME. Here, we discuss the promising potential of targeting the mutant Ras-induced immunomodulation mechanisms as a means to release the breaks on the host tumoricidal immune responses and thereby combat oncogenic Ras-driven cancers. Furthermore, we briefly describe the recent advances made in harnessing the power of the host immune system to target mutant Ras itself.

Setting Our Cytokines High

As previously described, oncogenic Ras-dependent processes that modulate the protumorigenic immune composition of the TME includes the secretion of a considerable number of cytokines and chemokines including CSF-1, IL-6, and CCL2 (Fukuda et al. 2011; Agalioti et al. 2017). Immunotherapeutic strategies aimed at neutralizing these secreted factors or blocking their cognate receptors may therefore be an efficient strategy in combinational therapies. As a proof of principle, Mace and colleagues showed antitumor activity of a combined PD-L1 and IL-6 blockade regimen in orthotopically transplanted Panc02 PDAC cells that also resulted in the induction of tumor regression and increased survival in KPC-BRCA2 mice (Mace et al. 2016). Similar results were obtained in a mouse model of mutant Kras-driven lung cancer on monotherapy with an IL-6-neutralizing antibody (Caetano et al. 2016). IL-6 inhibition in CCSP-Kras^{G12D} mice resulted in decreased M2 macrophages and MDSC levels in the TME and a skewing of the T-cell protumor Th17/Treg response to the antitumor Th1/CD8^{+} Tc response, resulting in tumor regression.

Along similar lines, individual administration of small molecule inhibitors of CCR2 and CSF1R tyrosine kinase in Kras^{G12D} orthotopic PDAC mice resulted in a significant reduction in tumor-infiltrating macrophages and monocytes (Mitchem et al. 2013). Furthermore, administration of either inhibitor with gemcitabine dramatically reduced tumor progression of both Pan02 and Kras^{G12D}-INK orthotopic transplants. Following up on these promising results, the CCR2 inhibitor PF-04136309 in combination with the chemotherapeutic regimen FOLFIRINOX is currently in clinical trials, with phase 1b results revealing 52% partial response and 47.8% of patients showing stable disease (Wang-Gillam et al. 2016). CXCR2 has also recently been shown as a potential effective target for combination therapy in PDAC. Pharmacological ablation of CXCR2 signaling in a preclinical PDAC mouse model not only abrogated metastasis but also enhanced T-cell infiltration, thereby sensitizing the tumor to anti-PD-1 checkpoint therapy and extending survival in combination therapy (Steele et al. 2016). Whereas these results are very encouraging, more extensive investigations in larger trials are needed for further evaluation and development.

A New Way to Target Mutant Kras?

Although the primary objective of this article is to provide an overview of oncogenic Kras-mediated immune modulation in cancer and its implications on disease progression and therapeutic targeting, we would be remiss to not discuss the potential of mutant Kras itself as a target for emerging immunotherapeutic strategies that are showing great promise.

Oncogenic mutation of Kras can generate neoantigenic peptides with the potential to generate a cytotoxic T lymphocyte (CTL) response against the mutated tumor cells. Activated CD8^{+} Tc cells express TCRs, which specifically recognize these neoantigens. They are typically 8- to 11-amino-acid-long mutated Ras peptides presented in complex with host-specific MHC class I molecules by antigen-presenting cells (APCs) or tumor cells (Maher and Davies 2004). Mutant Ras epitopes are attractive targets for CTL-based response because Ras mutations occur early in tumorigenesis and are inherited in almost all cells within a heterogeneous tumor. Although the overall levels of tumor-reactive CTLs in oncogenic Ras-driven tumors is low, multiple studies conducted earlier have detected the presence of such autoreactive CD8^{+} Tc populations in CRC, PDAC, melanoma, and lung cancer patients (Fossum et al. 1995; Linard et al. 2002; Trojan et al. 2003; Kubuschok et al. 2006). These observations have recently led to the first-ever successful targeting of metastatic lesions in a CRC patient by adoptive transfer of CTLs specifically against the mutant Kras^{G12D} isoform (Rosenberg et al. 2017). TIL cultures isolated from surgically resected secondary tumors in the lung containing active CD8^{+} Tc cells against the Kras^{G12D} neoantigen were identified, expanded, and transplanted back into the patient. Remarkably, all metastatic lung lesions underwent regression postadoptive transfer and the treatment had no adverse side effects. In the absence of direct mutant Kras inhibitors, the development of such alternative methods to target oncogenic Ras directly is an

 Cite this article as *Cold Spring Harb Perspect Med* doi: 10.1101/cshperspect.a031849

incredibly exciting prospect to pursue (Baines et al. 2011).

FUTURE PERSPECTIVES

In this review, we describe the emerging role of mutant Ras in shaping the immune microenvironment. This occurs through mutant Ras-dependent transcriptional induction of chemokines, cytokines, and growth factors that recruit and functionally regulate cells of both the innate and adaptive immune system. Although many of the pathways linking mutant Ras to immune modulation have been delineated, much remains unknown about how these pathways interact combinatorially and how their output is modulated by tissue and organ context. Furthermore, although significant progress has been made on broadly defining the immune composition of the tumor microenvironment, several challenges impede a detailed molecular understanding of the interactions between tumor and immune cells. These include a scarcity of model systems that faithfully recapitulate the immune characteristics of human tumors and barriers to accessing the full complement of immune-cell subpopulations within tumors. The development of tumor mouse models with genetically humanized immune systems and the rapid advancement of imaging and sequencing techniques that enable the interrogation of immune cells at the single-cell levels offer promising prospects for deeper characterization of the cross talk between mutant Ras cancers and immune cells in the tumor microenvironment. This knowledge will undoubtedly enhance our understanding of how to harness the immune system for the treatment of mutant Ras tumors.

ACKNOWLEDGMENTS

The authors thank L.J. Taylor for help with figure preparation. This work is supported by funding from National Institutes of Health (NIH)/National Cancer Institute Grant CA 210263, American Association for Cancer Research (AACR) PanCAN Grant 13-90-25-VOND, a Project Purple grant, and Stand Up To Cancer—The Lustgarten Foundation Pancreatic Cancer Convergence Dream Team Grant SU2C-AACR-DT14-14 (to D.B.-S.). Stand Up To Cancer is a program of the Entertainment Industry Foundation administered by the AACR. J.C. is supported by NIH Grants 5-T32 CA 009161-39 and 5-T32AI100853-04.

REFERENCES

Agalioti T, Giannou AD, Krontira AC, Kanellakis NI, Kati D, Vreka M, Pepe M, Spella M, Lilis I, Zazara DE, et al. 2017. Mutant KRAS promotes malignant pleural effusion formation. *Nat Commun* **8:** 15205.

Ahrendt SA, Decker PA, Alawi EA, Zhu Yr YR, Sanchez-Cespedes M, Yang SC, Haasler GB, Kajdacsy-Balla A, Demeure MJ, Sidransky D. 2001. Cigarette smoking is strongly associated with mutation of the K-*ras* gene in patients with primary adenocarcinoma of the lung. *Cancer* **92:** 1525–1530.

Ali K, Soond DR, Pineiro R, Hagemann T, Pearce W, Lim EL, et al. 2014. Inactivation of PI(3)K p110δ breaks regulatory T-cell-mediated immune tolerance to cancer. *Nature* **510:** 407–411.

Allen SJ, Crown SE, Handel TM. 2007. Chemokine: Receptor structure, interactions, and antagonism. *Annu Rev Immunol* **25:** 787–820.

Almand B, Clark JI, Nikitina E, van Beynen J, English NR, Knight SC, Carbone DP, Gabrilovich DI. 2001. Increased production of immature myeloid cells in cancer patients: A mechanism of immunosuppression in cancer. *J Immunol* **166:** 678–689.

Ancrile B, Lim KH, Counter CM. 2007. Oncogenic Ras-induced secretion of IL6 is required for tumorigenesis. *Genes Dev* **21:** 1714–1719.

Andoh A, Takaya H, Saotome T, Shimada M, Hata K, Araki Y, Nakamura F, Shintani Y, Fujiyama Y, Bamba T. 2000. Cytokine regulation of chemokine (IL-8, MCP-1, and RANTES) gene expression in human pancreatic periacinar myofibroblasts. *Gastroenterology* **119:** 211–219.

Apte MV, Pirola RC, Wilson JS. 2012. Pancreatic stellate cells: A starring role in normal and diseased pancreas. *Front Physiol* **3:** 344.

Bailey JM, Swanson BJ, Hamada T, Eggers JP, Singh PK, Caffery T, Oullette MM, Hollingsworth MA. 2008. Sonic hedgehog promotes desmoplasia in pancreatic cancer. *Clin Cancer Res* **14:** 5995–6004.

Baines AT, Xu D, Der CJ. 2011. Inhibition of Ras for cancer treatment: The search continues. *Future Med Chem* **3:** 1787–1808.

Baldwin GC, Gasson JC, Kaufman SE, Quan SG, Williams RE, Avalos BR, Gazdar AF, Golde DW, DiPersio JF. 1989. Nonhematopoietic tumor cells express functional GM-CSF receptors. *Blood* **73:** 1033–1037.

Baud V, Karin M. 2009. Is NF-κB a good target for cancer therapy? Hopes and pitfalls. *Nat Rev Drug Discov* **8:** 33–40.

Bayne LJ, Beatty GL, Jhala N, Clark CE, Rhim AD, Stanger BZ, Vonderheide RH. 2012. Tumor-derived granulocyte-macrophage colony-stimulating factor regulates myeloid inflammation and T cell immunity in pancreatic cancer. *Cancer Cell* **21:** 822–835.

Bellone G, Turletti A, Artusio E, Mareschi K, Carbone A, Tibaudi D, Robecchi A, Emanuelli G. 1999. Tumor-associated transforming growth factor β and interleukin-10 contribute to a systemic Th2 immune phenotype in pancreatic carcinoma patients. *Am J Pathol* **155:** 537–547.

Bellone G, Smirne C, Mauri FA, Tonel E, Carbone A, Buffolino A, Dughera L, Robecchi A, Pirisi M, Emanuelli G. 2006. Cytokine expression profile in human pancreatic carcinoma cells and in surgical specimens: Implications for survival. *Cancer Immunol Immunother* **55:** 684–698.

Busch SE, Hanke ML, Kargl J, Metz HE, MacPherson D, Houghton AM. 2016. Lung cancer subtypes generate unique immune responses. *J Immunol* **197:** 4493–4503.

Caetano MS, Zhang H, Cumpian AM, Gong L, Unver N, Ostrin EJ, Daliri S, Chang SH, Ochoa CE, Hanash S, et al. 2016. IL6 blockade reprograms the lung tumor microenvironment to limit the development and progression of K-ras-mutant lung cancer. *Cancer Res* **76:** 3189–3199.

Carpagnano GE, Resta O, Foschino-Barbaro MP, Gramiccioni E, Carpagnano F. 2002. Interleukin-6 is increased in breath condensate of patients with non-small cell lung cancer. *Int J Biol Markers* **17:** 141–145.

Chang CH, Hsiao CF, Yeh YM, Chang GC, Tsai YH, Chen YM, Huang MS, Chen HL, Li YJ, Yang PC, et al. 2013. Circulating interleukin-6 level is a prognostic marker for survival in advanced nonsmall cell lung cancer patients treated with chemotherapy. *Int J Cancer* **132:** 1977–1985.

Chang SH, Mirabolfathinejad SG, Katta H, Cumpian AM, Gong L, Caetano MS, Moghaddam SJ, Dong C. 2014. T helper 17 cells play a critical pathogenic role in lung cancer. *Proc Natl Acad Sci* **111:** 5664–5669.

Chen JJ, Yao PL, Yuan A, Hong TM, Shun CT, Kuo ML, Lee YC, Yang PC. 2003. Up-regulation of tumor interleukin-8 expression by infiltrating macrophages: Its correlation with tumor angiogenesis and patient survival in non-small cell lung cancer. *Clin Cancer Res* **9:** 729–737.

Clark CE, Beatty GL, Vonderheide RH. 2009. Immunosurveillance of pancreatic adenocarcinoma: Insights from genetically engineered mouse models of cancer. *Cancer Lett* **279:** 1–7.

Cortez-Retamozo V, Etzrodt M, Newton A, Rauch PJ, Chudnovskiy A, Berger C, Ryan RJ, Iwamoto Y, Marinelli B, Gorbatov R, et al. 2012. Origins of tumor-associated macrophages and neutrophils. *Proc Natl Acad Sci* **109:** 2491–2496.

Cox AD, Fesik SW, Kimmelman AC, Luo J, Der CJ. 2014. Drugging the undruggable RAS: Mission possible? *Nat Rev Drug Discov* **13:** 828–851.

Cullis J, Siolas D, Avanzi A, Barui S, Maitra A, Bar-Sagi D. 2017. Macropinocytosis of nab-paclitaxel drives macrophage activation in pancreatic cancer. *Cancer Immunol Res* **5:** 182–190.

Dai F, Liu L, Che G, Yu N, Pu Q, Zhang S, Ma J, Ma L, You Z. 2010. The number and microlocalization of tumor-associated immune cells are associated with patient's survival time in non-small cell lung cancer. *BMC Cancer* **10:** 220.

Daley D, Zambirinis CP, Seifert L, Akkad N, Mohan N, Werba G, Barilla R, Torres-Hernandez A, Hundeyin M, Mani VR, et al. 2016. γδ T cells support pancreatic oncogenesis by restraining αβ T cell activation. *Cell* **166:** 1485.

Delitto D, Black BS, Sorenson HL, Knowlton AE, Thomas RM, Sarosi GA, Moldawer LL, Behrns KE, Liu C, George TJ, et al. 2015. The inflammatory milieu within the pancreatic cancer microenvironment correlates with clinicopathologic parameters, chemoresistance and survival. *BMC Cancer* **15:** 783.

Dickson RB, Kasid A, Huff KK, Bates SE, Knabbe C, Bronzert D, Gelmann EP, Lippman ME. 1987. Activation of growth factor secretion in tumorigenic states of breast cancer induced by 17 β-estradiol or v-Ha-ras oncogene. *Proc Natl Acad Sci* **84:** 837–841.

Dong ZY, Zhong WZ, Zhang XC, Su J, Xie Z, Liu SY, Tu HY, Chen HJ, Sun YL, Zhou Q, et al. 2017. Potential predictive value of TP53 and KRAS mutation status for response to PD-1 blockade immunotherapy in lung adenocarcinoma. *Clin Cancer Res* **23:** 3012–3024.

Drake CG, Jaffee E, Pardoll DM. 2006. Mechanisms of immune evasion by tumors. *Adv Immunol* **90:** 51–81.

Duan M, Ning Z, Fu Z, Zhang J, Liu G, Wei Q, Zheng X. 2015. Decreased IL-27 negatively correlated with Th17 cells in non-small-cell lung cancer patients. *Mediators Inflamm* **2015:** 802939.

Ebrahimi B, Tucker SL, Li D, Abbruzzese JL, Kurzrock R. 2004. Cytokines in pancreatic carcinoma: Correlation with phenotypic characteristics and prognosis. *Cancer* **101:** 2727–2736.

Ene-Obong A, Clear AJ, Watt J, Wang J, Fatah R, Riches JC, Marshall JF, Chin-Aleong J, Chelala C, Gribben JG, et al. 2013. Activated pancreatic stellate cells sequester $CD8^+$ T cells to reduce their infiltration of the juxtatumoral compartment of pancreatic ductal adenocarcinoma. *Gastroenterology* **145:** 1121–1132.

Feig C, Jones JO, Kraman M, Wells RJ, Deonarine A, Chan DS, Connell CM, Roberts EW, Zhao Q, Caballero OL, et al. 2013. Targeting CXCL12 from FAP-expressing carcinoma-associated fibroblasts synergizes with anti-PD-L1 immunotherapy in pancreatic cancer. *Proc Natl Acad Sci* **110:** 20212–20217.

Feurino LW, Zhang Y, Bharadwaj U, Zhang R, Li F, Fisher WE, Brunicardi FC, Chen C, Yao Q, Min L. 2007. IL-6 stimulates Th2 type cytokine secretion and upregulates VEGF and NRP-1 expression in pancreatic cancer cells. *Cancer Biol Ther* **6:** 1096–1100.

Finco TS, Westwick JK, Norris JL, Beg AA, Der CJ, Baldwin AS Jr. 1997. Oncogenic Ha-Ras-induced signaling activates NF-κB transcriptional activity, which is required for cellular transformation. *J Biol Chem* **272:** 24113–24116.

Fortis C, Foppoli M, Gianotti L, Galli L, Citterio G, Consogno G, Gentilini O, Braga M. 1996. Increased interleukin-10 serum levels in patients with solid tumours. *Cancer Lett* **104:** 1–5.

Fossum B, Olsen AC, Thorsby E, Gaudernack G. 1995. $CD8^+$ T cells from a patient with colon carcinoma, specific for a mutant p21–Ras-derived peptide (Gly13→Asp), are cytotoxic towards a carcinoma cell line harbouring the same mutation. *Cancer Immunol Immunother* **40:** 165–172.

Frick VO, Rubie C, Wagner M, Graeber S, Grimm H, Kopp B, Rau BM, Schilling MK. 2008. Enhanced ENA-78 and IL-8 expression in patients with malignant pancreatic diseases. *Pancreatology* **8:** 488–497.

Fukuda A, Wang SC, Morris JP IV, Folias AE, Liou A, Kim GE, Akira S, Boucher KM, Firpo MA, Mulvihill SJ, et al. 2011. Stat3 and MMP7 contribute to pancreatic ductal adenocarcinoma initiation and progression. *Cancer Cell* **19:** 441–455.

Gabrilovich DI, Velders MP, Sotomayor EM, Kast WM. 2001. Mechanism of immune dysfunction in cancer mediated by immature Gr-1^{+} myeloid cells. *J Immunol* **166:** 5398–5406.

Gorelik L, Flavell RA. 2002. Transforming growth factor-β in T-cell biology. *Nat Rev Immunol* **2:** 46–53.

Granville CA, Memmott RM, Balogh A, Mariotti J, Kawabata S, Han W, Lopiccolo J, Foley J, Liewehr DJ, Steinberg SM, et al. 2009. A central role for Foxp3^{+} regulatory T cells in K-Ras-driven lung tumorigenesis. *PLoS ONE* **4:** e5061.

Guerra C, Collado M, Navas C, Schuhmacher AJ, Hernández-Porras I, Cañamero M, Rodriguez-Justo M, Serrano M, Barbacid M. 2011. Pancreatitis-induced inflammation contributes to pancreatic cancer by inhibiting oncogene-induced senescence. *Cancer Cell* **19:** 728–739.

Gunderson AJ, Kaneda MM, Tsujikawa T, Nguyen AV, Affara NI, Ruffell B, Gorjestani S, Liudahl SM, Truitt M, Olson P, et al. 2016. Bruton tyrosine kinase-dependent immune cell cross-talk drives pancreas cancer. *Cancer Discov* **6:** 270–285.

Hasegawa Y, Takanashi S, Kanehira Y, Tsushima T, Imai T, Okumura K. 2001. Transforming growth factor-β1 level correlates with angiogenesis, tumor progression, and prognosis in patients with nonsmall cell lung carcinoma. *Cancer* **91:** 964–971.

Hatanaka H, Abe Y, Kamiya T, Morino F, Nagata J, Tokunaga T, Oshika Y, Suemizu H, Kijima H, Tsuchida T, et al. 2000. Clinical implications of interleukin (IL)-10 induced by non-small-cell lung cancer. *Ann Oncol* **11:** 815–819.

Herter-Sprie GS, Koyama S, Korideck H, Hai J, Deng J, Li YY, Buczkowski KA, Grant AK, Ullas S, Rhee K, et al. 2016. Synergy of radiotherapy and PD-1 blockade in Kras-mutant lung cancer. *JCI Insight* **1:** e87415.

Holmer R, Goumas FA, Waetzig GH, Rose-John S, Kalthoff H. 2014. Interleukin-6: A villain in the drama of pancreatic cancer development and progression. *Hepatobiliary Pancreat Dis Int* **13:** 371–380.

Hsu P, Santner-Nanan B, Hu M, Skarratt K, Lee CH, Stormon M, Wong M, Fuller SM, Nanan R. 2015. IL-10 potentiates differentiation of human induced regulatory T cells via STAT3 and Foxo1. *J Immunol* **195:** 3665–3674.

Huang F, Wang XL, Geng Y, Li MX. 2005. Evaluation of IL-6 level in serum and bronchoalveolar lavage fluid of patients with non-small cell lung cancer (NSCLC). *Xi Bao Yu Fen Zi Mian Yi Xue Za ZhiEur* **21:** 507–509.

Ino Y, Yamazaki-Itoh R, Shimada K, Iwasaki M, Kosuge T, Kanai Y, Hiraoka N. 2013. Immune cell infiltration as an indicator of the immune microenvironment of pancreatic cancer. *Br J Cancer* **108:** 914–923.

Jess T, Rungoe C, Peyrin-Biroulet L. 2012. Risk of colorectal cancer in patients with ulcerative colitis: A meta-analysis of population-based cohort studies. *Clin Gastroenterol Hepatol* **10:** 639–645.

Ji H, Houghton AM, Mariani TJ, Perera S, Kim CB, Padera R, Tonon G, McNamara K, Marconcini LA, Hezel A, et al. 2006. K-ras activation generates an inflammatory response in lung tumors. *Oncogene* **25:** 2105–2112.

Kenkel JA, Tseng WW, Davidson MG, Tolentino L, Choi O, Bhattacharya N, Seeley ES, Winer DA, Reticker-Flynn NE, Engleman EG. 2017. An immunosuppressive dendritic cell subset accumulates at secondary sites and promotes metastasis in pancreatic cancer. *Cancer Res* **77:** 4158–4170.

Kim BY, Gaynor RB, Song K, Dritschilo A, Jung M. 2002. Constitutive activation of NF-κB in Ki-*ras*-transformed prostate epithelial cells. *Oncogene* **21:** 4490–4497.

Koizumi M, Hiasa Y, Kumagi T, Yamanishi H, Azemoto N, Kobata T, Matsuura B, Abe M, Onji M. 2013. Increased B-cell-activating factor promotes tumor invasion and metastasis in human pancreatic cancer. *PloS ONE* **8:** e71367.

Kordes C, Brookmann S, Haussinger D, Klonowski-Stumpe H. 2005. Differential and synergistic effects of platelet-derived growth factor-BB and transforming growth factor-β1 on activated pancreatic stellate cells. *Pancreas* **31:** 156–167.

Kubuschok B, Neumann F, Breit R, Sester M, Schormann C, Wagner C, Sester U, Hartmann F, Wagner M, Remberger K, et al. 2006. Naturally occurring T-cell response against mutated p21 ras oncoprotein in pancreatic cancer. *Clin Cancer Res* **12:** 1365–1372.

Kulkarni AB, Karlsson S. 1993. Transforming growth factor-β1 knockout mice. A mutation in one cytokine gene causes a dramatic inflammatory disease. *Am J Pathol* **143:** 3–9.

Kurahara H, Shinchi H, Mataki Y, Maemura K, Noma H, Kubo F, Sakoda M, Ueno S, Natsugoe S, Takao S. 2011. Significance of M2-polarized tumor-associated macrophage in pancreatic cancer. *J Surg Res* **167:** e211–e219.

Kusmartsev S, Gabrilovich DI. 2006. Effect of tumor-derived cytokines and growth factors on differentiation and immune suppressive features of myeloid cells in cancer. *Cancer Metastasis Rev* **25:** 323–331.

Lee J, Horuk R, Rice GC, Bennett GL, Camerato T, Wood WI. 1992. Characterization of two high affinity human interleukin-8 receptors. *J Biol Chem* **267:** 16283–16287.

Lee KE, Spata M, Bayne LJ, Buza EL, Durham AC, Allman D, Vonderheide RH, Simon MC. 2016. Hif1a deletion reveals pro-neoplastic function of B cells in pancreatic neoplasia. *Cancer Discov* **6:** 256–269.

Lehmann K, Janda E, Pierreux CE, Rytomaa M, Schulze A, McMahon M, Hill CS, Beug H, Downward J. 2000. Raf induces TGFβ production while blocking its apoptotic but not invasive responses: A mechanism leading to increased malignancy in epithelial cells. *Genes Dev* **4:** 2610–2622.

Lesina M, Kurkowski MU, Ludes K, Rose-John S, Treiber M, Klöppel G, Yoshimura A, Reindl W, Sipos B, Akira S, et al. 2011. Stat3/Socs3 activation by IL-6 transsignaling promotes progression of pancreatic intraepithelial neoplasia and development of pancreatic cancer. *Cancer Cell* **19:** 456–469.

Li MO, Wan YY, Sanjabi S, Robertson AK, Flavell RA. 2006. Transforming growth factor-β regulation of immune responses. *Annu Rev Immunol* **24:** 99–146.

Li A, King J, Moro A, Sugi MD, Dawson DW, Kaplan J, Li G, Lu X, Strieter RM, Burdick M, et al. 2011. Overexpression of CXCL5 is associated with poor survival in patients with pancreatic cancer. *Am J Pathol* **178:** 1340–1349.

Linard B, Bezieau S, Benlalam H, Labarrière N, Guilloux Y, Diez E, Jotereau F. 2002. A *ras*-mutated peptide targeted by CTL infiltrating a human melanoma lesion. *J Immunol* **168:** 4802–4808.

Liou GY, Storz P. 2015. Inflammatory macrophages in pancreatic acinar cell metaplasia and initiation of pancreatic cancer. *Oncoscience* **2:** 247–251.

Liou GY, Döppler H, Necela B, Krishna M, Crawford HC, Raimondo M, Storz P. 2013. Macrophage-secreted cytokines drive pancreatic acinar-to-ductal metaplasia through NF-κB and MMPs. *J Cell Biol* **202:** 563–577.

Liou GY, Doppler H, Necela B, Edenfield B, Zhang L, Dawson DW, Storz P. 2015. Mutant KRAS-induced expression of ICAM-1 in pancreatic acinar cells causes attraction of macrophages to expedite the formation of precancerous lesions. *Cancer Discov* **5:** 52–63.

Loetscher P, Pellegrino A, Gong JH, Mattioli I, Loetscher M, Bardi G, Baggiolini M, Clark-Lewis I. 2001. The ligands of CXC chemokine receptor 3, I-TAC, Mig, and IP10, are natural antagonists for CCR3. *J Biol Chem* **276:** 2986–2991.

Löhr M, Schmidt C, Ringel J, Kluth M, Müller P, Nizze H, Jesnowski R. 2001. Transforming growth factor-β1 induces desmoplasia in an experimental model of human pancreatic carcinoma. *Cancer Res* **61:** 550–555.

Löhr M, Kloppel G, Maisonneuve P, Lowenfels AB, Luttges J. 2005. Frequency of K-ras mutations in pancreatic intraductal neoplasias associated with pancreatic ductal adenocarcinoma and chronic pancreatitis: A meta-analysis. *Neoplasia* **7:** 17–23.

Lowenfels AB, Maisonneuve P, Cavallini G, Ammann RW, Lankisch PG, Andersen JR, Dimagno EP, Andrén-Sandberg A, Domellöf L. 1993. Pancreatitis and the risk of pancreatic cancer. International Pancreatitis Study Group. *N Engl J Med* **328:** 1433–1437.

Lunardi S, Jamieson NB, Lim SY, Griffiths KL, Carvalho-Gaspar M, Al-Assar O, Yameen S, Carter RC, McKay CJ, Spoletini G, et al. 2014. IP-10/CXCL10 induction in human pancreatic cancer stroma influences lymphocytes recruitment and correlates with poor survival. *Oncotarget* **5:** 11064–11080.

Lüttges J, Reinecke-Lüthge A, Möllmann B, Menke MA, Clemens A, Klimpfinger M, Sipos B, Klöppel G. 1999. Duct changes and K-ras mutations in the disease-free pancreas: Analysis of type, age relation and spatial distribution. *Virchows Arch* **435:** 461–468.

Mace TA, Ameen Z, Collins A, Wojcik S, Mair M, Young GS, Fuchs JR, Eubank TD, Frankel WL, Bekaii-Saab T, et al. 2013a. Pancreatic cancer-associated stellate cells promote differentiation of myeloid-derived suppressor cells in a STAT3-dependent manner. *Cancer Res* **73:** 3007–3018.

Mace TA, Bloomston M, Lesinski GB. 2013b. Pancreatic cancer-associated stellate cells: A viable target for reducing immunosuppression in the tumor microenvironment. *Oncoimmunology* **2:** e24891.

Mace TA, Shakya R, Pitarresi JR, Swanson B, McQuinn CW, Loftus S, Nordquist E, Cruz-Monserrate Z, Yu L, Young G, et al. 2016. IL-6 and PD-L1 antibody blockade combination therapy reduces tumour progression in murine models of pancreatic cancer. *Gut* doi: 10.1136/gutjnl-2016-311585.

Maher J, Davies ET. 2004. Targeting cytotoxic T lymphocytes for cancer immunotherapy. *Br J Cancer* **91:** 817–821.

Malka D, Hammel P, Maire F, Rufat P, Madeira I, Pessione F, Lévy P, Ruszniewski P. 2002. Risk of pancreatic adenocarcinoma in chronic pancreatitis. *Gut* **51:** 849–852.

Mantovani A, Sozzani S, Locati M, Allavena P, Sica A. 2002. Macrophage polarization: Tumor-associated macrophages as a paradigm for polarized M2 mononuclear phagocytes. *Trends Immunol* **23:** 549–555.

Mantovani A, Schioppa T, Porta C, Allavena P, Sica A. 2006. Role of tumor-associated macrophages in tumor progression and invasion. *Cancer Metastasis Rev* **25:** 315–322.

Markowitz J, Brooks TR, Duggan MC, Paul BK, Pan X, Wei L, Abrams Z, Luedke E, Lesinski GB, Mundy-Bosse B, et al. 2015. Patients with pancreatic adenocarcinoma exhibit elevated levels of myeloid-derived suppressor cells upon progression of disease. *Cancer Immunol Immunother* **64:** 149–159.

Marshall EA, Ng KW, Kung SH, Conway EM, Martinez VD, Halvorsen EC, Rowbotham DA, Vucic EA, Plumb AW, Becker-Santos DD, et al. 2016. Emerging roles of T helper 17 and regulatory T cells in lung cancer progression and metastasis. *Mol Cancer* **15:** 67.

Mathew E, Brannon AL, Del Vecchio A, Garcia PE, Penny MK, Kane KT, Vinta A, Buckanovich RJ, di Magliano MP. 2016. Mesenchymal stem cells promote pancreatic tumor growth by inducing alternative polarization of macrophages. *Neoplasia* **18:** 142–151.

McAllister F, Bailey JM, Alsina J, Nirschl CJ, Sharma R, Fan H, Rattigan Y, Roeser JC, Lankapalli RH, Zhang H, et al. 2014. Oncogenic Kras activates a hematopoietic-to-epithelial IL-17 signaling axis in preinvasive pancreatic neoplasia. *Cancer Cell* **25:** 621–637.

Means AL, Ray KC, Singh AB, Washington MK, Whitehead RH, Harris RC Jr, Wright CV, Coffey RJ Jr, Leach SD. 2003. Overexpression of heparin-binding EGF-like growth factor in mouse pancreas results in fibrosis and epithelial metaplasia. *Gastroenterology* **124:** 1020–1036.

Mielgo A, Schmid MC. 2013. Impact of tumour associated macrophages in pancreatic cancer. *BMB Rep* **46:** 131–138.

Mitchem JB, Brennan DJ, Knolhoff BL, Belt BA, Zhu Y, Sanford DE, Belaygorod L, Carpenter D, Collins L, Piwnica-Worms D, et al. 2013. Targeting tumor-infiltrating macrophages decreases tumor-initiating cells, relieves immunosuppression, and improves chemotherapeutic responses. *Cancer Res* **73:** 1128–1141.

Moghaddam SJ, Barta P, Mirabolfathinejad SG, Ammar-Aouchiche Z, Garza NT, Vo TT, Newman RA, Aggarwal BB, Evans CM, Tuvim MJ, et al. 2009. Curcumin inhibits COPD-like airway inflammation and lung cancer progression in mice. *Carcinogenesis* **30:** 1949–1956.

Moo-Young TA, Larson JW, Belt BA, Tan MC, Hawkins WG, Eberlein TJ, Goedegebuure PS, Linehan DC. 2009. Tumor-derived TGF-β mediates conversion of $CD4^{+}$*Foxp3*$^{+}$ regulatory T cells in a murine model of pancreas cancer. *J Immunother* **32:** 12–21.

Mroczko B, Groblewska M, Gryko M, Kedra B, Szmitkowski M. 2010. Diagnostic usefulness of serum interleukin 6 (IL-6) and C-reactive protein (CRP) in the differentiation between pancreatic cancer and chronic pancreatitis. *J Clin Lab Anal* **24:** 256–261.

Murphy PM. 1994. The molecular biology of leukocyte chemoattractant receptors. *Annu Rev Immunol* **12:** 593–633.

Nemunaitis J, Dillman RO, Schwarzenberger PO, Senzer N, Cunningham C, Cutler J, Tong A, Kumar P, Pappen B, Hamilton C, et al. 2006. Phase II study of belagenpumatucel-L, a transforming growth factor β-2 antisense gene-modified allogeneic tumor cell vaccine in non-small-cell lung cancer. *J Clin Oncol* **24:** 4721–4730.

Nemunaitis J, Nemunaitis M, Senzer N, Snitz P, Bedell C, Kumar P, et al. 2009. Phase II trial of belagenpumatucel-L, a TGF-β2 antisense gene modified allogeneic tumor vaccine in advanced non-small cell lung cancer (NSCLC) patients. *Cancer Gene Ther* **16:** 620–624.

O'Hayer KM, Brady DC, Counter CM. 2009. ELR$^+$ CXC chemokines and oncogenic Ras-mediated tumorigenesis. *Carcinogenesis* **30:** 1841–1847.

Okada S, Okusaka T, Ishii H, Kyogoku A, Yoshimori M, Kajimura N, Yamaguchi K, Kikizoe T. 1998. Elevated serum interleukin-6 levels in patients with pancreatic cancer. *Jpn J Clin Oncol* **28:** 12–15.

Omary MB, Lugea A, Lowe AW, Pandol SJ. 2007. The pancreatic stellate cell: A star on the rise in pancreatic diseases. *J Clin Invest* **117:** 50–59.

Ortegel JW, Staren ED, Faber LP, Warren WH, Braun DP. 2002. Modulation of tumor-infiltrating lymphocyte cytolytic activity against human non-small cell lung cancer. *Lung Cancer* **36:** 17–25.

Pauken KE, Nelson CE, Martinov T, Spanier JA, Heffernan JR, Sahli NL, Quarnstrom CF, Osum KC, Schenkel JM, Jenkins MK, et al. 2015. Cutting edge: Identification of autoreactive CD4$^+$ and CD8$^+$ T cell subsets resistant to PD-1 pathway blockade. *J Immunol* **194:** 3551–3555.

Principe DR, DeCant B, Diaz AM, Mangan RJ, Hwang R, Lowy A, Shetuni BB, Sreekumar BK, Chung C, Bentrem DJ, et al. 2016. PEDF inhibits pancreatic tumorigenesis by attenuating the fibro-inflammatory reaction. *Oncotarget* **7:** 28218–28234.

Purohit A, Varney M, Rachagani S, Ouellette MM, Batra SK, Singh RK. 2016. CXCR2 signaling regulates *KRAS*^{G12D}-induced autocrine growth of pancreatic cancer. *Oncotarget* **7:** 7280–7296.

Pylayeva-Gupta Y, Lee KE, Hajdu CH, Miller G, Bar-Sagi D. 2012. Oncogenic Kras-induced GM-CSF production promotes the development of pancreatic neoplasia. *Cancer Cell* **21:** 836–847.

Pylayeva-Gupta Y, Das S, Handler JS, Hajdu CH, Coffre M, Koralov SB, Bar-Sagi D. 2016. IL35-producing B cells promote the development of pancreatic neoplasia. *Cancer Discov* **6:** 247–255.

Qu Z, Sun F, Zhou J, Li L, Shapiro SD, Xiao G. 2015. Interleukin-6 prevents the initiation but enhances the progression of lung cancer. *Cancer Res* **75:** 3209–3215.

Redente EF, Higgins DM, Dwyer-Nield LD, Orme IM, Gonzalez-Juarrero M, Malkinson AM. 2010. Differential polarization of alveolar macrophages and bone marrow-derived monocytes following chemically and pathogen-induced chronic lung inflammation. *J Leukoc Biol* **88:** 159–168.

Remark R, Becker C, Gomez JE, Damotte D, Dieu-Nosjean MC, Sautès-Fridman C, Fridman WH, Powell CA, Altorki NK, Merad M, et al. 2015. The non-small cell lung cancer immune contexture. A major determinant of tumor characteristics and patient outcome. *Am J Respir Crit Care Med* **191:** 377–390.

Rishi A, Goggins M, Wood LD, Hruban RH. 2015. Pathological and molecular evaluation of pancreatic neoplasms. *Semin Oncol* **42:** 28–39.

Rosenberg SA, Tran E, Robbins PF. 2017. T-cell transfer therapy targeting mutant KRAS. *N Engl J Med* **376:** e11.

Rossi D, Zlotnik A. 2000. The biology of chemokines and their receptors. *Annu Rev Immunol* **18:** 217–242.

Schmall A, Al-Tamari HM, Herold S, Kampschulte M, Weigert A, Wietelmann A, Vipotnik N, Grimminger F, Seeger W, Pullamsetti SS, et al. 2015. Macrophage and cancer cell cross-talk via CCR2 and CX3CR1 is a fundamental mechanism driving lung cancer. *Am J Respir Crit Care Med* **191:** 437–447.

Seo JS, Ju YS, Lee WC, Shin JY, Lee JK, Bleazard T, Lee J, Jung YJ, Kim JO, Shin JY, et al. 2012. The transcriptional landscape and mutational profile of lung adenocarcinoma. *Genome Res* **22:** 2109–2119.

Sharma G, Whang EE, Ruan DT, Ito H. 2015. Efficacy of neoadjuvant versus adjuvant therapy for resectable pancreatic adenocarcinoma: A decision analysis. *Ann Surg Oncol* **22:** S1229–S1237.

Shull MM, Ormsby I, Kier AB, Pawlowski S, Diebold RJ, Yin M, Allen R, Sidman C, Proetzel G, Calvin D, et al. 1992. Targeted disruption of the mouse transforming growth factor-β1 gene results in multifocal inflammatory disease. *Nature* **359:** 693–699.

Sparmann A, Bar-Sagi D. 2004. Ras-induced interleukin-8 expression plays a critical role in tumor growth and angiogenesis. *Cancer Cell* **6:** 447–458.

Stambrook PJ, Maher J, Farzaneh F. 2017. Cancer immunotherapy: Whence and whither. *Mol Cancer Res* **15:** 635–650.

Steele CW, Karim SA, Leach JD, Bailey P, Upstill-Goddard R, Rishi L, Foth M, Bryson S, McDaid K, Wilson Z, et al. 2016. CXCR2 inhibition profoundly suppresses metastases and augments immunotherapy in pancreatic ductal adenocarcinoma. *Cancer Cell* **29:** 832–845.

Takeuchi S, Baghdadi M, Tsuchikawa T, Wada H, Nakamura T, Abe H, Nakanishi S, Usui Y, Higuchi K, Takahashi M, et al. 2015. Chemotherapy-derived inflammatory responses accelerate the formation of immunosuppressive myeloid cells in the tissue microenvironment of human pancreatic cancer. *Cancer Res* **75:** 2629–2640.

Talar-Wojnarowska R, Gasiorowska A, Smolarz B, Romanowicz-Makowska H, Kulig A, Malecka-Panas E. 2009. Clinical significance of interleukin-6 (IL-6) gene polymorphism and IL-6 serum level in pancreatic adenocarcinoma and chronic pancreatitis. *Dig Dis Sci* **54:** 683–689.

Tang D, Yuan Z, Xue X, Lu Z, Zhang Y, Wang H, Chen M, An Y, Wei J, Zhu Y, et al. 2012. High expression of galectin-1 in pancreatic stellate cells plays a role in the development and maintenance of an immunosuppressive microenvironment in pancreatic cancer. *Int J Cancer* **130:** 2337–2348.

Tang D, Gao J, Wang S, Yuan Z, Ye N, Chong Y, Xu C, Jiang X, Li B, Yin W, et al. 2015. Apoptosis and anergy of T cell induced by pancreatic stellate cells-derived galectin-1 in pancreatic cancer. *Tumour Biol* **36:** 5617–5626.

Tas F, Duranyildiz D, Oguz H, Camlica H, Yasasever V, Topuz E. 2006. Serum vascular endothelial growth factor (VEGF) and interleukin-8 (IL-8) levels in small cell lung cancer. *Cancer Invest* **24:** 492–496.

Tateishi M, Ishida T, Mitsudomi T, Sugimachi K. 1991. Prognostic implication of transforming growth factor α in adenocarcinoma of the lung—An immunohistochemical study. *Br J Cancer* **63:** 130–133.

Terhune PG, Phifer DM, Tosteson TD, Longnecker DS. 1998. K-ras mutation in focal proliferative lesions of human pancreas. *Cancer Epidemiol Biomarkers Prev* **7:** 515–521.

Tewari N, Zaitoun AM, Arora A, Madhusudan S, Ilyas M, Lobo DN. 2013. The presence of tumour-associated lymphocytes confers a good prognosis in pancreatic ductal adenocarcinoma: An immunohistochemical study of tissue microarrays. *BMC Cancer* **13:** 436.

Tian X, Chen G, Zhou S, Henne-Bruns D, Bachem M, Kornmann M. 2012. Interactions of pancreatic cancer and stellate cells are mediated by FGFR1-III isoform expression. *Hepatogastroenterology* **59:** 1604–1608.

Toomey D, Smyth G, Condron C, Kelly J, Byrne AM, Kay E, Conroy RM, Broe P, Bouchier-Hayes D. 2003. Infiltrating immune cells, but not tumour cells, express FasL in non-small cell lung cancer: No association with prognosis identified in 3-year follow-up. *Int J Cancer* **103:** 408–412.

Topalian SL, Hodi FS, Brahmer JR, Gettinger SN, Smith DC, McDermott DF, Powderly JD, Carvajal RD, Sosman JA, Atkins MB, et al. 2012. Safety, activity, and immune correlates of anti-PD-1 antibody in cancer. *N Engl J Med* **366:** 2443–2454.

Topalian SL, Wolchok JD, Chan TA, Mellman I, Palucka K, Banchereau J, Rosenberg SA, Dane Wittrup K. 2015. Immunotherapy: The path to win the war on cancer? *Cell* **161:** 185–186.

Trojan A, Urosevic M, Hummerjohann J, Giger R, Schanz U, Stahel RA. 2003. Immune reactivity against a novel HLA-A3-restricted influenza virus peptide identified by predictive algorithms and interferon-γ quantitative PCR. *J Immunother* **26:** 41–46.

Tsubaki M, Yamazo Y, Yanae M, Satou T, Itoh T, Kaneko J, Kidera Y, Moriyama K, Nishida S. 2011. Blockade of the Ras/MEK/ERK and Ras/PI3K/Akt pathways by statins reduces the expression of bFGF, HGF, and TGF-β as angiogenic factors in mouse osteosarcoma. *Cytokine* **54:** 100–107.

Vasiliades G, Kopanakis N, Vasiloglou M, Zografos G, Margaris H, Masselou K, Kokosi E, Liakakos T. 2012. Role of the hematopoietic cytokines SCF, IL-3, GM-CSF and M-CSF in the diagnosis of pancreatic and ampullary cancer. *Int J Biol Markers* **27:** e186–e194.

von Bernstorff W, Voss M, Freichel S, Schmid A, Vogel I, Jöhnk C, Henne-Bruns D, Kremer B, Kalthoff H. 2001. Systemic and local immunosuppression in pancreatic cancer patients. *Clin Cancer Res* **7:** 925s–932s.

Vonderheide RH, Bayne LJ. 2013. Inflammatory networks and immune surveillance of pancreatic carcinoma. *Curr Opin Immunol* **25:** 200–205.

Waghray M, Yalamanchili M, Dziubinski M, Zeinali M, Erkkinen M, Yang H, Schradle KA, Urs S, Pasca Di Magliano M, Welling TH, et al. 2016. GM-CSF mediates mesenchymal-epithelial cross-talk in pancreatic cancer. *Cancer Discov* **6:** 886–899.

Wang-Gillam A, Li CP, Bodoky G, Dean A, Shan YS, Jameson G, Macarulla T, Lee KH, Cunningham D, Blanc JF, et al. 2016. Nanoliposomal irinotecan with fluorouracil and folinic acid in metastatic pancreatic cancer after previous gemcitabine-based therapy (NAPOLI-1): A global, randomised, open-label, phase 3 trial. *Lancet* **387:** 545–557.

Watt J, Kocher HM. 2013. The desmoplastic stroma of pancreatic cancer is a barrier to immune cell infiltration. *Oncoimmunology* **2:** e26788.

Wenger FA, Jacobi CA, Zieren J, Docke W, Volk HD, Muller JM. 1999. Tumor size and lymph-node status in pancreatic carcinoma—Is there a correlation to the preoperative immune function? *Langenbecks Arch Surg* **384:** 473–478.

Wigmore SJ, Fearon KC, Sangster K, Maingay JP, Garden OJ, Ross JA. 2002. Cytokine regulation of constitutive production of interleukin-8 and -6 by human pancreatic cancer cell lines and serum cytokine concentrations in patients with pancreatic cancer. *Int J Oncol* **21:** 881–886.

Wislez M, Fujimoto N, Izzo JG, Hanna AE, Cody DD, Langley RR, Tang H, Burdick MD, Sato M, Minna JD, et al. 2006. High expression of ligands for chemokine receptor CXCR2 in alveolar epithelial neoplasia induced by oncogenic kras. *Cancer Res* **66:** 4198–4207.

Xue J, Sharma V, Hsieh MH, Chawla A, Murali R, Pandol SJ, Habtezion A. 2015. Alternatively activated macrophages promote pancreatic fibrosis in chronic pancreatitis. *Nat Commun* **6:** 7158.

Yamaji H, Iizasa T, Koh E, Suzuki M, Otsuji M, Chang H, Motohashi S, Yokoi S, Hiroshima K, Tagawa M, et al. 2004. Correlation between interleukin 6 production and tumor proliferation in non-small cell lung cancer. *Cancer Immunol Immunother* **53:** 786–792.

Yoon SH. 2014. Immunotherapy for non-small cell lung cancer. *Tuberc Respir Dis (Seoul)* **77:** 111–115.

Young RP, Hopkins RJ, Christmas T, Black PN, Metcalf P, Gamble GD. 2009. COPD prevalence is increased in lung cancer, independent of age, sex and smoking history. *Eur Respir J* **34:** 380–386.

Zambirinis CP, Levie E, Nguy S, Avanzi A, Barilla R, Xu Y, Seifert L, Daley D, Greco SH, Deutsch M, et al. 2015. TLR9 ligation in pancreatic stellate cells promotes tumorigenesis. *J Exp Med* **212:** 2077–2094.

Zdanov S, Mandapathil M, Abu Eid R, Adamson-Fadeyi S, Wilson W, Qian J, Carnie A, Tarasova N, Mkrtichyan M, Berzofsky JA, et al. 2016. Mutant KRAS conversion of conventional T cells into regulatory T cells. *Cancer Immunol Res* **4:** 354–365.

Zhang Y, Yan W, Collins MA, Bednar F, Rakshit S, Zetter BR, Stanger BZ, Chung I, Rhim AD, di Magliano MP. 2013. Interleukin-6 is required for pancreatic cancer progression by promoting MAPK signaling activation and oxidative stress resistance. *Cancer Res* **73:** 6359–6374.

Zhang Y, Yan W, Mathew E, Bednar F, Wan S, Collins MA, Evans RA, Welling TH, Vonderheide RH, di Magliano MP. 2014. $CD4^+$ T lymphocyte ablation prevents pancreatic carcinogenesis in mice. *Cancer Immunol Res* **2:** 423–435.

Zhou J, Min Z, Zhang D, Wang W, Marincola F, Wang X. 2014. Enhanced frequency and potential mechanism of B regulatory cells in patients with lung cancer. *J Transl Med* **12:** 304.

Zhu Y, Knolhoff BL, Meyer MA, Nywening TM, West BL, Luo J, Wang-Gillam A, Goedegebuure SP, Linehan DC, DeNardo DG. 2014a. CSF1/CSF1R blockade reprograms tumor-infiltrating macrophages and improves response to T-cell checkpoint immunotherapy in pancreatic cancer models. *Cancer Res* **74:** 5057–5069.

Zhu Z, Aref AR, Cohoon TJ, Barbie TU, Imamura Y, Yang S, Moody SE, Shen RR, Schinzel AC, Thai TC, et al. 2014b. Inhibition of KRAS-driven tumorigenicity by interruption of an autocrine cytokine circuit. *Cancer Discov* **4:** 452–465.

Synthetic Lethal Vulnerabilities in *KRAS*-Mutant Cancers

Andrew J. Aguirre[1,2,3] **and William C. Hahn**[1,2,3]

[1]Department of Medical Oncology, Dana-Farber Cancer Institute, Boston, Massachusetts 02215

[2]Broad Institute of Harvard and MIT, Cambridge, Massachusetts 02142

[3]Department of Medicine, Brigham and Women's Hospital and Harvard Medical School, Boston, Massachusetts 02115

Correspondence: william_hahn@dfci.harvard.edu

KRAS is the most commonly mutated oncogene in human cancer. Most *KRAS*-mutant cancers depend on sustained expression and signaling of KRAS, thus making it a high-priority therapeutic target. Unfortunately, development of direct small molecule inhibitors of KRAS function has been challenging. An alternative therapeutic strategy for *KRAS*-mutant malignancies involves targeting codependent vulnerabilities or synthetic lethal partners that are preferentially essential in the setting of oncogenic KRAS. KRAS activates numerous effector pathways that mediate proliferation and survival signals. Moreover, cancer cells must cope with substantial oncogenic stress conferred by mutant KRAS. These oncogenic signaling pathways and compensatory coping mechanisms of *KRAS*-mutant cancer cells form the basis for synthetic lethal interactions. Here, we review the compendium of previously identified codependencies in *KRAS*-mutant cancers, including the results of numerous functional genetic screens aimed at identifying KRAS synthetic lethal targets. Importantly, many of these vulnerabilities may represent tractable therapeutic opportunities.

RAS mutations occur in approximately 30% of human cancers, including the majority of pancreatic ductal adenocarcinoma (PDAC), half of colorectal cancers, and one-third of all lung cancers (Pylayeva-Gupta et al. 2011; Cox et al. 2014). The three RAS genes (*KRAS*, *NRAS*, and *HRAS*) have high sequence homology except for a carboxy-terminal hypervariable region (Cox et al. 2014), and RAS gene mutations are typically mutually exclusive in human cancer. Despite their similarity, *KRAS* mutations are far more common in human cancer than *NRAS* or *HRAS* mutations. Whereas much of the early studies of RAS signaling and biology focused on HRAS, KRAS has now become the dominant focus for cancer modeling and therapeutic development efforts (Stephen et al. 2014; Papke and Der 2017). The similarities and differences between RAS isoforms have been reviewed elsewhere (Pylayeva-Gupta et al. 2011; Cox et al. 2014; Stephen et al. 2014), and the remainder of this review will primarily focus on studies related to the KRAS oncogene.

Experiments in cell culture and animal models have shown that the majority of tumors that harbor *KRAS* mutations depend on sus-

Cite this article as *Cold Spring Harb Perspect Med* doi: 10.1101/cshperspect.a031518

tained expression of the oncogene for cell proliferation and viability (Podsypanina et al. 2008; Singh et al. 2009; Collins et al. 2012; Ying et al. 2012; Kapoor et al. 2014; Shao et al. 2014; Hayes et al. 2016), making oncogenic KRAS a high-priority therapeutic target. However, KRAS is a small, structurally dynamic protein that has not been particularly amenable to direct therapeutic targeting. Unfortunately, attempts to develop drugs that target mutant RAS proteins have thus far been unsuccessful (Stephen et al. 2014; Papke and Der 2017), leading investigators to explore alternative opportunities for targeting KRAS-driven cancers.

Oncogenic KRAS activates more than 10 different effector signaling pathways, and the therapeutic potential of inhibiting these effectors has been the focus of intensive investigation. The most well-studied and critical KRAS effector pathways include the mitogen-activated protein kinase (MAPK) signaling cascade (Moodie et al. 1993; Vojtek et al. 1993; Warne et al. 1993; Zhang et al. 1993), the phosphatidyl inositol 3-kinase (PI3K)-AKT-MTOR pathway (Sjolander et al. 1991; Rodriguez-Viciana et al. 1994), and the Ras-like (RAL)-guanine nucleotide exchange factor (GEF) family of GEFs for the RAL GTPases (Hofer et al. 1994; Kikuchi et al. 1994; Spaargaren and Bischoff 1994). Most genetic and functional studies have suggested that the MAPK pathway is the dominant oncogenic RAS signaling pathway, and *KRAS* mutations generally occur in a mutually exclusive manner with mutations affecting other MAPK pathway components (e.g., *BRAF*, *EGFR* mutation). RAS-MAPK signaling has been shown to mediate tumor cell proliferation and survival in a variety of in vitro and in vivo experiments (Blasco et al. 2011; Karreth et al. 2011; Collisson et al. 2012; Yuan et al. 2014), and numerous small molecule inhibitors of this pathway are in clinical development (Ryan et al. 2015). Although some KRAS-driven cancers such as pancreatic cancers depend on the PI3K pathway, other KRAS-mutant tumors such as some lung or colorectal cancers do not (Ebi et al. 2011; Eser et al. 2013). Thus, PI3K dependence in *KRAS*-mutant cancers is context-specific. Therapeutic targeting of PI3K isoforms in *KRAS*-mutant cancers remains an active area of study. RAL-GEFs activate multiple processes, including nuclear factor (NF)-κB signaling and cell motility and have also been shown to be important for transformation of human cells (Hamad et al. 2002; Lim et al. 2005). Therapeutic inhibition of this pathway may also hold promise for *KRAS*-mutant cancers (Barbie et al. 2009; Yan et al. 2014; Kitajima et al. 2016). Several other RAS effectors have been identified, including TIAM1 and PLCε, but their role in RAS signaling remains less well understood and their therapeutic value has not yet been proven.

An alternative approach to direct targeting of mutant cancer genes is to exploit the concept of synthetic lethality, in which gene products are identified that, when suppressed or inhibited, result in cell death only in the presence of the cancer-causing alteration (Fig. 1) (Kaelin 2005; Ngo et al. 2006; McLornan et al. 2014). Thus, in principle, targeting synthetic lethal vulnerabilities in cancer should reduce the potential for side effects, because cells harboring the biomarker oncogenic lesion should have more differential sensitivity to the perturbation than normal cells, which do not have the oncogenic lesion. The most salient example of this concept in cancer biology and therapeutics is the discovery that homologous-recombination-deficient *BRCA1* or *BRCA2* mutant cancers show profound sensitivity to inhibition of poly-(ADP-ribose) polymerase (PARP) (Bryant et al. 2005; Farmer et al. 2005). PARP inhibitors have subsequently shown therapeutic efficacy in BRCA1/2 mutant breast, ovarian, and some pancreatic cancers (McLornan et al. 2014; Lord and Ashworth 2017).

Given the relative intractability of KRAS itself as a therapeutic target, identification of synthetic lethal partners of oncogenic KRAS has been the focus of intense investigation by many groups. The fundamental premise underlying the concept of synthetic lethality in KRAS-mutant tumors is that oncogenic KRAS signaling establishes a distinct cell state, marked by altered KRAS effector signaling, adaptation to oncogenic stress, and transcriptional and metabolic reprogramming. Disruption of this KRAS-driven cell state may impair proliferation and

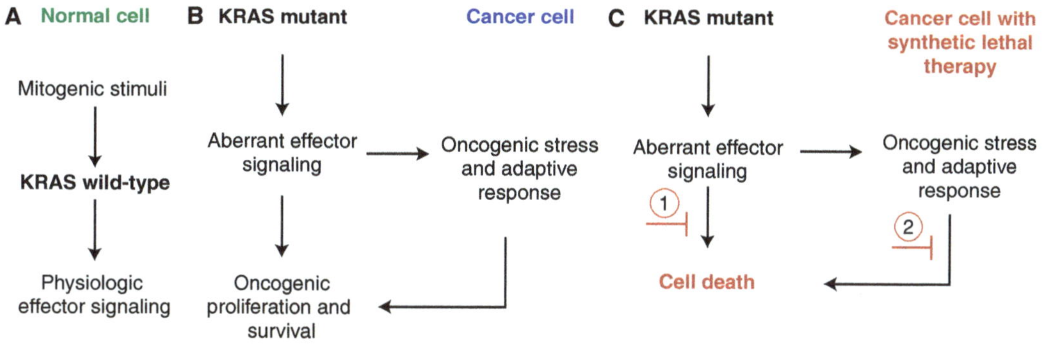

Figure 1. Synthetic lethality as a therapeutic paradigm in cancer. (*A*) In normal healthy cells, wild-type (WT) KRAS is activated by appropriate mitogenic stimuli such as receptor tyrosine kinase–mediated growth factor signaling. (*B*) In cancer cells with *KRAS* mutation, aberrant effector signaling mediates oncogenic proliferation and survival but also creates oncogenic stress to which cancer cells must adapt to sustain oncogenic growth. These downstream aberrant effector signaling pathways as well as parallel adaptive pathways that mitigate oncogenic stress represent unique features of *KRAS*-mutant cells that may include selective vulnerabilities and synthetic lethal targets. (*C*) Synthetic lethal targets refer to those targets whose inhibition results in cell death only in the presence of another mutation (i.e., *KRAS* mutation). Broadly applied, this concept may represent targets that act (1) downstream of aberrant effector signaling, or (2) in parallel adaptive pathways. In theory, synthetic lethal therapies result in cell death of the *KRAS*-mutant cancer cells, but not in normal cells (shown in *A*) because the aberrant oncogenic signals or parallel adaptive processes are not present in these *KRAS* WT cells.

viability of *KRAS*-mutant cells. Furthermore, many of these adaptive features may have partial selectivity or even unique specificity for the KRAS-driven oncogenic cell state over a KRAS wild-type (WT) state, and thus may represent nononcogene dependencies or vulnerabilities between these cell contexts. Because most cancer cells and normal cells use RAS effector signaling in some capacity for mitogenic processes, it should be expected that many of the oncogenic KRAS codependencies will not show unique specificity to the *KRAS*-mutant context and thus will not strictly conform to the genetic definition of synthetic lethality (Fig. 1). However, the term "synthetic lethal" has also been applied more liberally to denote such vulnerabilities that show some quantitative selectivity for the *KRAS*-mutant state over the *KRAS* WT state. Many recent studies, including hypothesis-driven focused studies as well as large small-molecule and genetic screens performed over the past decade, have identified several putative KRAS synthetic lethal partners of varying strength and specificity. Whereas the long-sought-after universal synthetic lethal target for all *KRAS*-mutant cancers has not yet been identified, these studies have informed our understanding of KRAS biology and have identified numerous context-selective vulnerabilities in *KRAS*-mutant cancers. Whether these synthetic lethal candidates identified in experimental systems will translate to effective therapeutic targets remains to be shown. Here, we review recent work on KRAS synthetic lethal vulnerabilities and provide a conceptual framework for how to interpret these observations and prioritize synthetic lethal targets for therapeutic development.

INHIBITING KRAS EFFECTOR PATHWAYS IN *KRAS*-MUTANT CANCERS

Numerous studies have shown that KRAS effectors and related receptor tyrosine kinase (RTK)-mediated signaling pathways may be preferential vulnerabilities in the *KRAS*-mutant context. Through large-scale small-molecule screens in many different cancer cell lines with a variety of genotypes and lineages, several groups have compared compound sensitivity in RAS-mutant and RAS WT cell lines. These studies have repeatedly identified inhibitors of the MAPK pathway as the most selective compounds for

KRAS-mutant cancers, including inhibitors of MAPK kinase 1 and 2 proteins (MEK1/2) and RAF kinases (Garnett et al. 2012; Basu et al. 2013; Molina-Arcas et al. 2013), although inhibition of these pathways leads primarily to cytostasis. These studies have also shown that *KRAS*-mutant cells harbor increased dependence on RTK signaling pathways, as small-molecule inhibitors of IGF1R and MET show some preferential selectivity for subsets of *KRAS*-mutant lines over *KRAS* WT lines (Ebi et al. 2011; Molina-Arcas et al. 2013). In addition, MAPK inhibition leads to increased dependence on RTK signaling, with combined targeting of MEK1/2 and IGF1R (Ebi et al. 2011; Molina-Arcas et al. 2013) or FGFR1 (Manchado et al. 2016), demonstrating enhanced differential impact on *KRAS*-mutant cells over WT cells. Moreover, KRAS-MAPK effector signaling ultimately leads to activation of AP-1 transcription factors and up-regulation of cell-cycle regulatory proteins, such as cyclin D1 (*CCND1*), which promotes G_1/S cell-cycle progression. Inhibition of CDK4/6 has been reported as a promising combination therapeutic strategy in *KRAS*-mutated malignancies (Puyol et al. 2010). Unfortunately, clinical trials testing single-agent inhibition of the critical KRAS effectors (e.g., MAPK or PI3K) in *KRAS*-mutant cancers have been disappointing to date (Adjei et al. 2008; Infante et al. 2012, 2013; Zhao and Adjei 2014). However, inhibition of these pathways will undoubtedly form the foundation for future combination therapy strategies in *KRAS*-mutated malignancies (Engelman et al. 2008; Britten 2013; Ebi et al. 2014; Ryan et al. 2015).

RAS signaling through both the MAPK and PI3K-AKT-MTOR pathways has been shown to mediate antiapoptotic signaling through down-regulation of proapoptotic mediators and up-regulation of antiapoptotic proteins (Pylayeva-Gupta et al. 2011). Engelman and colleagues have shown that disruption of RAS effector signaling unveils selective vulnerabilities in the *KRAS*-mutant context. MEK1/2 inhibition leads to induction of the proapoptotic protein BIM, but this protein is bound by antiapoptotic BCL-XL proteins (Corcoran et al. 2013). Exploiting this observation, they showed through genetic and pharmacologic means that combined MEK inhibition with genetic ablation or small-molecule inhibition of BCL-XL led to robust apoptosis and synthetic lethality in *KRAS*-mutant cells. This therapeutic strategy has now advanced to early-phase clinical trials (NCT02079740). Further work by the Engelman laboratory has also shown that *KRAS*- or *BRAF*-mutant colorectal cancer cells compared with WT cells harbor increased sensitivity to combined inhibition of the antiapoptotic proteins MCL-1, BCL-XL, and BCL-2 through dual treatment with an MTOR inhibitor and the BCL2/BCL-XL inhibitor ABT-263, with selectivity stemming from disruption of BIM/MCL-1 complexes in the *KRAS*- or *BRAF*-mutant context (Faber et al. 2014).

Several studies have shown that KRAS-associated inflammatory signaling mediates key proliferative and survival programs (Kitajima et al. 2016). NF-κB signaling has been shown to play an important role in *KRAS*-mutant cancers. As noted above, KRAS has been shown to activate NF-κB signaling through activation of RAL-GEFs. Genetic or pharmacologic suppression of NF-κB signaling inhibits tumorigenesis in the *Kras*^{G12D};*p53* (KP) mouse model of lung cancer (Meylan et al. 2009; Basseres et al. 2010). In a *Kras*^{G12D};*Ink4a/Arf* mutant PDAC model, oncogenic Kras signaling activates an interleukin (IL)-1a/p62 feedforward loop to induce NF-κB signaling (Ling et al. 2012). Small-molecule inhibitors of RAL GTPase proteins are under development (Yan et al. 2014), and such compounds could be used to directly abrogate KRAS-mediated activation of NF-κB signaling. Through an arrayed RNA-interference (RNAi)-based synthetic lethal screening (discussed below), the IκB kinase (IKK)–related kinase Tank-binding kinase-1 (TBK1) was identified as a synthetic lethal target with oncogenic KRAS (Barbie et al. 2009). This finding expanded upon the prior observation that the RALB GTPase activates TBK1, thus suggesting a mechanistic link to KRAS signaling through RAL-GEFs (Chien et al. 2006). Subsequent work has shown that TBK1 promotes KRAS-driven tumorigenesis by regulating an autocrine CCL5 and IL-6 cytokine signaling loop (Zhu et al.

2014). Numerous other studies have also shown an important role for Janus kinase (JAK)-signal transducers and activators of transcription (STAT) signaling in PDAC tumorigenesis (Corcoran et al. 2011; Fukuda et al. 2011; Lesina et al. 2011). MEK inhibition in *KRAS*-mutant cancer cells, as well as those driven by other oncogenes such as epidermal growth factor receptor (EGFR) or MET, leads to feedback activation of IL-6 signaling and STAT3 activation to drive resistance to this therapy (Lee et al. 2014). The multitargeted kinase inhibitor momelotinib inhibits both JAK1/2 and the TBK1 (and IKKe) kinases. Suppression of both JAK-STAT and TBK1/NF-κB signaling with momelotinib has shown strong preclinical efficacy in cell culture models and murine Kras-driven animal models (Zhu et al. 2014). Clinical trials of momelotinib in combination with chemotherapy or MEK inhibition in *KRAS*-mutant malignancies are currently underway. In addition, further development of more potent and specific inhibitors of TBK1 and JAK-STAT signaling is an area of current activity. Thus, KRAS-directed inflammatory signaling networks and compensatory cytokine signaling represent important vulnerabilities for therapeutic development.

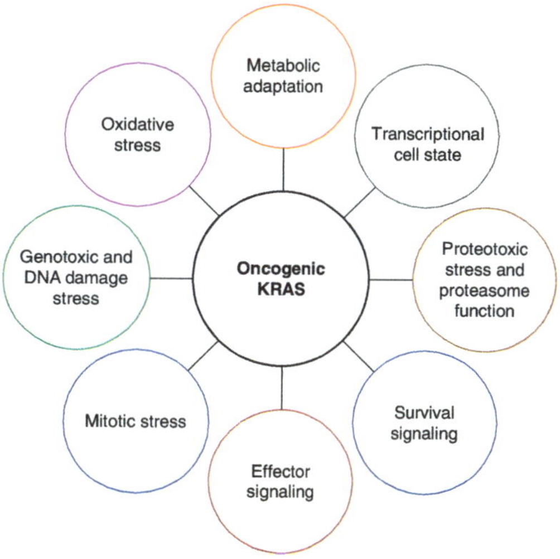

Figure 2. KRAS oncogenic stress and adaptation.

KRAS ONCOGENIC STRESS AND ADAPTATION

Cancer cells harbor a variety of oncogenic stresses, including DNA damage and replicative stress, proteotoxic stress, oxidative stress, and metabolic stress (Luo et al. 2009b). KRAS signaling establishes an oncogenic cell state that gives rise to adaptive changes to oncogenic stress that are necessary for cancer cell proliferation and survival. The myriad of parallel cell processes that are essential in the context of KRAS signaling have been termed nononcogene addictions (Luo et al. 2009b), and these processes form the basis for KRAS synthetic lethality (Fig. 2).

RAS signaling has been reported to directly lead to genotoxic stress stemming from generation of reactive oxygen species (ROS) and DNA hyperreplication (Grabocka et al. 2015). Cancer cells respond to this stress through activation of DNA damage repair (DDR) checkpoints and up-regulation of the function of DDR pathways (Grabocka et al. 2015). Bar-Sagi and colleagues have shown an important role for WT H-/N-Ras in the activation of the ATR/CHK1-mediated DDR and maintenance of genomic stability in *KRAS*-mutant cancers (Grabocka et al. 2014). They have shown that WT HRAS and NRAS negatively regulate MAPK and AKT signaling to control inhibitory phosphorylation of CHK1 on serine 280. Loss of WT HRAS or NRAS expression in *KRAS*-mutant cells resulted in impaired CHK1 activation and checkpoint failure, leading to increased genomic instability and sensitivity to DNA-damaging chemotherapy agents. These studies more generally suggest that inhibition of WT HRAS/NRAS signaling or inhibition of the ATR-CHK1 pathway may have important therapeutic value in combination with DNA damaging chemotherapy in *KRAS*-mutant cancers. Furthermore, combination small molecule screening studies in cancer-cell lines has identified synergistic interactions of DDR checkpoint inhibitors specifically in the *KRAS*-mutant context, with dual inhibition of the of CHK1 and MK2 cell-cycle checkpoint inhibitors having a potent effect in *KRAS*-mutant contexts (Dietlein et al. 2015). Whereas these therapeutic strategies involving DDR checkpoint blockade or DNA-damaging chemotherapy are by no means specific for the *KRAS*-mutant context, the above evidence suggests that there may be partial selectivity for *KRAS*-mutant cells that may offer a sufficient therapeutic window to enable therapeutic efficacy.

Using an integrative approach combining gene expression and RNAi screening data, White and colleagues recently described a specific dependency of *KRAS*-mutant non-small-cell lung cancer (NSCLC) cells to receptor-dependent nuclear export (Kim et al. 2016). Specifically, they showed that chemical inhibition of nuclear export with a clinically available inhibitor of XPO1 led to a synthetic lethal interaction with oncogenic KRAS. XPO1 inhibition was mechanistically linked to inhibition of NF-κB transcription factor activity. Thus, addiction to XPO1-dependent nuclear and cytoplasmic trafficking is a druggable liability in *KRAS*-mutant lung cancers.

Multiple groups have shown that *KRAS*-mutant cancers harbor proteotoxic or endoplasmic reticulum (ER) stress, resulting in activation of the unfolded protein response (UPR) signaling pathway (Denoyelle et al. 2006; De Raedt et al. 2011). Enhancement of this proteotoxic stress can overwhelm the cell's ability to compensate and leads to cell death. Cichowski and colleagues showed that *Kras/p53* mutant lung cancers are sensitive to agents that enhance proteotoxic stress (De Raedt et al. 2011). They showed that HSP90 inhibition with IPI-504 in combination with MTOR inhibition with rapamycin results in catastrophic ER stress and tumor regressions in murine lung tumor models. They argued that this combination is clinically feasible and should be evaluated in human clinical trials in *KRAS*-mutant lung cancers.

Downward (2015) has highlighted a fundamental role for protein synthesis machinery and proteasome function in *KRAS*-mutant cells. RNAi screening was performed on approximately 7000 druggable genes and identified the GATA2 transcription factor as a synthetic lethal target with preferential dependency in *KRAS*-mutant cells (Kumar et al. 2012; Steckel et al. 2012). GATA2 was shown to up-regulate proteasome components through the transcription factor Nrf1; however, loss of GATA2 impaired proteasome activity in both *KRAS*-mutant and WT lung cancer cells. GATA2 also up-regulates other essential pathways, including IL-1 and Rho signaling cascades, and the concurrent regulation of these three pathways was proposed to be the basis for partial selectivity of GATA2 essentiality in *KRAS*-mutant versus WT cells. Supporting these observations, proteasome components were also identified in other synthetic lethal screening efforts (Barbie et al. 2009; Luo et al. 2009a) discussed below. However, proteasome inhibitors such as bortezomib do not have selectivity for *KRAS*-mutant cells in culture (Garnett et al. 2012; Basu et al. 2013) and have a relatively narrow therapeutic window in human trials; thus, whether the current clinical grade proteasome inhibitors will show efficacy in KRAS-driven cancers remains uncertain.

Several studies suggest that RAS mutant cells may have enhanced sensitivity to oxidative stress. Stockwell and colleagues have pursued synthetic lethal small molecule screening in isogenic HRAS G12V mutant or WT fibroblasts and have identified a series of small molecules that induce ferroptosis, a nonapopototic cell death characterized by iron-dependent accumulation of lethal lipid ROS in oncogenic RAS-harboring cells (Dolma et al. 2003; Yagoda et al. 2007; Yang and Stockwell 2008). Furthermore, Shaw and colleagues (2011) have performed synthetic lethal screening of over 50,000 compounds in mouse embryonic fibroblasts (MEFs) isogenic for a $Kras^{G12D}$ allele and identified lanperisone, a compound that induces nonapoptotic cell death selectively in $\mathrm{Kras}^{\mathrm{G12D}}$ MEFs through induction of ROS leading to oxidative stress and cell death. Notably, lanperisone showed suppression of tumor growth in a xenograft model with $Kras^{G12D};p53^{-/-}$ transformed MEFs without significant toxicity; however, human *KRAS*-mutant lung cancer cell lines were relatively resistant to this compound, suggesting context specificity for this vulnerability (Shaw et al. 2011). Moreover, as noted above, the proteotoxic stress and *Kras*-mutant tumor regressions invoked by combined HSP90 and MTOR inhibition, described by Cichowski and colleagues, is thought to be fueled by oxidative stress (De Raedt et al. 2011). Last, work by Tuveson and colleagues has shown that KRAS and

Cite this article as *Cold Spring Harb Perspect Med* doi: 10.1101/cshperspect.a031518

other oncogenes promote detoxification of ROS through transcriptional up-regulation of an NFE2L2/NRF2-mediated antioxidant program, and genetic targeting of Nrf2 impairs Kras^{G12D}-driven tumorigenesis (DeNicola et al. 2011). Subsequent work has shown that loss of NRF2 impairs autocrine EGFR signaling and leads to oxidation of translational regulatory proteins that inhibit messenger RNA (mRNA) translation and pancreatic cancer cell proliferation (Chio et al. 2016). Targeting of these NRF2-mediated signaling pathways with combined inhibition of AKT and glutathione synthesis was shown to be a synthetic lethal strategy resulting in impaired pancreatic cancer growth in animal models.

Oncogenic KRAS drives altered cellular metabolism to support anabolic processes that enable proliferation and survival (Ying et al. 2012; Bryant et al. 2014). Studies of KRAS-driven metabolism have unveiled numerous potential synthetic lethal vulnerabilities. KRAS-mediated metabolic reprogramming results in increased utilization of autophagy for recycling of intracellular components to fuel biosynthetic pathways and promote KRAS-mutant pancreatic cancer cell proliferation (Guo et al. 2011; Lock et al. 2011; Yang et al. 2011, 2014). Moreover, rigorous studies in KRAS-mutant PDAC mouse models and human PDAC patient-derived xenografts (PDXs) showed that autophagy inhibition impairs PDAC tumor growth, thus highlighting this process as an important dependency that may be amenable to therapeutic targeting.

KRAS-driven tumors consume extracellular protein through the process of macropinocytosis (Commisso et al. 2013; Davidson et al. 2017). Macropinocytosis results in internalization of extracellular fluid and nutrients within vesicles, where proteins undergo degradation to yield amino acids that can enter central carbon metabolism. Pharmacologic inhibition of micropinocytosis results in growth inhibition of *KRAS*-mutant but not *KRAS*-WT PDAC xenografts (Commisso et al. 2013). Thus, the dependence of *KRAS*-mutant cancer cells on macropinocytosis is a metabolic vulnerability with therapeutic potential.

Kimmelman and colleagues have reported that KRAS directs reprogramming of glutamine metabolism in PDAC cells through transcriptional up-regulation of metabolic enzymes in a noncanonical pathway of glutamine utilization (Son et al. 2013). Through RNAi experiments, they showed that suppression of expression of multiple enzymes in this pathway was essential for the growth of PDAC cells in vitro and in vivo, but not essential for normal cell growth. Glutaminase is the initial enzyme in this pathway and is currently being investigated as a therapeutic target in *KRAS*-mutated PDAC. However, it is becoming increasingly appreciated that the cellular environment is a major determinant of metabolic phenotypes and that metabolic dependencies may be highly context specific. For example, whereas *Kras*-mutant lung cancer cells depend on glutaminase in culture, *Kras*-driven lung tumors do not depend on glutaminase in vivo (Davidson et al. 2016). Moreover, metabolic adaptations are not uniform in KRAS-driven cancers. Vander Heiden and colleagues have shown distinct utilization of branched-chain amino acids (BCAAs) between *Kras;Trp53* mutant mouse models of NSCLC and PDAC (Mayers et al. 2016). NSCLC incorporates BCAAs into protein and uses them as a nitrogen source; furthermore, loss of the BCAA-processing enzymes *Bcat1* and *Bcat2* impairs NSCLC growth. However, PDAC tumors have decreased BCAA uptake and do not depend on *Bcat1* or *Bcat2*. This context-selectivity of metabolic dependencies highlights the importance of studying *KRAS*-driven metabolic adaptations using in vivo model systems that most closely represent human tumor physiology.

Cantley and colleagues have recently harnessed the metabolic properties of *KRAS*-mutant cells and sensitivity of these cells to oxidative stress (Yun et al. 2015). They showed that high-dose vitamin C shows selective toxicity in *KRAS* (or *BRAF*) mutant tumor cells because of uptake of oxidized vitamin C via the GLUT1 transporter, leading to buildup of ROS and inactivation of glyceraldehyde 3-phosphate dehydrogenase (GAPDH). Notably, high GLUT1 expression is driven by the KRAS or BRAF oncogene-induced glycolytic addiction; thus, oxidative stress leads to an energetic crisis and cell death specifically in the KRAS- or BRAF-driven tumor cell context.

While there is not yet a clinically effective direct inhibitor of oncogenic KRAS, multiple groups have modeled KRAS inhibition through inducible suppression of KRAS expression in cell culture or mouse tumor models (Podsypanina et al. 2008; Collins et al. 2012; Ying et al. 2012; Kapoor et al. 2014; Shao et al. 2014). These studies have revealed important KRAS-independent escape mechanisms in the context of KRAS suppression. Using an inducible *Kras*^{G12D} (iKras) mouse model of pancreatic cancer, Draetta and colleagues have shown that metabolic reprogramming from a decreased reliance on glycolysis to a strong dependence on oxidative phosphorylation enables survival in the absence of KRAS signaling (Viale et al. 2014). This finding has prompted efforts to combine KRAS signaling blockade (e.g., MEK1/2) with inhibition of oxidative phosphorylation, although the efficacy and tolerability of this strategy needs to be evaluated in human clinical trials. Hahn and colleagues used a cDNA overexpression screen to identify the YAP1 transcription factor as a mediator of KRAS-independent growth (Shao et al. 2014). Using the iKras mouse model, DePinho and colleagues also showed that YAP1 could overcome Kras suppression to promote PDAC growth (Kapoor et al. 2014). YAP1 has also been shown to be essential for KRAS-induced cell transformation (Shao et al. 2014), as well as PDAC development in mice (Zhang et al. 2014). Moreover, Bivona and colleagues have described a role for YAP1 as a node of resistance to RAF and MEK inhibition, and therefore propose that YAP1 may be a synthetic lethal interactor that is an attractive candidate for combination therapy in *KRAS*- (or *BRAF*-) mutant malignancies (Lin et al. 2015).

RAS signaling is well known to lead to epithelial-mesenchymal transition (EMT) in certain epithelial cell contexts. Settleman and colleagues have previously described an important role for EMT in mediating KRAS-independent growth of *KRAS*-mutant cancer cells (Singh et al. 2009). Moreover, aggressive mesenchymal subpopulations have been shown to arise through Smarcb1-Myc network-driven reprogramming upon extinction of Kras^{G12D} expression in animal models of PDAC (Genovese et al. 2017). KRAS and YAP1 have also been shown to converge to activate an EMT transcriptional program (Shao et al. 2014). Additionally, the *SNAI2* gene encoding the SNAIL transcription factor, a known regulator of EMT, has been previously identified as a KRAS synthetic lethal target in an RNAi screen of an isogenic pair of colorectal cancer cell lines (Wang et al. 2010). Thus, the mesenchymal cell state itself, partially induced by KRAS signaling, may harbor synthetic lethal vulnerabilities for therapeutic targeting in a subset of KRAS-mutant cancers.

FUNCTIONAL GENETIC SCREENING FOR KRAS SYNTHETIC LETHAL TARGETS

Loss-of-function genetic screens provide a means to define the compendium of genes that are essential for cancer cell proliferation and viability in a context-selective manner (Boehm and Hahn 2011). Unbiased, genome-scale functional genetic screens hold promise to identify novel and unpredicted synthetic lethal relationships with oncogenic KRAS. Multiple different approaches have been used to screen for KRAS synthetic lethal relationships, including both arrayed/multiwell and pooled screening formats in both panels of *KRAS*-mutant or WT cell lines, as well as isogenic *KRAS*-mutant and WT cell systems. Most of these screens have been performed using RNAi, although the demonstration that CRISPR-Cas9 technology can be used in mammalian cells for genome scale screens has led several groups to begin using this technology. The strengths and disadvantages of these methodologic approaches to KRAS synthetic lethal screening have been recently reviewed elsewhere (Downward 2015). We summarize a wide range of KRAS synthetic lethal screens in Table 1, but focus on a few illustrative examples below.

A team led by Elledge and colleagues performed a pooled primary screen of a genome-scale lentivirally delivered short hairpin RNA (shRNA) library in an isogenic *KRAS*^{G13D} and WT DLD1 colorectal cancer cell line (Luo et al.

Table 1. RAS synthetic lethal functional genetic screens

Synthetic lethal genes or pathways	Library (assay and format)	Cells in primary screen	Drug inhibition	References
RAN, TPX2, SCD1	~3700 druggable genes, small interfering RNA (siRNA), arrayed cell death	NCIH1299 (*NRAS*^{Q6K} NSCLC)	Not tested	Morgan-Lappe et al. 2007
BIRC5 (survivinn), *CDK1, RBCK1,*	~4000 genes, siRNA, arrayed cell death	Isogenic DLD1 (CRC, *KRAS*^{G3D})	Not tested	Sarthy et al. 2007
PLK1, APC/C, proteosome	74,905 retroviral short hairpin RNAs (shRNAs) to 32,293 human transcripts (pooled proliferation screen)	Isogenic DLD1 (CRC, *KRAS*^{G3D})	BI-2536 (PLK)	Luo et al. 2009a
STK33, AKT3, CPNE1, CAMPK1, MLKL, FLT3LG, and *DGKZ*	~1000 druggable genes, shRNA, arrayed proliferation	Pan-cancer cell-line panel (four *KRAS*-mutant, four *KRAS*-wild-type [WT]) and two immortalized cell lines	STK33 kinase inhibitor, failed to suppress proliferation in *KRAS*-mutant cells (Babij et al. 2011; Luo et al. 2012)	Scholl et al. 2009
TBK1, PSKH2, PTCH2, CPNE1, MAP3K8, proteasome genes	~1000 druggable genes, shRNA, arrayed proliferation	Pan-cancer cell-line panel (seven KRAS-mutant, 10 *KRAS*-WT) and two immortalized cell lines	CYT387 (TBK1 and JAK inhibitor) assessed in Zhu et al. (2014)	Barbie et al. 2009
WT1, RAC1, PHB2	162 KRAS-related genes, shRNA, in vitro and in vivo pooled proliferation screens with bead array readout	LKR10 and LKR13 (*Kras;Trp53* mutant mouse lung-tumor-derived cell lines)	Not tested	Vicent et al. 2010
SNAI2 (SNAIL2)	~2500 druggable genes, shRNA, pooled proliferation	Isogenic HCT116 (CRC, *KRAS*^{G13D})	Not tested	Wang et al. 2010
GATA2, CDC6, proteasome	>7000 druggable genes, siRNA pools (arrayed apoptosis and cell proliferation)	Isogenic HCT116 (*KRAS*^{G13D}) and pan-cancer cell-line panel (14 *KRAS*-mutant, 12 KRAS-WT)	Bortezomib with fasudil (GATA2)	Kumar et al. 2012; Steckel et al. 2012
MAP3K7 (TAK1)	17 kinases highly expressed in *KRAS*-dependent CRC, shRNA, arrayed proliferation	KRAS-dependent SW620 and KRAS-independent SW837 (CRC, *KRAS*-mutant)	5Z-7-oxozeaenol	Singh et al. 2012

Continued

Table 1. *Continued*

Synthetic lethal genes or pathways	Library (assay and format)	Cells in primary screen	Drug inhibition	References
Cttnb1 (β-catenin), *Mllt6*, *Raf1*, *Akt3*	Genome scale, shRNA, pooled in vivo proliferation with NGS readout	Mouse keratinocytes (H-ras^{G12V})	Not tested	Beronja et al. 2013
COPI coatomer subunits ARCN1, COPB1, COPA dependent in KRAS;LKB1-mutant lung cancer cells	Genome scale, siRNA, arrayed proliferation	17 *KRAS*- and LKB1-mutant lung cancer cell lines, matched tumor (*KRAS*-mutant), and normal NSCLC cell-line pair	Saliphenylhalamide A	Kim et al. 2013
ARHGEF2 (GEFH1)	Genome scale, shRNA, pooled proliferation	72 human cancer cell lines	Not tested	Marcotte et al. 2012; Cullis et al. 2014
BCL2L1 (BCLXL)	~1200 druggable genes in presence of MEK inhibitor (selumetinib), shRNA, pooled proliferation screen for enhancement of MEK inhibitor effect	HCT116 and SW620 (CRC, *KRAS*-mutant)	Selumetinib and navitoclax	Corcoran et al. 2013
RAF1, *BRAF*	535 kinases and related genes, shRNA, pooled proliferation screen for enhancement with MEK inhibitor selumetinib	SW480 KRAS-mutant colon cancer cell line	RAF265 or AZ628 (RAF inhibitors) with selumetinib	Lamba et al. 2014
YAP1	5046 signaling components, shRNA, pooled proliferation screen for enhancement of BRAF-inhibitor response	HCC364 BRAF V600E human NSCLC cell line with validation in KRAS mutants with MEK inhibition	Not tested	Lin et al. 2015
CDK1	853 genes, primarily kinases, siRNA, arrayed cell viability	LIM1215 isogenic lines with KRAS WT, G12D, G12S, or G12V	Not tested	Costa-Cabral et al. 2016
FGFR1, *BRAF*, *RAF1*, *ERK2*	526 kinases, shRNA, pooled proliferation screen for enhancement with MEK inhibitor trametinib	H23 KRAS (G13C) mutant lung cancer cell line	Ponatinib (FGFR inhibitor)	Manchado et al. 2016
XPO1	Genome scale, siRNA arrayed toxicity screens	Panel of human NSCLC cell lines	KPT-330 (selinexor)	Kim et al. 2016

Continued

Table 1. *Continued*

Synthetic lethal genes or pathways	Library (assay and format)	Cells in primary screen	Drug inhibition	References
RAF1, SHOC2, PREX1, RCE1, ICMT	Genome-scale, CRISPR-Cas9, pooled proliferation screen	14 leukemia cell lines (six KRAS or NRAS, six RAS WT)	Not tested	Wang et al. 2017

Modeled after similar tables in Downward (2015) and Ebi et al. (2014).
NSCLC, Non-small-cell lung cancer; JAK, Janus kinase.

2009a). Their library contained 75,905 shRNAs targeting 32,293 unique human transcripts to approximately 12,000 genes. Stringent criteria identified a subset of 379 shRNAs targeting 368 genes that were candidate KRAS synthetic lethal interactors. Notably, the DLD1 *KRAS* WT cell line showed decreased MAPK signaling and slower in vitro proliferation than its *KRAS*-mutant counterpart. Follow-up studies were performed using a multicolor competition assay in both the DLD1 isogenic pair as well as a second isogenic HCT116 *KRAS*-mutant and WT pair. They identified 77 candidate KRAS synthetic lethal interactors across a variety of biological processes, including protein modification, nucleic acid metabolism, and cell-cycle and signal transduction. In particular, they described a striking number of genes involved in regulation of mitosis as being candidate KRAS synthetic lethals, including subunits of the anaphase-promoting complex/cyclosome (APC/C), the proteasome and the mitotic kinase polo-like kinase (PLK1). They showed that *KRAS*-mutant cells harbor heightened mitotic stress compared with WT cells, and both genetic and small molecule inhibition of PLK1 revealed greater sensitivity in the *KRAS*-mutant isogenic cancer cells over the *KRAS* WT cells. Moreover, inhibition of proteasome function with bortezomib also appeared to demonstrate selectivity in the KRAS isogenic cell systems. Mitotic stress is not unique to *KRAS*-mutant cancers; however, these studies suggest that *KRAS*-mutant cells may be selectively more sensitive to targeting mitotic regulation than WT cells.

Scholl et al. (2009) performed an arrayed format RNAi screen in four *KRAS*-mutant and four *KRAS*-WT cancer cell lines, as well as normal human fibroblasts and immortalized human mammary epithelial cells (HMECs). They used a library of 5024 lentivirally delivered shRNA constructs targeting 1011 human kinases, phosphatases, and known cancer genes. Comparing *KRAS*-mutant to WT cells, they identified the serine-threonine protein kinase STK33 as selectively essential in *KRAS*-mutant cells. Through evaluation of STK33 dependency in 25 additional cell lines, they determined that STK33 is preferentially required by KRAS mutants that also depend on sustained KRAS expression for viability and proliferation. STK33 has been subsequently investigated as a KRAS synthetic lethal target by multiple other groups, some with conflicting findings and interpretations (Babij et al. 2011; Frohling and Scholl 2011). Much of the excitement regarding STK33 stemmed from its kinase function and thus potential therapeutic tractability. Potent small molecule inhibitors of STK33 have been developed, and these molecules have conclusively shown that STK33 kinase activity is not required for KRAS-mutant cancers (Babij et al. 2011; Luo et al. 2012), although other strategies that result in loss of STK33 protein may still hold therapeutic potential (Azoitei et al. 2012).

Barbie et al. (2009) performed an arrayed format primary RNAi screen in 19 cancer cell lines, using a lentivirally delivered shRNA library targeting kinases, phosphatases, and oncogenes. They compared *KRAS*-mutant and WT cells to identify those genes that selectively killed *KRAS*-mutant cells and identified 45 candidates that were screened in a larger panel of mutant and WT cells. In addition to *KRAS* itself, they found RAF1 and the serine-threonine kinase TBK1 as top-scoring candidate KRAS synthetic lethals. They validated TBK1 as a synthetic lethal target in *KRAS*-mutant and -dependent cells and confirmed a link between RALB-mediated activation of TBK1 (Chien et al. 2006) to promote NF-κB survival signaling, in part, through c-Rel and BCL-XL. As discussed above, pharmacologic inhibition of the TBK1 induces cell death in preclinical animal models of KRAS-driven lung cancers (Zhu et al. 2014), and TBK1 inhibition is being tested in clinical studies as a potential therapeutic approach.

The lack of overlap in genes scoring from KRAS synthetic lethal screens has caused much consternation. However, the KRAS synthetic lethal screens performed to date have had several important differences and limitations. Each study used different experimental systems and were limited in scale in terms of the cell lines screened and the number of genes evaluated. None of the RNAi-based screens reached genetic saturation. Moreover, RNAi studies have previously suffered from an incomplete understanding of off-target effects. Advances in RNAi-screening efforts have improved upon these limitations. Recent technical advances in RNAi screening have enabled much higher density screens across many more cell lines (Luo et al. 2008; Cheung et al. 2011; Marcotte et al. 2012). Additionally, refinements in bioinformatic methodology have also improved the signal-to-noise resolution and distinction of off-target effects (Shao et al. 2013; Marcotte et al. 2016; Tsherniak et al. 2017). Furthermore, the advent of CRISPR-Cas9 screening technology as an alternative means of genetic perturbation has enabled a new approach to loss-of-function genetic screening (Shalem et al. 2014, 2015). CRISPR-Cas9 technology enables knockout of target genes, as compared to suppression of mRNA expression conferred by RNAi. Multiple groups have now shown that CRISPR-Cas9 genome-scale screens can effectively elucidate essential genes in cancer cell lines (Koike-Yusa et al. 2014; Wang et al. 2014, 2015; Hart et al. 2015; Parnas et al. 2015; Aguirre et al. 2016).

Sabatini and colleagues recently reported genome-scale CRISPR-Cas9 screens in 14 acute myeloid leukemia (AML) cell lines and compared six *KRAS* or *NRAS* mutant lines with six RAS WT cell lines to identify synthetic lethal genes (Wang et al. 2017). These studies identified a surprisingly small number of genes in the RAS processing (*RCE1* and *ICMT*) and MAPK signaling pathways (*RAF1*, *SHOC2*). A second pair of screens was performed in isogenic murine Ba/F3 cell lines, one of which was engineered to express oncogenic *NRAS*, and this screen also discovered the same set of genes as well as additional MAPK signaling components. PREX1 was identified as a novel RAS synthetic lethal partner in the human AML screen. PREX1 is a GEF for Rac GTPases, and the investigators showed that PREX1 is an AML-specific activator of MAPK signaling.

NEXT STEPS IN DISCOVERY AND UTILIZATION OF KRAS SYNTHETIC LETHAL TARGETS

While the results of Sabatini and colleagues suggest that in AML there are only a limited number of RAS synthetic lethal targets that are largely restricted to regulators of RAS itself or the MAPK signaling pathway, it is important to note that these findings reflect only one of many *KRAS*-mutant cancer contexts. Indeed, there is profound heterogeneity of *KRAS*-mutant cancers that complicates uniform identification of KRAS synthetic lethal targets. This multidimensional heterogeneity includes (1) variability in KRAS dependency across cancers; (2) differing lineage specificity across cancer types; (3) variation in effector signaling between different KRAS alleles (e.g., G12D vs. G12V or Q61H); and (4) variability in the co-occurring mutational and copy-number landscape across cancers. It may be unreasonable to expect that a universal KRAS synthetic lethal target will apply across all contexts. To adequately address this heterogeneity and to determine biomarker-driven predictive models of these synthetic lethal relationships, genome-scale screens will need to be performed in a large number of diverse cell lines. Toward this goal, the Achilles Consortium at the Broad Institute has now screened over 300 cell lines with genome-scale CRISPR-Cas9 knockout technology, including over 60 KRAS-mutant cell lines across a variety of contexts. These efforts have not found a single strong universal synthetic lethal target across all *KRAS*-mutant contexts. Instead, numerous strong vulnerabilities have emerged with preferential dependency in a subset of *KRAS*-mutant cell lines. Understanding these vulnerabilities and their context specificity across *KRAS*-mutant cell lines remains a top priority.

Besides standard CRISPR-Cas9 knockout screening, additional new screening approaches may prove useful for identifying KRAS synthetic lethal candidates. CRISPR-inhibitor (CRISPRi) technology targets a transcriptional repressor to specific genes, and CRISPRi approaches have been developed for genome-scale screening (Gilbert et al. 2014). Moreover, CRISPR-Cas9 libraries have been developed that enable simultaneous knockout of two different genes within the same cell, and will allow examination of pairwise genetic interactions in *KRAS*-mutant cancer cells (Rosenbluh et al. 2016; Han et al. 2017). Last, because most treatment strategies in *KRAS*-mutant cancers will require combination therapies, functional genetic screening in the context of small-molecule inhibition of key KRAS effector signaling pathways may prove most fruitful for the identification of high-priority KRAS synthetic lethal interactions. A number of screens have been performed using RNAi technologies to identify combination therapy targets with MAPK pathway inhibition, including FGFR1 (Manchado et al. 2016), BCL-XL (Corcoran et al. 2013), YAP1 (Lin et al. 2015), and PTPN11 (Prahallad et al. 2015).

In addition to performing screens in two-dimensional cell culture growth conditions for established cancer cell lines, it will also be important to evaluate for synthetic lethal candidates in early passage patient-derived cell lines as well as other potentially more physiologically relevant cancer model systems, such as three-dimensional organoid cultures (Sachs and Clevers 2014; Boj et al. 2015) or animal models of *KRAS*-mutant tumors (Zender et al. 2008; Zuber et al. 2011; Beronja et al. 2013; Carugo et al. 2016). Indeed, utilization of *KRAS*-mutant cancer models that recapitulate tumor–stroma interactions may enable elucidation of an entirely novel set of KRAS-driven non-cell-autonomous vulnerabilities that could impair tumor growth. Whether these alternative culture techniques or model systems will yield a different set of KRAS synthetic lethal interactors remains an open and important question. In collaboration with the National Cancer Institute's RAS Initiative, the RAS Synthetic Lethal Network (RSLN) of laboratories has been organized to examine these novel approaches to identify KRAS synthetic lethal targets and to share these data with the larger RAS scientific community to promote rapid translation of clinically useful therapeutic targets.

Beyond identification in functional genetic screens, extensive validation efforts will be required to rigorously demonstrate the value of novel KRAS synthetic lethal targets. Validation will require confirmation of the potency and specificity of the synthetic lethal relationship using multiple genetic and pharmacologic approaches in both established cell lines and patient-derived models. Moreover, although cross comparison of one tumor cell with a *KRAS* mutation to another that does not have a *KRAS* mutation may reveal vulnerabilities that are stronger in the *KRAS*-mutant cancer context, this approach does not guarantee tolerable toxicity in *KRAS* WT normal cells. Investigators will also need to consider the essentiality of targets within normal cells and tissues that would be most likely to lead to dose-limiting toxicities of targeted cancer therapies. Prioritizing synthetic lethal partners through patterns of expression of the dependent target or its biomarker in normal human tissues or through functional modeling in normal cell culture systems may help address this important issue.

Ultimately, clinical translation of novel synthetic lethal vulnerabilities of oncogenic KRAS into viable therapeutic approaches will require strategies for targeted inhibition of these proteins. Some candidates may possess protein domains that are therapeutically tractable, such as kinases or other enzymes; however, they most likely will not be inherently targetable. As with the oncogenic KRAS protein itself, novel approaches to target "undruggable" proteins are urgently needed. Phthalimide conjugation strategies for targeted protein degradation have recently been described and may be one viable path forward (Winter et al. 2015). It is expected that novel small molecules that directly target mutant KRAS will be soon be developed. Whereas one may argue that the advent of such therapies may obviate the need for targeting synthetic lethal partners, it is inevitable that combination treatments will be needed with direct KRAS inhibitors to overcome compensatory signaling and resistance mechanisms. It is likely that the catalog of synthetic lethal vulnerabilities discussed above as well as novel ones identified through future screening approaches will provide a critical framework for developing multifaceted combination therapy strategies in *KRAS*-mutant cancers.

ACKNOWLEDGMENTS

We thank Belinda Wang, PhD, for her assistance with the generation of some figures and tables used in this review. We also acknowledge the following funding sources: R01 CA130988 (W.C.H.), U01 CA199253 (W.C.H.), U01 CA 176058 (W.C.H.), P01 CA154303 (W.C.H), and P50 CA127003 (W.C.H. and A.J.A.). A.J.A. is supported by the Hope Funds for Cancer Research Postdoctoral Fellowship, Dana Farber Cancer Institute Hale Center for Pancreatic Cancer, Perry S. Levy Endowed Fellowship, and the Harvard Catalyst and Harvard Clinical and Translational Science Center (UL1 TR001102).

REFERENCES

Adjei AA, Cohen RB, Franklin W, Morris C, Wilson D, Molina JR, Hanson LJ, Gore L, Chow L, Leong S, et al. 2008. Phase I pharmacokinetic and pharmacodynamic study of the oral, small-molecule mitogen-activated protein kinasekinase 1/2 inhibitor AZD6244 (ARRY-142886) in patients with advanced cancers. *J Clin Oncol* **26:** 2139–2146.

Aguirre AJ, Meyers RM, Weir BA, Vazquez F, Zhang CZ, Ben-David U, Cook A, Ha G, Harrington WF, Doshi MB, et al. 2016. Genomic copy number dictates a gene-independent cell response to CRISPR/Cas9 targeting. *Cancer Discov* **6:** 914–929.

Azoitei N, Hoffmann CM, Ellegast JM, Ball CR, Obermayer K, Gosselé U, Koch B, Faber K, Genze F, Schrader M, et al. 2012. Targeting of KRAS mutant tumors by HSP90 inhibitors involves degradation of STK33. *J Exp Med* **209:** 697–711.

Babij C, Zhang Y, Kurzeja RJ, Munzli A, Shehabeldin A, Fernando M, Quon K, Kassner PD, Ruefli-Brasse AA, Watson VJ, et al. 2011. STK33 kinase activity is nonessential in KRAS-dependent cancer cells. *Cancer Res* **71:** 5818–5826.

Barbie DA, Tamayo P, Boehm JS, Kim SY, Moody SE, Dunn IF, Schinzel AC, Sandy P, Meylan E, Scholl C, et al. 2009. Systematic RNA interference reveals that oncogenic KRAS-driven cancers require TBK1. *Nature* **462:** 108–112.

Basseres DS, Ebbs A, Levantini E, Baldwin AS. 2010. Requirement of the NF-κB subunit p65/RelA for K-Ras-induced lung tumorigenesis. *Cancer Res* **70:** 3537–3546.

Basu A, Bodycombe NE, Cheah JH, Price EV, Liu K, Schaefer GI, Ebright RY, Stewart ML, Ito D, Wang S, et al. 2013. An interactive resource to identify cancer genetic and lineage dependencies targeted by small molecules. *Cell* **154:** 1151–1161.

Beronja S, Janki P, Heller E, Lien WH, Keyes BE, Oshimori N, Fuchs E. 2013. RNAi screens in mice identify physiological regulators of oncogenic growth. *Nature* **501:** 185–190.

Blasco RB, Francoz S, Santamaria D, Canamero M, Dubus P, Charron J, Baccarini M, Barbacid M. 2011. c-Raf, but not B-Raf, is essential for development of K-Ras oncogene-driven non-small-cell lung carcinoma. *Cancer Cell* **19:** 652–663.

Boehm JS, Hahn WC. 2011. Towards systematic functional characterization of cancer genomes. *Nat Rev Genet* **12:** 487–498.

Boj SF, Hwang CI, Baker LA, Chio II, Engle DD, Corbo V, Jager M, Ponz-Sarvise M, Tiriac H, Spector MS, et al. 2015. Organoid models of human and mouse ductal pancreatic cancer. *Cell* **160:** 324–338.

Britten CD. 2013. PI3K and MEK inhibitor combinations: Examining the evidence in selected tumor types. *Cancer Chemother Pharmacol* **71:** 1395–1409.

Bryant HE, Schultz N, Thomas HD, Parker KM, Flower D, Lopez E, Kyle S, Meuth M, Curtin NJ, Helleday T. 2005. Specific killing of BRCA2-deficient tumours with inhibitors of poly(ADP-ribose) polymerase. *Nature* **434:** 913–917.

Bryant KL, Mancias JD, Kimmelman AC, Der CJ. 2014. KRAS: Feeding pancreatic cancer proliferation. *Trends Biochem Sci* **39:** 91–100.

Carugo A, Genovese G, Seth S, Nezi L, Rose JL, Bossi D, Cicalese A, Shah PK, Viale A, Pettazzoni PF, et al. 2016. In vivo functional platform targeting patient-derived xenografts identifies WDR5-Myc association as a critical determinant of pancreatic cancer. *Cell Rep* **16:** 133–147.

Cheung HW, Cowley GS, Weir BA, Boehm JS, Rusin S, Scott JA, East A, Ali LD, Lizotte PH, Wong TC, et al. 2011. Systematic investigation of genetic vulnerabilities across cancer cell lines reveals lineage-specific dependencies in ovarian cancer. *Proc Natl Acad Sci* **108:** 12372–12377.

Chien Y, Kim S, Bumeister R, Loo YM, Kwon SW, Johnson CL, Balakireva MG, Romeo Y, Kopelovich L, Gale M Jr, et al. 2006. RalB GTPase-mediated activation of the IκB family kinase TBK1 couples innate immune signaling to tumor cell survival. *Cell* **127:** 157–170.

Chio II, Jafarnejad SM, Ponz-Sarvise M, Park Y, Rivera K, Palm W, Wilson J, Sangar V, Hao Y, Ohlund D, et al. 2016. NRF2 promotes tumor maintenance by modulating mRNA translation in pancreatic cancer. *Cell* **166:** 963–976.

Collins MA, Bednar F, Zhang Y, Brisset JC, Galban S, Galban CJ, Rakshit S, Flannagan KS, Adsay NV, Pasca di Magliano M. 2012. Oncogenic Kras is required for both the initiation and maintenance of pancreatic cancer in mice. *J Clin Invest* **122:** 639–653.

Collisson EA, Trejo CL, Silva JM, Gu S, Korkola JE, Heiser LM, Charles RP, Rabinovich BA, Hann B, Dankort D, et al. 2012. A central role for RAF→MEK→ERK signaling in the genesis of pancreatic ductal adenocarcinoma. *Cancer Discov* **2:** 685–693.

Commisso C, Davidson SM, Soydaner-Azeloglu RG, Parker SJ, Kamphorst JJ, Hackett S, Grabocka E, Nofal M, Drebin JA, Thompson CB, et al. 2013. Macropinocytosis of protein is an amino acid supply route in Ras-transformed cells. *Nature* **497:** 633–637.

Corcoran RB, Contino G, Deshpande V, Tzatsos A, Conrad C, Benes CH, Levy DE, Settleman J, Engelman JA, Bardeesy N. 2011. STAT3 plays a critical role in KRAS-induced pancreatic tumorigenesis. *Cancer Res* **71:** 5020–5029.

Corcoran RB, Cheng KA, Hata AN, Faber AC, Ebi H, Coffee EM, Greninger P, Brown RD, Godfrey JT, Cohoon TJ, et al. 2013. Synthetic lethal interaction of combined BCL-XL and MEK inhibition promotes tumor regressions in KRAS mutant cancer models. *Cancer Cell* **23:** 121–128.

Costa-Cabral S, Brough R, Konde A, Aarts M, Campbell J, Marinari E, Riffell J, Bardelli A, Torrance C, Lord CJ, et al. 2016. CDK1 is a synthetic lethal target for KRAS mutant tumours. *PLoS ONE* **11:** e0149099.

Cox AD, Fesik SW, Kimmelman AC, Luo J, Der CJ. 2014. Drugging the undruggable RAS: Mission possible? *Nat Rev Drug Discov* **13:** 828–851.

Cullis J, Meiri D, Sandi MJ, Radulovich N, Kent OA, Medrano M, Mokady D, Normand J, Larose J, Marcotte R, et al. 2014. The RhoGEF GEF-H1 is required for oncogenic RAS signaling via KSR-1. *Cancer Cell* **25:** 181–195.

Davidson SM, Papagiannakopoulos T, Olenchock BA, Heyman JE, Keibler MA, Luengo A, Bauer MR, Jha AK, O'Brien JP, Pierce KA, et al. 2016. Environment impacts the metabolic dependencies of Ras-driven non-small-cell lung cancer. *Cell Metab* **23:** 517–528.

Davidson SM, Jonas O, Keibler MA, Hou HW, Luengo A, Mayers JR, Wyckoff J, Del Rosario AM, Whitman M, Chin CR, et al. 2017. Direct evidence for cancer-cell-autonomous extracellular protein catabolism in pancreatic tumors. *Nat Med* **23:** 235–241.

DeNicola GM, Karreth FA, Humpton TJ, Gopinathan A, Wei C, Frese K, Mangal D, Yu KH, Yeo CJ, Calhoun ES, et al. 2011. Oncogene-induced Nrf2 transcription promotes ROS detoxification and tumorigenesis. *Nature* **475:** 106–109.

Denoyelle C, Abou-Rjaily G, Bezrookove V, Verhaegen M, Johnson TM, Fullen DR, Pointer JN, Gruber SB, Su LD, Nikiforov MA, et al. 2006. Anti-oncogenic role of the endoplasmic reticulum differentially activated by mutations in the MAPK pathway. *Nat Cell Biol* **8:** 1053–1063.

De Raedt T, Walton Z, Yecies JL, Li D, Chen Y, Malone CF, Maertens O, Jeong SM, Bronson RT, Lebleu V, et al. 2011. Exploiting cancer cell vulnerabilities to develop a combination therapy for ras-driven tumors. *Cancer Cell* **20:** 400–413.

Dietlein F, Kalb B, Jokic M, Noll EM, Strong A, Tharun L, Ozretic L, Kunstlinger H, Kambartel K, Randerath WJ, et al. 2015. A synergistic interaction between Chk1- and MK2 inhibitors in KRAS-mutant cancer. *Cell* **162:** 146–159.

Dolma S, Lessnick SL, Hahn WC, Stockwell BR. 2003. Identification of genotype-selective antitumor agents using synthetic lethal chemical screening in engineered human tumor cells. *Cancer Cell* **3:** 285–296.

Downward J. 2015. RAS synthetic lethal screens revisited: Still seeking the elusive prize? *Clin Cancer Res* **21:** 1802–1809.

Ebi H, Corcoran RB, Singh A, Chen Z, Song Y, Lifshits E, Ryan DP, Meyerhardt JA, Benes C, Settleman J, et al. 2011. Receptor tyrosine kinases exert dominant control over PI3K signaling in human KRAS mutant colorectal cancers. *J Clin Invest* **121:** 4311–4321.

Ebi H, Faber AC, Engelman JA, Yano S. 2014. Not just gRASping at flaws: Finding vulnerabilities to develop novel therapies for treating KRAS mutant cancers. *Cancer Sci* **105:** 499–505.

Engelman JA, Chen L, Tan X, Crosby K, Guimaraes AR, Upadhyay R, Maira M, McNamara K, Perera SA, Song Y, et al. 2008. Effective use of PI3K and MEK inhibitors to treat mutant Kras G12D and PIK3CA H1047R murine lung cancers. *Nat Med* **14:** 1351–1356.

Eser S, Reiff N, Messer M, Seidler B, Gottschalk K, Dobler M, Hieber M, Arbeiter A, Klein S, Kong B, et al. 2013. Selective requirement of PI3K/PDK1 signaling for Kras oncogene-driven pancreatic cell plasticity and cancer. *Cancer cell* **23:** 406–420.

Faber AC, Coffee EM, Costa C, Dastur A, Ebi H, Hata AN, Yeo AT, Edelman EJ, Song Y, Tam AT, et al. 2014. mTOR inhibition specifically sensitizes colorectal cancers with KRAS or BRAF mutations to BCL-2/BCL-XL inhibition by suppressing MCL-1. *Cancer Discov* **4:** 42–52.

Farmer H, McCabe N, Lord CJ, Tutt AN, Johnson DA, Richardson TB, Santarosa M, Dillon KJ, Hickson I, Knights C, et al. 2005. Targeting the DNA repair defect in BRCA mutant cells as a therapeutic strategy. *Nature* **434:** 917–921.

Frohling S, Scholl C. 2011. STK33 kinase is not essential in KRAS-dependent cells—Letter. *Cancer Res* **71:** 7716; author reply 7717.

Fukuda A, Wang SC, Morris JPt, Folias AE, Liou A, Kim GE, Akira S, Boucher KM, Firpo MA, Mulvihill SJ, et al. 2011. Stat3 and MMP7 contribute to pancreatic ductal adenocarcinoma initiation and progression. *Cancer Cell* **19:** 441–455.

Garnett MJ, Edelman EJ, Heidorn SJ, Greenman CD, Dastur A, Lau KW, Greninger P, Thompson IR, Luo X, Soares J, et al. 2012. Systematic identification of genomic markers of drug sensitivity in cancer cells. *Nature* **483:** 570–575.

Genovese G, Carugo A, Tepper J, Robinson FS, Li L, Svelto M, Nezi L, Corti D, Minelli R, Pettazzoni P, et al. 2017. Synthetic vulnerabilities of mesenchymal subpopulations in pancreatic cancer. *Nature* **542:** 362–366.

Gilbert LA, Horlbeck MA, Adamson B, Villalta JE, Chen Y, Whitehead EH, Guimaraes C, Panning B, Ploegh HL, Bassik MC, et al. 2014. Genome-scale CRISPR-mediated control of gene repression and activation. *Cell* **159:** 647–661.

Grabocka E, Pylayeva-Gupta Y, Jones MJ, Lubkov V, Yemanaberhan E, Taylor L, Jeng HH, Bar-Sagi D. 2014. Wild-type H- and N-Ras promote mutant K-Ras-driven tumorigenesis by modulating the DNA damage response. *Cancer cell* **25:** 243–256.

Grabocka E, Commisso C, Bar-Sagi D. 2015. Molecular pathways: Targeting the dependence of mutant RAS cancers on the DNA damage response. *Clin Cancer Res* **21:** 1243–1247.

Guo JY, Chen HY, Mathew R, Fan J, Strohecker AM, Karsli-Uzunbas G, Kamphorst JJ, Chen G, Lemons JM, Karantza V, et al. 2011. Activated Ras requires autophagy to maintain oxidative metabolism and tumorigenesis. *Genes Dev* **25:** 460–470.

Hamad NM, Elconin JH, Karnoub AE, Bai W, Rich JN, Abraham RT, Der CJ, Counter CM. 2002. Distinct requirements for Ras oncogenesis in human versus mouse cells. *Genes Dev* **16:** 2045–2057.

Han K, Jeng EE, Hess GT, Morgens DW, Li A, Bassik MC. 2017. Synergistic drug combinations for cancer identified in a CRISPR screen for pairwise genetic interactions. *Nat Biotechnol* **35:** 463–474.

Hart T, Chandrashekhar M, Aregger M, Steinhart Z, Brown KR, MacLeod G, Mis M, Zimmermann M, Fradet-Turcotte A, Sun S, et al. 2015. High-resolution CRISPR screens reveal fitness genes and genotype-specific cancer liabilities. *Cell* **163:** 1515–1526.

Hayes TK, Neel NF, Hu C, Gautam P, Chenard M, Long B, Aziz M, Kassner M, Bryant KL, Pierobon M, et al. 2016. Long-term ERK inhibition in KRAS-mutant pancreatic cancer is associated with MYC degradation and senescence-like growth suppression. *Cancer Cell* **29:** 75–89.

Hofer F, Fields S, Schneider C, Martin GS. 1994. Activated Ras interacts with the Ral guanine nucleotide dissociation stimulator. *Proc Natl Acad Sci* **91:** 11089–11093.

Infante JR, Fecher LA, Falchook GS, Nallapareddy S, Gordon MS, Becerra C, DeMarini DJ, Cox DS, Xu Y, Morris SR, et al. 2012. Safety, pharmacokinetic, pharmacodynamic, and efficacy data for the oral MEK inhibitor trametinib: A phase 1 dose-escalation trial. *Lancet Oncol* **13:** 773–781.

Infante JR, Papadopoulos KP, Bendell JC, Patnaik A, Burris HA III, Rasco D, Jones SF, Smith L, Cox DS, Durante M, et al. 2013. A phase 1b study of trametinib, an oral Mitogen-activated protein kinase kinase (MEK) inhibitor, in combination with gemcitabine in advanced solid tumours. *Eur J Cancer* **49:** 2077–2085.

Kaelin WG Jr. 2005. The concept of synthetic lethality in the context of anticancer therapy. *Nat Rev Cancer* **5:** 689–698.

Kapoor A, Yao W, Ying H, Hua S, Liewen A, Wang Q, Zhong Y, Wu CJ, Sadanandam A, Hu B, et al. 2014. Yap1 activation enables bypass of oncogenic Kras addiction in pancreatic cancer. *Cell* **158:** 185–197.

Karreth FA, Frese KK, DeNicola GM, Baccarini M, Tuveson DA. 2011. C-Raf is required for the initiation of lung cancer by K-Ras^{G12D}. *Cancer Discov* **1:** 128–136.

Kikuchi A, Demo SD, Ye ZH, Chen YW, Williams LT. 1994. ralGDS family members interact with the effector loop of ras p21. *Mol Cell Biol* **14:** 7483–7491.

Kim HS, Mendiratta S, Kim J, Pecot CV, Larsen JE, Zubovych I, Seo BY, Kim J, Eskiocak B, Chung H, et al. 2013. Systematic identification of molecular subtype-selective vulnerabilities in non-small-cell lung cancer. *Cell* **155:** 552–566.

Kim J, McMillan E, Kim HS, Venkateswaran N, Makkar G, Rodriguez-Canales J, Villalobos P, Neggers JE, Mendiratta S, Wei S, et al. 2016. XPO1-dependent nuclear export is a druggable vulnerability in KRAS-mutant lung cancer. *Nature* **538:** 114–117.

Kitajima S, Thummalapalli R, Barbie DA. 2016. Inflammation as a driver and vulnerability of KRAS mediated oncogenesis. *Semin Cell Dev Biol* **58:** 127–135.

Koike-Yusa H, Li Y, Tan EP, Velasco-Herrera Mdel C, Yusa K. 2014. Genome-wide recessive genetic screening in mammalian cells with a lentiviral CRISPR-guide RNA library. *Nat Biotechnol* **32:** 267–273.

Kumar MS, Hancock DC, Molina-Arcas M, Steckel M, East P, Diefenbacher M, Armenteros-Monterroso E, Lassailly F, Matthews N, Nye E, et al. 2012. The GATA2 transcriptional network is requisite for RAS oncogene-driven non-small-cell lung cancer. *Cell* **149:** 642–655.

Lamba S, Russo M, Sun C, Lazzari L, Cancelliere C, Grernrum W, Lieftink C, Bernards R, Di Nicolantonio F, Bardelli A. 2014. RAF suppression synergizes with MEK inhibition in KRAS mutant cancer cells. *Cell Rep* **8:** 1475–1483.

Lee HJ, Zhuang G, Cao Y, Du P, Kim HJ, Settleman J. 2014. Drug resistance via feedback activation of Stat3 in oncogene-addicted cancer cells. *Cancer Cell* **26:** 207–221.

Lesina M, Kurkowski MU, Ludes K, Rose-John S, Treiber M, Kloppel G, Yoshimura A, Reindl W, Sipos B, Akira S, et al. 2011. Stat3/Socs3 activation by IL-6 transsignaling promotes progression of pancreatic intraepithelial neoplasia and development of pancreatic cancer. *Cancer Cell* **19:** 456–469.

Lim KH, Baines AT, Fiordalisi JJ, Shipitsin M, Feig LA, Cox AD, Der CJ, Counter CM. 2005. Activation of RalA is critical for Ras-induced tumorigenesis of human cells. *Cancer Cell* **7:** 533–545.

Lin L, Sabnis AJ, Chan E, Olivas V, Cade L, Pazarentzos E, Asthana S, Neel D, Yan JJ, Lu X, et al. 2015. The Hippo effector YAP promotes resistance to RAF- and MEK-targeted cancer therapies. *Nat Genet* **47:** 250–256.

Ling J, Kang Y, Zhao R, Xia Q, Lee DF, Chang Z, Li J, Peng B, Fleming JB, Wang H, et al. 2012. KrasG12D-induced IKK2/β/NF-κB activation by IL-1α and p62 feedforward loops is required for development of pancreatic ductal adenocarcinoma. *Cancer Cell* **21:** 105–120.

Lock R, Roy S, Kenific CM, Su JS, Salas E, Ronen SM, Debnath J. 2011. Autophagy facilitates glycolysis during Ras-mediated oncogenic transformation. *Mol Biol Cell* **22:** 165–178.

Lord CJ, Ashworth A. 2017. PARP inhibitors: Synthetic lethality in the clinic. *Science* **355:** 1152–1158.

Luo B, Cheung HW, Subramanian A, Sharifnia T, Okamoto M, Yang X, Hinkle G, Boehm JS, Beroukhim R, Weir BA, et al. 2008. Highly parallel identification of essential genes in cancer cells. *Proc Natl Acad Sci* **105:** 20380–20385.

Luo J, Emanuele MJ, Li D, Creighton CJ, Schlabach MR, Westbrook TF, Wong KK, Elledge SJ. 2009a. A genome-wide RNAi screen identifies multiple synthetic lethal interactions with the Ras oncogene. *Cell* **137:** 835–848.

Luo J, Solimini NL, Elledge SJ. 2009b. Principles of cancer therapy: Oncogene and non-oncogene addiction. *Cell* **136:** 823–837.

Luo T, Masson K, Jaffe JD, Silkworth W, Ross NT, Scherer CA, Scholl C, Frohling S, Carr SA, Stern AM, et al. 2012. STK33 kinase inhibitor BRD-8899 has no effect on KRAS-dependent cancer cell viability. *Proc Natl Acad Sci* **109:** 2860–2865.

Manchado E, Weissmueller S, Morris JPt, Chen CC, Wullenkord R, Lujambio A, de Stanchina E, Poirier JT, Gainor JF, Corcoran RB, et al. 2016. A combinatorial strategy for treating KRAS-mutant lung cancer. *Nature* **534:** 647–651.

Marcotte R, Brown KR, Suarez F, Sayad A, Karamboulas K, Krzyzanowski PM, Sircoulomb F, Medrano M, Fedyshyn Y, Koh JL, et al. 2012. Essential gene profiles in breast, pancreatic, and ovarian cancer cells. *Cancer Discov* **2:** 172–189.

Marcotte R, Sayad A, Brown KR, Sanchez-Garcia F, Reimand J, Haider M, Virtanen C, Bradner JE, Bader GD, Mills GB, et al. 2016. Functional genomic landscape of human breast cancer drivers, vulnerabilities, and resistance. *Cell* **164:** 293–309.

Mayers JR, Torrence ME, Danai LV, Papagiannakopoulos T, Davidson SM, Bauer MR, Lau AN, Ji BW, Dixit PD, Hosios AM, et al. 2016. Tissue of origin dictates branched-chain amino acid metabolism in mutant Kras-driven cancers. *Science* **353:** 1161–1165.

McLornan DP, List A, Mufti GJ. 2014. Applying synthetic lethality for the selective targeting of cancer. *N Engl J Med* **371:** 1725–1735.

Meylan E, Dooley AL, Feldser DM, Shen L, Turk E, Ouyang C, Jacks T. 2009. Requirement for NF-κB signalling in a mouse model of lung adenocarcinoma. *Nature* **462:** 104–107.

Molina-Arcas M, Hancock DC, Sheridan C, Kumar MS, Downward J. 2013. Coordinate direct input of both KRAS and IGF1 receptor to activation of PI3 kinase in KRAS-mutant lung cancer. *Cancer Discov* **3:** 548–563.

Moodie SA, Willumsen BM, Weber MJ, Wolfman A. 1993. Complexes of Ras.GTP with Raf-1 and mitogen-activated protein kinase kinase. *Science* **260:** 1658–1661.

Morgan-Lappe SE, Tucker LA, Huang X, Zhang Q, Sarthy AV, Zakula D, Vernetti L, Schurdak M, Wang J, Fesik SW. 2007. Identification of Ras-related nuclear protein, targeting protein for xenopus kinesin-like protein 2, and stearoyl-CoA desaturase 1 as promising cancer targets from an RNAi-based screen. *Cancer Res* **67:** 4390–4398.

Ngo VN, Davis RE, Lamy L, Yu X, Zhao H, Lenz G, Lam LT, Dave S, Yang L, Powell J, et al. 2006. A loss-of-function RNA interference screen for molecular targets in cancer. *Nature* **441:** 106–110.

Papke B, Der CJ. 2017. Drugging RAS: Know the enemy. *Science* **355:** 1158–1163.

Parnas O, Jovanovic M, Eisenhaure TM, Herbst RH, Dixit A, Ye CJ, Przybylski D, Platt RJ, Tirosh I, Sanjana NE, et al. 2015. A genome-wide CRISPR screen in primary immune cells to dissect regulatory networks. *Cell* **162:** 675–686.

Podsypanina K, Politi K, Beverly LJ, Varmus HE. 2008. Oncogene cooperation in tumor maintenance and tumor recurrence in mouse mammary tumors induced by Myc and mutant Kras. *Proc Natl Acad Sci* **105:** 5242–5247.

Prahallad A, Heynen GJ, Germano G, Willems SM, Evers B, Vecchione L, Gambino V, Lieftink C, Beijersbergen RL, Di Nicolantonio F, et al. 2015. PTPN11 Is a central node in intrinsic and acquired resistance to targeted cancer drugs. *Cell Rep* **12:** 1978–1985.

Puyol M, Martin A, Dubus P, Mulero F, Pizcueta P, Khan G, Guerra C, Santamaria D, Barbacid M. 2010. A synthetic lethal interaction between K-Ras oncogenes and Cdk4 unveils a therapeutic strategy for non-small-cell lung carcinoma. *Cancer Cell* **18:** 63–73.

Pylayeva-Gupta Y, Grabocka E, Bar-Sagi D. 2011. RAS oncogenes: Weaving a tumorigenic web. *Nat Rev Cancer* **11:** 761–774.

Rodriguez-Viciana P, Warne PH, Dhand R, Vanhaesebroeck B, Gout I, Fry MJ, Waterfield MD, Downward J. 1994. Phosphatidylinositol-3-OH kinase as a direct target of Ras. *Nature* **370:** 527–532.

Rosenbluh J, Mercer J, Shrestha Y, Oliver R, Tamayo P, Doench JG, Tirosh I, Piccioni F, Hartenian E, Horn H, et al. 2016. Genetic and proteomic interrogation of lower confidence candidate genes reveals signaling networks in β-catenin-active cancers. *Cell Syst* **3:** 302–316 e304.

Ryan MB, Der CJ, Wang-Gillam A, Cox AD. 2015. Targeting RAS-mutant cancers: Is ERK the key? *Trends Cancer* **1:** 183–198.

Sachs N, Clevers H. 2014. Organoid cultures for the analysis of cancer phenotypes. *Curr Opin Genet Dev* **24:** 68–73.

Sarthy AV, Morgan-Lappe SE, Zakula D, Vernetti L, Schurdak M, Packer JC, Anderson MG, Shirasawa S, Sasazuki T, Fesik SW. 2007. Survivin depletion preferentially reduces the survival of activated K-Ras-transformed cells. *Mol Cancer Therapeut* **6:** 269–276.

Scholl C, Frohling S, Dunn IF, Schinzel AC, Barbie DA, Kim SY, Silver SJ, Tamayo P, Wadlow RC, Ramaswamy S, et al. 2009. Synthetic lethal interaction between oncogenic KRAS dependency and STK33 suppression in human cancer cells. *Cell* **137:** 821–834.

Shalem O, Sanjana NE, Hartenian E, Shi X, Scott DA, Mikkelsen TS, Heckl D, Ebert BL, Root DE, Doench JG, et al. 2014. Genome-scale CRISPR-Cas9 knockout screening in human cells. *Science* **343:** 84–87.

Shalem O, Sanjana NE, Zhang F. 2015. High-throughput functional genomics using CRISPR-Cas9. *Nat Rev Genet* **16:** 299–311.

Shao DD, Tsherniak A, Gopal S, Weir BA, Tamayo P, Stransky N, Schumacher SE, Zack TI, Beroukhim R, Garraway LA, et al. 2013. ATARiS: Computational quantification of gene suppression phenotypes from multisample RNAi screens. *Genome Res* **23:** 665–678.

Shao DD, Xue W, Krall EB, Bhutkar A, Piccioni F, Wang X, Schinzel AC, Sood S, Rosenbluh J, Kim JW, et al. 2014. KRAS and YAP1 converge to regulate EMT and tumor survival. *Cell* **158:** 171–184.

Shaw AT, Winslow MM, Magendantz M, Ouyang C, Dowdle J, Subramanian A, Lewis TA, Maglathin RL, Tolliday N, Jacks T. 2011. Selective killing of K-ras mutant cancer cells by small molecule inducers of oxidative stress. *Proc Natl Acad Sci* **108:** 8773–8778.

Singh A, Greninger P, Rhodes D, Koopman L, Violette S, Bardeesy N, Settleman J. 2009. A gene expression signature associated with "K-Ras addiction" reveals regulators of EMT and tumor cell survival. *Cancer Cell* **15:** 489–500.

Singh A, Sweeney MF, Yu M, Burger A, Greninger P, Benes C, Haber DA, Settleman J. 2012. TAK1 inhibition promotes apoptosis in KRAS-dependent colon cancers. *Cell* **148:** 639–650.

Sjolander A, Yamamoto K, Huber BE, Lapetina EG. 1991. Association of p21ras with phosphatidylinositol 3-kinase. *Proc Natl Acad Sci* **88:** 7908–7912.

Son J, Lyssiotis CA, Ying H, Wang X, Hua S, Ligorio M, Perera RM, Ferrone CR, Mullarky E, Shyh-Chang N, et al. 2013. Glutamine supports pancreatic cancer growth through a KRAS-regulated metabolic pathway. *Nature* **496:** 101–105.

Spaargaren M, Bischoff JR. 1994. Identification of the guanine nucleotide dissociation stimulator for Ral as a putative effector molecule of R-ras, H-ras, K-ras, and Rap. *Proc Natl Acad Sci* **91:** 12609–12613.

Steckel M, Molina-Arcas M, Weigelt B, Marani M, Warne PH, Kuznetsov H, Kelly G, Saunders B, Howell M, Downward J, et al. 2012. Determination of synthetic lethal interactions in KRAS oncogene-dependent cancer cells reveals novel therapeutic targeting strategies. *Cell Res* **22:** 1227–1245.

Stephen AG, Esposito D, Bagni RK, McCormick F. 2014. Dragging Ras back in the ring. *Cancer Cell* **25:** 272–281.

Tsherniak A, Vazquez F, Montgomery PG, Weir BA, Kryukov G, Cowley GS, Gill S, Harrington WF, Pantel S, Krill-Burger JM, et al. 2017. Defining a cancer dependency map. *Cell* **170:** 564-576.e16.

Viale A, Pettazzoni P, Lyssiotis CA, Ying H, Sanchez N, Marchesini M, Carugo A, Green T, Seth S, Giuliani V, et al. 2014. Oncogene ablation-resistant pancreatic cancer cells depend on mitochondrial function. *Nature* **514:** 628–632.

Vicent S, Chen R, Sayles LC, Lin C, Walker RG, Gillespie AK, Subramanian A, Hinkle G, Yang X, Saif S, et al. 2010. Wilms tumor 1 (WT1) regulates KRAS-driven oncogenesis and senescence in mouse and human models. *J Clin Invest* **120:** 3940–3952.

Vojtek AB, Hollenberg SM, Cooper JA. 1993. Mammalian Ras interacts directly with the serine/threonine kinase Raf. *Cell* **74:** 205–214.

Wang Y, Ngo VN, Marani M, Yang Y, Wright G, Staudt LM, Downward J. 2010. Critical role for transcriptional repressor Snail2 in transformation by oncogenic RAS in colorectal carcinoma cells. *Oncogene* **29:** 4658–4670.

Wang T, Wei JJ, Sabatini DM, Lander ES. 2014. Genetic screens in human cells using the CRISPR-Cas9 system. *Science* **343:** 80–84.

Wang T, Birsoy K, Hughes NW, Krupczak KM, Post Y, Wei JJ, Lander ES, Sabatini DM. 2015. Identification and characterization of essential genes in the human genome. *Science* **350:** 1096–1101.

Wang T, Yu H, Hughes NW, Liu B, Kendirli A, Klein K, Chen WW, Lander ES, Sabatini DM. 2017. Gene essentiality profiling reveals gene networks and synthetic lethal interactions with oncogenic Ras. *Cell* **168:** 890–903 e815.

Warne PH, Viciana PR, Downward J. 1993. Direct interaction of Ras and the amino-terminal region of Raf-1 in vitro. *Nature* **364:** 352–355.

Winter GE, Buckley DL, Paulk J, Roberts JM, Souza A, Dhe-Paganon S, Bradner JE. 2015. DRUG DEVELOPMENT. Phthalimide conjugation as a strategy for in vivo target protein degradation. *Science* **348:** 1376–1381.

Yagoda N, von Rechenberg M, Zaganjor E, Bauer AJ, Yang WS, Fridman DJ, Wolpaw AJ, Smukste I, Peltier JM, Boniface JJ, et al. 2007. RAS-RAF-MEK-dependent oxidative cell death involving voltage-dependent anion channels. *Nature* **447:** 864–868.

Yan C, Liu D, Li L, Wempe MF, Guin S, Khanna M, Meier J, Hoffman B, Owens C, Wysoczynski CL, et al. 2014. Discovery and characterization of small molecules that target the GTPase Ral. *Nature* **515:** 443–447.

Yang WS, Stockwell BR. 2008. Synthetic lethal screening identifies compounds activating iron-dependent, nonapoptotic cell death in oncogenic-RAS-harboring cancer cells. *Chem Biol* **15:** 234–245.

Yang S, Wang X, Contino G, Liesa M, Sahin E, Ying H, Bause A, Li Y, Stommel JM, Dell'antonio G, et al. 2011. Pancreatic cancers require autophagy for tumor growth. *Genes Dev* **25:** 717–729.

Yang A, Rajeshkumar NV, Wang X, Yabuuchi S, Alexander BM, Chu GC, Von Hoff DD, Maitra A, Kimmelman AC. 2014. Autophagy is critical for pancreatic tumor growth and progression in tumors with p53 alterations. *Cancer Discov* **4:** 905–913.

Ying H, Kimmelman AC, Lyssiotis CA, Hua S, Chu GC, Fletcher-Sananikone E, Locasale JW, Son J, Zhang H, Coloff JL, et al. 2012. Oncogenic Kras maintains pancreatic tumors through regulation of anabolic glucose metabolism. *Cell* **149:** 656–670.

Yuan TL, Fellmann C, Lee CS, Ritchie CD, Thapar V, Lee LC, Hsu DJ, Grace D, Carver JO, Zuber J, et al. 2014. Development of siRNA payloads to target KRAS-mutant cancer. *Cancer Discov* **4:** 1182–1197.

Yun J, Mullarky E, Lu C, Bosch KN, Kavalier A, Rivera K, Roper J, Chio II, Giannopoulou EG, Rago C, et al. 2015. Vitamin C selectively kills KRAS and BRAF mutant colorectal cancer cells by targeting GAPDH. *Science* **350:** 1391–1396.

Zender L, Xue W, Zuber J, Semighini CP, Krasnitz A, Ma B, Zender P, Kubicka S, Luk JM, Schirmacher P, et al. 2008. An oncogenomics-based in vivo RNAi screen identifies tumor suppressors in liver cancer. *Cell* **135:** 852–864.

Zhang XF, Settleman J, Kyriakis JM, Takeuchi-Suzuki E, Elledge SJ, Marshall MS, Bruder JT, Rapp UR, Avruch J. 1993. Normal and oncogenic p21ras proteins bind to the amino-terminal regulatory domain of c-Raf-1. *Nature* **364:** 308–313.

Zhang W, Nandakumar N, Shi Y, Manzano M, Smith A, Graham G, Gupta S, Vietsch EE, Laughlin SZ, Wadhwa M, et al. 2014. Downstream of mutant KRAS, the transcription regulator YAP is essential for neoplastic progression to pancreatic ductal adenocarcinoma. *Sci Signal* **7:** ra42.

Zhao Y, Adjei AA. 2014. The clinical development of MEK inhibitors. *Nat Rev Clin Oncol* **11:** 385–400.

Zhu Z, Aref AR, Cohoon TJ, Barbie TU, Imamura Y, Yang S, Moody SE, Shen RR, Schinzel AC, Thai TC, et al. 2014. Inhibition of KRAS-driven tumorigenicity by interruption of an autocrine cytokine circuit. *Cancer Discov* **4:** 452–465.

Zuber J, Shi J, Wang E, Rappaport AR, Herrmann H, Sison EA, Magoon D, Qi J, Blatt K, Wunderlich M, et al. 2011. RNAi screen identifies Brd4 as a therapeutic target in acute myeloid leukaemia. *Nature* **478:** 524–528.

Efforts to Develop KRAS Inhibitors

Matthew Holderfield

NCI-Ras Initiative, Cancer Research Technology Program, Frederick National Laboratory for Cancer Research, Leidos Biomedical Research, Frederick, Maryland 21702

Correspondence: matthew.holderfield@nih.gov

The high prevalence of KRAS mutations in human cancers and the lack of effective treatments for patients ranks KRAS among the most highly sought-after targets for preclinical oncologists. Pharmaceutical companies and academic laboratories have tried for decades to identify small molecule inhibitors of oncogenic KRAS proteins, but little progress has been made and many have labeled KRAS undruggable. However, recent progress in in silico screening, fragment-based drug design, disulfide tethered screening, and some emerging themes in RAS biology have caused the field to reconsider previously held notions about targeting KRAS. This review will cover some of the historical efforts to identify RAS inhibitors, and some of the most promising efforts currently being pursued.

The age of whole-genome sequencing has paved the road for "personalized" medicine in oncology. The Cancer Genome Atlas Research Network (2017), the Cancer Cell Line Encyclopedia (Barretina et al. 2012), the Genomics of Drug Sensitivity in Cancer (Yang et al. 2013), and similar efforts aim to catalog driver mutations with actionable and effective treatment regimens. While significant progress has been made in some notable areas, including BRAF-mutated melanomas (Davies et al. 2002; Chapman et al. 2011) and BRCA-mutated breast and ovarian cancers (Fong et al. 2009; The Cancer Genome Atlas Research Network 2011), many of the most common oncogenes were identified decades ago and remain the most deadly and elusive oncology drug targets. Perhaps chief among these is KRAS.

Interest in KRAS as a cancer target stems from work with rodent sarcoma viruses in the late 1970s. First, DNAs from viruses and later from human tumor cells were shown to be sufficient to transform NIH-3T3 cells in culture (Shih et al. 1979; Shih et al. 1981). It was not until the early 1980s that KRAS was identified as the causative oncogene (Der et al. 1982; Parada and Weinberg 1983). By 1987, KRAS oncogenes were cloned from multiple tumor biopsies of lung, colon, and pancreas cancers (Santos et al. 1984; Bos et al. 1987; Forrester et al. 1987; Rodenhuis et al. 1987; Almoguera et al. 1988). By this time, all three RAS isoforms, H-, N-, and K-RAS were known and mutations at codons 12, 13, and 61 of a RAS isoform were shown to be sufficient to induce cell-cycle progression. Finally, the HRAS crystal structure revealed that the cancer-associated mutations at codons G12, G13, and Q61 are each located at a site of guanosine triphosphate (GTP) binding, suggesting a common mechanism of oncogenesis (Pai et al. 1989).

Cite this article as *Cold Spring Harb Perspect Med* doi: 10.1101/cshperspect.a031864

It was later determined that RAS proteins are prenylated small GTPases. H-, N-, and K-RAS are all farnesylated at the carboxyl terminus. H- and N-RAS as well as one of the two KRAS splice variants (KRAS 4A) are also palmitoylated, whereas KRAS 4B contains a series of charged lysine residues near the carboxyl terminus that aid in membrane localization (Hancock 2003). Fully processed RAS proteins localize to the plasma membrane where they are activated upon growth factor stimulus. Guanosine exchange factors (GEFs) promote RAS–GTP association to offset the high rate of GTP hydrolysis in cells. Hydrolysis is stimulated by GTPase-activating proteins (GAPs) (Trahey and McCormick 1987). Among the most well-characterized RAS GAP proteins are p120 and NF1, which increase GTP hydrolysis >10,000-fold over the intrinsic rate. GAP proteins all contain a catalytic arginine residue that fits into the active site of RAS to stimulate hydrolysis of the terminal GTP phosphate. Mutations at RAS G12, G13, or Q61 prevent the GAP arginine from accessing GTP, which stabilizes RAS in the GTP-bound state (Scheffzek et al. 1997).

In the GTP-bound form, RAS proteins associate with and activate a variety of effector molecules including phosphatidylinositol-3 kinase (PI3K) and the RAF kinases. RAF kinases activate the mitogen-activated protein kinase (MAPK) pathway by phosphorylating MEK1 and MEK2, which in turn phosphorylate ERK1 and ERK2, which phosphorylate and activate many transcription factors that promote cell-cycle progression. Alternatively, PI3K activates AKT and mechanistic target of rapamycin (mTOR) signaling to promote protein translation. Unlike RAF kinases, PI3K can be activated independent of RAS by receptor tyrosine kinases (RTKs) or indirectly through G-protein-coupled receptor (GPCR) and integrin signaling, but both effector pathways are strongly implicated in cancer progression and have been evaluated as therapeutic targets for KRAS-mutated cancers. Genetically engineered mouse models demonstrate that KRAS G12D-driven non-small-cell lung cancer (NSCLC) requires a single RAS isoform, CRAF, as well as the RAF client proteins, MEK1/MEK2, and the MEK substrates, ERK1/ERK2 (Blasco et al. 2011). Likewise, KRAS activation of PI3K is also essential in a similar NSCLC mouse model (Castellano et al. 2013). Unfortunately, KRAS-mutated cancers and many of the available cell lines have proven unresponsive to MEK, RAF, or PI3K inhibitors administered as single agents (Barretina et al. 2012). Therefore, cotargeting the MAPK and PI3K pathways may be required to treat KRAS-driven cancers (Engelman et al. 2008). Unfortunately, coadministration of PI3K and MEK inhibitors is not well tolerated. Phase I clinical trials reported grade 3 and 4 dose-limiting toxicities, which is thought to limit clinical efficacy (Bedard et al. 2015). Unless dosing regimens can be modified to reduce general toxicity, cotargeting MAPK and PI3K does not appear to be a viable therapeutic option, which further illustrates the unmet clinical need for effective treatment strategies for KRAS-mutated cancers. This has, in part, nucleated a resurgence for basic science and drug discovery efforts to target oncogenic KRAS proteins and to identify new targets required for KRAS-driven oncogenesis.

DIRECT TARGETING OF KRAS PROTEINS

The difficulties drugging RAS stem from the small size of the protein, just 21 kD, with a lack of an apparent "pocket" for a small molecule to bind apart from the nucleotide substrate-binding site. However, the high picomolar affinity reported for nucleotide substrates paired with the high millimolar cellular concentration of GTP suggests that this strategy is thermodynamically and pharmacologically unattainable. As such, RAS proteins are commonly referred to as "undruggable," but a renewed interest in targeting KRAS may change this paradigm (Stephen et al. 2014). This review will focus on many of the strategies to target KRAS in the past, as well as some promising efforts that are currently being pursued.

Targeting Nucleotide Exchange

SCH54292 was described in the first report of a small molecule that binds directly to RAS (Taveras et al. 1997). The series was developed from a biochemical screen designed to identify inhibitors of nucleotide exchange, and nuclear magnetic resonance (NMR) studies showed that it interacts with the Switch II domain of RAS in the guanosine diphosphate (GDP)-bound conformation. Similar studies have since been performed by other groups using different nucleotide exchange assays, but all nucleotide exchange inhibitors typically suffer from low affinity for RAS (Cox et al. 2014) and often lack activity in cells. None of the compounds identified have been developed for clinical use.

More recently, three groups have used NMR to identify small molecule fragments that bind to RAS (Maurer et al. 2012; Sun et al. 2014; Winter et al. 2015). This strategy seeks to identify novel compound-binding pockets using low molecular weight chemical matter and to improve affinity by increasing the inhibitor size through a rational structure-based drug design or by chemically linking multiple fragments with low affinity for the target protein. Many of the chemical fragments identified bind to RAS in a shallow pocket near the α2 helix, which is also the site for SOS binding (Boriack-Sjodin et al. 1998). Indeed, the fragments inhibit SOS binding and SOS-dependent nucleotide exchange in vitro. In theory, this method could be used to develop a small molecule with high enough affinity for RAS to inhibit activity in cells. The α2 compound binding site is not apparent in the apocrystal structures, and it is not clear to what extent the pocket can be exploited with a larger compound. It is also not clear whether such a strategy will have activity in KRAS-mutated cells because KRAS oncoproteins are not affected by epidermal growth factor receptor (EGFR) or other RTK inhibitors (Khambata-Ford et al. 2007) and are presumably not responsive to upstream perturbations, including inhibition of SOS activity.

Targeting RAS–Effector Interactions

Perhaps a more direct mechanism of action for a RAS inhibitor might be to inhibit RAS–effector interactions. Protein–protein interactions are not always easy to target as proteins often interact through multiple residues along flat surfaces, which makes inhibition thermodynamically unfavorable with a small molecule. However, a single-point mutation in the RAS-binding domain of RAF kinase or RAL-GDS is sufficient to abolish RAS-RAF or RAS-RAL-GDS binding (Fabian et al. 1994), suggesting that even a minor disruption of the RAS-effector binding site may prevent binding. Thus, assays that measure KRAS–effector interactions in vitro and in cell systems have been used to screen many compound libraries. One such compound, MCP1, was identified in an HRAS-CRAF yeast two-hybrid screen, and demonstrated activity in NRAS-mutated HT1080, KRAS-mutated PANC-1, and A549 cells in the 10- to 20-μM range, but no inhibition of BRAF V600E-mutated A2058 cells was observed (Kato-Stankiewicz et al. 2002). It is not clear whether MCP1 binds directly to RAS or CRAF, and the compound series is not potent enough for therapeutic activity. However, the proposed mechanism of action remains a viable approach for inhibiting oncogenic KRAS signal transduction.

Peptide chemistries are often used to increase compound library diversity and they are well suited to target protein–protein interactions. A large combinatorial library of cyclic peptides was screened for the ability to inhibit KRAS-RAF1 binding in vitro. The cyclic peptide, cyclorasin, was shown to bind RAS only in the GTP-bound state and compete directly with RAS effectors (Wu et al. 2013). The peptide was further optimized for cell permeability, and the authors report submicromolar IC50s in KRAS-mutated cancer cell lines (Trinh et al. 2016). However, the large molecular weight of the compounds may present significant pharmacokinetic challenges for clinical development.

In Silico Screening

Computer "docking" studies offer a cost-effective alternative to wet-bench screening efforts to identify potential chemical leads for difficult targets. Computational methods use structural information obtained from crystallography and NMR to virtually dock small molecules into putative binding pockets. Molecules predicted to dock are then synthesized and tested in functional biological assays. Although in silico docking does not always correlate perfectly with biological activity, in silico screens can start with a large virtual library and enrich for compounds likely to bind to the target of interest. The library size is limited only by computing time needed for the docking experiments. Virtual screening methods can accommodate libraries orders of magnitude larger than what can be screened in biochemical assays. This allows for a greater diversity of chemical space that can be tested and reduces the size of the library to be tested in subsequent biochemical or biophysical experiments.

The first-reported effort to target the effector–binding interface on RAS revealed Kobe 2602, which inhibited SOS-mediated exchange in vitro and inhibited RAS signal transduction pathways and proliferation of HRAS G12V-transformed NIH-3T3 cells (Shima et al. 2013). The compound also demonstrated modest activity in a xenograft model implanted with KRAS G12V SW480 cells. Whereas the effort demonstrated proof-of-concept for an in silico–based screening approach, the 2602 chemical series itself has not been pursued for further development because of instability and general toxicity.

Improved methods and increased computing power continue to advance in silico drug-screening efforts. A recent publication sought to target KRAS–RAF1 interactions using iterative in silico docking to develop 3144, a compound with activity on KRAS-mutated cancer cell lines in the 1- to 10-μM range (Shima et al. 2013). The authors went on to validate the in silico docking to KRAS by microscale thermophoresis, isothermal calorimetry, and NMR, and reported a dissociation constant also in the low micromolar range. Although not yet validated by crystallography, the NMR structure confirmed the binding site to be directly on the Switch I and II loops, and the binding is predicted to be competitive with effector proteins. Perhaps most impressively, 3144 had a dramatic response in KRAS-mutated cell lines and patient-derived xenograft mouse models. With a molecular weight of 717.66, 3144 is already quite large for a small molecule, and it is not yet clear whether the series can be further developed to improve potency for clinical development.

Disulfide Tethering

A variation of fragment-based screening was recently used to identify a previously unrecognized small molecule-binding pocket on KRAS. Disulfide fragment libraries were designed to identify low-affinity small-molecule fragments that bind with low affinity to shallow pockets and react with nearby cysteine residues. The small molecule–protein complex can then be detected by mass spectrometry to identify hit compounds. Shokat and colleagues used such a library to identify fragments that bind to the oncogenic KRAS G12C protein in a pocket adjacent to the nucleotide-binding site within the Switch II region (Ostrem et al. 2013). This pocket was not apparent in the apocrystal structures, but appears to be induced or stabilized by the compound. The disulfide reactive groups were then converted to electrophilic acrylamides or vinyl sulfonamides, which irreversibly attach to cysteine residues in the reducing environment of a cell. The binding pocket was subsequently expanded to accommodate larger small molecules with higher affinity and the current generation of inhibitors is active in cells in the nanomolar range (Lito et al. 2016; Patricelli et al. 2016). Furthermore, by targeting the oncogenic mutation itself (which happens to be a cysteine), the compounds potentially will benefit from an extraordinary therapeutic index because the compounds will only react with the KRAS G12C protein, but not the wild-type RAS molecules that harbor a glycine at codon 12.

INHIBITING MEMBRANE LOCALIZATION

In addition to directly targeting KRAS proteins, several efforts have been made to target KRAS-driven cancers by inhibiting cellular processes required for KRAS function. One of the first and most extensive efforts was to inhibit farnesyl transferase, which is required for H- N- and KRAS farnesylation (End et al. 2001). The small molecule, Tipifarnib was developed and tested in clinical trials for this purpose. Although the compounds were quite effective at inhibiting farnesyl transferase, it was discovered that KRAS itself does not require farnesylation for membrane localization. Although HRAS-dependent cells are quite sensitive to farnesyl transferase inhibitors, geranylgeranyl transferase prenylates the carboxyl terminus of KRAS (but not HRAS) upon farnesyl transferase inhibition, which restores membrane localization and full biochemical activity in the presence of Tipifarnib. Unfortunately, a dual inhibitor of farnesyl transferase and geranylgeranyl transferase reached dose-limiting toxicity but did not inhibit KRAS prenylation in human patients (Lobell et al. 2002). Although farnesyl transferase inhibitors are not effective in KRAS-mutated cancers, they are currently being evaluated for use in HRAS-mutated cancers, which represent 5% of bladder and thyroid cancers.

ICMT and RCE1

After prenylation, the final posttranslational modifications before KRAS is transported to the membrane are proteolysis of the three carboxy-terminal resides and then methylation of the prenylated cysteine. These are done by RCE1 (Ras converting CAAX endopeptidase 1) (Boyartchuk et al. 1997) and ICMT (isoprenylcysteine carboxyl methyltransferase) (Bergo et al. 2000). Inhibitors of both enzymes have been developed and are reported to mislocalize RAS molecules in cells, but RCE1 inhibitors have not been fully evaluated in vivo, and the antiproliferative effects of ICMT inhibition are not RAS specific (Bergo et al. 2004), suggesting that ICMT may be required for normal cellular function in KRAS wild-type cells.

PDE6-Delta

It was discovered that the KRAS splice forms 4A and 4B, which differ in the last 25 amino acids, are transported throughout the cell by different mechanisms (Wright and Philips 2006), and that KRAS 4B may require a chaperone, PDE6d (Chandra et al. 2011; Schmick et al. 2014). PDE6d is a prenyl-binding protein that is thought to solubilize prenylated proteins by shielding the hydrophobic farnesyl chain when KRAS is not bound to a membrane surface (Dharmaiah et al. 2016). A small molecule inhibitor, deltarasin, was developed to prevent PDE6d from binding to KRAS by occupying the prenyl-binding pocket on PDE6d (Zimmermann et al. 2013). Cells treated with the compound cause a mislocalization of KRAS 4B proteins, which accumulate on endomembranes. Cancer cells are also quite sensitive to deltarasin, though it is unclear whether KRAS mislocalization inhibits oncogenic KRAS 4B signal transduction or whether cellular toxicity is a result of inhibition of another PDE6d substrate. A second generation of PDE6d inhibitors was subsequently published with improved selectivity (Papke et al. 2016). Further investigation into PDE6d and similar mechanisms that regulate KRAS localization is likely to reveal additional cancer-relevant biology that can be exploited for therapeutic intervention.

DIMERIZATION AND MULTIMERIZATION

One emerging area of interest is the idea that activated RAS molecules may dimerize at the plasma membrane. Because RAF activation is thought to require dimerization, RAS dimerization would be a logical prerequisite, and there is some evidence that RAS molecules form high-order complexes. Superresolution single-molecule imaging studies of PA-mCherry-tagged KRAS proteins expressed in mammalian cells revealed the presences of apparent KRAS dimers (Nan et al. 2015). Artificially induced dimerization of KRAS molecules also activated the MAPK pathway, suggesting that KRAS dimerization is sufficient for signal transduction, yet it remains unclear whether RAS dimerization is required for oncogenic or wild-type KRAS signal transduction. One study of purified HRAS proteins in supported lipid bilayers identified a putative dimerization interface and a point mutation at Y64C that abolished dimerization (Lin et al. 2014). However, it is dimerization in cells that appears to be more complicated. The proposed dimerization interface is also within the effector-binding site and Y64 is required for binding to the RAS GEF, SOS. Other studies suggest that RAS may form higher-order "nanoclusters" in cells that are associated with and may be required for signal transduction (Zhou and Hancock 2015), and cellular expression of a "monobody" protein engineered to bind directly to RAS was shown to disrupt RAS nanoclusters and perturb signal transduction (Spencer-Smith et al. 2017). These studies offer some optimism that disrupting RAS–RAS interactions could be a viable therapeutic mechanism of action, but the underlying biology is not fully understood.

CONCLUDING REMARKS

Over 35 years of investigation has yet to produce an effective treatment for the large patient population with KRAS-mutated cancers. All the essential signal transduction pathways feeding into and emanating from KRAS, including multiple RTKs, RAF, MEK, PI3K, AKT, mTOR, and a host of ancillary proteins, have all been successfully targeted with small molecules. Inhibitors have been evaluated in cancer patients, and many have been approved by regulatory bodies, yet none of the treatments have proven effective in KRAS-mutated cancers. It is therefore understandable why KRAS is frequently referred to as one of the "undruggable" oncology targets. However, recent advances in in silico docking, covalent tethering, and fragment-based drug screening offer new hope for developing direct inhibitors of RAS proteins. These and other technologies are likely to yield small molecules that bind to KRAS with sufficient affinity for activity in cells and whole animals. Additionally, new biology has shed some light onto mechanisms required for RAS trafficking, membrane localization, and membrane dynamics, which may reveal new targets or targetable mechanisms.

ACKNOWLEDGMENTS

This project is funded in part with federal funds from the National Cancer Institute, National Institutes of Health (NIH), under Contract No. HHSN261200800001E. The content of this publication does not necessarily reflect the views or policies of the Department of Health and Human Services, and the mention of trade names, commercial products, or organizations does not imply endorsement by the U.S. Government.

REFERENCES

Barretina J, Caponigro G, Stransky N, Venkatesan K, Margolin AA, Kim S, Wilson CJ, Lehar J, Kryukov GV, Sonkin D, et al. 2012. The Cancer Cell Line Encyclopedia enables predictive modelling of anticancer drug sensitivity. *Nature* **483:** 603–607.

Bedard PL, Tabernero J, Janku F, Wainberg ZA, Paz-Ares L, Vansteenkiste J, Van Cutsem E, Perez-Garcia J, Stathis A, Britten CD, et al. 2015. A phase Ib dose-escalation study of the oral pan-PI3K inhibitor buparlisib (BKM120) in combination with the oral MEK1/2 inhibitor trametinib (GSK1120212) in patients with selected advanced solid tumors. *Clin Cancer Res* **21:** 730–738.

Almoguera C, Shibata D, Forrester K, Martin J, Arnheim N, Perucho M. 1988. Most human carcinomas of the exocrine pancreas contain mutant c-K-ras genes. *Cell* **53:** 549–554.

Bergo MO, Leung GK, Ambroziak P, Otto JC, Casey PJ, Young SG. 2000. Targeted inactivation of the isoprenylcysteine carboxyl methyltransferase gene causes mislocalization of K-Ras in mammalian cells. *J Biol Chem* **275:** 17605–17610.

Bergo MO, Gavino BJ, Hong C, Beigneux AP, McMahon M, Casey PJ, Young SG. 2004. Inactivation of *Icmt* inhibits transformation by oncogenic K-Ras and B-Raf. *J Clin Invest* **113:** 539–550.

Blasco RB, Francoz S, Santamaria D, Canamero M, Dubus P, Charron J, Baccarini M, Barbacid M. 2011. c-Raf, but not B-Raf, is essential for development of K-Ras oncogene-driven non-small-cell lung carcinoma. *Cancer Cell* **19:** 652–663.

Boriack-Sjodin PA, Margarit SM, Bar-Sagi D, Kuriyan J. 1998. The structural basis of the activation of Ras by Sos. *Nature* **394:** 337–343.

Bos JL, Fearon ER, Hamilton SR, Verlaan-de Vries M, van Boom JH, van der Eb AJ, Vogelstein B. 1987. Prevalence of ras gene mutations in human colorectal cancers. *Nature* **327:** 293–297.

Boyartchuk VL, Ashby MN, Rine J. 1997. Modulation of Ras and a-factor function by carboxyl-terminal proteolysis. *Science* **275:** 1796–1800.

Castellano E, Sheridan C, Thin MZ, Nye E, Spencer-Dene B, Diefenbacher ME, Moore C, Kumar MS, Murillo MM, Gronroos E, et al. 2013. Requirement for interaction of PI3-kinase p110α with RAS in lung tumor maintenance. *Cancer Cell* **24:** 617–630.

Chandra A, Grecco HE, Pisupati V, Perera D, Cassidy L, Skoulidis F, Ismail SA, Hedberg C, Hanzal-Bayer M, Venkitaraman AR, et al. 2011. The GDI-like solubilizing factor PDEδ sustains the spatial organization and signalling of Ras family proteins. *Nat Cell Biol* **14:** 148–158.

Chapman PB, Hauschild A, Robert C, Haanen JB, Ascierto P, Larkin J, Dummer R, Garbe C, Testori A, Maio M, et al. 2011. Improved survival with vemurafenib in melanoma with BRAF V600E mutation. *N Engl J Med* **364:** 2507–2516.

Cox AD, Fesik SW, Kimmelman AC, Luo J, Der CJ. 2014. Drugging the undruggable RAS: Mission possible? *Nat Rev Drug Discov* **13:** 828–851.

Davies H, Bignell GR, Cox C, Stephens P, Edkins S, Clegg S, Teague J, Woffendin H, Garnett MJ, Bottomley W, et al. 2002. Mutations of the BRAF gene in human cancer. *Nature* **417:** 949–954.

Der CJ, Krontiris TG, Cooper GM. 1982. Transforming genes of human bladder and lung carcinoma cell lines are homologous to the *ras* genes of Harvey and Kirsten sarcoma viruses. *Proc Natl Acad Sci* **79:** 3637–3640.

Dharmaiah S, Bindu L, Tran TH, Gillette WK, Frank PH, Ghirlando R, Nissley DV, Esposito D, McCormick F, Stephen AG, et al. 2016. Structural basis of recognition of farnesylated and methylated KRAS4b by PDEδ. *Proc Natl Acad Sci* **113:** E6766–E6775.

End DW, Smets G, Todd AV, Applegate TL, Fuery CJ, Angibaud P, Venet M, Sanz G, Poignet H, Skrzat S, et al. 2001. Characterization of the antitumor effects of the selective farnesyl protein transferase inhibitor R115777 in vivo and in vitro. *Cancer Res* **61:** 131–137.

Engelman JA, Chen L, Tan X, Crosby K, Guimaraes AR, Upadhyay R, Maira M, McNamara K, Perera SA, Song Y, et al. 2008. Effective use of PI3K and MEK inhibitors to treat mutant Kras G12D and PIK3CA H1047R murine lung cancers. *Nat Med* **14:** 1351–1356.

Fabian JR, Vojtek AB, Cooper JA, Morrison DK. 1994. A single amino acid change in Raf-1 inhibits Ras binding and alters Raf-1 function. *Proc Natl Acad Sci* **91:** 5982–5986.

Fong PC, Boss DS, Yap TA, Tutt A, Wu P, Mergui-Roelvink M, Mortimer P, Swaisland H, Lau A, O'Connor MJ, et al. 2009. Inhibition of poly(ADP-ribose) polymerase in tumors from BRCA mutation carriers. *N Engl J Med* **361:** 123–134.

Forrester K, Almoguera C, Han K, Grizzle WE, Perucho M. 1987. Detection of high incidence of K-ras oncogenes during human colon tumorigenesis. *Nature* **327:** 298–303.

Hancock JF. 2003. Ras proteins: Different signals from different locations. *Nat Rev Mol Cell Biol* **4:** 373–384.

Kato-Stankiewicz J, Hakimi I, Zhi G, Zhang J, Serebriiskii I, Guo L, Edamatsu H, Koide H, Menon S, Eckl R, et al. 2002. Inhibitors of Ras/Raf-1 interaction identified by two-hybrid screening revert Ras-dependent transformation phenotypes in human cancer cells. *Proc Natl Acad Sci* **99:** 14398–14403.

Khambata-Ford S, Garrett CR, Meropol NJ, Basik M, Harbison CT, Wu S, Wong TW, Huang X, Takimoto CH, Godwin AK, et al. 2007. Expression of epiregulin and amphiregulin and K-ras mutation status predict disease control in metastatic colorectal cancer patients treated with cetuximab. *J Clin Oncol* **25:** 3230–3237.

Lin WC, Iversen L, Tu HL, Rhodes C, Christensen SM, Iwig JS, Hansen SD, Huang WY, Groves JT. 2014. H-Ras forms dimers on membrane surfaces via a protein–protein interface. *Proc Natl Acad Sci* **111:** 2996–3001.

Lito P, Solomon M, Li LS, Hansen R, Rosen N. 2016. Allele-specific inhibitors inactivate mutant KRAS G12C by a trapping mechanism. *Science* **351:** 604–608.

Lobell RB, Liu D, Buser CA, Davide JP, DePuy E, Hamilton K, Koblan KS, Lee Y, Mosser S, Motzel SL, et al. 2002. Preclinical and clinical pharmacodynamic assessment of L-778,123, a dual inhibitor of farnesyl:protein transferase and geranylgeranyl:protein transferase type-I. *Mol Cancer Ther* **1:** 747–758.

Maurer T, Garrenton LS, Oh A, Pitts K, Anderson DJ, Skelton NJ, Fauber BP, Pan B, Malek S, Stokoe D, et al. 2012. Small-molecule ligands bind to a distinct pocket in Ras and inhibit SOS-mediated nucleotide exchange activity. *Proc Natl Acad Sci* **109:** 5299–5304.

Nan X, Tamguney TM, Collisson EA, Lin LJ, Pitt C, Galeas J, Lewis S, Gray JW, McCormick F, Chu S. 2015. Ras-GTP dimers activate the mitogen-activated protein kinase (MAPK) pathway. *Proc Natl Acad Sci* **112:** 7996–8001.

Ostrem JM, Peters U, Sos ML, Wells JA, Shokat KM. 2013. K-Ras^{G12C} inhibitors allosterically control GTP affinity and effector interactions. *Nature* **503:** 548–551.

Pai EF, Kabsch W, Krengel U, Holmes KC, John J, Wittinghofer A. 1989. Structure of the guanine-nucleotide-binding domain of the Ha-ras oncogene product p21 in the triphosphate conformation. *Nature* **341:** 209–214.

Papke B, Murarka S, Vogel HA, Martin-Gago P, Kovacevic M, Truxius DC, Fansa EK, Ismail S, Zimmermann G, Heinelt K, et al. 2016. Identification of pyrazolopyridazinones as PDEδ inhibitors. *Nat Commun* **7:** 11360.

Parada LF, Weinberg RA. 1983. Presence of a Kirsten murine sarcoma virus ras oncogene in cells transformed by 3-methylcholanthrene. *Mol Cell Biol* **3:** 2298–2301.

Patricelli MP, Janes MR, Li LS, Hansen R, Peters U, Kessler LV, Chen Y, Kucharski JM, Feng J, Ely T, et al. 2016. Selective inhibition of oncogenic KRAS output with small molecules targeting the inactive state. *Cancer Discov* **6:** 316–329.

Rodenhuis S, van de Wetering ML, Mooi WJ, Evers SG, van Zandwijk N, Bos JL. 1987. Mutational activation of the K-ras oncogene. A possible pathogenetic factor in adenocarcinoma of the lung. *N Engl J Med* **317:** 929–935.

Santos E, Martin-Zanca D, Reddy EP, Pierotti MA, Della Porta G, Barbacid M. 1984. Malignant activation of a K-ras oncogene in lung carcinoma but not in normal tissue of the same patient. *Science* **223:** 661–664.

Scheffzek K, Ahmadian MR, Kabsch W, Wiesmuller L, Lautwein A, Schmitz F, Wittinghofer A. 1997. The Ras–RasGAP complex: Structural basis for GTPase activation and its loss in oncogenic Ras mutants. *Science* **277:** 333–338.

Schmick M, Vartak N, Papke B, Kovacevic M, Truxius DC, Rossmannek L, Bastiaens PIH. 2014. KRas localizes to the plasma membrane by spatial cycles of solubilization, trapping and vesicular transport. *Cell* **157:** 459–471.

Shih C, Shilo BZ, Goldfarb MP, Dannenberg A, Weinberg RA. 1979. Passage of phenotypes of chemically transformed cells via transfection of DNA and chromatin. *Proc Natl Acad Sci* **76:** 5714–5718.

Shih C, Padhy LC, Murray M, Weinberg RA. 1981. Transforming genes of carcinomas and neuroblastomas introduced into mouse fibroblasts. *Nature* **290:** 261–264.

Shima F, Yoshikawa Y, Ye M, Araki M, Matsumoto S, Liao J, Hu L, Sugimoto T, Ijiri Y, Takeda A, et al. 2013. In silico discovery of small-molecule Ras inhibitors that display antitumor activity by blocking the Ras–effector interaction. *Proc Natl Acad Sci* **110:** 8182–8187.

Spencer-Smith R, Koide A, Zhou Y, Eguchi RR, Sha F, Gajwani P, Santana D, Gupta A, Jacobs M, Herrero-Garcia E, et al. 2017. Inhibition of RAS function through targeting an allosteric regulatory site. *Nat Chem Biol* **13:** 62–68.

Stephen AG, Esposito D, Bagni RK, McCormick F. 2014. Dragging Ras back in the ring. *Cancer Cell* **25:** 272–281.

Sun Q, Phan J, Friberg AR, Camper DV, Olejniczak ET, Fesik SW. 2014. A method for the second-site screening of K-Ras in the presence of a covalently attached first-site ligand. *J Biomol NMR* **60:** 11–14.

Taveras AG, Remiszewski SW, Doll RJ, Cesarz D, Huang EC, Kirschmeier P, Pramanik BN, Snow ME, Wang YS, del Rosario JD, et al. 1997. Ras oncoprotein inhibitors: The discovery of potent, ras nucleotide exchange inhibitors and the structural determination of a drug–protein complex. *Bioorg Med Chem* **5:** 125–133.

The Cancer Genome Atlas Research Network. 2011. Integrated genomic analyses of ovarian carcinoma. *Nature* **474:** 609–615.

The Cancer Genome Atlas Research Network. 2017. National Cancer Institute and the National Human Genome Research Institute, National Institutes of Health, Bethesda, MD.

Trahey M, McCormick F. 1987. A cytoplasmic protein stimulates normal N-ras p21 GTPase, but does not affect oncogenic mutants. *Science* **238:** 542–545.

Trinh TB, Upadhyaya P, Qian Z, Pei D. 2016. Discovery of a direct ras inhibitor by screening a combinatorial library of cell-permeable bicyclic peptides. *ACS Comb Sci* **18:** 75–85.

Winter JJ, Anderson M, Blades K, Brassington C, Breeze AL, Chresta C, Embrey K, Fairley G, Faulder P, Finlay MR, et al. 2015. Small molecule binding sites on the Ras:SOS complex can be exploited for inhibition of Ras activation. *J Med Chem* **58:** 2265–2274.

Wright LP, Philips MR. 2006. Thematic review series: Lipid posttranslational modifications. CAAX modification and membrane targeting of Ras. *J Lipid Res* **47:** 883–891.

Wu X, Upadhyaya P, Villalona-Calero MA, Briesewitz R, Pei D. 2013. Inhibition of ras–effector interaction by cyclic peptides. *Med Chem Commun* **4:** 378–382.

Yang W, Soares J, Greninger P, Edelman EJ, Lightfoot H, Forbes S, Bindal N, Beare D, Smith JA, Thompson IR, et al. 2013. Genomics of drug sensitivity in cancer (GDSC): A resource for therapeutic biomarker discovery in cancer cells. *Nucleic Acids Res* **41:** D955–D961.

Zhou Y, Hancock JF. 2015. Ras nanoclusters: Versatile lipid-based signaling platforms. *Biochim Biophys Acta* **1853:** 841–849.

Zimmermann G, Papke B, Ismail S, Vartak N, Chandra A, Hoffmann M, Hahn SA, Triola G, Wittinghofer A, Bastiaens PI, et al. 2013. Small molecule inhibition of the KRAS-PDEδ interaction impairs oncogenic KRAS signalling. *Nature* **497:** 638–642.

Targeting Ras with Macromolecules

Dehua Pei, Kuangyu Chen, and Hui Liao

Department of Chemistry and Biochemistry, The Ohio State University, Columbus, Ohio 43210

Correspondence: pei.3@osu.edu

Activating Ras mutations are associated with ~30% of all human cancers and the four Ras isoforms are highly attractive targets for anticancer drug discovery. However, Ras proteins are challenging targets for conventional drug discovery because they function through intracellular protein–protein interactions and their surfaces lack major pockets for small molecules to bind. Over the past few years, researchers have explored a variety of approaches and modalities, with the aim of specifically targeting oncogenic Ras mutants for anticancer treatment. This perspective will provide an overview of the efforts on developing "macromolecular" inhibitors against Ras proteins, including peptides, macrocycles, antibodies, nonimmunoglobulin proteins, and nucleic acids.

Ras is a small GTPase, acting as a molecular switch in many signaling pathways and regulating cell proliferation, differentiation, and survival, among other functions (Young et al. 2009). Its four isoforms, H-Ras, N-Ras, K-Ras4A, and K-Ras4B, are identical within the amino-terminal 85 amino acids and differ primarily in the carboxyl-termini (amino acids 165–189). Wild-type Ras oscillates between the inactive guanosine diphosphate (GDP)-bound form (Ras-GDP) and the active guanosine triphosphate (GTP)-bound form (Ras-GTP) (Fig. 1A). Ras-GTP interacts with and activates multiple effector proteins, including kinases Raf and phosphoinositide 3-kinase (PI3K), turning cells on for proliferation and survival. Somatic mutations at Gly-12, Gly-13, or Gln-61, which are all located within the GTPase active site, impair GTP hydrolysis, resulting in an excessive Ras-GTP population leading to uncontrolled cell growth. Ras mutations are found in ~30% of all human cancers and are well-established cancer drivers (Prior et al. 2012; Singh et al. 2015). Mutations in K-Ras are particularly prevalent in some of the most deadly cancers, including pancreatic (90% prevalence), colon (35% prevalence), and lung cancers (16% prevalence). Disruption of Ras function genetically (i.e., by gene mutations or small-interfering RNA [siRNA]) inhibits the proliferation of Ras-mutant cancer cells and induces apoptosis, validating Ras as one of the most compelling cancer drug targets (Gupta et al. 2007; Singh et al. 2009; Castellano et al. 2013; Khvalevsky et al. 2013).

Over the past three decades, researchers have explored several approaches to inhibit Ras function (Wang et al. 2012; Spiegel et al. 2014; Stephen et al. 2014). The ideal approach would be direct binding to the GTPase active site by a small molecule. However, this approach was quickly found to be impractical, because Ras binds to GTP (and GDP) with pM affinity and GTP is present in the cytosol at mM concentrations. The next-best option is to block the interaction between Ras and its partner proteins and/or its nucleotide exchange activity.

Cite this article as *Cold Spring Harb Perspect Med* doi: 10.1101/cshperspect.a031476

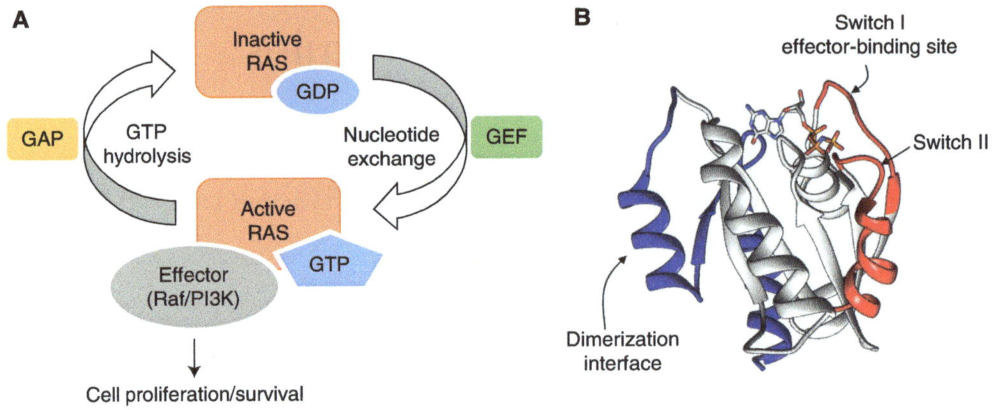

Figure 1. Ras structure and function. (*A*) Interconversion between inactive Ras-guanosine diphosphate (GDP) and active Ras-guanosine triphosphate (GTP) mediated by guanine nucleotide exchange factor (GEF) and GTPase-activating protein (GAP). The active Ras-GTP interacts with effector proteins and activates downstream cell signaling. (*B*) Structure of K-Ras bound with GppNp, a nonhydrolyzable analog of GTP. The switch I/II regions are shown in red, the dimerization interface in blue, and the bound nucleotide analog is shown as sticks. (Figure was recreated from PDB data [pdb5p21].)

Unfortunately, Ras has been a very challenging target for direct inhibition, because its surface has no obvious pocket for small molecules to bind. Consequently, most of the early efforts were focused on targeting the signaling steps upstream or downstream of Ras. Small-molecule inhibitors against the posttranslational modification of Ras proteins (e.g., inhibitors against farnesyl and geranylgeranyl transferases) ultimately proved to be ineffective clinically, because of either a lack of efficacy or unacceptable toxicity (Whyte et al. 1997; Konstantinopoulos et al. 2007; Berndt et al. 2011). Another indirect approach is inhibition of the protein kinases downstream of Ras. Several MEK and PI3K inhibitors are currently undergoing clinical evaluation, although the clinical effectiveness of this approach remains to be seen (Engelman et al. 2008). It appears that simultaneous inhibition of at least two different pathways (e.g., by both MEK and PI3K inhibitors) is necessary to effectively block Ras signaling in Ras-mutant tumors (Britten et al. 2013); however, blocking multiple signaling pathways may result in extensive toxicity.

Driven by both urgent unmet medical needs and remarkable technological breakthroughs, there has been a recent "renaissance" of the direct Ras inhibition approach (Ledford 2015), which attacks Ras-driven cancers at the source and provides an opportunity for selective inhibition of Ras-mutant cancers. Development of direct Ras inhibitors is challenging because Ras performs its biological functions by engaging in intracellular protein–protein interactions (PPIs). PPIs are in general a challenging class of drug targets for conventional small molecules (defined as molecules of ≤500 molecular weight), because small molecules typically do not make a sufficient number of contacts with the large flat PPI interfaces to impart high affinity or specificity. To effectively compete with the protein partners, a PPI inhibitor should also possess a binding surface comparable in size to that of the partner proteins. Recognizing this, researchers have increasingly turned their attention to larger molecules, including peptides, macrocycles, and proteins, as PPI inhibitors. Powerful combinatorial library technologies for discovering peptide—especially macrocyclic peptide—ligands and novel approaches to improving their membrane permeability have recently been developed (Dougherty et al. 2017). Application of these technologies to Ras has resulted in a variety of Ras inhibitors, ranging from small molecules, peptides, macrocycles, proteins, to nucleic acids over the past few years (Fig. 2). In this review, we provide a summary of the "macromolecular" approaches

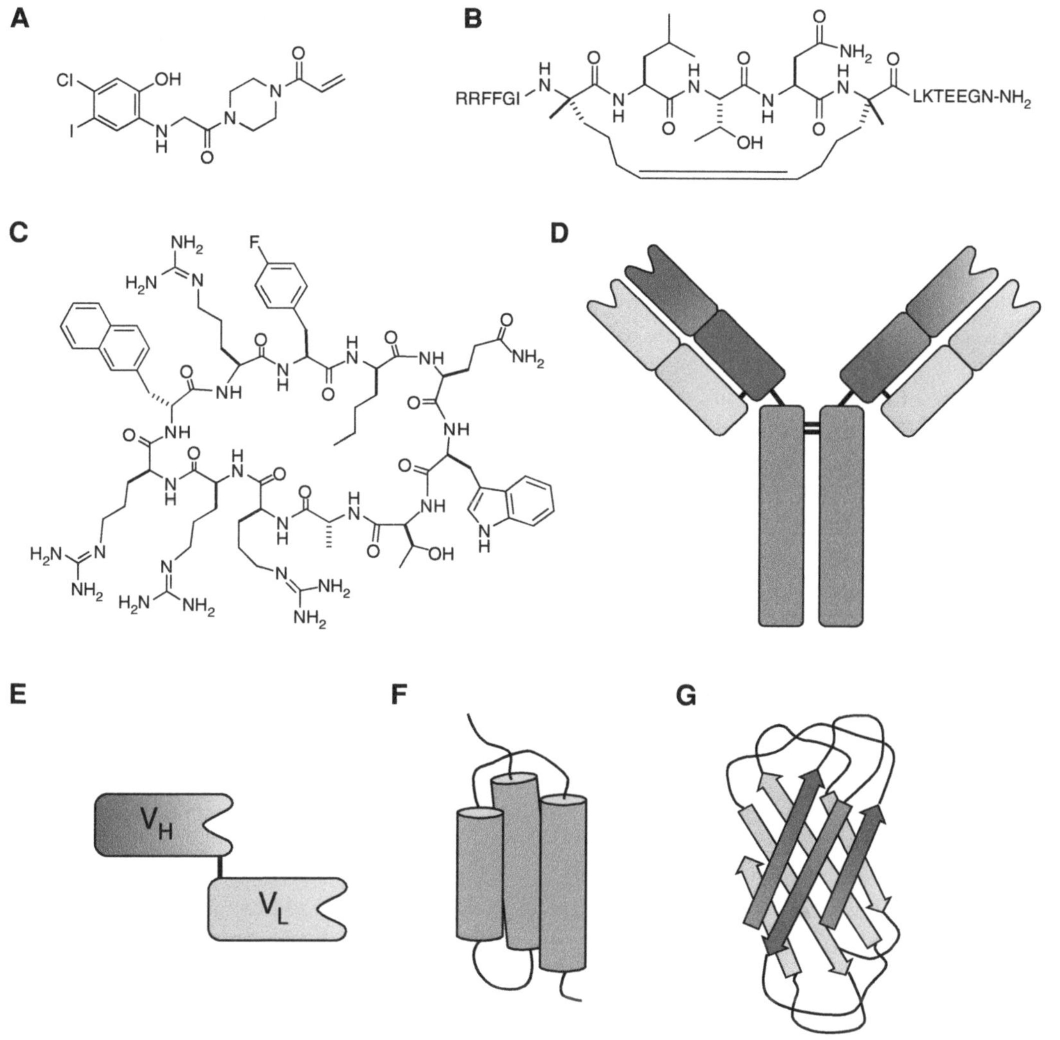

Figure 2. Molecular modalities that have been used as direct Ras inhibitors. (*A*) Small molecules (example shown is the K-Ras^{G12C}-selective inhibitor by Ostrem et al. 2013); (*B*) stapled peptides (example shown is SAH-SOS1); (*C*) macrocyclic peptides (example shown is cyclorasin 9A5); (*D*) monoclonal antibodies; (*E*) intrabodies (the V_H and V_L regions of an antibody linked by a polypeptide or disulfide bond); (*F*) affibodies; and (*G*) monobodies (the tenth fibronectin type III domain of human fibronectin).

to Ras inhibitor development (as opposed to the small-molecule approach). For an updated account on small-molecule Ras inhibitors, readers are referred to Zhou et al. (2017).

PEPTIDE-BASED INHIBITORS

Linear Peptides

The first peptidyl Ras inhibitors were discovered by Clark et al. (1996), who compared the sequences of several Ras effector proteins and identified a consensus Ras-binding sequence shared among a subset of the Ras effectors. They showed that peptides containing this sequence from Raf-1 (Table 1, compound 1) and NF1-GAP (Table 1, compound 2) blocked NF1-GAP stimulation of Ras GTPase activity and Ras-mediated activation of mitogen-activated protein kinases, with IC_{50} values of ~45 μM. Similarly, Barnard et al. (1998) systematically tested short peptides corresponding to different regions of H-Ras and Raf-1 for in vitro inhibition of the Ras-Raf association and found that two of the Raf-1 peptides, E^{94}CCAVFR^{100} (Table 1, compound 3) and C^{95}CAVFRL^{101} (Table 1, compound 4), were significant inhibitors of Ras-Raf binding. Peptide C^{95}CAVFRL^{101} had an IC_{50} value of 7 µM and also inhibited Ras-RalGDS binding. Xu and Luo (2002) screened 3.5×10^7 peptide aptamers and isolated two peptides (Table 1, compounds 5 and 6) that blocked the Ras–Raf interaction in vitro with IC_{50} values of 100 and 500 nM, respectively. Impressively, the peptides bound selectively to Ras^{G12V} over wild-type Ras and, when expressed in COS cells, inhibited Raf kinase activation in cell culture. More recently, Gareiss et al. (2010) identified a moderately potent dodecapeptide ligand (Table 1, compound 7) that binds to both H-Ras and K-Ras. However, the peptide did not alter the intrinsic GTPase activity of Ras or compete for Ras binding with Raf and showed no cellular activity.

Stapled Peptides

Stapled peptides have emerged as a powerful tool for disrupting protein–protein interactions by mimicking the secondary structure of crucial binding motifs (Walensky and Bird 2014). The preformed α-helical structure minimizes the entropic penalty associated with ligand–protein interaction, greatly improves the proteolytic stability of the peptides, and in some cases increases the membrane permeability of the peptides. Based on the observation that the guanine nucleotide exchange factor SOS binds to Ras primarily via an α-helical motif, Patgiri et al. (2011) designed a hydrogen bond surrogate (HBS) helical peptide (Table 1, compound 8), a special variant of stapled peptides, which had 56% helical content and bound to Ras-GDP with moderate affinity ($K_D = 158$ µM). ^1H-^{15}N heteronuclear single quantum coherence (HSQC) nuclear magnetic resonance (NMR) studies indicated that the peptide interacts with the switch I/II region on Ras, overlapping with the SOS-binding site (Fig. 1B). Remarkably, the HBS peptide was cell-permeable and reduced epidermal growth factor (EGF)-stimulated Ras activation and downstream signaling events in HeLa cells. More recently, Leshchiner et al. (2015) designed a hydrocarbon-stapled α-helical peptide with much greater Ras-binding affinity, SAH-SOS1 (K_D <200 nM) (Table 1, compound 9 and Fig. 2B), also corresponding to the Ras-binding helical motif in SOS. SAH-SOS1 was insensitive to the nucleotide-bound states of Ras. NMR and docking studies showed that SAH-SOS1 interacts with the SOS-binding site on Ras and inhibits nucleotide association of K-Ras. Importantly, SAH-SOS1 down-regulated AKT/MEK/ERK phosphorylation and reduced the viability of different K-Ras mutant cancer cells with IC_{50} values of 5–15 µM. Furthermore, SAH-SOS1 reduced the AKT and ERK phosphorylation levels in *Drosophila* tissues.

Macrocyclic Peptides

Because Ras-effector interactions are not mediated by α-helical motifs, rational design of stapled peptides against the effector-binding site is not an option. As described above, linear peptide ligands against the effector-binding site have been discovered, but they are generally weak binders. In addition, linear peptides face other challenges, including proteolytic degradation and the lack of membrane permeability. Meanwhile, macrocyclic peptides have emerged as an effective modality for inhibition of protein–protein interactions over the past decade (Dougherty et al. 2017). With molecular mass generally in the range of 500–2000, macrocyclic peptides are 3–5 times larger than conventional small-molecule drugs and possess binding surfaces similar in size to that of PPI interfaces. They also have a balanced conformational flexibility/rigidity that is conducive to recognizing the large flat protein surfaces. In addition, macrocyclic peptides have greatly increased proteolytic stability, especially when unnatural amino acids (e.g., D-amino acids) are incorporated into their structures.

In an attempt to generalize the mode of action by rapamycin, Wu et al. (2013) constructed a one-bead–two-compound (OBTC)

Table 1. A list of peptide-based Ras inhibitors and their inhibition mechanisms

No.	Peptide sequence/name	Mechanism	K_D or IC_{50}	References
1	FARKTFLKLAF	Blocks Ras-effector interaction in vitro	44.4 ± 6.6 μM	Clark et al. 1996
2	ARRFFLDIAD		86.5 ± 8.9 μM	
3	Ac-E^{94}CCAVFR^{100}-NH_2	Blocks Ras-effector interaction in vitro	10 μM	Barnard et al. 1998
4	Ac-C^{95}CAVFRL^{101}-NH_2		7 μM	
5	DMDWFFRFYASVSRLFRHLH	Blocks Ras-effector interaction in vitro	0.5 μM	Xu and Luo 2002
6	WWGYLGKLFQTETDRSQGAL-GPRPMCFILVEAVFASAACD SC	Cellular activity	0.1 μM	
7	NH_2-HYPWFKARLYPL	Direct binding to Ras G domain	5–10 μM	Gareiss et al. 2010
8	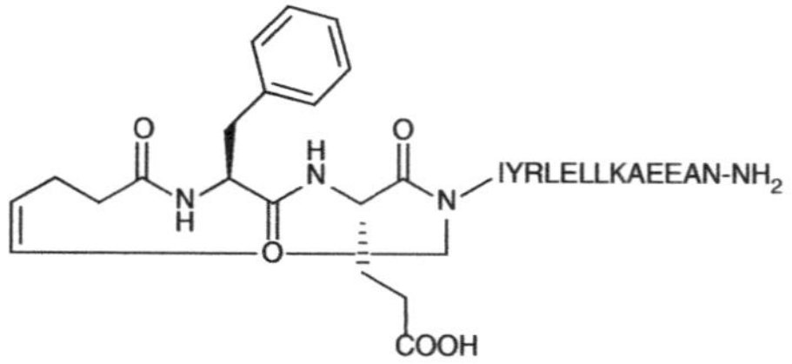	Interferes with Ras–SOS interaction Cellular activity	158 ± 16 μM	Patgiri et al. 2011
9	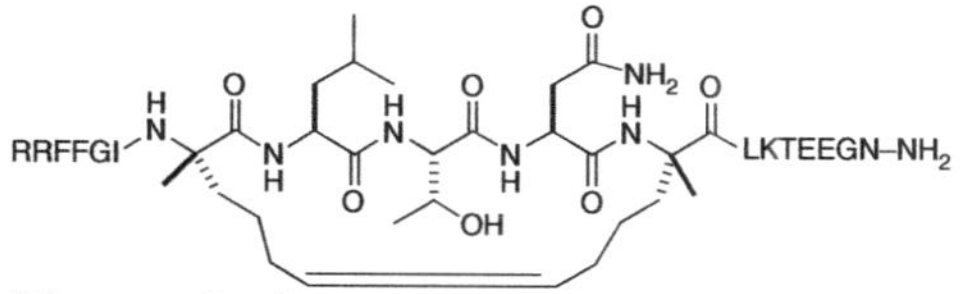	Inhibits SOS-mediated nucleotide association Cellular activity Active in fly	60–160 nM	Leshchiner et al. 2015
10	"Compound 12"	Blocks Ras-effector interaction in vitro Cell-impermeable	0.5 μM	Wu et al. 2013
11	Cyclorasin 9A5	Blocks Ras-effector interaction in vitro Cellular activity	120 nM	Upadhyaya et al. 2015
12	Cyclorasin B3	Blocks Ras-effector interaction in vitro Cellular activity	2.6 μM	Upadhyaya et al. 2014
13	"Compound 49"	Blocks Ras-effector interaction in vitro Cellular activity	3.4 μM	Trinh et al. 2016
14	Ac-RRRR-Cys-PLYISYDPV-Cys-RRRR-NH_2 (S–S bridged Cys)	Inhibits SOS-mediated nucleotide exchange Cellular activity	42 nM (wild-type K-Ras) 1.6 nM (K-Ras^{G12D})	Sakamoto et al. 2017

Structures of compounds 10–13 are shown in Figure 3.

10

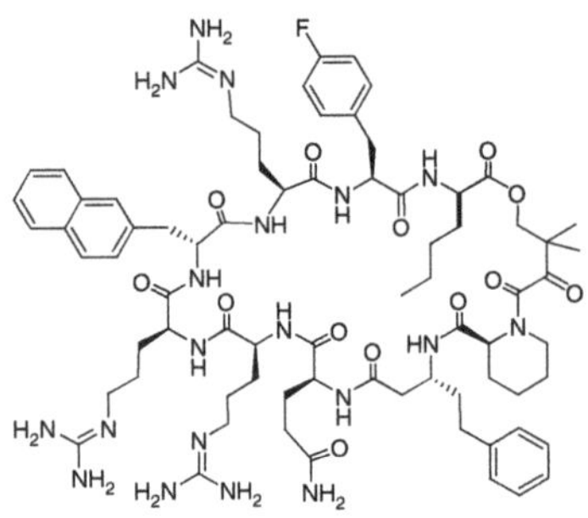

11

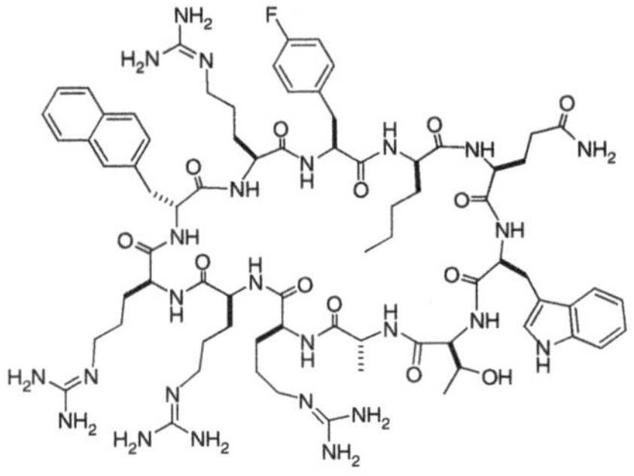

12

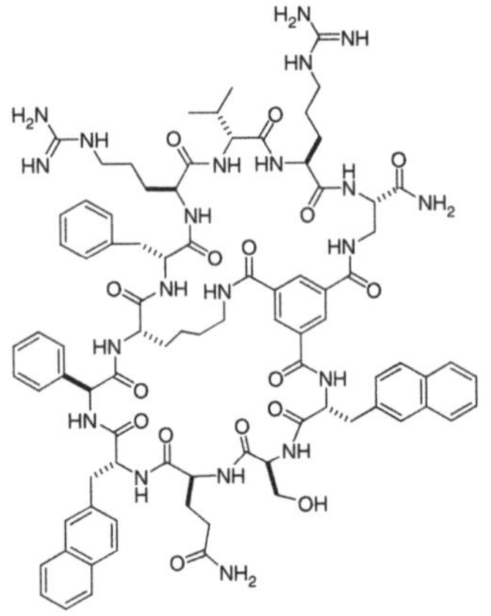

13

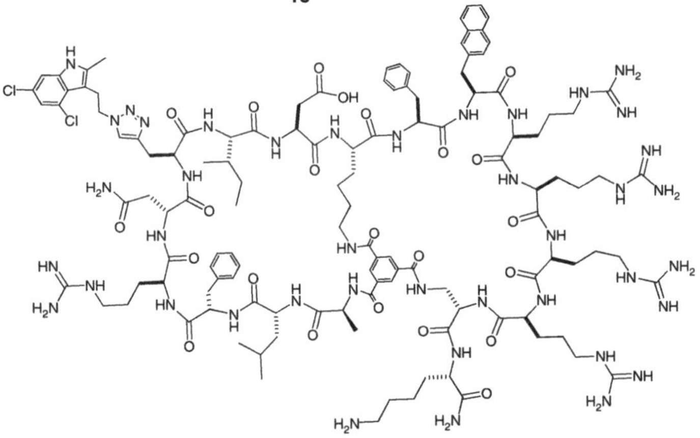

Figure 3. Structures of macrocyclic peptidyl Ras inhibitors.

cyclic peptide library containing a fixed FKBP-binding motif fused with a randomized peptide sequence made of 25 different natural and unnatural amino acid building blocks. Screening of the library against K-Ras^{G12V} identified a hit peptide, "compound 12" (Table 1, compound 10 and Fig. 3), which inhibited the Ras–Raf RBD interaction with an IC_{50} value of 0.5 μM, in the absence of FKBP. As expected from its large size and hydrophilic structure, compound 10 was impermeable to the cell membrane and had no cellular activity. Interestingly, the structure of compound 10 contained a pentapeptide motif, Arg-Arg-nal-Arg-Fpa (where Fpa is L-4-fluorophenylalanine and nal is D-β-naphthylalanine), which bears a similarity to a family of cyclic cell–penetrating peptides (CPPs) (e.g., cyclo (Phe-Nal-Arg-Arg-Arg-Arg-Gln), where Nal is L-β-naphthylalanine) (Qian et al. 2013, 2016). The investigators subsequently constructed a second-generation library, in which the CPP-like motif was retained, while the remaining structure was replaced with a randomized peptide sequence of 0–5 amino acids (Upadhyaya et al. 2015). Screening of the second-generation library against K-Ras^{G12V} identified two hits that blocked the Ras-Raf association in vitro, were cell-permeable, and showed modest antiproliferative activity against cancer cells. One of the hits, cyclorasin 9A, was further optimized through medicinal chemistry efforts to produce an improved compound, cyclorasin 9A5 (Table 1, compound 11 and Fig. 3), which orthosterically blocked the Ras–Raf interaction in vitro (IC_{50} = 120 nM), was readily cell-permeable, and showed a sixfold selectivity for Ras-GTP over Ras-GDP. NMR studies suggested that cyclorasin 9A5 binds to an extended area between the switch I and II loops, which overlaps with the effector-binding site (Fig. 1B).

When tested against Ras mutant lung cancer cells, cyclorasin 9A5 dose dependently inhibited the Ras–Raf interaction and abolished MEK, ERK, and AKT phosphorylation at low µM concentrations. It showed moderately potent antiproliferative activity against a panel of lung cancer cell lines (e.g., EC_{50} ~3 µM against H1299 cells) and induced apoptosis of both wild-type and mutant Ras cancer cells at low micromolar concentrations, consistent with its ability to simultaneously block both the RAF/MEK/ERK and the PI3K/AKT/mTOR pathways. Transfection of H1299 cells with a constitutively active form of AKT rendered the cancer cells substantially less sensitive to cyclorasin 9A5, providing further support that the observed anticancer activity was caused by specific inhibition of Ras signaling by the cyclic peptide.

In a separate study, Upadhyaya et al. (2014) synthesized a highly structurally constrained bicyclic peptide library in the OBTC format and screened the library against K-Ras^{G12V}. They discovered two different classes of Ras ligands, with one class of ligands (e.g., cyclorasin B3) apparently bound at or near the effector-binding site and blocked the Ras-Raf association (K_D = 2.6 µM). Cyclorasin B3 (Table 1, compound 12 and Fig. 3) competed with monocyclic peptide 10 and Raf-1 RBD for binding to K-Ras^{G12V}. Interestingly, the second class of Ras ligands (e.g., cyclorasin B2) apparently bound to a yet-unidentified site, different from the effector-binding site (K_D = 0.49 µM), as it did not inhibit the Ras–Raf interaction. As expected, both classes of cyclic peptides were impermeable to the cell membrane and had no significant activity during cellular assays. Efforts to improve both the potency and cell-permeability of the library hits are currently ongoing in the authors' laboratory.

To develop a general approach to discovering cell-permeable macrocyclic inhibitors against intracellular targets, Trinh et al. (2016) designed a bicyclic peptide library in which one ring consisted of a fixed CPP motif (e.g., Phe-Nal-Arg-Arg-Arg-Arg), while the second ring featured 5.7 million different peptide sequences. To increase the probability of high-affinity binding to K-Ras, the investigators included propargylglycine as a building block for the degenerate sequence and attached 4,6-dichloro-2-methyl-3-aminoethylindole (DCAI), a previously reported weak ligand for K-Ras (Maurer et al. 2012), to the alkyne side-chain by click chemistry. It was anticipated that most of the 5.7 million peptides should be cell permeable. Indeed, screening of the library against K-Ras^{G12V} followed by medicinal chemistry gave a cell-permeable and moderately potent K-Ras inhibitor ("compound 49") (Table 1, compound 13 and Fig. 3), which inhibited the Ras–Raf interaction with an IC_{50} value of 3.4 µM. Compound 13 competed with DCAI for Ras binding and a control peptide without the DCAI moiety showed 80-fold lower affinity to K-Ras^{G12V}, indicating that the DCAI moiety plays a critical role in Ras binding and the bicyclic peptide binds to an area that overlaps with the switch I/II regions and the DCAI-binding pocket (Fig. 1B). Compound 13 (but not the control peptide without DCAI) inhibited Akt and MEK phosphorylation in cancer cells in a dose-dependent manner and induced apoptosis of lung cancer cells at 10 µM concentration.

Very recently, Sakamoto et al. (2017) discovered a highly potent and K-Ras^{G12D}-selective cyclic peptide inhibitor by screening a disulfide-cyclized peptide library displayed on T7 phage against the recombinant K-Ras^{G12D}. To obtain mutant-selective inhibitors, the researchers used both positive and negative screening steps, with thorough subtraction of phage bound to wild-type K-Ras. One of the library hits, KRpep-2 (Ac-RRCPLYISYDPVCRR-NH_2), showed 14-fold selectivity for K-Ras^{G12D} over wild-type K-Ras (K_D values of 51 and 700 nM, respectively). Subsequent sequence optimization resulted in cyclic peptide KRpep-2d (Table 1, compound 14), which inhibited SOS-mediated nucleotide exchange activity of wild-type and G12D K-Ras with IC_{50} values of 42 and 1.6 nM, respectively. At 30 µM peptide concentration, KRpep-2d inhibited the phosphorylation levels of ERK1/2 and the proliferation of A427 lung cancer cells (which harbor a K-Ras^{G12D} mutation), but not A549 lung cancer cells (which contain a K-Ras^{G12C} mutation). The high peptide concentration needed to cause cellular effects (30 μM, relative to IC_{50} of 1.6 nM for inhibition of nucleotide exchange) was attributed to limited cellular entry and inactivation of KRpep-2d on reduction of its disulfide bond inside the mammalian cytosol.

PROTEIN-BASED INHIBITORS

Anti-Ras Antibodies

Shortly after the discovery of the *ras* oncogene, researchers explored the possibility of using anti-Ras antibodies as potential anticancer therapeutics. Feramisco et al. (1985) reported that microinjection of antibodies specific for a mutant K-Ras protein into NRK cells transformed by *V-K-ras* gene caused a transient reversion of the cells to a normal phenotype. Similarly, Kung et al. (1986) found that microinjection of a neutralizing monoclonal antibody against H-Ras (Y13-259) neutralized the transforming activity of a coinjected bacterially synthesized H-Ras protein and that the neutralization effect was blocked by coinjection of excess H-Ras protein. In addition, microinjection of Y13-259 into transformed NIH3T3 cells (obtained by DNA transfection of NIH3T3 cells with a mutant *H-ras* gene) reversed their transformed phenotypes. Whereas these early studies showed the validity of using anti-Ras antibodies for treatment of Ras-driven cancers, their clinical application was hampered by two technical challenges: (1) how to effectively deliver the antibodies into cancer cells, and (2) how to maintain the structural integrity of the antibodies, whose multiple disulfide bonds would be cleaved inside the reducing environment of the mammalian cytosol.

To overcome these challenges, researchers prepared antibody fragments, consisting of single-chain variable region domains (scFv) derived from Y13-259 and other anti-Ras antibodies (Fig. 2E). These antibody fragments are more stable intracellularly (because of their reduced number of disulfide bonds) and are also easier to produce recombinantly, especially in bacteria. Cochet et al. (1998) showed that intracellular expression of an scFv of Y13-259 led to the specific inhibition of the Ras signaling pathway in *Xenopus laevis* oocytes and NIH3T3 fibroblasts. Moreover, neutralizing Ras with the scFv specifically promoted apoptosis in vitro in human cancer cells but not in untransformed cells. As a step toward cancer gene therapy, they showed that intratumor transduction of HCT116 colon carcinoma cells with the anti-Ras scFv using an adenoviral vector elicited sustained tumor regression in nude mice. Tanaka and Rabbitts (2003) developed an intracellular antibody capture (IAC) technology to isolate antibody fragments against protein targets of interest. The antibody fragments were further engineered to generate "intrabodies" that can be stably expressed intracellularly. By screening a phage display library of 7×10^9 intrabodies against H-Ras^{G12V}, the investigators discovered a potent binder, iDab#6, which interacted selectively with mutant H-Ras^{G12V} loaded with GTPγS ($K_D = 6.2$ nM) over wild-type H-Ras and Ras-GDP and blocked Ras-GTP from interacting with effector proteins as well as downstream signaling pathways (Tanaka et al. 2007). Mutagenesis and X-ray crystallographic studies revealed that iDab#6 bound to the switch I/II regions on Ras (Fig. 1B). Expression of iDab#6 in mutant Ras-transformed mouse and human cells reverted them back to the untransformed phenotype. Furthermore, retroviral delivery of a membrane-anchored iDab#6 variant into tumor cell lines prevented Ras-dependent tumorigenesis in a mouse xenograft model. Interestingly, by using a transgenic mouse model, the same investigators later found that blocking the mutant Ras-effector protein interactions by iDab#6 was effective in preventing tumor initiation and controlling tumor growth, but was not sufficient to cause tumor regression (Tanaka and Rabbitts 2010). Very recently, Yang et al. constructed a recombinant adenovirus KGHV300 that carried an anti-p21Ras intrabody and that could replicate in tumor cell lines but not in normal cell lines (Yang et al. 2016a,b; Pan et al. 2017). In KGHV300, the expression levels of the essential replication genes E1a and E1b, were controlled by the human telomerase reverse transcriptase promoter and the hypoxia response element, respectively, and the anti-p21Ras intrabody gene was controlled by the cytomegalovirus promoter. The conditional replication of KGHV300 and its antitumor efficacy were shown in several tumor cell lines in vitro and in xenograft models of human breast cancer in nude mice. The recombinant adenovirus KGHV300 provides a promising and potentially safer antitumor therapeutic for Ras-driven cancers. Finally, Shin et al. (2017) engineered a cell-permeable antibody, RT11, which enters the cytosol of mammalian cells by endocytosis, selectively binds to the Ras-GTP form of various oncogenic Ras mutants, and blocks the Ras-effector interactions. RT11 suppressed Ras signaling and exerted antiproliferative effects in a variety of Ras mutant tumor cells. When systemically administered, an RT11 variant containing a tumor-targeting moiety significantly inhibited the in vivo growth of Ras-mutant tumor xenografts in mice, but not wild-type Ras-harboring tumors.

Nonimmunoglobulin Protein Inhibitors

Compared to antibodies and their fragments, nonimmunoglobulin protein scaffolds containing no disulfide bonds are more stable under reducing conditions and easier to produce recombinantly, offering alternative modalities as macromolecular therapeutics (Löfblom et al. 2010). Affibodies are ~6-kDa three-helix bundles containing flexible regions for sequence randomization (Fig. 2F). By screening a combinatorial affibody library displayed on bacteriophage, Grimm et al. (2010) isolated affibody molecules that bound H-Ras and Raf with 79 nM and 1.9 μM affinities, respectively. Interestingly, in a real-time biospecific interaction analysis, the two Ras targeting affibodies failed to inhibit the Ras–Raf interaction, suggesting that they bind a different site(s) from the effector protein-binding region. On the other hand, the Raf targeting affibody was able to block the Ras–Raf interaction in the same assay.

Cetin et al. (2017) developed a protein inhibitor against K-Ras using a miniprotein scaffold derived from the 10th fibronection type III domain of human fibronection (Fig. 2G). These 10-kD domains have an immunoglobulin-like fold but do not contain disulfide bonds. They contain two flexible loop regions (BC and FG) similar to the CDRH1 and CDRH3 loops at the antibody-combining site. These investigators generated a biased messenger RNA (mRNA) display library on the fibronection scaffold by incorporating the previously reported Ras-binding sequence from iDab#6 into the BC loop and a completely randomized sequence into the FG loop. Screening of the library followed by a second round of cell-based selection resulted in a hit molecule, RasIn1, which bound selectively to K-Ras^{G12V}-GTP (K_D = 2.1 μM) over K-Ras^{G12V}-GDP and wild-type K-Ras and blocked the Ras–Raf interaction. Mutational analysis showed the dependence of the BC and FG loops for Ras binding. When expressed inside mammalian cells, RasIn1 colocalized with the activated Ras^{G12V}. To further increase the potency of the hit molecule, a second-generation library was constructed based on the sequence of Rasln1 and screened to give Rasln2, which showed both improved binding affinity (K_D = 120 nM for K-Ras^{G12V}-GTP) and selectivity. ELISA assays suggested that Rasln2 binds to Ras^{G12V} with significantly higher affinity than the Ras-binding domain (RBD) of Raf. Similarly, Tamaskovic et al. (2016) developed designed ankyrin repeat proteins (DARPins) against Ras, which specifically blocked the Ras–RBD interactions. Breast cancer cells (which carry wild-type Ras) expressing the anti-Ras DARPins showed significantly higher susceptibility to anti-HER2 treatment (trastuzumab) than cells without RAS interference.

In addition to Ras-effector interaction blockers, proteins that allosterically inhibit Ras function have also been reported. Spencer-Smith et al. (2017) isolated a monobody molecule (NS1) from a phage display library. NS1 bound to H-Ras and K-Ras with K_D values of 14 and 67 nM, respectively, regardless of the GTP/GDP-bound state, but did not bind to N-Ras. NS1 effectively inhibited growth-factor-induced signaling and oncogenic H-Ras-/K-Ras-mediated transformation. Interestingly, NS1 had no effect on SOS-mediated nucleotide exchange of H-Ras, suggesting that its binding site also differs from that of SOS. X-ray crystal structure of the NS1/H-Ras complex revealed that NS1 binds to the dimerization interface of activated H-Ras, which is located on the back side of the protein with respect to the effector- and SOS-binding sites and divergent among the different Ras isoforms (Fig. 1B). Electron microscopic analysis showed that NS1 disrupted the dimerization and nanoclustering of H-Ras and K-Ras inside cells. NS1 inhibited Ras-induced Raf activation, consistent with the notion that Ras dimerization is required for recruitment of Raf to the plasma membrane.

NUCLEIC ACID–BASED INHIBITORS

siRNA

RNA interference (RNAi) provides an alternative to small-molecule and antibody-based therapeutics against challenging targets such as Ras. RNAi can, in principle, be applied to reversibly silence any target gene, including oncogenic *ras* mutants (as opposed to wild-type *ras* gene). Using short hairpin RNAs (shRNAs) to deplete K-Ras in lung and pancreatic cancer cell lines harboring K-ras mutations, Singh et al. (2009) identified two classes of cancer cell lines, those that do or do not require K-Ras to maintain viability. It was found that epithelial differentiation and tumor cell viability are associated, and that epithelial-mesenchymal transformation (EMT) regulators in "K-Ras-addicted" cancers represent candidate therapeutic targets. Since then, several recent studies have shown the validity of targeting mutant K-Ras by siRNAs as therapeutics. Khvalevsky et al. (2013) showed that siRNA targeting K-Ras mutations with a local prolonged release system knocks down K-Ras expression in vitro and in vivo (mouse models), leading to an antitumor effect. Treatment of pancreatic cancer cells with siRNA against K-Ras^{G12D} resulted in a significant decrease in K-Ras levels, leading to inhibition of proliferation and EMT. In vivo, intratumorally implanted siRNA impeded the growth of human pancreatic tumor cells and prolonged mouse survival. Pecot et al. (2014) subsequently showed that systemic delivery of K-Ras siRNAs with nanoparticles was effective for treatment of lung and colon cancers in mouse models. siRNA against K-Ras has also been chemically attached to anti-epidermal growth factor receptor (EGFR) antibodies for cancer-specific delivery to K-Ras-mutated EGFR-positive cancer cells (Bäumer et al. 2015). Finally, siRNA against mutant K-Ras has been used in combination with siRNAs against other Ras signaling pathway targets (e.g., Raf and PI3K) to treat colorectal cancer in xenograft models (Yuan et al. 2014).

Other Nucleic Acid–Based Approaches

Yu et al. (2009) reported an interesting approach to treating K-Ras mutant cancers with an oligodeoxyribonucleotide-based ribonuclease, 10-23 DNAzyme. The DNAzyme consists of a 15-nucleotide catalytic domain flanked by two target-binding sequences, and is capable of cleaving specific target mRNAs at a purine-pyrimidine dinucleotide. A K-Ras^{G12V}-specific DNAzyme was designed to cleave the mRNA of K-Ras^{G12V} (GGU→GUU) at the GU dinucleotide while leaving the wild-type K-Ras mRNA untouched. Transfection of SW480 cells (which carry homozygous K-Ras^{G12V} mutation) with the DNAzyme reduced K-Ras^{G12V} at both mRNA and protein levels, but not in HEK cells (wild-type K-Ras). Although the DNAzyme alone did not inhibit proliferation of SW480 or HEK cells, pretreatment with this DNAzyme sensitized the K-Ras^{G12V} mutant cells to anticancer agents such as doxorubicin and radiation. These results suggest the potential of using allele-specific DNAzymes in combination with other cancer therapies for more effective cancer treatment.

Another interesting approach involved direct targeting of mutant K-Ras DNA using pyrrole-imidazole polyamide bearing a reactive chloromethyl moiety (KR12) (Hiraoka et al. 2015). KR12 selectively recognizes oncogenic codon 12 K-Ras mutations and alkylates adenine N3 at the target sequence, causing strand cleavage and growth suppression in human colon cancer cells with G12D or G12V mutations. In xenograft models, KR12 infusions induced significant tumor growth suppression, with low host toxicity in K-Ras-mutated but not wild-type tumors.

CONCLUDING REMARKS AND OUTLOOK

Renewed efforts on the development of direct Ras inhibitors over the past few years have resulted in a large number of small-molecule, peptide, protein, and nucleic acid–based Ras inhibitors. Although achieving high potency and selectivity for small-molecule Ras inhibitors remains challenging (Zhou et al. 2017), highly potent and selective macromolecular Ras inhibitors have already been obtained. Some of the latter compounds (e.g., macrocyclic peptides and proteins) have shown remarkable selectivity for oncogenic Ras mutants over the wild-type protein, suggesting that selective targeting of mutant Ras proteins pharmacologically is feasible. When properly delivered into cancer cells and tissues (e.g., by intratumoral injection), some of the macromolecular Ras inhibitors have shown in vivo efficacy for preventing tumor initiation/growth and in some cases causing cancer cell apoptosis/tumor regression. Nucleic acid–based therapeutics have also shown in vivo efficacy in animal models and can, in principle, selectively target mutant Ras genes.

However, clinical development of these macromolecular Ras inhibitors still faces significant challenges. Because Ras is an intracellular target, any anti-Ras therapeutics must be effectively delivered into the cytosol of cancer cells. Larger molecules, such as the Ras inhibitors described in this review, violate almost every rule of Lipinski's rule of five (Lipinski et al. 2001) and generally cannot cross the plasma membrane by passive diffusion. The currently used methods for protein and nucleic acid delivery typically require endocytic uptake of the drug/vehicle complex, resulting in their initial localization in the early endosome (Bareford and Swaan 2007). To reach a cytosolic target (e.g., Ras), the drug must escape from the endosome into the cytosol by physically crossing the endosomal membrane. Endosomal escape is poorly understood and believed to be the limiting factor for all current nonviral protein and nucleic acid delivery technologies (Varkouhi et al. 2011). Protein therapeutics additionally face the challenge of immunogenicity if administered repeatedly. Fortunately, highly active cell-penetrating peptides have recently been discovered (Qian et al. 2013, 2016) and have been applied to efficiently deliver linear and macrocyclic peptides (Lian et al. 2014; Qian et al. 2015, 2017; Upadhyaya et al. 2015), proteins (Qian et al. 2014), and nucleic acids (D Pei, unpubl.) into the cytosol of mammalian cell in vitro and in vivo. Integration of the macromolecular Ras inhibitors with the latest delivery technologies may provide macromolecular therapeutics for clinical treatment of Ras mutant cancers.

ACKNOWLEDGMENTS

The Ras-related work in the Pei group is supported by grants from the National Institutes of Health (GM062820, GM110208, and GM122459).

REFERENCES

**Reference is also in this collection.*

Bareford LM, Swaan PW. 2007. Endocytic mechanisms for targeted drug delivery. *Adv Drug Deliv Rev* **59**: 748–758.

Barnard D, Sun H, Baker L, Marshall MS. 1998. In vitro inhibition of Ras-Raf association by short peptides. *Biochem Biophys Res Commun* **247**:176–180.

Bäumer S, Bäumer N, Appel N, Terheyden L, Fremerey J, Schelhaas S, Wardelmann E, Buchholz F, Berdel WE, Müller-Tidow C. 2015. Antibody-mediated delivery of anti-KRAS-siRNA in vivo overcomes therapy resistance in colon cancer. *Clin Cancer Res* **21**: 1383–1394.

Berndt N, Hamilton AD, Sebti SM. 2011. Targeting protein prenylation for cancer therapy. *Nat Rev Cancer* **11**: 775–791.

Britten CD. 2013. PI3K and MEK inhibitor combinations: examining the evidence in selected tumor types. *Cancer Chemother Pharmacol* **71**: 1395–1409.

Castellano E, Downward J. 2011. Ras interaction with PI3K: More than just another effector pathway. *Genes Cancer* **2**: 261–274.

Castellano E, Sheridan C, Thin MZ, Nye E, Spencer-Dene B, Diefenbacher ME, Moore C, Kumar MS, Murillo MM, Gronroos E, et al. 2013. Requirement for interaction of PI3-kinase p110α with RAS in lung tumor maintenance. *Cancer Cell* **24**: 617–630.

Cetin M, Evenson WE, Gross GG, Jalali-Yazdi F, Krieger D, Arnold D, Takahashi TT, Roberts RW. 2017. RasIns: Genetically encoded intrabodies of activated Ras proteins. *J Mol Biol* **429**: 562–573.

Chandra A, Grecco HE, Pisupati V, Perera D, Cassidy L, Skoulidis F, Ismail SA, Hedberg C, Hanzal-Bayer M, Venkitaraman AR, et al. 2012. The GDI-like solubilizing factor PDEδ sustains the spatial organization and signaling of Ras family proteins. *Nat Cell Biol* **14**: 148–158.

Clark GJ, Drugan JK, Terrell RS, Bradham C, Der CJ, Bell RM, Campbell S. 1996. Peptides containing a consensus Ras binding sequence from Raf-1 and the GTPase activating protein NF1 inhibit Ras function. *Proc Natl Acad Sci* **93**:1577–1581.

Cochet O, Kenigsberg M, Delumeau I, Virone-Oddos A, Multon MC, Fridman WH, Schweighoffer F, Teillaud JL, Tocqué B. 1998. Intracellular expression of an antibody fragment-neutralizing p21 ras promotes tumor regression. *Cancer Res* **58**: 1170–1176.

Dougherty PG, Qian Z, Pei D. 2017. Macrocycles as protein–protein interaction inhibitors. *Biochem J* **474**: 1109–1125.

Duncan JS, Whittle MC, Nakamura K, Abell AN, Midland AA, Zawistowski JS, Johnson NL, Granger DA, Jordan NV, Darr DB, et al. 2012. Dynamic reprogramming of the kinome in response to targeted MEK inhibition in triple-negative breast cancer. *Cell* **149**: 307–321.

Engelman JA, Chen L, Tan X, Crosby K, Guimaraes AR, Upadhyay R, Maira M, McNamara K, Perera SA, Song Y, et al. 2008. Effective use of PI3K and MEK inhibitors to treat mutant Kras G12D and PIK3CA H1047R murine lung cancers. *Nat Med* **14**: 1351–1356.

Feramisco J, Clark R, Wong G, Arnheim N, Milley R, McComick F. 1985. Transient reversion of ras oncogene-induced cell transformation by antibodies specific for amino acid 12 of ras protein. *Nature* **314**: 639–642.

Gareiss PC, Schneekloth AR, Salcius MJ, Seo SY, Crews CM. 2010. Identification and characterization of a peptidic ligand for Ras. *ChemBioChem* **11**: 517–522.

Grimm S, Lundberg E, Yu F, Shibasaki S, Vernet E, Skogs M, Nygren PÅ, Gräslund T. 2010. Selection and characterization of affibody molecules inhibiting the interaction between Ras and Raf in vitro. *N Biotechnol* **27**: 766–773.

Gupta S, Ramjaun AR, Haiko P, Wang Y, Warne PH, Nicke B, Nye E, Stamp G, Alitalo K, Downward J. 2007. Binding of Ras to phosphoinositide 3-kinase p110α is required for Ras-driven tumorigenesis in mice. *Cell* **129**: 957–968.

Hatzivassiliou G, Haling JR, Chen H, Song K, Price S, Heald R, Hewitt JFM, Zak M, Peck A, Orr C, et al. 2013. Mechanism of MEK inhibition determines efficacy in mutant KRAS- versus BRAF- driven cancers. *Nature* **501**: 232–236.

Hiraoka K, Inoue T, Taylor RD, Watanabe T, Koshikawa N, Yoda H, Shinohara K, Takatori A, Sugimoto H, Maru Y, et al. 2015. Inhibition of KRAS codon 12 mutants using a novel DNA-alkylating pyrrole-imidazole polyamide conjugate. *Nat Commun* **6**: 6706.

Khvalevsky EZ, Gabai R, Rachmut IH, Horwitz E, Brunschwig Z, Orbach A, Shemi A, Golan T, Domb AJ, Yavin E, et al. 2013. Mutant KRas is a druggable target for pancreatic cancer. *Proc Natl Acad Sci* **110**: 20723–20728.

Konstantinopoulos PA, Karamouzis MV, Papavassiliou AG. 2007. Post-translational modifications and regulation of the RAS superfamily of GTPases as anticancer targets. *Nat Rev Drug Discov* **6**: 541–555.

Kung HF, Smith MR, Bekesi E, Manne V, Stacey DW. 1986. Reversal of transformed phenotype by monoclonal antibodies against Ha-*ras* p21 proteins. *Exp Cell Res* **162**: 363–371.

Ledford H. 2015. The Ras renaissance. *Nature* **520**: 278–280.

Leshchiner ES, Parkhitko A, Bird GH, Luccarelli J, Bellairs JA, Escudero S, Opoku-Nsiah K, Godes M, Perrimon N, Walensky LD. 2015. Direct inhibition of oncogenic KRAS by hydrocarbon-stapled SOS1 helices. *Proc Natl Acad Sci* **112**: 1761–1766.

Lian W, Jiang B, Qian Z, Pei D. 2014. Cell-permeable bicyclic peptide inhibitors against intracellular proteins. *J Am Chem Soc* **136**: 9830–9833.

Lipinski CA, Lombardo F, Dominy BW, Feeney PJ. 2001. Experimental and computational approaches to estimate solubility and permeability in drug discovery and development settings. *Adv Drug Deliv Rev* **46**: 3–26.

Löfblom J, Feldwisch J, Tolmachev V, Carlsson J, Ståhl S, Frejd FY. 2010. Affibody molecules: engineered proteins for therapeutic, diagnostic and biotechnological applications. *FEBS Lett* **584**: 2670–2680.

Maurer T, Garrenton LS, Oh A, Pitts K, Anderson DJ, Skelton NJ, Fauber BP, Pan B, Malek S, Stokoe D, et al. 2012. Small-molecule ligands bind to a distinct pocket in Ras and inhibit SOS-mediated nucleotide exchange activity. *Proc Natl Acad Sci* **109**: 5299–5304.

Ostrem JM, Peters U, Sos ML, Wells JA, Shokat KM. 2013. K-Ras (G12C) inhibitors allosterically control GTP affinity and effector interactions. *Nature* **503**: 548–551.

Pan XY, Liu XJ, Li J, Zhen SJ, Liu DX, Feng Q, Zhao WX, Luo Y, Zhang YL, Li HW, et al. 2017. The antitumor efficacy of anti-p21Ras scFv mediated by the dual-promoter-regulated recombinant adenovirus KGHV300. *Gene Therapy* **24**: 40–48.

Patgiri A, Yadav KK, Arora PS, Bar-Sagi D. 2011. An orthosteric inhibitor of the Ras-Sos interaction. *Nat Chem Biol* **7**: 585–587.

Pecot CV, Wu SY, Bellister S, Filant J, Rupaimoole R, Hisamatsu T, Bhattacharya R, Maharaj A, Azam S, Rodriguez-Aguayo C, et al. 2014. Therapeutic silencing of KRAS using systemically delivered siRNAs. *Mol Cancer Ther* **13**: 2876–2885.

Poulikakos PI, Zhang C, Bollag G, Shokat KM, Rosen N. 2010. RAF inhibitors transactivate RAF dimers and ERK signaling in cells with wild-type BRAF. *Nature* **464**: 427–430.

Prior IA, Lewis PD, Mattos C. 2012. A comprehensive survey of Ras mutations in cancer. *Cancer Res* **72**: 2457–2467.

Qian Z, Liu T, Liu YY, Briesewitz R, Barrios AM, Jhiang SM, Pei D. 2013. Efficient delivery of cyclic peptides into mammalian cells with short sequence motifs. *ACS Chem Biol* **8**: 423–431.

Qian Z, LaRochelle JR, Jiang B, Lian W, Hard RL, Selner NG, Luechapanickhul R, Barrios AM, Pei D. 2014. Early endosomal escape of a cyclic cell-penetrating peptide allows effective cytosolic cargo delivery. *Biochemistry* **53**: 4034–4046.

Qian Z, Xu X, Amacher JF, Madden DR, Cormet-Boyaka E, Pei D. 2015. Intracellular delivery of peptidyl ligands by reversible cyclization: Discovery of a PDZ domain inhibitor that rescues CFTR activity. *Angew Chem Int Ed* **54**: 5874–5878.

Qian Z, Martyna A, Hard RL, Wang J, Appiah-Kubi G, Coss C, Phelps MA, Rossman JS, Pei D. 2016. Discovery and mechanism of highly efficient cyclic cell-penetrating peptides. *Biochemistry* **55**: 2601–2612.

Qian Z, Rhodes CA, McCroskey LC, Wen J, Appiah-Kubi G, Wang DJ, Guttridge DC, Pei D. 2017. Enhancing the cell permeability and metabolic stability of peptidyl drugs by reversible bicyclization. *Angew Chem Int Ed* **56**: 1525–1529.

Sakamoto K, Kamada Y, Sameshima T, Yaguchi M, Niida A, Sasaki S, Miwa M, Ohkubo S, Sakamoto J, Kamaura M, et al. 2017. K-Ras (G12D)-selective inhibitory peptides generated by random peptide T7 phage display technology. *Biochem Biophys Res Commun* **484**: 605–611.

Shin SM, Choi DK, Jung K, Bae J, Kim JS, Park SW, Song KH, Kim YS. 2017. Antibody targeting intracellular oncogenic Ras mutants exerts anti-tumour effects after systemic administration. *Nat Commun* doi: 10.1038/ncomms15090.

Singh A, Greninger P, Rhodes D, Koopman L, Violette S, Bardeesy N, Settleman J. 2009. A gene expression signature associated with "K-Ras addiction" reveals regulators of EMT and tumor cell survival. *Cancer Cell* **15**: 489–500.

Singh H, Longo DL, Chalmer BA. 2015. Improving prospects for targeting Ras. *J Clin Oncol* **33**: 3650–3659.

Spencer-Smith R, Koide A, Zhou Y, Eguchi RR, Sha F, Gajwani P, Santana D, Gupta A, Jacobs M, Herrero-Garcia E, et al. 2017. Inhibition of RAS function through targeting an allosteric regulatory site. *Nat Chem Biol* **13**: 62–68.

Spiegel J, Cromm PM, Zimmermann G, Grossmann TN, Waldmann H. 2014. Small-molecule modulation of Ras signaling. *Nat Chem Biol* **10**: 613–622.

Stephen AG, Esposito D, Bagni RK, McCormick F. 2014. Dragging ras back in the ring. *Cancer Cell* **25**: 272–281.

Tamaskovic R, Schwill M, Nagy-Davidescu G, Jost C, Schaefer DC, Verdurmen WPR, Schaefer JV, Honegger A, Plückthun A. 2016. Intermolecular biparatopic trapping of ErbB2 prevents compensatory activation of PI3K/AKT via RAS–p110 crosstalk. *Nat Commun* **7**: 11672.

Tanaka T, Rabbitts TH. 2003. Intrabodies based on intracellular capture frameworks that bind the RAS protein with high affinity and impair oncogenic transformation. *EMBO J* **22**: 1025–1035.

Tanaka T, Rabbitts TH. 2010. Interfering with RAS-effector protein interactions prevent RAS-dependent tumour initiation and causes stop–start control of cancer growth. *Oncogene* **29**: 6064–6070.

Tanaka T, Williams R, Rabbitts TH. 2007. Tumour prevention by a single antibody domain targeting the interaction of signal transduction proteins with RAS. *EMBO J* **6**: 3250–3259.

Trinh TB, Upadhyaya P, Qian Z, Pei D. 2016. Discovery of a direct Ras inhibitor by screening a combinatorial library of cell-permeable bicyclic peptides. *ACS Comb Sci* **18**: 75–85.

Upadhyaya P, Qian Z, Habir NA, Pei D. 2014. Direct Ras inhibitors identified from a structurally rigidified bicyclic peptide library. *Tetrahedron* **70**: 7714–7720.

Upadhyaya P, Qian Z, Selner NG, Clippinger SR, Wu Z, Briesewitz R, Pei D. 2015. Inhibition of Ras signaling by blocking Ras-effector interactions with cyclic peptides. *Angew Chem Int Ed* **54**: 7602–7606.

Varkouhi AK, Scholte M, Storm G, Haisma HJ. 2011. Endosomal escape pathways for delivery of biologicals. *J Control Release* **151**: 220–228.

Walensky LD, Bird GH. 2014. Hydrocarbon-stapled peptides: Principles, practice, and progress. *J Med Chem* **57**: 6275–6288.

Wang W, Fang G, Rudolph J. 2012. Ras inhibition via direct Ras binding—Is there a path forward? *Bioorg Med Chem Lett* **22**: 5766–5776.

Whyte DB, Kirschmeier P, Hockenberry TN, Nunez-Oliva I, James L, Catino JJ, Bishop WR, Pai JK. 1997. K- and N-Ras are geranylgeranylated in cells treated with farnesyl protein transferase inhibitors. *J Biol Chem* **272**: 14459–14464.

Winter-Vann AM, Baron RA, Wong W, dela Cruz J, York JD, Gooden DM, Bergo MO, Young SG, Toone EJ, Casey PJ. 2005. A small-molecule inhibitor of isoprenylcysteine carboxyl methyltransferase with antitumor activity in cancer cells. *Proc Natl Acad Sci* **102**: 4336–4341.

Wu X, U padhyaya P, Villalona-Calero MA, Briesewitz R, Pei D. 2013. Inhibition of Ras-effector interaction by cyclic peptides. *Medchemcomm* **4**: 378–382.

Xu CW, Luo Z. 2002. Inactivation of Ras function by allele-specific peptide aptamers. *Oncogene* **21**: 5753–5757.

Yang JL, Liu DX, Zhen SJ, Zhou YG, Zhang DJ, Yang LY, Chen HB, Feng Q. 2016a. A novel anti-p21Ras scFv antibody reacting specifically with human tumour cell lines and primary tumour tissues. *BMC Cancer* **16**: 131.

Yang J, Pan X, Zhao W, Hu Q, Ding F, Feng Q, Li G, Luo Y. 2016b. The antitumor efficacy of a novel adenovirus-mediated anti-p21Ras single chain fragment variable antibody on human cancers in vitro and in vivo. *Int J Oncol* **48**: 1218–1228.

Young A, Lyons J, Miller AL, Phan VT, Alarcón IR, McCormick F. 2009. Ras signaling and therapies. *Adv Cancer Res* **102**: 1–17.

Yu SH, Wang TH, Au LC. 2009. Specific repression of mutant K-RAS by 10-23 DNAzyme: Sensitizing cancer cell to anti-cancer therapies. *Biochem Biophys Res Commun* **378**: 230–234.

Yuan T, Fellmann C, Lee CS, Ritchie CD, Thapar V, Lee LC, Hsu DJ, Grace DO, Carver J, Zuber J, et al. 2014. Development of siRNA payloads to target KRAS-mutant cancer. *Cancer Discov* **4**:1182–1197.

* Zhou Y, Prakash P, Gorfe AA, Hancock JF. 2017. Ras and the plasma membrane: A complicated relationship. *Cold Spring Harb Perspect Med* doi: 10.1101/cshperspect.a031831.

Zimmermann G, Papke B, Ismail S, Vartak N, Chandra A, Hoffmann M, Hahn SA, Triola G, Wittinghofer A, Bastiaens PI, et al. 2013. Small molecule inhibition of the KRAS-PDEδ interaction impairs oncogenic KRAS signaling. *Nature* **497**: 638–642.

Targeting the MAPK Pathway in RAS Mutant Cancers

Sarah G. Hymowitz[1] and Shiva Malek[2]

[1]Department of Structural Biology, Genentech Inc., South San Francisco, California 94080

[2]Department of Discovery Oncology, Genentech Inc., South San Francisco, California 94080

Correspondence: shivam@gene.com

Despite decades of extensive drug discovery efforts, there are currently no targeted therapies approved to treat KRAS mutant cancers. In this review, we highlight the challenges and opportunities in targeting KRAS mutant tumors through inhibition of mitogen-activated protein kinase (MAPK) signaling with conformation-specific kinase inhibitors. Through structural analysis and mechanistic studies with BRAF and mitogen-activated protein kinase (MEK) inhibitors, we describe how kinase-dependent and -independent functions of MAPK signaling components regulate KRAS-driven tumorigenesis and how these insights can be used to treat RAS mutant cancers with small molecule kinase inhibitors.

MECHANISM OF NORMAL RAS/MAPK SIGNALING

HRAS, KRAS, and NRAS comprise a highly conserved family of small GTPases that relay signals from upstream growth factor receptors to downstream effector pathways, including mitogen-activated protein kinase (MAPK), phosphoinositide 3-kinase (PI3K), and RAL guanine nucleotide dissociation stimulator (RAL-GDS). Ligand stimulation results in activation of the guanine nucleotide exchange factor son of sevenless (SOS) and facilitates exchange of the inactive guanosine diphosphate (GDP)-bound state of RAS to an active guanosine triphosphate (GTP)-bound state (Fig. 1A). This switch between inactive and active states involves structural rearrangement of the Switch I (SW1) and Switch II (SW2) regions in RAS, enabling RAS to adopt a conformation that interacts with the RAS-binding domain (RBD) of its downstream effectors (Fig. 1B) (Erijman and Shifman 2016). In the context of MAPK signaling, GTP-bound RAS recruits rapidly accelerated fibrosarcoma (RAF) family members (ARAF, BRAF, CRAF) from the cytosol to the plasma membrane through a direct and high affinity interaction with the RAF RBD (Fig. 1B). Structurally, the RBD has a ubiquitin-like fold and interacts with RAS by forming an extended β sheet in the vicinity of the switch regions (Nassar et al. 1995). This binding mode is common to most RAS–effector interactions.

More recent studies have suggested that RAS dimerizes via a region distal to the SW1 and SW2 regions (the α4–α5 region) (Fig. 1C). These dimers form at the membrane and may also be required for proper signaling activity,

Cite this article as *Cold Spring Harb Perspect Med* doi: 10.1101/cshperspect.a031492

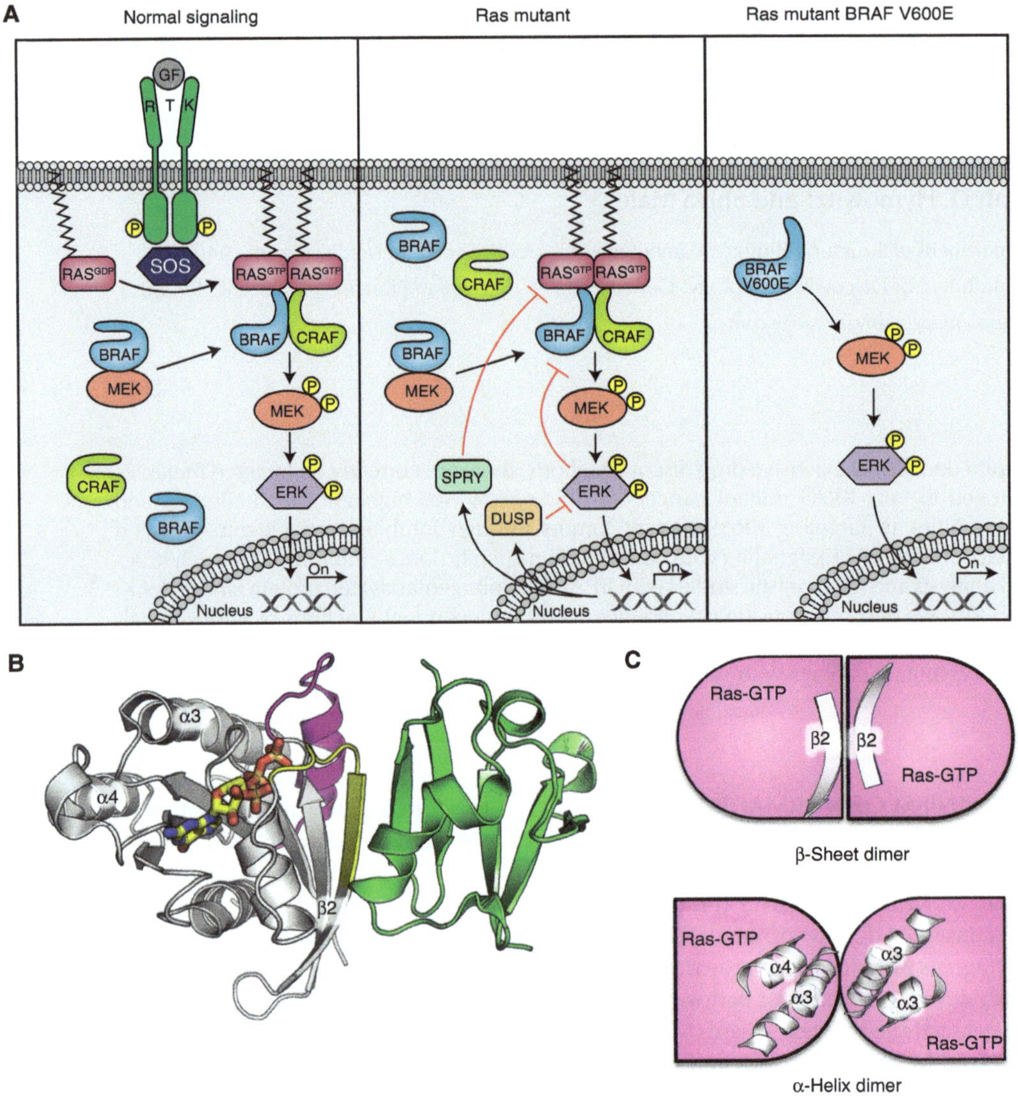

Figure 1. RAS/mitogen-activated protein kinase (MAPK) pathway signaling and protein–protein interactions. (*A*) Pathway signaling showing normal, RAS mutant, and BRAF V600E mutant signaling. (*B*) Structure of the Ras-like small GTPase Rap1A (white) with Switch I (yellow) and Switch II (magenta) bound to the RAS-binding domain (RBD) domain (green) of CRAF (Protein Data Bank [PDB] code 1C1Y). (*C*) Cartoon of two proposed RAS dimers proposed by Muratcioglu et al. (2015) formed either by the β2 strand from Switch I, which forms a dimer analogous to the Ras–RBD interactions, or by mediated α3/α4.

including RAF dimerization and activation (Lin et al. 2014; Spencer-Smith et al. 2017). Indeed, it has been proposed that upon GTP binding, the α4–α5 regions of RAS are positioned in close proximity and may promote RAF dimerization and activation. Lending support to this model is the fact that NS1, a monobody that selectively recognizes the α4–α5 region of RAS, prevents RAS dimerization and RAF activation (Spencer-Smith et al. 2017). Dimerization of RAF family members has been shown to induce kinase activity likely through a conformational change releasing the negative regulatory region (NRR) within the amino-terminal regions of RAF (Weber et al. 2001; Udell et al. 2011) and exposing the carboxy-terminal kinase domains. This results in phosphorylation and activation of RAF through an asymmetric allosteric mechanism (Garnett et al. 2005; Rajakulendran et al. 2009; Hu et al. 2013) that triggers mitogen-activated protein kinase (MEK) and extracellular signal-related kinase (ERK) phosphorylation.

Under physiological conditions, the duration and amplitude of MAPK signaling is regulated through a network of negative feedback loops mediated by ERK. ERK suppresses pathway signaling either directly through phosphorylation of upstream components such as SOS or CRAF or transcriptionally through the regulation of expression of MAPK phosphatases (dual-specificity phosphatase [DUSP] family members) and Sprouty family of proteins (Fig. 1A) (Dougherty et al. 2005; Owens and Keyse 2007; Pratilas et al. 2009). These negative feedback loops enable the pathway output to be tightly controlled under basal and stressed conditions. Pathway signaling is also tightly regulated by a variety of additional mechanisms, including phosphorylation and interaction with 14-3-3 proteins (both of which have been described extensively in previous reviews) as well as interaction with the scaffolding protein kinase suppressor of (activated) RAS ([KSR], see more below; Freeman et al. 2013; Lavoie and Therrien 2015).

RAF FAMILY DOMAIN ARCHITECTURE AND MECHANISM OF ASYMMETRIC ALLOSTERIC ACTIVATION

The protein domains of ARAF, BRAF, and CRAF are highly conserved, with each containing three conserved regions: CR1, CR2, and CR3 (Fig. 2A). CR1 contains the RBD and cysteine-rich domains (CRDs). CR2 contains a stretch of serine/threonine residues implicated in activation of RAF. Hence, both CR1 and CR2 together are commonly referred to as the NRR. CR3 is composed of the kinase domain and is held inactive via its interaction with the amino-terminal regulatory region (Lavoie and Therrien 2015). Interaction with RAS-GTP relieves the self-regulatory interaction between the NRR and kinase domains, such that RAF can dimerize with a second protomer and potentiate downstream signaling through the MEK-ERK pathway.

Although the molecular details of the interaction between the NRR and the RAF kinase domain are yet to be determined, the structure of the RAF kinase domain has been studied in detail both to better understand the mechanism of normal and oncogenic kinase activation as well as to support structure-based drug design of pathway-specific inhibitors. Structures of BRAF and CRAF kinase domains show a typical bi-lobed architecture of kinases with an adenosine triphosphate (ATP) (or inhibitor)-binding site at the interface between the amino (N-lobes) and carboxy-terminal lobes (C-lobes) (Wan et al. 2004; Hatzivassiliou et al. 2010). Key features include the activation loop (A-loop), the αC-helix, the catalytic salt bridge between K^{483} and E^{501} (in BRAF), the DFG motif, and the hinge region between the N- and C-lobes. All of these features contribute to the normal catalytic role of kinases and most are affected by the transition between the active and inactive conformation of RAF. On activation of RAF, the A-loop forms an extended platform for substrate binding while the αC-helix moves "in" toward the active site leading the formation of the catalytic K^{483}–E^{501} salt bridge while the DFG motif is positioned such that the hydrophobic phenylalanine contributes to the hydrophobic core (Fig. 2B).

Immediately before the kinase domain is a short stretch of amino acids termed the amino-terminal acidic region (NtA; residues ^{446}SSDD^{449} in BRAF). This region is constitutively phosphorylated in BRAF in addition to possessing two negatively charged residues (Lavoie and Therrien 2015). In CRAF, the sequence (^{338}SSYY^{341}) lacks the negatively charged residues and is only phosphorylated on recruitment to the plasma membrane. Phosphorylation of these residues in CRAF has been shown to enhance dimerization and RAF kinase activity (Hu et al. 2013). This amino acid sequence difference may account for the increased intrinsic activity of BRAF relative to CRAF (Emuss et al. 2005). The phosphorylated residues in the NtA are typically disordered in structures of BRAF, indicating that the mechanism of activation may rely on

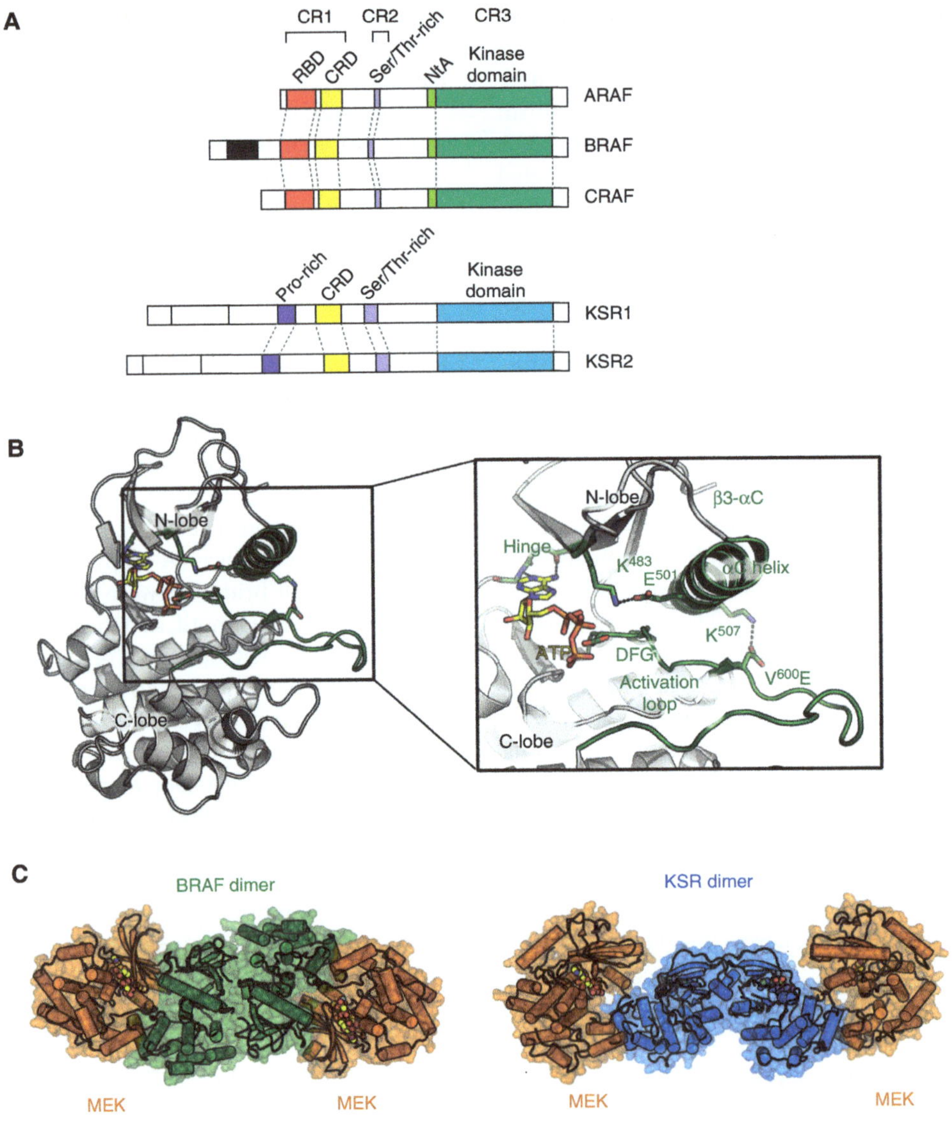

Figure 2. Domain organization and structure of RAF family members. (*A*) Domain organization of ARAF, BRAF, CRAF, and kinase suppressor of (activated) RAS (KSR) with features of interest labeled. (*B*) Model of BRAF kinase domain (based on chain B form PDB code 4MNE) with adenosine triphosphate (ATP) docked in and V600E mutation modeled. Regions of interest are labeled and colored green. (*C*) Comparison of BRAF-mitogen-activated protein kinase (MEK) heterotetramer (*left*, in green and orange, respectively; PDB code 1MNE) and the KSR-MEK heterotetramer (*right*, in blue and orange, respectively; PDB code 2Y4I).

destabilizing the repressive effects of the NRR. When they are resolved, they reside close to R509 and have been suggested to help form the proper dimerization interface. On release of the NRR region by interaction with RAS, the kinase domains dimerize stabilizing the active conformation of BRAF.

BRAF and CRAF dimers can exist as homodimers (Farrar et al. 1996; Luo et al. 1996) and heterodimers (Mizutani et al. 2001), with heterodimers contributing to the majority of kinase activity within the cell (Rushworth et al. 2006). Recent work has shown that RAF activity is allosterically regulated via the side-to-side dimer

 Cite this article as *Cold Spring Harb Perspect Med* doi: 10.1101/cshperspect.a031492

configuration that positions the αC-helix in a catalytically active orientation (Rajakulendran et al. 2009). Removal of the regulatory amino terminus results in a constitutively active kinase, as does the induction of RAF dimerization through mutation of an interface residue (Rajakulendran et al. 2009). Furthermore, kinase-dead BRAF can activate CRAF in *trans* (Wan et al. 2004; Rushworth et al. 2006; Heidorn et al. 2010). A similar transactivation mechanism of CRAF by BRAF has been reported with certain ATP-competitive BRAF inhibitors (Hatzivassiliou et al. 2010; Heidorn et al. 2010; Poulikakos et al. 2010).

Further arguing for an asymmetric dimer state, the distant RAF family member KSR can also activate RAF activity. KSR was originally identified through RAS synthetic lethal genetic screens in *Drosophila melanogaster* and *Caenorhabditis elegans* (Kornfeld et al. 1995; Sundaram and Han 1995; Therrien et al. 1995). In these studies, KSR was shown to function downstream from RAS regulating RAS/MAPK signaling through direct interactions with RAF, MEK, and ERK family members. Interestingly, the KSR proteins are most closely related to the RAF kinase family, with the carboxy-terminal region of KSR containing a pseudokinase domain, a region similar to canonical kinase domains yet lacking a critical catalytic lysine residue (Fig. 2A). Hence, KSR lacks kinase activity but has been shown to regulate RAF activity through a dimerization-dependent mechanism involving its pseudokinase domain (Roy et al. 2002; Rajakulendran et al. 2009; Hu et al. 2011). Given its role as a scaffolding protein, KSR expression levels have been shown to modulate the duration and amplitude of MAPK signaling (Therrien et al. 1996). Moderate levels of KSR expression enhance MAPK signaling, whereas high levels of KSR expression appear to function in a dominant negative manner by dampening pathway signaling likely through a sequestering mechanism (Kortum and Lewis 2004). Biochemical and structural studies have shown that KSR can dimerize with RAF and MEK proteins through its pseudokinase domain (Rajakulendran et al. 2009; Brennan et al. 2011). Dimerization of KSR with BRAF through a side-to-side dimer interface has been shown to enhance RAF kinase activity (Brennan et al. 2011).

MECHANISM OF ONCOGENIC RAS/MAPK SIGNALING

Oncogenic mutations in BRAF and KRAS circumvent the highly regulated activation process through a variety of mechanisms. The precise mechanism of MAPK pathway activation is dependent on the specific underlying genetic mutation. For example, the activating BRAF-V600E mutation or BRAF-β3-αC deletion, which is found in nearly 70% of melanomas and a small subset of pancreatic adenocarcinomas, respectively, predisposes BRAF to an active αC-helix-"in" conformation that is independent of RAS activation and BRAF dimerization for activity (Wan et al. 2004; Poulikakos et al. 2011; Foster et al. 2016b). In contrast, noncanonical BRAF mutations are not necessarily highly activating and, in some cases, abrogate kinase activity completely (i.e., BRAF-D594V). In this context, the mutant BRAF requires dimerization with CRAF for MAPK activity (Wan et al. 2004; Yao et al. 2015). Interestingly, noncanonical BRAF mutations are enriched in lung cancers (~5%–7% of lung adenocarcinomas) and are mutually exclusive with KRAS mutations, suggesting that there may be a tissue-specific preference for this alteration (Foster et al. 2016a).

Similar to the mechanism of noncanonical BRAF mutations, activating KRAS mutations drive MAPK signaling through CRAF and BRAF dimerization and CRAF activation (Garnett et al. 2005; Dumaz et al. 2006; Rushworth et al. 2006). Mutations within the KRAS family member represent the second most frequent alteration in cancer, found in 20% of all patients, and 90% of pancreatic, 43% of colorectal, and 32% of lung adenocarcinomas (Cerami et al. 2012; Gao et al. 2013) with mutations typically occurring at codons 12, 13, or 61. These mutations result in reduced intrinsic and GTPase-activating protein (GAP)-mediated GTPase activity, rendering KRAS in a constitutively active, GTP-bound form. Activated RAS can then signal to its downstream effector pathways driving cell growth, proliferation, and survival in the absence of external signaling cues.

More recently, it has been reported that in the context of KRAS mutant tumors, BRAF is in a preassembled, inactive complex with MEK (Fig. 1A) (Haling et al. 2014). MEK phosphorylation results in rapid dissociation of the complex and kinetic studies with epidermal growth factor (EGF) stimulation (in wild-type cells) show that the complex can be reformed on accumulation of inactive MEK. Intriguingly, a cocrystal structure of the active, heterotetrameric complex of BRAF with MEK has been determined in the presence of a MEK inhibitor G-573 and ATP (Haling et al. 2014). In this structure, BRAF side-to-side dimers reside at the heart of the tetrameric complex, with BRAF adopting an active conformation and mediating a face-to-face interaction with MEK (Fig. 2C). Small molecule inhibitors of RAF or MEK can perturb this complex depending on their binding mode. Interestingly, unlike the KSR-MEK complex previously reported, the BRAF-MEK complex is high in affinity and appears to favor BRAF in its active conformation. Furthermore, structural studies show that the BRAF-MEK and KSR-MEK interfaces are largely overlapping such that MEK and KSR cannot interact with the BRAF kinase domain simultaneously (Haling et al. 2014). This suggests that KSR is more likely to function as a scaffolding protein that may transactivate RAF family members through an allosteric, side-to-side dimerization mechanism rather than through recruitment of MEK as a substrate for RAF.

These mechanistic insights into BRAF and KRAS mutant tumors underscore the fact that BRAF has both catalytic and noncatalytic functions in regulating MAPK signaling. In the context of KRAS mutant tumors, BRAF resides in the cytosol in a dormant complex with MEK and additional signaling cues are required to form a transient, active heterotetrameric complex with CRAF that results in MEK phosphorylation. In contrast, BRAF-V600E and BRAF deletions activate BRAF in the absence of dimerization (Fig. 1). Finally, noncanonical BRAF mutations (including kinase-dead oncogenic mutations) constitutively dimerize with CRAF and allosterically activate CRAF through a side-to-side dimerization mechanism. Taken together, regulation of RAF dimerization is the rate-limiting step in pathway activity and oncogenic mutations in BRAF either abrogate the requirement for dimerization for kinase activity or result in constitutive dimers independent of KRAS as in the case of the noncanonical BRAF mutations. In KRAS mutant tumors, additional studies need to be conducted to understand how the preassembled MEK-BRAF complex is primed for activity.

TARGETING THE MAPK PATHWAY TO TREAT RAS MUTANT TUMORS

Several lines of evidence support the notion that targeting the MAPK effector pathway may be beneficial for KRAS mutant patients. First, CRAF but not BRAF has been shown to be required for KRAS driven signaling (Dumaz et al. 2006) as well as for the initiation of KRAS driven lung tumorigenesis in a genetically engineered KRAS driven lung cancer mouse model (Blasco et al. 2011; Karreth et al. 2011). In this KRAS genetically engineered mouse model, depletion of CRAF and not BRAF increased survival and decreased tumor incidence. Second, of the effector pathways downstream from RAS, KRAS and BRAF mutations appear to be mutually exclusive of one another, suggesting redundant functions. In contrast, mutations in other effector pathways such as PI3K can co-occur significantly with KRAS mutations (as exemplified in colorectal cancer [CRC]), suggesting nonoverlapping functions. Third, a recent genome-wide clustered regularly interspaced short palindromic repeat (CRISPR) dropout screen (Wang et al. 2017) of 14 acute myeloid leukemia (AML) cell lines with a subset of KRAS mutant and KRAS wild-type lines identified CRAF as a top KRAS synthetic lethal hit and not an essential gene, again suggesting that of the effector pathways downstream from KRAS, the MAPK pathway, namely, the RAF family members, remain critical for KRAS-driven tumorigenesis. Thus, of the downstream effector pathways, the MAPK pathway continues to be the most compelling therapeutic target in the context of RAS mutant cancers.

Cite this article as *Cold Spring Harb Perspect Med* doi: 10.1101/cshperspect.a031492

USING MAPK INHIBITORS IN BRAF AND RAS MUTANT CANCERS

Significant efforts have been undertaken to find small molecule inhibitors to selectively target the MAPK pathway in malignancies. BRAF and MEK inhibitors have been approved to treat BRAF mutant cancers with dozens of other RAF, MEK, and ERK inhibitors either still in preclinical research stage or early stage clinical trials (Table 1). These drugs bind to a diverse array of kinase conformational states and have been shown to exhibit efficacy in mutation-specific contexts. Despite extensive clinical efforts, the only approved MAPK therapies to-date are the BRAF inhibitors vemurafenib and dabrafenib and the MEK inhibitors trametinib and cobimetinib, all of which have been approved only for BRAF V600E/K metastatic melanomas (Kim et al. 2014).

CLASSES OF RAF INHIBITORS

RAF inhibitors can be grouped into several classes based on their molecular mechanism of action, all of which are ATP-competitive (Fig. 3). Type I inhibitors such as GDC-0879, which bind to a similar site as ATP, make interactions with the hinge region and favor an active conformation of the A-loop, αC-helix, DFG motif and permit formation of the K^{483}–E^{501} catalytic salt bridge (Hatzivassiliou et al. 2010). In contrast, the DFG-"out" (also known as type II) inhibitors such as sorafenib, AZ-628, and LY3009120 bind to RAF such that the F within the DFG motif is oriented toward solvent as opposed to being buried in the hydrophobic core of the protein (Wan et al. 2004; Peng et al. 2015; Foster et al. 2016b). However, the αC-helix remains in the "in" position with formation of the catalytic salt bridge. Type I.5 inhibitors have an intermediate effect. They typically occupy both the ATP-binding site as well as extend toward the αC-helix, perturbing its conformation and forcing it into varying degrees of an "out" conformation in which the catalytic salt bridge is disrupted.

All of these inhibitors prevent the ability of RAF to phosphorylate MEK. However, the differences in RAF conformation, affinity for ATP, and interactions with partners results in different cellular consequences. For instance, all classes of RAF inhibitors preferentially inhibit BRAF-V600E in cells. This is attributed to the fact that the V600E mutations result in a ~10-fold decrease in ATP K_M (reduced affinity for ATP), similar to what has been reported for epidermal growth factor receptor (EGFR) mutations, thereby sensitizing this specific mutant oncoprotein to ATP-competitive inhibitors (Hatzivassiliou et al. 2010). In contrast, type II (but not type I.5) inhibitors are able to inhibit tumors harboring noncanonical BRAF mutations that are constitutively dimerized with CRAF (Yao et al. 2015). In this context, BRAF-CRAF dimerization conformationally restricts the αC-helix from shifting to accommodate the type I.5 inhibitor-binding mode, and hence RAF dimer inhibitors (which are typically type II inhibitors) have been shown to be effective toward RAF dimers (Foster et al. 2016b; Karoulia et al. 2016). In yet another context, BRAF-β3-αC deletions are resistant to type I.5 inhibitors because the αC-helix is conformationally restricted from moving as a result of the deletion in the β3-αC loop (Foster et al. 2016b). However, because the conformation caused by these deletions is similar to conformation induced by di-

Table 1. BRAF, MEK, and ERK inhibitors with their binding modes defined with structures shown below

Molecule	Target	Binding mode	αC-helix "in" or "out"
GDC-0879	BRAF	Type I	"in"
Vemurafenib	BRAF	Type I.5	"out"
Dabrafenib	BRAF	Type I.5	"out"
PLX-8394	BRAF	Type I.5	"out"
Sorafenib	BRAF	Type II	"in"
AZ-628	BRAF	Type II	"in"
Trametinib	MEK	Class I	"out"
Cobimetinib	MEK	Class I	"out"
GDC-0623	MEK	Class II	"out"
G-573	MEK	Class II	"out"
CH5126766	MEK	Class II	"out"
GDC-0994	ERK	Type I	"in"
SCH772984	ERK	Type I	"in"

MEK, Mitogen-activated protein kinase; ERK, extracellular signal-related kinase.

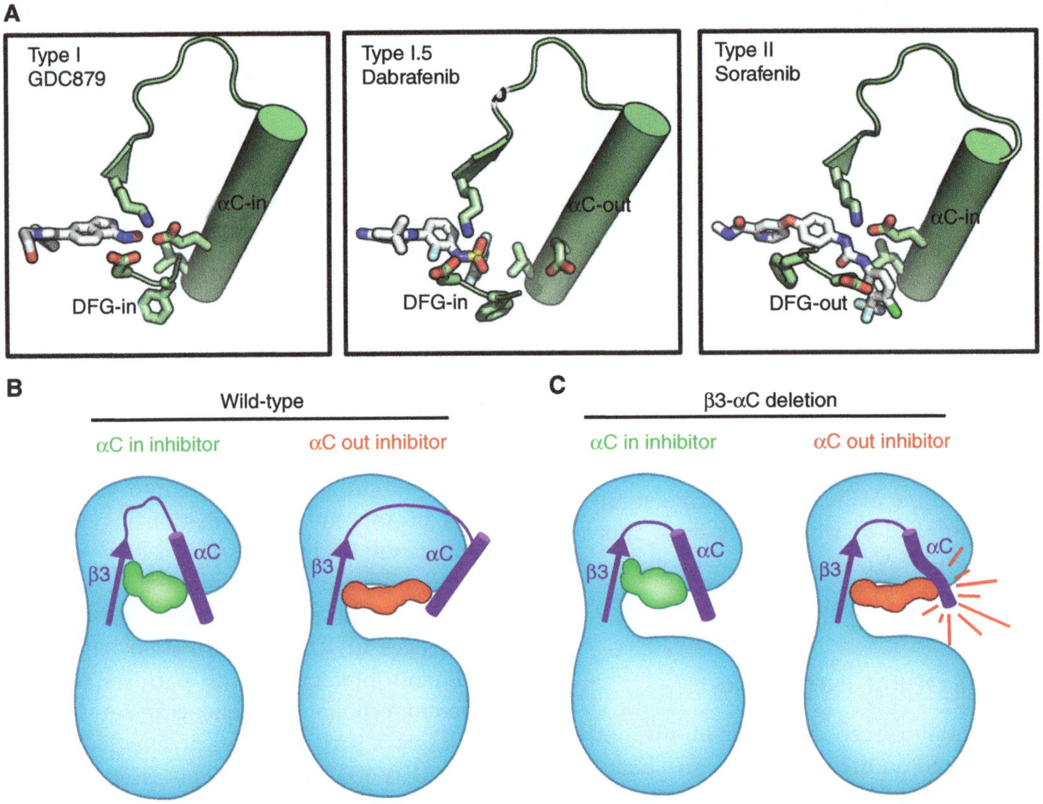

Figure 3. BRAF inhibitor-binding modes. (*A*) Cartoon images of BRAF wild-type and β3-αC deletion mutants bound to type I, type I.5, and type II inhibitors, including the effect on αC-helix and the catalytic salt bridge. (*B*) Cartoon representations of BRAF bound to type I (*left*), type I.5 (*center*), and type II inhibitors (*right*), including the effect on the αC-helix and the catalytic salt bridge.

merization (shift of the αC-helix "in"), this class of mutation is sensitive to type II inhibitors. Interestingly, whereas both the BRAF-V600E and BRAF-β3-αC deletions do not require dimerization for activity, they have distinct requirements for inhibition. In the case of BRAF-V600E, a type I.5 inhibitor can inhibit its activity effectively because the C-helix remains conformationally flexible and able to "shift" on inhibitor binding; however, deletion of the β3-αC loop restricts the C-helix from this same movement with type I.5 inhibitors (Fig. 3).

The approved BRAF inhibitors vemurafenib and dabrafenib are both type I.5 inhibitors, although the extent of the interaction with the αC-helix results in differences in their activity (Joseph et al. 2010; Zhang et al. 2015; Foster et al. 2016b). Vemurafenib was derived from a fragment-based drug discovery approach and, along with the B-cell lymphoma 2 (Bcl-2) inhibitor venetoclax, are examples of this approach leading to approved therapeutics (Erlanson et al. 2016). In particular, the fragment lead for vemurafinib was found in a high-concentration biochemical assay-based screen of a 20,000 library of small (150–300 Da) scaffold-like compounds. Extensive structure-based drug design was used in the process of discovering vemurafenib. A highly engineered version of BRAF with improved expression, purification, and crystallization properties was developed to support this effort (Bollag et al. 2012).

Most type I and type I.5 inhibitors, including the clinically approved drugs, have been shown preclinically to induce paradoxical activation of MAPK signaling in RAS mutant cells or cells with elevated RTK signaling (Hatzivassiliou et al. 2010; Poulikakos et al. 2010). In some contexts, these inhibitors have been shown to increase cellular proliferation both in vitro and in vivo (Hatzivassiliou et al. 2010). Paradoxical activation is believed to be the consequence of drug-induced dimerization of the drug-bound BRAF protomer with a second RAF protomer (i.e., CRAF), which results in increased CRAF kinase activity (which is RAS-GTP dependent) and downstream signaling through MEK and ERK. At high concentrations of the drugs, both protomers of the dimer (i.e., BRAF and CRAF) are occupied and the paradoxical activation is inhibited. It has been hypothesized that the significant incidence of cutaneous squamous cell carcinoma and keratoacanthomas (~20% of patients) observed clinically with BRAF-V600E melanoma patients treated with BRAF inhibitors is attributed to paradoxical activation (Anforth et al. 2012a,b; Su et al. 2012). In support of this, these paradox-induced skin tumors show a high incidence of HRAS mutations, consistent with the preclinical observations demonstrating enhanced growth of RAS mutant tumors on treatment with vemurafenib (Anforth et al. 2012a; Callahan et al. 2012; Oberholzer et al. 2012). These clinical findings have raised concerns that secondary RAS-driven tumors may be accelerated by this class of drugs.

Two main approaches have been explored to address the deleterious effects of paradoxical activation. Type I.5 binders have been discovered that induce even greater effects on the conformation of the αC-helix such that BRAF/CRAF dimerization can be more effectively inhibited (Zhang et al. 2015). This class of compounds, also called "paradox busters," still effectively inhibits BRAF-V600E tumors but do not induce paradoxical activation. The best characterized paradox buster PLX-8394 has been shown to block BRAF-V600E activity without inducing paradoxical activation in RAS mutant tumors (Zhang et al. 2015). Although the structural basis of this compared with other type I.5 inhibitors is not clear, structural studies suggest that these sulfonamide-containing type I.5 inhibitors may interact with Leu505 within the C-helix, which may result in greater dynamics in this region, thus disfavoring the dimeric structure. Unfortunately, despite the lack of paradoxical activation, paradox busters still show no efficacy in KRAS mutant tumors as they are not able to bind to preexisting RAF dimers. Alternatively, type II pan-RAF inhibitors have been shown to bind to BRAF/CRAF heterodimers and thereby prevent induction of MAPK signaling while remaining effective at inhibiting BRAF-V600E-dependent signaling. In this context, the type II inhibitors will still induce BRAF and CRAF dimerization in RAS mutant tumors, but paradoxical activation is not observed owing to occupancy of both promoters simultaneously (Nakamura et al. 2013; Peng et al. 2015).

Despite the significant progress that has been made in developing inhibitors that selectively target BRAF, several of which have been clinically approved for treating BRAF-V600 mutant cancers, these approaches have resulted in little success with KRAS mutant cancers. Some hope may lie in recent reports of type II RAF dimer inhibitors demonstrating efficacy in KRAS mutant tumors; however, the dose required for efficacy in KRAS mutant tumors is significantly greater than that required for efficacy in BRAF-V600E models, raising concerns around the therapeutic utility of such an approach as a single agent (Peng et al. 2015). Several recent reports have also described the combination of RAF inhibitors, with MEK inhibitors showing efficacy in RAS mutant tumors (Lamba et al. 2014; Lito et al. 2014). In this context, it has been shown that MEK inhibitors disable the negative feedback loops upstream of RAF, resulting in RAF kinase activation and increased sensitivity to RAF kinase inhibition. It remains to be seen whether the combination of RAF and MEK inhibitors will be successful clinically in KRAS mutant tumors.

CLASSES OF MEK INHIBITORS

In addition to approaches to targeting RAF itself, MEK inhibitors have also been approved for BRAF-V600E/K melanomas. Phosphorylation by BRAF at two sites in the MEK activation loop (Ser218 and 222 in MEK1) correlates with release of MEK from BRAF, MEK activation, and downstream signaling to ERK. MEK1 and MEK2 are unusual dual-specificity kinases that phosphorylate ERK on both serine and tyrosine residues within the ERK activation loop. Unlike RAF, MEK does not possess extensive accessory domains analogous to the RAF NRR. In contrast, MEK is primarily comprised of a kinase domain with short flanking amino- and carboxy-terminal regions with sequences that modulate MEK activity by mechanisms, including facilitating interactions with binding partners and dictating subcellular localization (Fig. 4). MEK interacts with RAF by forming a face-to-face heterodimer leaving RAF free to dimerize with another RAF protomer, such that dimeric BRAF can bind two copies of MEK simultaneously (Haling et al. 2014). Unlike most kinases, MEK binds ATP with high affinity, which makes development of non-ATP competitive inhibitors the more preferred approach for this kinase. Accordingly, the majority of MEK inhibitors developed to date, including the two clinically approved inhibitors cobimetinib and trametinib, both bind to MEK in the presence of ATP within an allosteric binding pocket. The affinity of these inhibitors is increased in the presence of ATP, supporting a noncompetitive mechanism of action (Fig. 3) (Ohren et al. 2004; Caunt et al. 2015).

In contrast to the clinically approved BRAF inhibitors, trametinib and cobimetinib do not induce paradoxical activation. However, similar to BRAF inhibitors, the mechanism of action of these MEK inhibitors modulates their ability to blunt MAPK signaling effectively in KRAS mutant tumors. Class I MEK inhibitors such as cobimetinib show efficacy in BRAF-V600E tumors, but have limited efficacy in KRAS mutant tumors (Hatzivassiliou et al. 2013). Studies have shown that this class of MEK inhibitor disables negative feedback in KRAS mutant tumors and, over time, results in pathway reactivation as observed with induction of phosphorylated MEK and phosphorylated ERK. Multiple studies have shown that this pathway reactivation is highly dependent on RAF kinase activation and the combination of RAF and MEK inhibition results in more robust efficacy in KRAS mutant tumors (Hatzivassiliou et al. 2013; Lito et al. 2014). In contrast, class II MEK inhibitors such as G-573 have shown efficacy in KRAS mutant tumors, demonstrating sustained inhibition of the pathway, despite disabled negative feedback (Hatzivassiliou et al. 2013; Lito et al. 2014). This activity has been attributed to the dual role of this class as RAF-MEK inhibitors through trapping and sequestering an inactive RAF-MEK complex. Therefore, these MEK inhibitors do not induce phosphorylated MEK and ERK over time and effectively inhibit MAPK signaling pathways in KRAS mutant tumors. These differences in behavior of class I and class II inhibitors likely derive from subtle differences of their interaction with the MEK A-loop (in particular the strength of the hydrogen bond to Ser212) (Hatzivassiliou et al. 2013). This effectively alters the stability of the inactive BRAF-MEK complex and contributes to the extent of MEK phosphorylation suppression.

ERK INHIBITORS

ERK inhibitors have also been investigated for use in both BRAF-V600E and KRAS mutant tumors. As ERK is the terminal kinase in the pathway, it has numerous substrates, and inhibition of its activity can have a broad downstream effect. As such, ERK has been targeted for small molecule drug discovery for inhibitors to work as single agents or in conjunction with upstream kinase inhibitors. Like MEK, the overall ERK structure is simpler than that of RAF, suggesting less complicated regulation. Unlike MEK, well-characterized ERK inhibitors to date are ATP competitive and bind to αC-"in" conformation. This includes the ERK inhibitor GDC-0994, which has been tested clinically (Fig. 4) (Blake et al. 2016). More recently, a novel binding mode with the ERK inhibitor SCH772984 has been reported. High-resolution crystal structures of SCH772984 bound to ERK1 and ERK2 reveal that the extended structure of this inhibitor resulted in discrete interactions with an induced binding pocket located between the αC-helix and the phosphate-binding loop (P-loop). This binding mode maintains the DFG and catalytic salt bridge in an active conformation while modestly tilting the C-helix

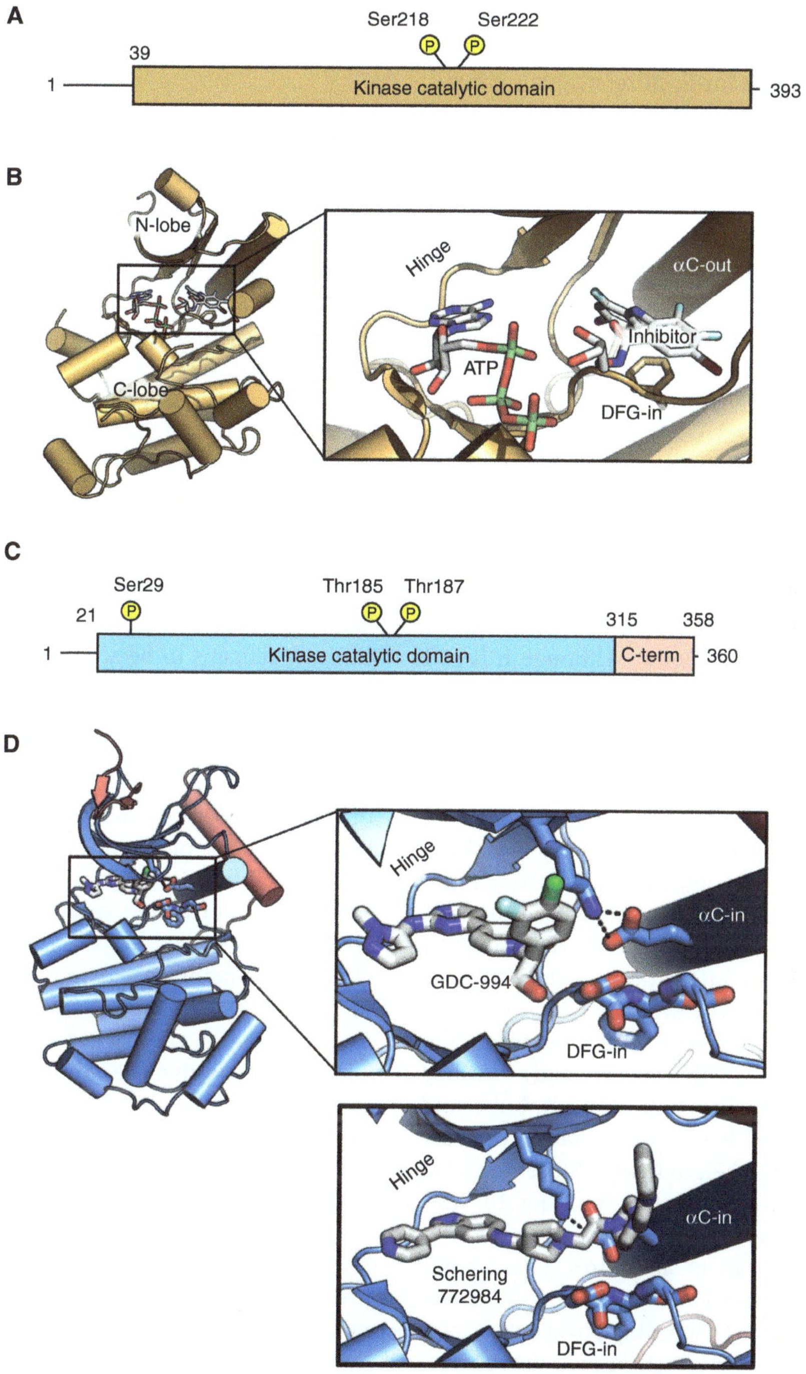

Figure 4. Domain organization and structure of mitogen-activated protein kinase (MEK) and extracellular signal-related kinase (ERK). (*A*) Domain organization of MEK1. (*B*) Structure of the MEK kinase domain (PDB 1S9J). *Inset* shows features of interest, including the adenosine triphosphate (ATP) site, allosteric inhibitor site, and the αC-helix that adopts an "out" conformation, including disruption of the catalytic salt bridge. (*C*) Domain structure of ERK. (*D*) Structure of ERK kinase bound to GDC-0994 (blue with additions to the canonical kinase domain in red; PDB 5K4I). C-term, Carboxy terminal.

and slightly distorting the P-loop. It remains to be seen whether these differential binding conformations result in differences in clinical responses.

KSR INHIBITORS

In addition to catalytic components, the RAF pathway is modulated by the pseudokinase KSR (KSR1/2). KSR possesses an overall domain structure analogous to that of BRAF but lacks the critical catalytic lysine residues thereby rendering it a pseudokinase. KSR has been described as having a scaffolding function as it can associate with many of the components of the MAPK pathway to potentiate signaling (Raabe and Rapp 2002). In particular, the kinase domain of KSR can homodimerize and interact with MEK (Brennan et al. 2011) or can heterodimerize and hence activate BRAF (and CRAF) (Rajakulendran et al. 2009). Although it is still unclear whether residual KSR enzymatic activity is essential for its biological function (Brennan et al. 2011), the ability of KSR to activate the pathway appears to depend in part on the conformation of the pseudokinase domain, suggesting that small molecules, which bind to KSR in specific conformational states, might be able to regulate its function. For instance, blocking the KSR active site by the mutation A587F stabilizes interactions with RAF and MEK, decreases pathway activity, and has been shown to prevent RAS transformation (Hu et al. 2011). This was originally ascribed to a latent catalytic activity of KSR, but could also be attributable to a sequestration of MEK by a KSR/RAF heterodimer. More recently, ATP-competitive inhibitors binding to KSR have been described and cocrystal structures reported (Dhawan et al. 2016). These compounds, exemplified by APS-2-79, bind to the KSR within the ATP-binding pocket and stabilize an inactive conformation such that heterodimerization with BRAF is blocked. Although the kinase selectivity of these tool compounds is not yet optimal and the target conformational state not clearly defined, they serve as good tool compounds to probe the effect of KSR conformation states and better understand how KSR regulates MAPK signaling downstream from KRAS.

CONCLUDING REMARKS

Identifying approaches to target RAS mutant tumors remains a critical unmet need in cancer therapy. As described in this review, there is strong rationale to target the MAPK signaling pathway downstream from KRAS; however, inhibition of any individual target in the pathway has had little single activity in KRAS mutant tumors to date. Some of this is dependent on the nature of the inhibitor used as described for the various classes of RAF and MEK inhibitors. Depending on the genetic context, molecules binding to the same target can either inhibit or activate the pathway in a drug-induced, conformation-specific manner. In other cases, efficacy has been limited by dose-limiting toxicities in the clinic. It remains to be seen whether combinations of agents targeting the MAPK pathway will show efficacy in RAS mutant tumors; many of those trials are currently underway.

However, given the profound differences in efficacy observed with subtle differences in binding modes of various MAPK inhibitors, the possibility exists that other classes of allosteric or conformation-specific inhibitors will show efficacy in this context. Discovery of such classes of inhibitors may require novel approaches such as phenotypic screens and novel fragment screening approaches directly targeting allosteric sites rather than the ATP-competitive binding sites. Allosteric binding sites for other kinases have been reported in the literature, including for oncogenic kinases such as EGFR and BCR-ABL (Zhang et al. 2010; Jia et al. 2016). Further, novel biological, biochemical, and structural data characterizing scaffolding proteins that regulate this pathway may provide new directions for either single agent or combination therapies to treat KRAS mutant cancers. Although much as been learned about the RAF pathway and its regulation in recent years, much remains to be discovered especially with regard to the development of therapeutics with transformative tolerability and durability targeting MAPK pathway-driven cancers.

 Cite this article as *Cold Spring Harb Perspect Med* doi: 10.1101/cshperspect.a031492

REFERENCES

Anforth R, Tembe V, Blumetti T, Fernandez-Penas P. 2012a. Mutational analysis of cutaneous squamous cell carcinomas and verrucal keratosis in patients taking BRAF inhibitors. *Pigment Cell Melanoma Res* **25:** 569–572.

Anforth RM, Blumetti TC, Kefford RF, Sharma R, Scolyer RA, Kossard S, Long GV, Fernandez-Penas P. 2012b. Cutaneous manifestations of dabrafenib (GSK2118436): A selective inhibitor of mutant BRAF in patients with metastatic melanoma. *Br J Dermatol* **167:** 1153–1160.

Blake JF, Burkard M, Chan J, Chen H, Chou KJ, Diaz D, Dudley DA, Gaudino JJ, Gould SE, Grina J, et al. 2016. Discovery of (*S*)-1-(1-(4-Chloro-3-fluorophenyl)-2-hydroxyethyl)-4-(2-((1-methyl-1*H*-pyrazol-5-yl)amino) pyrimidin-4-yl)pyridin-2(1*H*)-one (GDC-0994), an extracellular signal-regulated kinase 1/2 (ERK1/2) inhibitor in early clinical development. *J Med Chem* **59:** 5650–5660.

Blasco RB, Francoz S, Santamaria D, Canamero M, Dubus P, Charron J, Baccarini M, Barbacid M. 2011. C-Raf, but not B-Raf, is essential for development of K-Ras oncogene-driven non-small cell lung carcinoma. *Cancer Cell* **19:** 652–663.

Bollag G, Tsai J, Zhang J, Zhang C, Ibrahim P, Nolop K, Hirth P. 2012. Vemurafenib: The first drug approved for BRAF-mutant cancer. *Nat Rev Drug Discov* **11:** 873–886.

Brennan DF, Dar AC, Hertz NT, Chao WC, Burlingame AL, Shokat KM, Barford D. 2011. A Raf-induced allosteric transition of KSR stimulates phosphorylation of MEK. *Nature* **472:** 366–369.

Callahan MK, Rampal R, Harding JJ, Klimek VM, Chung YR, Merghoub T, Wolchok JD, Solit DB, Rosen N, Abdel-Wahab O, et al. 2012. Progression of RAS-mutant leukemia during RAF inhibitor treatment. *N Engl J Med* **367:** 2316–2321.

Caunt CJ, Sale MJ, Smith PD, Cook SJ. 2015. MEK1 and MEK2 inhibitors and cancer therapy: The long and winding road. *Nat Rev Cancer* **15:** 577–592.

Cerami E, Gao J, Dogrusoz U, Gross BE, Sumer SO, Aksoy BA, Jacobsen A, Byrne CJ, Heuer ML, Larsson E, et al. 2012. The cBio cancer genomics portal: An open platform for exploring multidimensional cancer genomics data. *Cancer Discov* **2:** 401–404.

Dhawan NS, Scopton AP, Dar AC. 2016. Small molecule stabilization of the KSR inactive state antagonizes oncogenic Ras signalling. *Nature* **537:** 112–116.

Dougherty MK, Muller J, Ritt DA, Zhou M, Zhou XZ, Copeland TD, Conrads TP, Veenstra TD, Lu KP, Morrison DK. 2005. Regulation of Raf-1 by direct feedback phosphorylation. *Mol Cell* **17:** 215–224.

Dumaz N, Hayward R, Martin J, Ogilvie L, Hedley D, Curtin JA, Bastian BC, Springer C, Marais R. 2006. In melanoma, RAS mutations are accompanied by switching signaling from BRAF to CRAF disrupted cyclic AMP signaling. *Cancer Res* **66:** 9483–9491.

Emuss V, Garnett M, Mason C, Marais R. 2005. Mutations of C-RAF are rare in human cancer because C-RAF has a low basal kinase activity compared with B-RAF. *Cancer Res* **65:** 9719–9726.

Erijman A, Shifman JM. 2016. RAS/effector interactions from structural and biophysical perspective. *Mini Rev Med Chem* **16:** 370–375.

Erlanson DA, Fesik SW, Hubbard RE, Jahnke W, Jhoti H. 2016. Twenty years on: The impact of fragments on drug discovery. *Nat Rev Drug Discov* **15:** 605–619.

Farrar MA, Alberol-Ila J, Perlmutter RM. 1996. Activation of the Raf-1 kinase cascade by coumermycin-induced dimerization. *Nature* **383:** 178–181.

Foster SA, Klijn C, Malek S. 2016a. Tissue specific mutations in BRAF and EGFR necessitate unique therapeutic approaches. *Trends Cancer* **2:** 699–701.

Foster SA, Whalen DM, Ozen A, Wongchenko MJ, Yin J, Yen I, Schaefer G, Mayfield JD, Chmielecki J, Stephens PJ, et al. 2016b. Activation mechanism of oncogenic deletion mutations in BRAF, EGFR, and HER2. *Cancer Cell* **29:** 477–493.

Freeman AK, Ritt DA, Morrison DK. 2013. The importance of Raf dimerization in cell signaling. *Small GTPases* **4:** 180–185.

Gao J, Aksoy BA, Dogrusoz U, Dresdner G, Gross B, Sumer SO, Sun Y, Jacobsen A, Sinha R, Larsson E, et al. 2013. Integrative analysis of complex cancer genomics and clinical profiles using the cBioPortal. *Sci Signal* **6:** pl1.

Garnett MJ, Rana S, Paterson H, Barford D, Marais R. 2005. Wild-type and mutant B-RAF activate C-RAF through distinct mechanisms involving heterodimerization. *Mol Cell* **20:** 963–969.

Haling JR, Sudhamsu J, Yen I, Sideris S, Sandoval W, Phung W, Bravo BJ, Giannetti AM, Peck A, Masselot A, et al. 2014. Structure of the BRAF-MEK complex reveals a kinase activity independent role for BRAF in MAPK signaling. *Cancer Cell* **26:** 402–413.

Hatzivassiliou G, Song K, Yen I, Brandhuber BJ, Anderson DJ, Alvarado R, Ludlam MJ, Stokoe D, Gloor SL, Vigers G, et al. 2010. RAF inhibitors prime wild-type RAF to activate the MAPK pathway and enhance growth. *Nature* **464:** 431–435.

Hatzivassiliou G, Haling JR, Chen H, Song K, Price S, Heald R, Hewitt JF, Zak M, Peck A, Orr C, et al. 2013. Mechanism of MEK inhibition determines efficacy in mutant KRAS- versus BRAF-driven cancers. *Nature* **501:** 232–236.

Heidorn SJ, Milagre C, Whittaker S, Nourry A, Niculescu-Duvas I, Dhomen N, Hussain J, Reis-Filho JS, Springer CJ, Pritchard C, et al. 2010. Kinase-dead BRAF and oncogenic RAS cooperate to drive tumor progression through CRAF. *Cell* **140:** 209–221.

Hu J, Yu H, Kornev AP, Zhao J, Filbert EL, Taylor SS, Shaw AS. 2011. Mutation that blocks ATP binding creates a pseudokinase stabilizing the scaffolding function of kinase suppressor of Ras, CRAF and BRAF. *Proc Natl Acad Sci* **108:** 6067–6072.

Hu J, Stites EC, Yu H, Germino EA, Meharena HS, Stork PJ, Kornev AP, Taylor SS, Shaw AS. 2013. Allosteric activation of functionally asymmetric RAF kinase dimers. *Cell* **154:** 1036–1046.

Jia Y, Yun CH, Park E, Ercan D, Manuia M, Juarez J, Xu C, Rhee K, Chen T, Zhang H, et al. 2016. Overcoming EGFR (T790M) and EGFR(C797S) resistance with mutant-selective allosteric inhibitors. *Nature* **534:** 129–132.

Joseph EW, Pratilas CA, Poulikakos PI, Tadi M, Wang W, Taylor BS, Halilovic E, Persaud Y, Xing F, Viale A, et al. 2010. The RAF inhibitor PLX4032 inhibits ERK signaling and tumor cell proliferation in a V600E BRAF-selective manner. *Proc Natl Acad Sci* **107:** 14903–14908.

Karoulia Z, Wu Y, Ahmed TA, Xin Q., Bollard J, Krepler C, Wu X, Zhang C, Bollag G, Herlyn M, et al. 2016. An integrated model of RAF inhibitor action predicts inhibitor activity against oncogenic BRAF signaling. *Cancer Cell* **30:** 485–498.

Karreth FA, Frese KK, DeNicola GM, Baccarini M, Tuveson DA. 2011. C-Raf is required for the initiation of lung cancer by K-Ras^{G12D}. *Cancer Discov* **1:** 128–136.

Kim G, McKee AE, Ning YM, Hazarika M, Theoret M, Johnson JR, Xu QC, Tang S, Sridhara R, Jiang X, et al. 2014. FDA approval summary: Vemurafenib for treatment of unresectable or metastatic melanoma with the BRAFV600E mutation. *Clin Cancer Res* **20:** 4994–5000.

Kornfeld K, Hom DB, Horvitz HR. 1995. The *ksr-1* gene encodes a novel protein kinase involved in Ras-mediated signaling in *C. elegans*. *Cell* **83:** 903–913.

Kortum RL, Lewis RE. 2004. The molecular scaffold KSR1 regulates the proliferative and oncogenic potential of cells. *Mol Cell Biol* **24:** 4407–4416.

Lamba S, Russo M, Sun C, Lazzari L, Cancelliere C, Grernrum W, Lieftink C, Bernards R, Di Nicolantonio F, Bardelli A. 2014. RAF suppression synergizes with MEK inhibition in KRAS mutant cancer cells. *Cell Rep* **8:** 1475–1483.

Lavoie H, Therrien M. 2015. Regulation of RAF protein kinases in ERK signalling. *Nature Rev Mol Cell Biol* **16:** 281–298.

Lin WC, Iversen L, Tu HL, Rhodes C, Christensen SM, Iwig JS, Hansen SD, Huang WY, Groves JT. 2014. H-Ras forms dimers on membrane surfaces via a protein–protein interface. *Proc Natl Acad Sci* **111:** 2996–3001.

Lito P, Saborowski A, Yue J, Solomon M, Joseph E, Gadal S, Saborowski M, Kastenhuber E, Fellmann C, Ohara K, et al. 2014. Disruption of CRAF-mediated MEK activation is required for effective MEK inhibition in KRAS mutant tumors. *Cancer Cell* **25:** 697–710.

Luo Z, Tzivion G, Belshaw PJ, Vavvas D, Marshall M, Avruch J. 1996. Oligomerization activates c-Raf-1 through a Ras-dependent mechanism. *Nature* **383:** 181–185.

Mizutani S, Inouye K, Koide H, Kaziro Y. 2001. Involvement of B-Raf in Ras-induced Raf-1 activation. *FEBS Lett* **507:** 295–298.

Muratcioglu S, Chavan TS, Freed BC, Jang H, Khavrutskii L, Freed RN, Dyba MA, Stefanisko K, Tarasov SG, Gursoy A, et al. 2015. GTP-dependent K-Ras dimerization. *Structure* **23:** 1325–1335.

Nakamura A, Arita T, Tsuchiya S, Donelan J, Chouitar J, Carideo E, Galvin K, Okaniwa M, Ishikawa T, Yoshida S. 2013. Antitumor activity of the selective pan-RAF inhibitor TAK-632 in BRAF inhibitor-resistant melanoma. *Cancer Res* **73:** 7043–7055.

Nassar N, Horn G, Herrmann C, Scherer A, McCormick F, Wittinghofer A. 1995. The 2.2 A crystal structure of the Ras-binding domain of the serine/threonine kinase c-Raf1 in complex with Rap1A and a GTP analogue. *Nature* **375:** 554–560.

Oberholzer PA, Kee D, Dziunycz P, Sucker A, Kamsukom N, Jones R, Roden C, Chalk CJ, Ardlie K, Palescandolo E, et al. 2012. RAS mutations are associated with the development of cutaneous squamous cell tumors in patients treated with RAF inhibitors. *J Clin Oncol* **30:** 316–321.

Ohren JF, Chen H, Pavlovsky A, Whitehead C, Zhang E, Kuffa P, Yan C, McConnell P, Spessard C, Banotai C, et al. 2004. Structures of human MAP kinase kinase 1 (MEK1) and MEK2 describe novel noncompetitive kinase inhibition. *Nat Struct Mol Biol* **11:** 1192–1197.

Owens DM, Keyse SM. 2007. Differential regulation of MAP kinase signalling by dual-specificity protein phosphatases. *Oncogene* **26:** 3203–3213.

Peng SB, Henry JR, Kaufman MD, Lu WP, Smith BD, Vogeti S, Rutkoski TJ, Wise S, Chun L, Zhang Y, et al. 2015. Inhibition of RAF isoforms and active dimers by LY3009120 leads to anti-tumor activities in RAS or BRAF mutant cancers. *Cancer Cell* **28:** 384–398.

Poulikakos PI, Zhang C, Bollag G, Shokat KM, Rosen N. 2010. RAF inhibitors transactivate RAF dimers and ERK signalling in cells with wild-type BRAF. *Nature* **464:** 427–430.

Poulikakos PI, Persaud Y, Janakiraman M, Kong X, Ng C, Moriceau G, Shi H, Atefi M, Titz B, Gabay MT, et al. 2011. RAF inhibitor resistance is mediated by dimerization of aberrantly spliced BRAF(V600E). *Nature* **480:** 387–390.

Pratilas CA, Taylor BS, Ye Q, Viale A, Sander C, Solit DB, Rosen N. 2009. (V600E)BRAF is associated with disabled feedback inhibition of RAF-MEK signaling and elevated transcriptional output of the pathway. *Proc Natl Acad Sci* **106:** 4519–4524.

Raabe T, Rapp UR. 2002. KSR—A regulator and scaffold protein of the MAPK pathway. *Science STKE* **2002:** pe28.

Rajakulendran T, Sahmi M, Lefrancois M, Sicheri F, Therrien M. 2009. A dimerization-dependent mechanism drives RAF catalytic activation. *Nature* **461:** 542–545.

Roy F, Laberge G, Douziech M, Ferland-McCollough D, Therrien M. 2002. KSR is a scaffold required for activation of the ERK/MAPK module. *Genes Dev* **16:** 427–438.

Rushworth LK, Hindley AD, O'Neill E, Kolch W. 2006. Regulation and role of Raf-1/B-Raf heterodimerization. *Mol Cell Biol* **26:** 2262–2272.

Spencer-Smith R, Koide A, Zhou Y, Eguchi RR, Sha F, Gajwani P, Santana D, Gupta A, Jacobs M, Herrero-Garcia E, et al. 2017. Inhibition of RAS function through targeting an allosteric regulatory site. *Nat Chem Biol* **13:** 62–68.

Su F, Viros A, Milagre C, Trunzer K, Bollag G, Spleiss O, Reis-Filho JS, Kong X, Koya RC, Flaherty KT, et al. 2012. RAS mutations in cutaneous squamous-cell carcinomas in patients treated with BRAF inhibitors. *N Engl J Med* **366:** 207–215.

Sundaram M, Han M. 1995. The *C. elegans ksr-1* gene encodes a novel Raf-related kinase involved in Ras-mediated signal transduction. *Cell* **83:** 889–901.

Therrien M, Chang HC, Solomon NM, Karim FD, Wassarman DA, Rubin GM. 1995. KSR, a novel protein kinase required for RAS signal transduction. *Cell* **83:** 879–888.

Therrien M, Michaud NR, Rubin GM, Morrison DK. 1996. KSR modulates signal propagation within the MAPK cascade. *Genes Dev* **10:** 2684–2695.

Udell CM, Rajakulendran T, Sicheri F, Therrien M. 2011. Mechanistic principles of RAF kinase signaling. *Cell Mol Life Sci* **68:** 553–565.

Wan PT, Garnett MJ, Roe SM, Lee S, Niculescu-Duvaz D, Good VM, Jones CM, Marshall CJ, Springer CJ, Barford D, et al. 2004. Mechanism of activation of the RAF-ERK signaling pathway by oncogenic mutations of B-RAF. *Cell* **116:** 855–867.

Wang T, Yu H, Hughes NW, Liu B, Kendirli A, Klein K, Chen WW, Lander ES, Sabatini DM. 2017. Gene essentiality profiling reveals gene networks and synthetic lethal interactions with oncogenic ras. *Cell* **168:** 890–903 e815.

Weber CK, Slupsky JR, Kalmes HA, Rapp UR. 2001. Active Ras induces heterodimerization of cRaf and BRaf. *Cancer Res* **61:** 3595–3598.

Yao Z, Torres NM, Tao A, Gao Y, Luo L, Li Q, de Stanchina E, Abdel-Wahab O, Solit DB, Poulikakos PI, et al. 2015. BRAF mutants evade ERK-dependent feedback by different mechanisms that determine their sensitivity to pharmacologic inhibition. *Cancer Cell* **28:** 370–383.

Zhang J, Adrian FJ, Jahnke W, Cowan-Jacob SW, Li AG, Iacob RE, Sim T, Powers J, Dierks C, Sun F, et al. 2010. Targeting Bcr-Abl by combining allosteric with ATP-binding-site inhibitors. *Nature* **463:** 501–506.

Zhang C, Spevak W, Zhang Y, Burton EA, Ma Y, Habets G, Zhang J, Lin J, Ewing T, Matusow B, et al. 2015. RAF inhibitors that evade paradoxical MAPK pathway activation. *Nature* **526:** 583–586.

Ras-Specific GTPase-Activating Proteins—Structures, Mechanisms, and Interactions

Klaus Scheffzek and Giridhar Shivalingaiah

Division of Biological Chemistry (Biocenter), Medical University of Innsbruck, A-6020 Innsbruck, Austria

Correspondence: Klaus.Scheffzek@i-med.ac.at

Ras-specific GTPase-activating proteins (RasGAPs) down-regulate the biological activity of Ras proteins by accelerating their intrinsic rate of GTP hydrolysis, basically by a transition state stabilizing mechanism. Oncogenic Ras is commonly not sensitive to RasGAPs caused by interference of mutants with the electronic or steric requirements of the transition state, resulting in up-regulation of activated Ras in respective cells. RasGAPs are modular proteins containing a helical catalytic RasGAP module surrounded by smaller domains that are frequently involved in the subcellular localization or contributing to regulatory features of their host proteins. In this review, we summarize current knowledge about RasGAP structure, mechanism, regulation, and dual-substrate specificity and discuss in some detail neurofibromin, one of the most important negative Ras regulators in cellular growth control and neuronal function.

GTPase-activating proteins (GAPs) are commonly thought to act as negative regulators of guanine nucleotide-binding proteins (GNBPs) (Bourne et al. 1990, 1991; Vetter and Wittinghofer 2001). They function by accelerating GNBP-mediated GTP hydrolysis thereby down-regulating the biological activity of their targets (Scheffzek and Ahmadian 2005; Mishra and Lambright 2016). Although they appear to be family specific in the sense that each GNBP family has its cognate GAP module, substrate promiscuous variants are known (Krapivinsky et al. 2004; Kupzig et al. 2006). Currently there are 14 Ras-specific GTPase-activating protein (RasGAP) homologous sequences known, the majority of which result in RasGAP active proteins (Bernards 2003). The discovery of p120GAP/RASA1 by Trahey and McCormick (1987; Trahey et al. 1988; Vogel et al. 1988) not only marked the first and Ras-directed GTPase-activating protein (RasGAP), but most importantly opened up a new concept of how G proteins are regulated in the cell (Boguski and McCormick 1993). Today we know that virtually every G protein has a cognate GAP as well as a cognate guanine nucleotide exchange factor (GEF) (Bos et al. 2007; Cherfils and Zeghouf 2013). In this review, we look back over 30 years of RasGAP research, here focusing on structural and biochemical aspects. For functional and physiological aspects, the reader is referred to respective excellent review articles published over the last decade (Donovan et al. 2002; Yarwood et al. 2006; Iwashita and Song 2008; Grewal et al. 2011; Ligeti et al. 2012; King et al. 2013; Maertens and Cichowski 2014).

Cite this article as *Cold Spring Harb Perspect Med* doi: 10.1101/cshperspect.a031500

RasGAPs typically display a modular architecture of variable size, typically ranging from 100 to 300 kDa. They share a catalytic RasGAP module (35 kDa) (see Figs. 1 and 2) commonly flanked by C2 (Cho and Stahelin 2006), pleckstrin homology (PH) (Lemmon 2004; Scheffzek and Welti 2012), calponin homology (CH) (Banuelos et al. 1998), or Sec14-like domains (Bankaitis et al. 2010; Ghosh and Bankaitis 2011) with variable positioning in the coding sequence (Bernards 2003). Some canonical RasGAPs show at least dual-target specificity inactivating Ras and Rap family members (Krapivinsky et al. 2004; Kupzig et al. 2006). Members of the IQGAP group show no detectable RasGAP activity but bind to Rho family members (Kurella et al. 2009; LeCour et al. 2016). Generally, accessory domains have been implicated in protein localization as well as in regulatory features (Pamonsinlapatham et al. 2009).

From a systematic viewpoint, six groups of mammalian RasGAP homologous sequences are commonly distinguished (see Fig. 2):

1. P120GAP/RASA1 (120 kDa) contains a carboxy-terminal RasGAP module (Marshall et al. 1989) preceded by a PH and C2 domain and an SH2/SH3/SH2 motif (Pamonsinlapatham et al. 2009). It is ubiquitously expressed and is essential for embryonic development (Henkemeyer et al. 1995). Germline mutations in the RASA1 gene are associated with the genetic disorder capillary vascular malformation-arteriovenous malformation (CVM-AM) with most of the lesions implying loss of function of the protein (Boon et al. 2005).
2. Neurofibromin (320 kDa), the protein product of the neurofibromatosis type 1 (NF1) tumor suppressor gene *NF1*, contains a central GAP-related domain (GRD) (Ballester et al. 1990; Martin et al. 1990; Xu et al. 1990b) followed by a bipartite glycerophospholipid-binding Sec14-PH-like module (D'Angelo et al. 2006; Welti et al. 2007). Neurofibromin is ubiquitously expressed with highest levels in neuronal cells in the adult. It is essential for embryonic development and mice lacking a functional *NF1* gene die early from cardiovascular defects (Henkemeyer et al. 1995). NF1 patients have germline alterations in the *NF1* gene causative for the disease (Riccardi 1992;

Figure 1. Sculpture of the canonical Ras-specific GTPase-activating protein (RasGAP) module showing the catalyitic portion central domain (GAPc) in red and the extra domain (GAPex) in green. (Artwork, by G. Haeusser, was created using foam rubber strips and acrylic on hard board [original size: 48″ × 35.4″].)

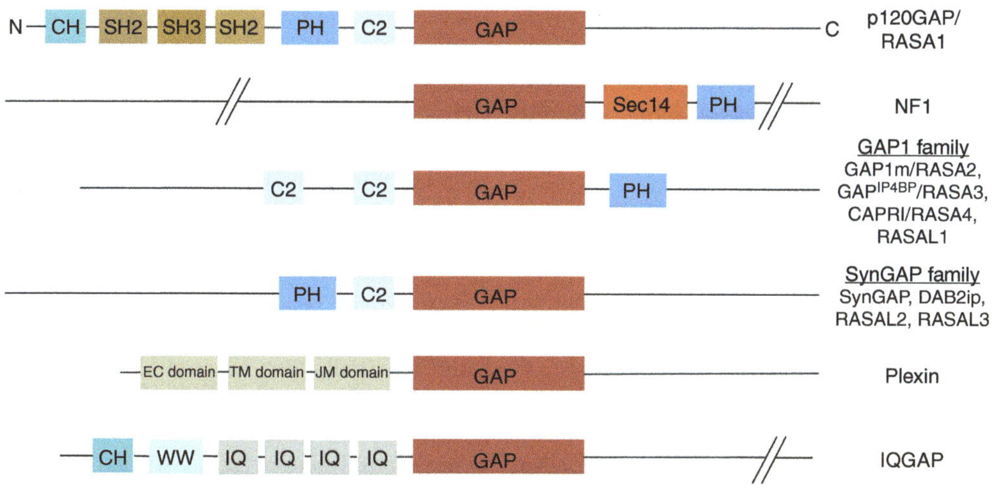

Figure 2. Domain architecture of Ras-specific GTPase-activating proteins (RasGAPs). Representative RasGAPs are schematically shown with emphasis on their domain structure. Red boxes indicate GAP-(related) domains (GRDs). Sec14 represented in orange, src homology (SH, bronze), pleckstrin homology (PH, blue). Other domains like Ca^{2+}-dependent phospholipid-binding/conserved region 2 (C2), isoleucine-glutamine repeat region (IQ), WW (conserved tryptophan), calponin homology (CH), juxtamembrane (JM), transmembrane (TM), and extracellular (EC) domains are also shown.

Cichowski and Jacks 2001; Zhu and Parada 2001) contributing to a rather complex clinical picture that include growth regulatory as well as brain function defects. The *NF1* gene is also genetically altered in several sporadic human malignancies, unrelated to NF1 (Philpott et al. 2017). See separate section below for a more detailed description of neurofibromin.

3. GAP1 subfamily members (ca. 95 kDa) contain a canonical RasGAP module preceded by two C2 domains followed by a PH domain. The group comprises GAP1m/RASA2, GAP^{IP4BP}/RASA3, CAPRI/RASA4 and RASAL1. Dual-target specificity inactivating Ras and Rap family members has been shown for GAP^{IP4BP}, CAPRI, and RASAL1 (Kupzig et al. 2006), Ca^{2+}-dependent membrane translocation has been seen in CAPRI and RASAL representatives (Yarwood et al. 2006).

4. SynGAP family members (ca. 140 kDa) contain a canonical RasGAP module preceded by a C2 and PH domain, comprising the representatives SynGAP, DAB2ip, RASAL2, and RASAL3. SynGAP is present only in neuronal tissues (Chen et al. 1998; Kim et al. 1998; Jeyabalan and Clement 2016). Different from p120GAP/RASA1, the PH-C2-RasGAP triple modules are located at the amino-terminus of the protein. Germline mutations in SynGAP encoding genes have been associated with intellectual disabilities and autism (Jeyabalan and Clement 2016).

5. Plexins are transmembrane receptors (70 kDa) for semaphorins, small proteins triggering neuronal axon guidance and other processes, after receptor binding (Hota and Buck 2012). In these proteins, a transmembrane span separates an extracellular receptor from an intracellular RasGAP-like domain regulating the nucleotide-binding state of Rap proteins, mechanistically potentially similar to dual-target specificity RasGAPs (Wang et al. 2012; Pascoe et al. 2015).

6. IQGAPs (108–190 kDa) are cytoskeletal scaffolding proteins that share a carboxy-terminal structural RasGAP module accompanied by amino-terminal CH, a WW domain, and calmodulin-binding IQ motifs (Brown and Sacks 2006). Although IQGAPs are

devoid of RasGAP activity (Kurella et al. 2009), they bind Rho family GNBPs (LeCour et al. 2016), presumably to coordinate processes involving cytoskeletal architecture and signaling (Hedman et al. 2015; Smith et al. 2015).

RasGAP STRUCTURE AND MECHANISM OF GTPase ACTIVATION

The structure of the canonical RasGAP module as originally identified in p120GAP (Marshall et al. 1989) shows a largely helical protein that is composed of two subdomains: a central domain (GAPc) that carries the catalytic amino acid equipment required for RasGAP function and an extra domain (GAPex) (see Fig. 3) (Scheffzek et al. 1996, 1998a; Pena et al. 2008) that is dispensable for GAP activity (Ahmadian et al. 1996). It is conserved in canonical RasGAP structures (Scheffzek et al. 1996, 1998a; Scheffzek and Ahmadian 2005; Pena et al. 2008) and also in the GAP inactive IQGAP (Kurella et al. 2009) but its function has been largely unclear (see below).

A seminal contribution to the elucidation of the RasGAP mechanism was the discovery that GDP-bound Ras forms a stable complex with RasGAP in the presence of aluminum fluoride (Mittal et al. 1996), a compound that when bound to GDP can be viewed as a transition state mimic of the phosphoryl transfer reaction (Wittinghofer 1997). Its binding to GDP-bound Ras in the presence of RasGAP suggested a transition state-stabilizing mechanism of the GAP-catalyzed GTP hydrolysis reaction (Mittal et al. 1996). Such a complex was not formed with representative oncogenic Ras mutants in which GTPase activity is impaired or in the presence of Ras effectors (Mittal et al. 1996; Gremer et al. 2008). In the Ras–RasGAP complex derived from the catalytic domain of p120GAP/RASA1 (Scheffzek et al. 1996) bound to the putative transition state mimic GDP-aluminum fluoride (GDP-AlF_3), Ras binds to a shallow grove in the GAPc module that is lined by conserved residues that interact with conserved regions in Ras, primarily the switch and P-loop regions (Scheffzek et al. 1997b). A major feature of the nucleotide-binding site situation is the presence of a conserved arginine (Arg789 in p120GAP/RASA1, Arg1276 in neurofibromin), termed arginine finger, contributed from RasGAP that contacts the β/γ phosphate region and is believed to stabilize the transition state of the phosphoryl transfer reaction (see Fig. 3) (Mittal et al. 1996; Ahmadian et al. 1997c; Sermon et al. 1998). The arginine-containing finger loop is stabilized by residues derived from the FLRV (Phe-Leu-Arg-Val) motif of RasGAPs (Hettich and Marshall 1994; Miao et al. 1996). Its site-directed mutagenesis confirms the impact of the catalytic arginine, as even conservative substitution to lysine reduces GAP activity by three orders of magnitude (Ahmadian et al. 1997c; Sermon et al. 1998). Additional features include the stabilization of the switch regions that are found frequently flexible in isolated Ras structures (Pai et al. 1989; Milburn et al. 1990; Wittinghofer and Vetter 2011). Along this line, particularly, the critical Gln61 is oriented to stabilize a water molecule for nucleophilic attack (see Fig. 3) (Scheffzek et al. 1997a). A so-called variable loop has been implicated in mediating substrate specificity (Scheffzek et al. 1997a; Ahmadian et al. 2003).

The structure of the Ras–RasGAP complex provided an explanation why transforming Gly12/Gln61-substitutions in Ras (Seeburg et al. 1984; Der et al. 1986) render Ras insensitive to RasGAPs; they would interfere with the geometric or electronic requirements of the transition state as represented by the complex bound to GDP-AlF_3 (Scheffzek et al. 1997b). The Gly12Pro substitution is nontransforming, has a slightly increased rate of GTP hydrolysis with respect to cellular Ras, and is yet insensitive to RasGAP (Franken et al. 1993), suggesting that potentially a slight increase in the GTPase rate with respect to oncogenic Ras might be sufficient to overcome the Ras-activating effects believed to be prevalent in oncogenic Ras (Wittinghofer and Waldmann 2000).

The RasGAP mechanism has been investigated in detail using Fourier transform infrared (FTIR) spectroscopic methods and computer simulation approaches (Martin-Garcia et al. 2012; Gerwert et al. 2017). In these studies, the

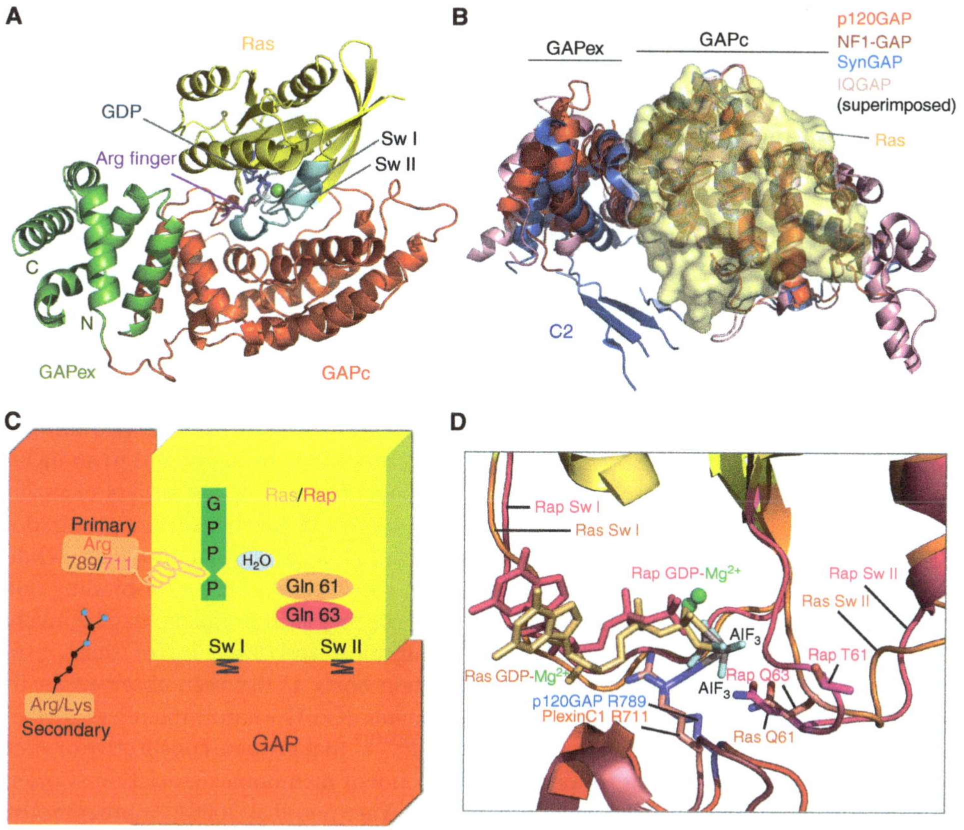

Figure 3. Ras-specific GTPase-activating protein (RasGAP) structure and mechanism. (*A*) Ribbon representation of the Ras–RasGAP complex as derived from the structure of the catalytic domain of p120GAP/RASA1 (red and green) bound to Ras (yellow) and GDP-aluminum fluoride (AlF_3) (blue) (Scheffzek et al. 1997a), switch regions are shown in cyan, and arginine finger motif in violet. (*B*) Ribbon representation of different RasGAPs superimposed and modeled onto Ras shown in transparent yellow. View is rotated approximately 90° about a horizontal axis as compared to RasGAP in panel *A* representing a *top* view on the modeled complex. C2 domain fragment of SynGAP (blue) appears in proximity of the Ras-binding region in SynGAP. (*C*) Cartoon representation of the Ras–RasGAP complex including dual-specificity functionalities. Ras/Rap is shown in yellow and GAP in red. Upon binding of Ras/Rap to GAP, the GTPase activity is strongly enhanced by the complementation of the active site, delivering the catalytic arginine, 789 (p120GAP), or 711(PlexinC1). This is further stabilized by interaction with a secondary Arg/Lys in the GAP domain. Glutamine 61 of Ras or Glutamine 63 of Rap contributes to the catalysis reaction by positioning phosphate accepting water molecule for nucleophilic attack. Additional residues and motifs in GAP stabilize the switch I and switch II regions of Ras/Rap, supporting a conformation that is favorable for efficient GTP hydrolysis. (*D*) Close-up view of the Ras/Rap active site bound to GAP showing conformational concordance or variations in the switch regions on Ras-p120GAP and Rap-PlexinC1GAP. Components of the nucleotide-binding site are shown for Ras (beige) and Rap (pink). The catalytic arginines in PlexinC1GAP and p120GAP are included.

authors define three rate constants in the catalytic complex. In a first step, the protein adopts the GTP-bound ON-state; in the second step, the catalytic arginine side chain moves into the active site to complete the active site for GTP hydrolysis occurring simultaneously and forming a protein-bound P_i intermediate; the last step is characterized by release of P_i into the solvent, switch regions returning to the OFF-state conformation, and the catalytic arginine finger leaving the nucleotide-binding pocket (Kotting et al. 2008; Rudack et al. 2012; reviewed in Gerwert et al. 2017).

The GAP mechanism outlined above appears to be essentially valid also for Rho/RhoGAP systems (Rittinger et al. 1997; Nassar et al. 1998) that also share similar structural folds (Bax 1998; Gamblin and Smerdon 1998; Scheffzek et al. 1998b), although individual effects of mutations on the formation of $AlF_{3/4}$ complexes and thus catalysis may be variable (Gremer et al. 2008). It is interesting to note that the arginine finger mechanism appears to be shared also by bacterial toxins that unfold their toxic activity by acting as Rho family–specific GAPs but seem to have a minimal and entirely different three-dimensional structure (Stebbins and Galan 2000; Wurtele et al. 2001; Evdokimov et al. 2002). Different from initial considerations, the arginine finger mechanism does not seem to be generally valid for GAPs targeting other small GNBPs (Scheffzek et al. 1998b; Scheffzek and Ahmadian 2005; Mishra and Lambright 2016), although the concept of transition state stabilization is supported in most of the structures.

The Ras-related Rap protein is generally regulated by its cognate Rap1GAP (Rubinfeld et al. 1991; Bos et al. 2007). Structural and biochemical investigations have revealed that these Rap-specific GAPs do not use a catalytic arginine "finger," like in RasGAPs. Instead, they provide an asparagine "thumb" as catalytic residue, presumably to replace the GNBP conserved Gln 61 that structurally corresponds to a noncatalytic threonine (Thr 61) in Rap proteins (Daumke et al. 2004; Scrima et al. 2008). Dual Ras/Rap-specific RasGAPs (sharing the canonical RasGAP fold as shown in Figs. 1 and 3) seem to follow a mechanism employing a catalytic arginine (see Fig. 3) (Pena et al. 2008; Kupzig et al. 2009) associated with a glutamine residue (Gln 63) of the switch II region (Sot et al. 2010) (see below). A dual *in trans* mechanism employing catalytic arginine and glutamine fingers contributed by the GAP component seems to be implemented in the RabGAP Gyp1p and potentially in other RabGAPs (Pan et al. 2006; Gavriljuk et al. 2012).

MECHANISMS OF MEMBRANE TRANSLOCATION

The subcellular localization of RasGAPs is certainly influenced and modulated by their requirement to meet Ras at cellular membranes. Post-translational regulation of Ras signaling commonly requires translocation of respective protein components to membrane compartments where they can interact with membrane-associated Ras species to regulate their biological activity (i.e., the level of their output signals). Translocation to the membrane environment is thought to increase the local two-dimensional concentration to be equivalent to an increase in binding constants of five orders of magnitude compared with the cytosolic concentration (Simanshu et al. 2017). The modular architecture of RasGAPs containing lipid-binding domains suggests that such domains may be involved in localization and targeting of their respective protein hosts, as has been observed for the PH domains of GAP1 family members GAP1m and GAP^{IP4BP} that mediate constitutive or stimulus-dependent membrane association (Cozier et al. 2000; Yarwood et al. 2006). Another mechanism of GAP localization may be via adaptor proteins such as Annexins (Gerke and Moss 2002). Annexin A6 has been shown to promote membrane binding of p120GAP/RASA1 in vitro and membrane localization in living cells (Grewal et al. 2005; Grewal and Enrich 2006). The C2 domains of CAPRI/RASA4 and RASAL1 mediate membrane association after increase of intracellular calcium concentration (Liu et al. 2005). In the case of CAPRI/RASA4 this association is stable. RASAL1 senses calcium oscillations shuttling between membrane and cytoplasm. In the case of CAPRI/RASA4 and RASAL1, detection of RasGAP activity requires calcium-dependent interaction of the C2 domain with membrane structures and glycerophospholipids (Sot et al. 2013). Translocation to the sites of GAP regulation may also be mediated by protein–protein interaction of GAP and receptor tyrosine kinase components (Kaplan et al. 1990; Fantl et al. 1992; Kashishian et al. 1992; Cooper and Kashishian 1993), and in the case of p120GAP/RASA1 may involve SH2/SH3 domains (Rojas et al. 2007; Pamonsinlapatham et al. 2009).

Interestingly, membrane translocation of neurofibromin is mediated by binding of the membrane targeting Sprouty-related protein (Bundschu et al. 2007) Spred1 (Stowe et al. 2012) to the GAP-related domain (Dunzendorfer-Matt et al. 2016; Hirata et al. 2016) (see separate section below). Sec14- and PH-like domains have been detected in a variety of signal regulatory proteins that harbor GEF or GAP domains (Lemmon 2004; Bankaitis et al. 2009; Ghosh and Bankaitis 2011; Scheffzek and Welti 2012). They have been found to control cellular localization of their host proteins, presumably by protein–protein interaction mechanisms (Fig 4) (Kostenko et al. 2004; Sirokmany et al. 2005). With neurofibromin, such localization has not been reported but cannot be excluded at the moment.

REGULATION OF GAP ACTIVITY

Regulatory mechanisms other than protein translocation have been investigated with GAPs in general, including protein–protein interaction, phosphorylation, proteasomal degradation, and regulation by small molecules such as lipids, as reviewed earlier (Bernards and Settleman 2004).

The fact that RasGAPs apparently have to interact with membrane-associated GNBP/Ras targets (Simanshu et al. 2017) raises the question whether membrane lipids may affect GAP activity via competitive or noncompetitive mechanisms. Early studies reported modulation of p120GAP/RASA1 or neurofibromin GAP activity by lipid compounds, including arachidonic acid, eicosanoids, fatty acids, and phospholipids with conflicting results (Tsai et al. 1989; Bollag and McCormick 1991; Han et al. 1991; Serth et al. 1991; Sermon et al. 1996) potentially reflecting challenges in defining the experimental conditions under which measurements were performed. The structural Ras–RasGAP interface does not offer a particularly hydrophobic amino acid distribution (Ahmadian et al. 2003) to suggest lipid binding in the Ras-binding region.

A GAP regulatory mechanism proposed for neurofibromin has been proteasomal degradation (followed by resynthesis) upon growth factor stimulation (Cichowski et al. 2003) mediated by the Cullin 3–associated complex (Hollstein and Cichowski 2013) or in a developmental context by the SAG/RBX2/ROC2 protein, an essential RING component of SCF E3 ubiquitin ligase (Dai et al. 2011). Little is known about the mechanistic details of the underlying protein interactions and in some cell types or conditions growth factor–dependent degradation (and resynthesis) may be inhibited (McClatchey and Cichowski 2012). Another study showed ETEA/UBXD8-mediated ubiquitination of the GRD of neurofibromin in vitro, correlating with increased levels of activated Ras after silencing of ETEA expression (Phan et al. 2010).

Protein phosphorylation is an important regulatory post-translational modification and has been reported to regulate GAP activity with some GNBP systems (Bernards and Settleman 2004). Protein kinase A (PKA) has been shown to phosphorylate the carboxy-terminal region of neurofibromin. Phosphorylation promoted its association with a 14-3-3 protein followed by down-regulation of neurofibromin-mediated GAP activity (Feng et al. 2004). Phosphorylation of SynGAP by Ca^{2+}/calmodulin-dependent protein kinase II (CamKII) was reported to increase RasGAP activity by ~90% potentially by an allosteric mechanism because phosphorylated residues are located outside the GAP domain (Oh et al. 2004). Another phosphorylation process involving cyclin-dependent kinase 5 (CDK5) and CamKII appears to control the ratio of Ras- and RapGAP activities of SynGAP (Walkup et al. 2015) (see below). Allosteric positive regulation of neurofibromin GAP activity after protein kinase C (PKC)-mediated phosphorylation of the cysteine serine-rich domain (CSRD) was proposed to act like a molecular switch in epidermal growth factor (EGF) receptor signaling (Mangoura et al. 2006).

Protein–protein interaction may represent another strategy/mechanism of RasGAP regulation (Bernards and Settleman 2004, 2005). Along this line, the isolated PH domain of p120GAP/RASA1 has been reported to interact specifically with its GAP domain and to inhibit interaction between activated Ras and p120GAP/RASA1

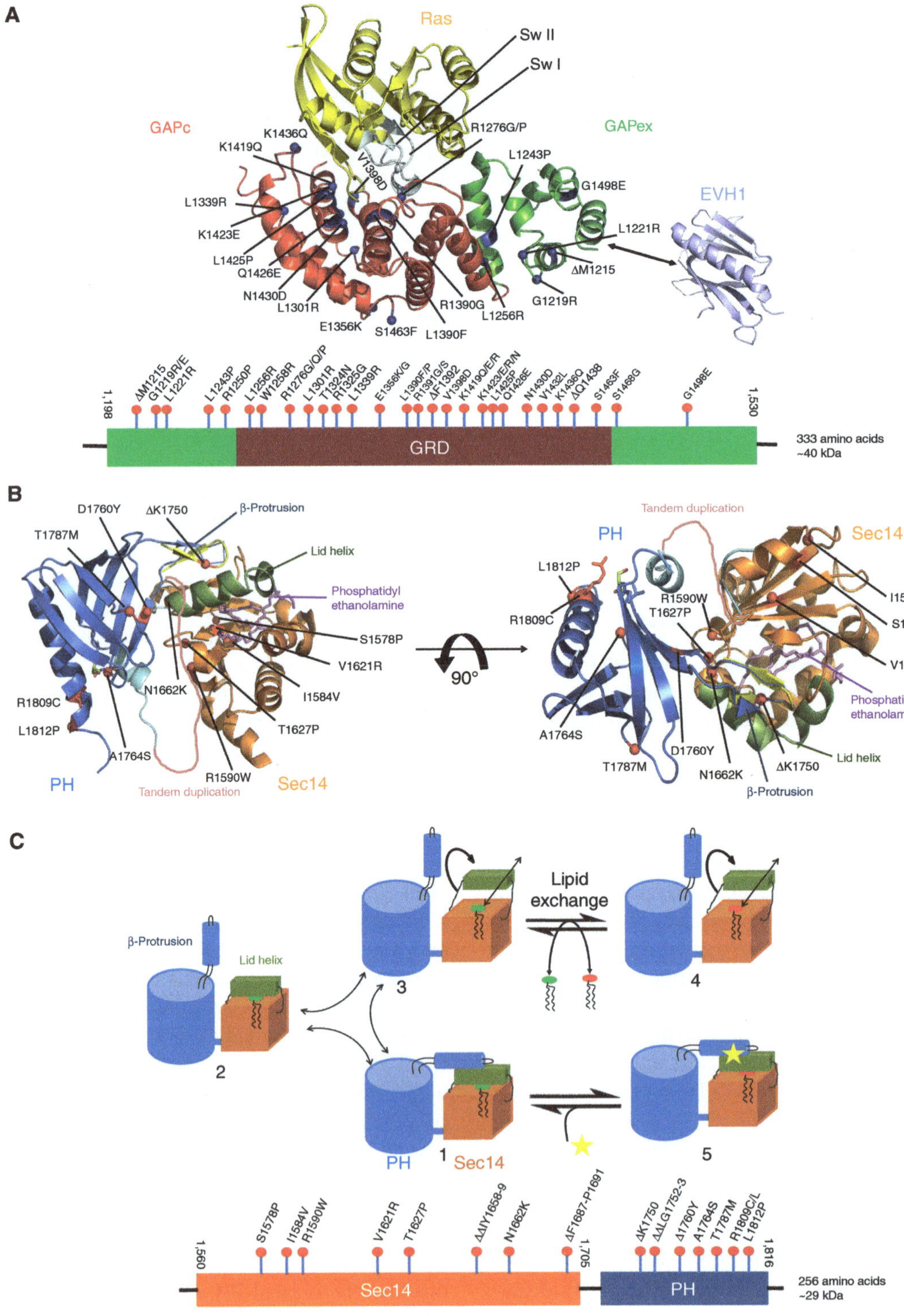

Figure 4. (*Legend on facing page.*)

Cite this article as *Cold Spring Harb Perspect Med* doi: 10.1101/cshperspect.a031500

(Drugan et al. 2000). The physiological relevance of this finding remained unclear as such an interaction would have to be shown also in the full-length context of p120GAP/RASA1.

Neurofibromin has been shown to directly interact with $G_{\beta/\gamma}$ subunits linking opioid receptors to Ras signaling in the brain. $G_{\beta/\gamma}$ interacts with the Sec14-like domain and reduces GAP activity of neurofibromin in a construct composed of GRD and the Sec14-PH modules (Xie et al. 2016). To fully evaluate this observation the measurement would have to be performed also with full-length neurofibromin.

DUAL-TARGET SPECIFICITY AND MECHANISM VARIATION

The GAP mechanism implemented in the canonical RasGAP module is found modified in systems showing dual Ras/Rap specificity as in members of the GAP1 (GAP^{IP4BP}, RASAL1, CAPRI) and SynGAP families. Although the isolated RasGAP module commonly mediates RasGAP activity, surrounding domains are required for RapGAP activity (Krapivinsky et al. 2004; Kupzig et al. 2006). The specificity switch may be achieved by different yet complementary mechanisms.

In the case of CAPRI/RASA4, Ca^{2+}-dependent monomer/dimer formation has been shown to switch the protein between RasGAP and RapGAP activities (Dai et al. 2011), although a detailed mechanism or structures were not presented in the respective paper. SynGAP requires the RasGAP module preceding C2 domain to unfold RapGAP activity (Pena et al. 2008) and has otherwise very poor RasGAP activity (Krapivinsky et al. 2004; Pena et al. 2008). In the structure of a C2-GAP construct of SynGAP the C2 domain is partly disordered with only a few β-strands recognizable. Their positions are in the vicinity of the putative Ras/Rap-binding sites but did not allow mechanistic conclusions regarding RapGAP activity (see Fig. 3) (Pena et al. 2008). It has been reported that CamKII increases SynGAP activity (Oh et al. 2004) and along with CDK5-mediated phosphorylation of SynGAP modulates the ratio of its GAP activities toward Ras and Rap targets (see above) (Walkup et al. 2015).

The semaphorin receptor plexin transduces signals regulating axon guidance in neurons and other systems (Pascoe et al. 2015). It contains an extracellular receptor domain binding the protein semaphorin and an intracellular RasGAP-like module that acts on Rap proteins upon dimerization (Wang et al. 2012) has been reported to exhibit GAP activity toward R- and M-Ras in cellular systems (Oinuma et al. 2004; Saito et al. 2009), and in pure preparations only as a RapGAP (Wang et al. 2012). Upon semaphorin binding, the intracellular GAP module dimerizes thereby inducing a GAP active conformation that interacts with activated Rap proteins to accelerate GTP hydrolysis via an arginine finger-dependent mechanism in which the role of Gln61 of Ras is taken by Gln63 of Rap (see Fig. 3). A similar mechanism has been proposed by

Figure 4. Structural components of neurofibromin. (*A*) Ribbon representation of a putative Ras–GRD complex, modeled on the basis of the GAP-related domain (GRD) structure (Scheffzek et al. 1998a) and the Ras-p120GAP complex (Scheffzek et al. 1997a). Blue dots indicate the location of missense mutations found in neurofibromatosis type 1 (NF1) patients. The EVH1 domain of Spredl shown to interact with extra domain (GAPex) is depicted in cyan. Below is the domain scheme of GRD of neurofibromin with amino-terminal and carboxy-terminal extra domains with nontruncating patient-derived mutations included (see text). (*B*) Ribbon diagram of the Sec14-PH module bound to lipid with approximately 90° rotation about a horizontal axis. Orange ribbon represents Sec14-like domain bound to lipid (phosphatidyl ethanolamine) (in violet). Pleckstrin homology (PH)-like domain is shown in blue. Helix represented in green is a part of Sec14 that forms lid helix. Red dots indicate the location of missense mutations found in NF1 patients. A tandem duplication and ΔK1750 are also shown. Below is the domain scheme of Sec14-PH of neurofibromin with nontruncating mutations identified in NF1 patients (see text). (*C*) Cartoon of the Sec14-PH module, illustrating the proposed structural changes associated with the exchange of glycerophospholipid ligands. Starting from the observed structure (1), lid helix (in green), and β-protrusion helix can move either stepwise (1 > 2 > 3) or in a concerted fashion (1 > 3) into an open conformation, which allows the exchange of lipid molecules between Sec14-PH and membrane (4). Binding of modulator to the lid helix or the β-protrusion appears to prevent the lipid exchange reaction (5).

Sot et al. (2010) for GAP^{IP4BP} using biophysical approaches that did not include a three-dimensional structure of a respective Rap-RasGAP complex. It appears that the RapGAP mechanisms implemented in canonical RasGAP modules require an arginine finger for catalysis (Pena et al. 2008; Kupzig et al. 2009). Interestingly, the Rab family–specific RabGAP Gyp1p employs a dual-finger mechanism comprising a catalytic arginine along with a catalytic glutamine to promote catalysis. Importantly, the RasGln61 analogous glutamine does not stabilize the transition state (Pan et al. 2006; Gavriljuk et al. 2012). Recently, the topology of RASAL1 has been presented using negative stain electron microscopy. The authors show the relative organization of the individual domains but higher resolution will be required to obtain detailed information about the mechanism of dual-target specificity (Cuellar et al. 2017).

RasGAPS IN HUMAN DISEASES

Given the importance of Ras signaling pathways in numerous cellular functions and their association with various diseases, most importantly with malignant cell transformation and cancer, it is not surprising that molecular damages do occur not only in Ras itself but also in Ras-regulating protein components. The most prominent example in this context is neurofibromin and will be described in more detail below. Its encoding gene is altered in the germline of NF1 patients (Upadhyaya and Cooper 2012) and also in a large number of sporadic malignancies unrelated to NF1 (Philpott et al. 2017). Up-regulation of Ras is believed to represent at least one mechanism of pathogenicity in respective cases. Genetic alterations in the p120GAP/RASA1 encoding gene have been reported to be responsible for capillary malformation-arteriovenous malformation (CVM-AM), a genetic disorder characterized by vascular development defects (Boon et al. 2005; Revencu et al. 2008, 2013). Most alterations would likely be incompatible with a folded protein suggesting loss of RasGAP activity as at least one mechanism of pathogenicity. Tumor-suppressive functions have been shown for DAB2ip (McLaughlin et al. 2013; Bellazzo et al. 2017) and for members of the GAP1 family, suggesting mechanisms potentially involving transcriptional deregulation (Jin et al. 2007) including DNA hypermethylation. Germline mutations in SynGAP encoding genes have been associated with intellectual disabilities and autism (Jeyabalan and Clement 2016). Semaphorin signaling via neuropilin and plexin receptors has been implicated in tumor progression, including metastasis and tumor angiogenesis (Neufeld and Kessler 2008).

NEUROFIBROMIN—A MAJOR RasGAP IN GROWTH CONTROL AND NEURONAL FUNCTION

NF1 (von Recklinghausen 1882; Riccardi 1992) was the first RASopathy identified (Rauen 2013; Ratner and Miller 2015). NF1 patients have an increased risk to develop the disease-typical neurofibroma, benign and malignant tumors of the nervous system. Additional features include pigment anomalies of the skin, bone deformations, and learning disabilities in more than 10% of adolescent patients. Further medical complications contribute to a clinical picture that is far from being well understood. Although the disease was described in 1882 (von Recklinghausen 1882; Riccardi 1992; Upadhyaya and Cooper 2012), the gene responsible was identified about 100 years later (Cawthon et al. 1990; Viskochil et al. 1990; Wallace et al. 1990). Alterations in the tumor suppressor gene *NF1* are responsible for the pathology of NF1 (Riccardi 1992; Cichowski and Jacks 2001; Zhu and Parada 2001) and are found also in a large number of aggressive sporadic malignancies unrelated to NF1 (Cancer Genome Atlas Research Network 2008, 2014; Sangha et al. 2008; Patch et al. 2015; Philpott et al. 2017). The *NF1* gene shows sequence similarity to the previously discovered p120GAP/RASA1 and the related IRA1/IRA2 proteins in yeast and encodes a large cytoplasmic protein, termed neurofibromin (320 kDa) (DeClue et al. 1991; Gutmann et al. 1991). Neurofibromin is highly conserved in vertebrates with sequence identity of >90% in the evolutionarily old Fugu and Zebrafish proteins (Kehrer-Sawatzki et al. 1998; Padmanabhan et al. 2009)

 Cite this article as *Cold Spring Harb Perspect Med* doi: 10.1101/cshperspect.a031500

and thus seems to be an essential vertebrate protein. Homologous sequences have been found in fly (The et al. 1997) and can be detected in honey bee (Honeybee Genome Sequencing 2006), hydra (Chapman et al. 2010), and slime mold (Eichinger et al. 2005). In yeast strains lacking IRA1/2 neurofibromin can complement a missing IRA function (Xu et al. 1990a).

A central region, GRD, shows sequence similarity to the carboxy-terminal catalytic domain of p120GAP/RASA1 (Marshall et al. 1989) and displayed RasGAP activity (Ballester et al. 1990; Martin et al. 1990; Xu et al. 1990b), defining neurofibromin as a RasGAP. The cellular impact of neurofibromin on Ras regulation is evident from studies in which loss of neurofibromin is associated with increased intracellular levels of activated Ras (Basu et al. 1992; DeClue et al. 1992; Bollag et al. 1996). Conversely, expression of the GRD in otherwise $NF1^{-/-}$ cells can restore the cellular phenotypes (Hiatt et al. 2001; Thomas et al. 2006). Neuronal studies suggest that neurofibromin contributes to more than 90% of Ras inactivation in dendritic spines (Oliveira and Yasuda 2014). Nontruncating gene alterations (currently >400 identified) cover the whole coding sequence suggesting important functions represented in regions outside the GRD (Kaufmann 2008). The GRD is preceded by a 90-residue peptide segment that has been reported to be important for the interaction with the cytoskeletal protein tubulin (Bollag et al. 1993). Mutations in the GRD region abolish tubulin binding (Xu and Gutmann 1997). At its carboxy-terminal end, it is followed by a bipartite glycerophospholipid-binding module that is composed of an amino-terminal Sec14- and a carboxy-terminal PH-like domain (see below) (D'Angelo et al. 2006; Welti et al. 2007). The so-called cysteine serine-rich domain (CSRD) (Izawa et al. 1996) appears to cluster a number of mutations found in NF1 patients (see Fig. 5) (Fahsold et al. 2000; Mattocks et al. 2004), but has not been structurally validated so far.

Structure of Neurofibromin

Bioinformatics sequence analysis indicates a highly helical protein with few flexible segments in the carboxy-terminal regions. From a structural perspective, the domain composition/architecture is poorly defined and include only the GRD (Scheffzek et al. 1998a) and the neighboring Sec14-PH module (D'Angelo et al. 2006; Welti et al. 2007). These account for approximately 25% of the total molecular weight. In the following sections, we describe the structures of the respective modules and try to address potential functional implications also considering mutational studies.

The GAP-Related Domain (GRD)

The classic GRD corresponding to the catalytic domain of p120GAP comprises a segment ranging from residues 1198 to 1530, which for crystallographic structure determination has been proteolytically treated to facilitate crystallization. The structure of the GRD module (see Fig. 4A) (Scheffzek et al. 1998a) confirms the canonical helical RasGAP fold with few differences to the respective module of p120GAP/RASA1 (Scheffzek et al. 1996). They can be located essentially in loop regions connecting helices.

The differentially spliced and GAP activity modulating exon 23a (Andersen et al. 1993) would insert a 21-residue peptide segment in a helix that is in a structural neighborhood near the active site region (Scheffzek et al. 1998a). A structure derived from this type II transcript has not been reported so far. GAP-accelerated GTP hydrolysis has been characterized in detail using kinetic approaches (Ahmadian et al. 1997a,b, 2003; Sermon et al. 1998). Nontruncating mutations found in NF1 patients have different effects on the catalytic properties of the GRD (see Fig. 4). A most striking alteration has been found earlier in a patient with severe tumor burden where Arg1276 (the arginine finger of neurofibromin) was replaced by a proline residue. Structural modeling of this mutation indicates loss of a catalytic residue for which there is no functionally alternative residue in the vicinity of the site of GTP hydrolysis, confirmed by biochemical investigation of the associated enzyme kinetics yielding an 8000-fold decrease in catalytic efficiency for the mutated protein (Klose et al. 1998). Numerous mutations have been identified in the GRD. Several are indicated in Figure 4A.

Those and other mutations include ΔM1215, G1219R/E, L1221R, L1243P, R1250P, L1256R, W1258R, R1276G/Q/P, L1301R, T1324N, R1325G, L1339R, E1356K/G, L1390F/P, R1391G/S, ΔF1392, V1398D, K1419Q/E/R, K1423E/R/N, L1425P, Q1426E, N1430D, V1432L, K1436Q, ΔQ1438, S1463F, S1468G, and G1498E (see Fig. 4A) (Li et al. 1992; Upadhyaya et al. 1997b; Fahsold et al. 2000; Han et al. 2001; Ars et al. 2003; De Luca et al. 2003; Mattocks et al. 2004; Lee et al. 2006; Kaufmann 2008; Thomas et al. 2012; Bianchessi et al. 2015). The HGMD (Institute of Medical Genetics, Cardiff, Wales, UK, www.hgmd.cf.ac.uk/ac/index.php) and LOVD (Leiden Open Variation Database, www.LOVD.nl/NF1) (Fokkema et al. 2011; van Minkelen et al. 2014) databases were interrogated to summarize patient-derived mutations.

The Sec14-PH Module

Using structural biology approaches we have discovered earlier a bipartite structural module composed of an amino-terminal Sec14- and a carboxy-terminal PH-like domain (D'Angelo et al. 2006), the former binding glycerophospholipids (Welti et al. 2007). The structure of the Sec14-PH module in neurofibromin allows definition of glycerophospholipid binding in a hydrophobic cage of the Sec14-like portion that appears to be modulated/regulated by a hairpin-like protrusion derived from the PH-like portion (see Fig. 4B). This protrusion appears to keep the lid structure closed in the respective crystal structure thereby preventing access of lipid ligands to the interior of the Sec14 portion. In a hypothetical mechanistic model, two ligands would be involved in Sec14-PH function/regulation. Ligand 1 would induce conformational changes in the hairpin region thereby releasing structural constraints of the lid helix to allow binding of ligand 2 (a glycerophospholipid) to the lipid-binding cage (see Fig. 4B and C). Given the abundance of phosphatidyl ethanolamine (PE) in eukaryotic cells, it appears likely that PE represents at least one physiological ligand of the Sec14-like portion in neurofibromin (Welti et al. 2007). Binding of immobilized PIPs in overlay assays has been shown for the Sec14-PH module and is modulated by mutations of conserved residues in the vicinity of the hairpin protrusion (D'Angelo et al. 2006; Welti et al. 2007). The isolated Sec14-PH module does not seem to localize to specific cellular structures/compartments (S Welti and K Scheffzek, unpubl.). Structural consequences of prominent alterations have been investigated, suggesting structural rearrangements in at least one of the three cases investigated (Welti et al. 2011). We speculate that they affect interactions between neurofibromin and/or small molecules or proteins.

A particularly interesting alteration has been identified at the carboxy-terminal end of the PH-like domain. Arg1809 was found substituted by cysteine in a set of six NF1 patients in a cohort of patients presenting with a mild NF1 phenotype but without neurofibroma (Ekvall et al. 2014; Rojnueangnit et al. 2015; Santoro et al. 2015). The genetic analysis of the mutation along with the common symptoms in patients gives one of the rare examples for a genotype–phenotype relationship of which only two have been reported previously (Cnossen et al. 1997; Upadhyaya et al. 2007). In the Sec14-PH structure representing the cellular protein sequence Arg1809 interacts with Ser1738 in the amino-terminal region of the PH β-sheet (D'Angelo et al. 2006). Mutation of Arg1809 to cysteine (or other amino acids observed) is likely to abolish the interaction between the two structural elements of the PH-like portion. The structural impact of this interaction is difficult to assess because in contrast to Arg1809, Ser1738 shows little conservation in *NF1*-related sequences (D'Angelo et al. 2006). To evaluate the cellular consequences of the observed alteration, its investigation in the full-length context of neurofibromin within relevant cell lines is required, potentially comparing protein mass spectrometry-guided interaction profiles with those of the wild-type protein. In any case, one would expect that alterations in the respective sequence position would converge on a common biochemical/cellular feature of neurofibromin function.

Other nontruncating mutations in the Sec14-PH module include S1578P, I1584V,

R1590W, V1621R, T1627P, ΔΔIY1658-9, N1662K, ΔF1687-P1691, ΔK1750, ΔΔLG1752-3, Δ1760Y, A1764S, T1787M, R1809C/L, and L1812P (Tassabehji et al. 1993; Upadhyaya et al. 1997a; Wu et al. 1999; Fahsold et al. 2000; Jeong et al. 2006; Welti et al. 2011; Bianchessi et al. 2015). The HGMD (Institute of Medical Genetics, Cardiff, Wales, UK, www.hgmd.cf.ac.uk/ac/index.php) and LOVD (Leiden Open Variation Database, www.LOVD.nl/NF1) (Fokkema et al. 2011; van Minkelen et al. 2014) databases were interrogated to summarize patient mutations.

What is the function of the Sec14-PH module? The module has been discovered in search for novel soluble domains of neurofibromin (D'Angelo et al. 2006; Bonneau et al. 2009). While the Sec14-like portion has been anticipated based on bioinformatics analyses (Aravind et al. 1999), the PH-like portion was entirely unexpected in the structure of neurofibromin as was the seemingly mechanistic interaction between the two domains (D'Angelo et al. 2006). PH-like domains commonly share a sandwich-like fold with halves formed by two β-sheets closed by a carboxy-terminal α-helix (Scheffzek and Welti 2012). Depending on the detailed amino acid equipment, they may bind various ligand including phospholipids, phosphotyrosine peptides, or proteins including Rho family GNBPs (Lemmon 2004; Scheffzek and Welti 2012). Sec14- and PH-like domains occur as stand-alone modules or as parts of signal regulatory proteins, such as RhoGEFs, RhoGAPs, or phosphotyrosine phosphatases (PTPs) (Saito et al. 2007; Bankaitis et al. 2010). Their function in these proteins is not entirely understood. Some studies nevertheless suggest targeting functions in p50RhoGAP and the RhoGEF Dbs (Kostenko et al. 2004; Sirokmany et al. 2005). It is so far unclear whether this function is associated with lipid-binding properties of the proteins.

Full-Length Neurofibromin

The structure of the entire neurofibromin is not known so far. The classic method for structure determination of biomolecular species is X-ray crystallography. However, this method requires the availability of high-quality protein crystals to yield detailed information of the crystalline protein. To obtain such crystals the protein has to be available in pure and homogeneous quality (Shi 2014). Obtaining high-quality neurofibromin for structural investigation has long been a rate-limiting step in neurofibromin structural biology. With the routine availability of gene synthesis, including sequence optimization for protein expression (Bonneau et al. 2009), neurofibromin overexpression in insect cells became a powerful method to obtain pure and at least partially homogeneous neurofibromin from such preparations (Dunzendorfer-Matt et al. 2016). Recombinant neurofibromin has been used in protein transduction experiments demonstrating down-regulation of the Ras/MAPK pathway in *NF1*-depleted cells, as shown by monitoring phospho-ERK decrease (Mellert et al. 2018). To our knowledge, crystalline neurofibromin has so far not been obtained. An alternative method with increasingly high potential is cryoelectron microscopy (cryoEM) (Carroni and Saibil 2016). With this method, the molecular species is preserved in a native-like vitrified state that does not require crystallinity for structural analysis (Earl et al. 2017). Many large macromolecular and signaling species have been analyzed by cryoEM over the past years (Merk et al. 2016). Initial and preliminary analyses of such specimens show neurofibromin at low resolution (>20 Å), forming compact dimeric particles of unclear molecular oligomerization state (J Briggs, H Stark, and K Scheffzek, unpubl.). Further efforts will be necessary to improve sample and data quality of cryoEM approaches to obtain a high-resolution view of full-length neurofibromin.

Nontruncating mutations can be found along the entire coding sequence of neurofibromin (Figs. 4 and 5), suggesting the whole sequence to be essential for protein function. Their assessment in terms of impact on protein function may be difficult depending on structural and biochemical details. From a structural perspective, nonconservative substitutions are likely to have a more severe effect on the structure or function of the protein than conservative

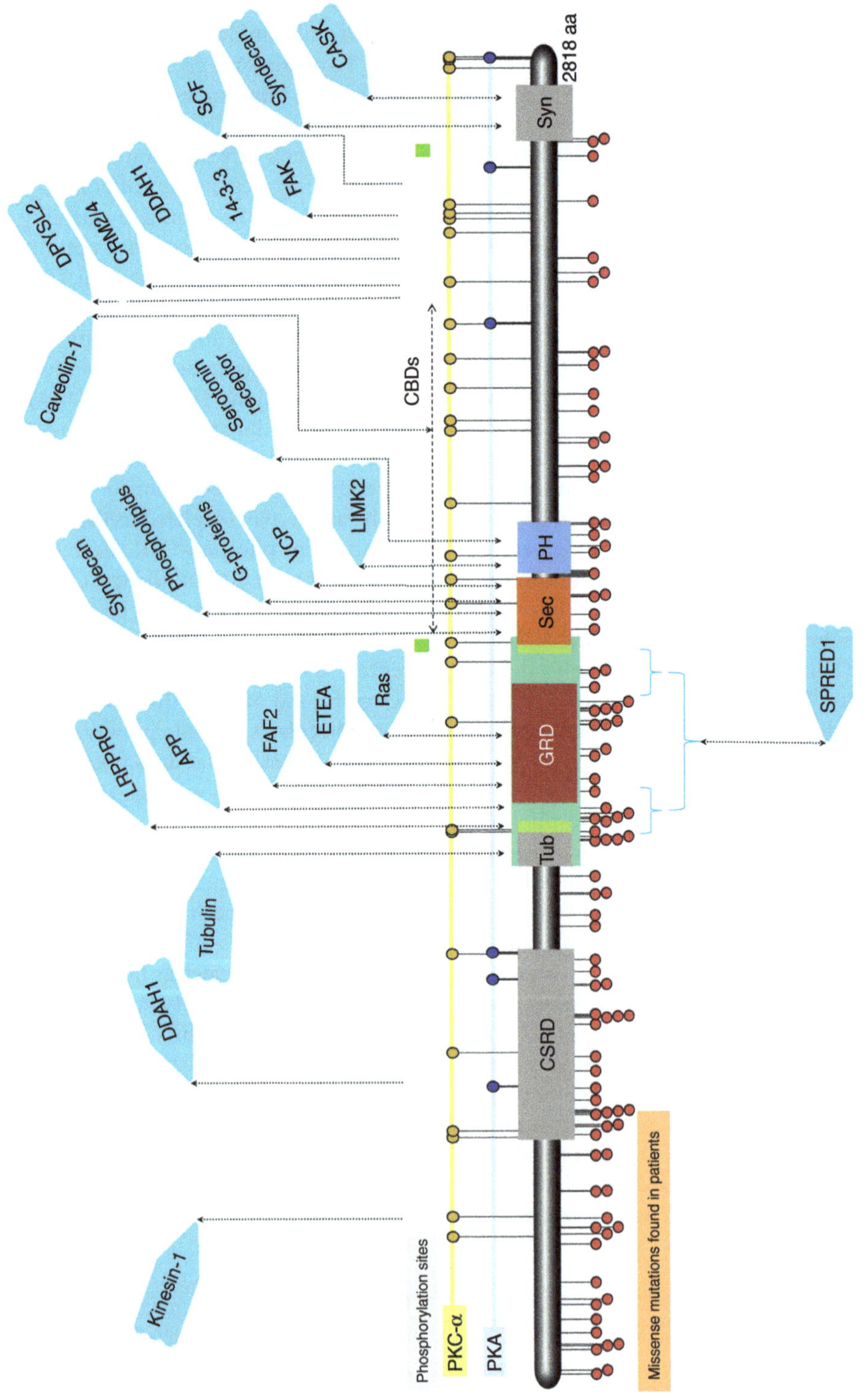

Figure 5. *(Legend on facing page.)*

 Cite this article as *Cold Spring Harb Perspect Med* doi: 10.1101/cshperspect.a031500

exchanges. Such substitutions are likely to change the local environment in a manner that may not be compatible with the native neurofibromin structure. In general, the availability of 3D structural information contributes significantly to the assessment of nontruncating mutations found in patients.

Interaction Partners of Neurofibromin

Meanwhile, a large number of interaction partners beyond activated Ras have been reported (see Fig. 5) (Ratner and Miller 2015). They include kinesin-1 (Hakimi et al. 2002), DDAH (Tokuo et al. 2001), PKA (Feng et al. 2004), tubulin (Bollag et al. 1993), LRPPRC (Arun et al. 2013), amyloid precursor protein (APP) (De Schepper et al. 2006), Spred1 (Stowe et al. 2012; Dunzendorfer-Matt et al. 2016; Hirata et al. 2016), syndecans (Hsueh et al. 2001), phospholipids (D'Angelo et al. 2006; Welti et al. 2007), G proteins or components thereof (Tong et al. 2002; Xie et al. 2016), p97/vcp (Wang et al. 2011), LIMK2 (Vallee et al. 2012), serotonin receptor (Deraredj Nadim et al. 2016), caveolin-1 (Boyanapalli et al. 2006), CRMP2/4 (Lin and Hsueh 2008; Patrakitkomjorn et al. 2008), 14-3-3 (Feng et al. 2004), focal adhesion kinase (FAK) (Kweh et al. 2009), SCF (Tan et al. 2011), Cullin-3 (Hollstein and Cichowski 2013), and CASK (Volta et al. 2010). Technically, they were identified in a variety of approaches including yeast two-hybrid screening, coimmunoprecipitation or pull-down-like experiments. The biological significance of the underlying protein–protein interactions has not been clearly established in most of the reported interactions. We describe here in more detail selected examples where direct interaction and/or a functional link appear explored or established.

Spred1

A most insightful observation came from experiments uncovering the interaction between neurofibromin and Sprouty-related proteins with an EVI1l domain 1 (Spred1) (Stowe et al. 2012).

SPRED domain proteins (Bundschu et al. 2007) have been reported to act as negative regulators of the Ras/MAPK pathway. Although interaction with Ras has been reported, the mechanism of Ras inhibition remains unclear (Wakioka et al. 2001). Genetic alterations in the Spred1 gene are associated with Legius syndrome (LS), a rare genetic disorder that shares mild symptoms with NF1 (Brems et al. 2007) in the absence of severe clinical phenotypes. Stowe et al. (2012) have found that Spred1 down-regulates the Ras/MAPK pathway by interaction with neurofibromin. In a respective model, Spred1 uses its amino-terminal EVH1 domain to bind neurofibromin and is recruited to the cellular membrane by the carboxy-terminal SPY-domain of Spred1. Membrane recruitment enables neurofibromin to interact with Ras, and by its RasGAP activity, down-regulate the bio-

Figure 5. Interaction partners of neurofibromin. Domain scheme of neurofibromin with reported interaction partners and structurally validated domains colored. Red dots at the *bottom* part of the scheme indicate nontruncating mutations found in patients and blue and yellow dots on the *top* part of the scheme represent potential phosphorylation sites for protein kinase A (PKA) and protein kinase C (PKC), respectively. Two ubiquitination sites are indicated as green boxes. Several potential caveolin-binding domains (CBDs) are also indicated. The GAP-related domain (GRD) part indicated in dark red is the minimal GAP domain required for the GAP activity and binds to Ras. Amino-terminal and carboxy-terminal extensions, indicated in light green are overlapping with the tubulin-binding region and the Sec14-PH domain, respectively, and interact with the Sprouty-related protein with an EVH1 domain 1 (Spred1). DDAH1, dimethylarginine dimethylaminohydrolase 1; LRPPRC, leucine-rich pentatricopeptide repeat-containing protein; APP, amyloid precursor protein; FAF2, Fas-associated factor 2; ETEA, expressed in T cells and eosinophils in atopic dermatitis; VCP, valosin-containing protein; LIMK2, LIM domain kinase 2; DPYSL2, dihydropyrimidinase-related protein 2; CRMP2/4, collapsin response mediator protein 2/4; SCF, Skp-Cullin-F box-containing complex; FAK, focal adhesion kinase; CASK, calcium/calmodulin-dependent serine protein kinase (see text).

logical activity of Ras at the cellular membrane. These observations provide a molecular link between NF1 and LS both being associated with deregulation of the Ras/MAPK signaling pathway (McClatchey and Cichowski 2012; Stowe et al. 2012). In a collaborative follow-up study with the laboratory of F. McCormick, we were able to map the EVH1-binding site to the GRD of neurofibromin (Dunzendorfer-Matt et al. 2016). EVH1 binding to the GRD was compatible with binding of activated Ras and did not influence RasGAP activity. Refined mapping experiments finally allowed us to identify the extra domain of the GRD (GAPex) as the EVH1-binding site on the GRD (Dunzendorfer-Matt et al. 2016). Importantly, EVH1 binding to the GRD did not interfere with binding of activated Ras or with GAP activity or with the formation of a transition state–like complex (Dunzendorfer-Matt et al. 2016). Pathogenic mutations found in LS patients impaired the EVH1–GRD binding as has also been found by the group of A. Yoshimura who mapped the EVH1-binding site based on yeast two-hybrid experiments (Hirata et al. 2016). In addition, a pathogenic single residue deletion of the GAPex retained RasGAP activity but failed to translocate to the membrane for interaction with Ras (Dunzendorfer-Matt et al. 2016). Although it is tempting to speculate that GAPex provides a general docking platform for proteins targeting RasGAPs to membrane compartments, it has to be noted that at least p120GAP/RASA1 is not membrane targeted by Spred proteins (Stowe et al. 2012).

P97/VCP

P97/VCP is a cytoplasmic abundant AAA (pronounced “triple A”) ATPase that is associated with various cellular activities including proteasome-mediated protein degradation. Based on immunoprecipitation studies, Wang et al. (2011) identified the protein as a neurofibromin-binding protein and mapped the binding region to a portion comprising Sec14-PH and amino- and carboxy-terminal extensions thereof. Employing pull-down experiments from mouse brain lysates using glutathione-S-transferase (GST)-fused Sec14-PH as bait we identified p97/VCP as an interaction partner of the protein (M Schwarz, F Bonneau, and K Scheffzek, unpubl.) indicating that the Sec14-PH module is sufficient to mediate the underlying protein–protein interaction. The physiological relevance of the p97/VCP neurofibromin interaction was proposed and experimentally supported to negatively regulate dendritic spine density. Mutations found in NF1 patients indeed interfered with spine density (Wang et al. 2011). Further studies are necessary to investigate the mechanistic aspects of this protein–protein interaction.

LIM Kinase2

Using yeast two-hybrid approaches to screen for interaction partners of the Sec14-PH module, Vallee et al. (2012) identified LIMK2 to interact with Sec14-PH. Sec14-PH binds to LIMK2 but not to the closely related LIMK1 and inhibits its kinase activity with respect to cofilin phosphorylation by preventing ROCK activation. Little information has been presented about the specific mechanism that causes ROCK activation to be inhibited (Vallee et al. 2012).

Cross Talk with GPCR Signaling

Potential cross talk with G protein–coupled receptor signaling (GPCR) has been identified with neurofibromin (Tong et al. 2002). The serotonin 5-hydroxytryptamin 6 (5-HT$_6$) receptor constitutively activates the Gs/adenylyl cyclase pathway in several cell types including neurons. Silencing the expression of neurofibromin strongly reduces the constitutive activity of the receptor (Deraredj Nadim et al. 2016). This process apparently requires the interaction with the PH-like domain of neurofibromin. Interestingly, disease-associated mutations found in NF1 patients blocked the interaction with the receptor.

Another GPCR cross talk has been described recently when Xie et al. (2016) described a mechanism linking opioid receptors to Ras signaling in the brain. Most importantly, $G_{\beta/\gamma}$ subunits of the μ-opioid receptor interact directly with neurofibromin. The authors mapped a $G_{\beta/\gamma}$-binding site to the Sec14-PH portion of neurofibromin and using a GST-fused GRD-Sec14-PH construct showed that $G_{\beta/\gamma}$ inhibits neurofibromin-mediated GAP activity (Xie et al. 2016). It will be important to perform GAP assays in the context of full-length neurofibromin.

CONCLUDING REMARKS

Thirty years after the discovery of the first RasGAP, the concept of G protein regulation by GEFs, and GTPase-activating proteins is well established and a plethora of research papers investigating the details of RasGAP-mediated signal regulation has accumulated in the scientific literature and beyond. Given that Ras and its major regulator neurofibromin play major roles in cellular growth control and cancer development, understanding the GAP mechanism and the roles of individual RasGAPs including neurofibromin in more detail, and also for each functionally relevant mutant, will be among the important challenges in the future.

ACKNOWLEDGMENTS

We thank our colleagues for sharing research enthusiasm, scientific results, and stimulating discussions over decades of RasGAP investigation. We are grateful to Gabi Haeusser (SAP-SE, Walldorf, Germany) for the artwork in Figure 1.

REFERENCES

Ahmadian MR, Wiesmuller L, Lautwein A, Bischoff FR, Wittinghofer A. 1996. Structural differences in the minimal catalytic domains of the GTPase-activating proteins p120GAP and neurofibromin. *J Biol Chem* **271:** 16409–16415.

Ahmadian MR, Hoffmann U, Goody RS, Wittinghofer A. 1997a. Individual rate constants for the interaction of Ras proteins with GTPase-activating proteins determined by fluorescence spectroscopy. *Biochemistry* **36:** 4535–4541.

Ahmadian MR, Mittal R, Hall A, Wittinghofer A. 1997b. Aluminum fluoride associates with the small guaninc nucleotide binding proteins. *FEBS Lett* **408:** 315–318.

Ahmadian MR, Stege P, Scheffzek K, Wittinghofer A. 1997c. Confirmation of the arginine-finger hypothesis for the GAP-stimulated GTP-hydrolysis reaction of Ras. *Nat Struct Biol* **4:** 686–689.

Ahmadian MR, Kiel C, Stege P, Scheffzek K. 2003. Structural fingerprints of the Ras-GTPase activating proteins neurofibromin and p120GAP. *J Mol Biol* **329:** 699–710.

Andersen LB, Ballester R, Marchuk DA, Chang E, Gutmann DH, Saulino AM, Camonis J, Wigler M, Collins FS. 1993. A conserved alternative splice in the von Recklinghausen neurofibromatosis (NF1) gene produces two neurofibromin isoforms, both of which have GTPase-activating protein activity. *Mol Cell Biol* **13:** 487–495.

Aravind L, Neuwald AF, Ponting CP. 1999. Sec14p-like domains in NF1 and Dbl-like proteins indicate lipid regulation of Ras and Rho signaling. *Curr Biol* **9:** R195–R197.

Ars E, Kruyer H, Morell M, Pros E, Serra E, Ravella A, Estivill X, Lazaro C. 2003. Recurrent mutations in the NF1 gene are common among neurofibromatosis type 1 patients. *J Med Genet* **40:** e82.

Arun V, Wiley JC, Kaur H, Kaplan DR, Guha A. 2013. A novel neurofibromin (NF1) interaction with the leucine-rich pentatricopeptide repeat motif-containing protein links neurofibromatosis type 1 and the French Canadian variant of Leigh's syndrome in a common molecular complex. *J Neurosci Res* **91:** 494–505.

Ballester R, Marchuk D, Boguski M, Saulino A, Letcher R, Wigler M, Collins F. 1990. The NF1 locus encodes a protein functionally related to mammalian GAP and yeast IRA proteins. *Cell* **63:** 851–859.

Bankaitis VA, Mousley CJ, Schaaf G. 2009. The Sec14 superfamily and mechanisms for crosstalk between lipid metabolism and lipid signaling. *Trends Biochem Sci* **35:** 150–160.

Bankaitis VA, Mousley CJ, Schaaf G. 2010. The Sec14 superfamily and mechanisms for crosstalk between lipid metabolism and lipid signaling. *Trends Biochem Sci* **35:** 150–160.

Banuelos S, Saraste M, Djinovic Carugo K. 1998. Structural comparisons of calponin homology domains: Implications for actin binding. *Structure* **6:** 1419–1431.

Basu TN, Gutmann DH, Fletcher JA, Glover TW, Collins FS, Downward J. 1992. Aberrant regulation of ras proteins in malignant tumour cells from type 1 neurofibromatosis patients. *Nature* **356:** 713–715.

Bax B. 1998. Domains of rasGAP and rhoGAP are related. *Nature* **392:** 447–448.

Bellazzo A, Di Minin G, Collavin L. 2017. Block one, unleash a hundred. Mechanisms of DAB2IP inactivation in cancer. *Cell Death Differ* **24:** 15–25.

Bernards A. 2003. GAPs galore! A survey of putative Ras superfamily GTPase activating proteins in man and *Drosophila*. *Biochim Biophys Acta* **1603:** 47–82.

Bernards A, Settleman J. 2004. GAP control: Regulating the regulators of small GTPases. *Trends Cell Biol* **14:** 377–385.

Bernards A, Settleman J. 2005. GAPs in growth factor signalling. *Growth Factors* **23:** 143–149.

Bianchessi D, Morosini S, Saletti V, Ibba MC, Natacci F, Esposito S, Cesaretti C, Riva D, Finocchiaro G, Eoli M. 2015. 126 novel mutations in Italian patients with neurofibromatosis type 1. *Mol Genet Genomic Med* **3:** 513–525.

Boguski MS, McCormick F. 1993. Proteins regulating Ras and its relatives. *Nature* **366:** 643–654.

Bollag G, McCormick F. 1991. Differential regulation of rasGAP and neurofibromatosis gene product activities. *Nature* **351:** 576–579.

Bollag G, McCormick F, Clark R. 1993. Characterization of full-length neurofibromin: Tubulin inhibits Ras GAP activity. *EMBO J* **12:** 1923–1927.

Bollag G, Adler F, elMasry N, McCabe PC, Conner E Jr, Thompson P, McCormick F, Shannon K. 1996. Biochemical characterization of a novel KRAS insertion mutation from a human leukemia. *J Biol Chem* **271:** 32491–32494.

Bonneau F, Lenherr ED, Pena V, Hart DJ, Scheffzek K. 2009. Solubility survey of fragments of the neurofibromatosis type 1 protein neurofibromin. *Protein Expr Purif* **65:** 30–37.

Boon LM, Mulliken JB, Vikkula M. 2005. RASA1: Variable phenotype with capillary and arteriovenous malformations. *Curr Opin Genet Dev* **15:** 265–269.

Bos JL, Rehmann H, Wittinghofer A. 2007. GEFs and GAPs: Critical elements in the control of small G proteins. *Cell* **129:** 865–877.

Bourne HR, Sanders DA, McCormick F. 1990. The GTPase superfamily: A conserved switch for diverse cell functions. *Nature* **348:** 125–132.

Bourne HR, Sander DA, McCormick F. 1991. The GTPase superfamily: Conserved structure and molecular mechanism. *Nature* **349:** 117–127.

Boyanapalli M, Lahoud OB, Messiaen L, Kim B, Anderle de Sylor MS, Duckett SJ, Somara S, Mikol DD. 2006. Neurofibromin binds to caveolin-1 and regulates ras, FAK, and Akt. *Biochem Biophys Res Commun* **340:** 1200–1208.

Brems H, Chmara M, Sahbatou M, Denayer E, Taniguchi K, Kato R, Somers R, Messiaen L, De Schepper S, Fryns JP, et al. 2007. Germline loss-of-function mutations in SPRED1 cause a neurofibromatosis 1-like phenotype. *Nat Genet* **39:** 1120–1126.

Brown MD, Sacks DB. 2006. IQGAP1 in cellular signaling: Bridging the GAP. *Trends Cell Biol* **16:** 242–249.

Bundschu K, Walter U, Schuh K. 2007. Getting a first clue about SPRED functions. *Bioessays* **29:** 897–907.

Cancer Genome Atlas Research Network. 2008. Comprehensive genomic characterization defines human glioblastoma genes and core pathways. *Nature* **455:** 1061–1068.

Cancer Genome Atlas Research Network. 2014. Comprehensive molecular profiling of lung adenocarcinoma. *Nature* **511:** 543–550.

Carroni M, Saibil HR. 2016. Cryo electron microscopy to determine the structure of macromolecular complexes. *Methods* **95:** 78–85.

Cawthon RM, O'Connell P, Buchberg AM, Viskochil D, Weiss RB, Culver M, Stevens J, Jenkins NA, Copeland NG, White R. 1990. Identification and characterization of transcripts from the neurofibromatosis 1 region: The sequence and genomic structure of EVI2 and mapping of other transcripts. *Genomics* **7:** 555–565.

Chapman JA, Kirkness EF, Simakov O, Hampson SE, Mitros T, Weinmaier T, Rattei T, Balasubramanian PG, Borman J, Busam D, et al. 2010. The dynamic genome of Hydra. *Nature* **464:** 592–596.

Chen HJ, Rojas-Soto M, Oguni A, Kennedy MB. 1998. A synaptic Ras-GTPase activating protein (p135 SynGAP) inhibited by CaM kinase II. *Neuron* **20:** 895–904.

Cherfils J, Zeghouf M. 2013. Regulation of small GTPases by GEFs, GAPs, and GDIs. *Physiol Rev* **93:** 269–309.

Cho W, Stahelin RV. 2006. Membrane binding and subcellular targeting of C2 domains. *Biochim Biophys Acta* **1761:** 838–849.

Cichowski K, Jacks T. 2001. NF1 tumor suppressor gene function: Narrowing the GAP. *Cell* **104:** 593–604.

Cichowski K, Santiago S, Jardim M, Johnson BW, Jacks T. 2003. Dynamic regulation of the Ras pathway via proteolysis of the NF1 tumor suppressor. *Genes Dev* **17:** 449–454.

Cnossen MH, van der Est MN, Breuning MH, van Asperen CJ, Breslau-Siderius EJ, van der Ploeg AT, de Goede-Bolder A, van den Ouweland AM, Halley DJ, Niermeijer MF. 1997. Deletions spanning the neurofibromatosis type 1 gene: Implications for genotype-phenotype correlations in neurofibromatosis type 1? *Hum Mutat* **9:** 458–464.

Cooper JA, Kashishian A. 1993. In vivo binding properties of SH2 domains from GTPase-activating protein and phosphatidylinositol 3-kinase. *Mol Cell Biol* **13:** 1737–1745.

Cozier GE, Lockyer PJ, Reynolds JS, Kupzig S, Bottomley JR, Millard TH, Banting G, Cullen PJ. 2000. GAP1IP4BP contains a novel group I pleckstrin homology domain that directs constitutive plasma membrane association. *J Biol Chem* **275:** 28261–28268.

Cuellar J, Valpuesta JM, Wittinghofer A, Sot B. 2017. Domain topology of human Rasal. *Biol Chem* **399:** 63–72.

Dai Y, Walker SA, de Vet E, Cook S, Welch HC, Lockyer PJ. 2011. Ca^{2+}-dependent monomer and dimer formation switches CAPRI Protein between Ras GTPase-activating protein (GAP) and RapGAP activities. *J Biol Chem* **286:** 19905–19916.

D'Angelo I, Welti S, Bonneau F, Scheffzek K. 2006. A novel bipartite phospholipid-binding module in the neurofibromatosis type 1 protein. *EMBO Rep* **7:** 174–179.

Daumke O, Weyand M, Chakrabarti PP, Vetter IR, Wittinghofer A. 2004. The GTPase-activating protein Rap1-GAP uses a catalytic asparagine. *Nature* **429:** 197–201.

DeClue JE, Cohen BD, Lowy DR. 1991. Identification and characterization of the neurofibromatosis type 1 protein product. *Proc Natl Acad Sci* **88:** 9914–9918.

DeClue JE, Papageorge AG, Fletcher JA, Diehl SR, Ratner N, Vass WC, Lowy DR. 1992. Abnormal regulation of mammalian p21ras contributes to malignant tumor growth in von Recklinghausen (type 1) neurofibromatosis. *Cell* **69:** 265–273.

De Luca A, Buccino A, Gianni D, Mangino M, Giustini S, Richetta A, Divona L, Calvieri S, Mingarelli R, Dallapiccola B. 2003. NF1 gene analysis based on DHPLC. *Hum Mutat* **21:** 171–172.

Der CJ, Finkel T, Cooper GM. 1986. Biological and biochemical properties of human rasH genes mutated at codon 61. *Cell* **44:** 167–176.

Deraredj Nadim W, Chaumont-Dubel S, Madouri F, Cobret L, De Tauzia ML, Zajdel P, Benedetti H, Marin P, Morisset-Lopez S. 2016. Physical interaction between neurofibromin and serotonin 5-HT6 receptor promotes receptor constitutive activity. *Proc Natl Acad Sci* **113:** 12310–12315.

De Schepper S, Boucneau JM, Westbroek W, Mommaas M, Onderwater J, Messiaen L, Naeyaert JM, Lambert JL.

2006. Neurofibromatosis type 1 protein and amyloid precursor protein interact in normal human melanocytes and colocalize with melanosomes. *J Invest Dermatol* **126:** 653–659.

Donovan S, Shannon KM, Bollag G. 2002. GTPase activating proteins: Critical regulators of intracellular signaling. *Biochim Biophys Acta* **1602:** 23–45.

Drugan JK, Rogers-Graham K, Gilmer T, Campbell S, Clark GJ. 2000. The Ras/p120 GTPase-activating protein (GAP) interaction is regulated by the p120 GAP pleckstrin homology domain. *J Biol Chem* **275:** 35021–35027.

Dunzendorfer-Matt T, Mercado EL, Maly K, McCormick F, Scheffzek K. 2016. The neurofibromin recruitment factor Spred1 binds to the GAP related domain without affecting Ras inactivation. *Proc Natl Acad Sci* **113:** 7497–7502.

Earl LA, Falconieri V, Milne JL, Subramaniam S. 2017. Cryo-EM: Beyond the microscope. *Curr Opin Struct Biol* **46:** 71–78.

Eichinger L, Pachebat JA, Glockner G, Rajandream MA, Sucgang R, Berriman M, Song J, Olsen R, Szafranski K, Xu Q, et al. 2005. The genome of the social amoeba *Dictyostelium discoideum*. *Nature* **435:** 43–57.

Ekvall S, Sjors K, Jonzon A, Vihinen M, Anneren G, Bondeson ML. 2014. Novel association of neurofibromatosis type 1-causing mutations in families with neurofibromatosis-Noonan syndrome. *Am J Med Genet A* **164A:** 579–587.

Evdokimov AG, Tropea JE, Routzahn KM, Waugh DS. 2002. Crystal structure of the *Yersinia pestis* GTPase activator YopE. *Protein Sci* **11:** 401–408.

Fahsold R, Hoffmeyer S, Mischung C, Gille C, Ehlers C, Kucukceylan N, Abdel-Nour M, Gewies A, Peters H, Kaufmann D, et al. 2000. Minor lesion mutational spectrum of the entire NF1 gene does not explain its high mutability but points to a functional domain upstream of the GAP-related domain. *Am J Hum Genet* **66:** 790–818.

Fantl WJ, Escobedo JA, Martin GA, Turck CW, del Rosario M, McCormick F, Williams LT. 1992. Distinct phosphotyrosines on a growth factor receptor bind to specific molecules that mediate different signaling pathways. *Cell* **69:** 413–423.

Feng L, Yunoue S, Tokuo H, Ozawa T, Zhang D, Patrakitkomjorn S, Ichimura T, Saya H, Araki N. 2004. PKA phosphorylation and 14-3-3 interaction regulate the function of neurofibromatosis type I tumor suppressor, neurofibromin. *FEBS Lett* **557:** 275–282.

Fokkema IF, Taschner PE, Schaafsma GC, Celli J, Laros JF, den Dunnen JT. 2011. LOVD v.2.0: The next generation in gene variant databases. *Hum Mutat* **32:** 557–563.

Franken SM, Scheidig AJ, Krengel U, Rensland H, Lautwein A, Geyer M, Scheffzek K, Goody RS, Kalbitzer HR, Pai EF, et al. 1993. Three-dimensional structures and properties of a transforming and a nontransforming glycine-12 mutant of p21H-ras. *Biochemistry* **32:** 8411–8420.

Gamblin SJ, Smerdon SJ. 1998. GTPase-activating proteins and their complexes. *Curr Opin Struct Biol* **8:** 195–201.

Gavriljuk K, Gazdag EM, Itzen A, Kotting C, Goody RS, Gerwert K. 2012. Catalytic mechanism of a mammalian Rab·RabGAP complex in atomic detail. *Proc Natl Acad Sci* **109:** 21348–21353.

Gerke V, Moss SE. 2002. Annexins: From structure to function. *Physiol Rev* **82:** 331–371.

Gerwert K, Mann D, Kotting C. 2017. Common mechanisms of catalysis in small and heterotrimeric GTPases and their respective GAPs. *Biol Chem* **398:** 523–533.

Ghosh R, Bankaitis VA. 2011. Phosphatidylinositol transfer proteins: Negotiating the regulatory interface between lipid metabolism and lipid signaling in diverse cellular processes. *Biofactors* **37:** 290–308.

Gremer L, Gilsbach B, Ahmadian MR, Wittinghofer A. 2008. Fluoride complexes of oncogenic Ras mutants to study the Ras–RasGap interaction. *Biol Chem* **389:** 1163–1171.

Grewal T, Enrich C. 2006. Molecular mechanisms involved in Ras inactivation: The annexin A6-p120GAP complex. *Bioessays* **28:** 1211–1220.

Grewal T, Evans R, Rentero C, Tebar F, Cubells L, de Diego I, Kirchhoff MF, Hughes WE, Heeren J, Rye KA, et al. 2005. Annexin A6 stimulates the membrane recruitment of p120GAP to modulate Ras and Raf-1 activity. *Oncogene* **24:** 5809–5820.

Grewal T, Koese M, Tebar F, Enrich C. 2011. Differential regulation of RasGAPs in cancer. *Genes Cancer* **2:** 288–297.

Gutmann DH, Wood DL, Collins FS. 1991. Identification of the neurofibromatosis type 1 gene product. *Proc Natl Acad Sci* **88:** 9658–9662.

Hakimi MA, Speicher DW, Shiekhattar R. 2002. The motor protein kinesin-1 links neurofibromin and merlin in a common cellular pathway of neurofibromatosis. *J Biol Chem* **277:** 36909–36912.

Han JW, McCormick F, Macara IG. 1991. Regulation of Ras-GAP and the neurofibromatosis-1 gene product by eicosanoids. *Science* **252:** 576–579.

Han SS, Cooper DN, Upadhyaya MN. 2001. Evaluation of denaturing high performance liquid chromatography (DHPLC) for the mutational analysis of the neurofibromatosis type 1 (*NF1*) gene. *Hum Genet* **109:** 487–497.

Hedman AC, Smith JM, Sacks DB. 2015. The biology of IQGAP proteins: Beyond the cytoskeleton. *EMBO Rep* **16:** 427–446.

Henkemeyer M, Rossi DJ, Holmyard DP, Puri MC, Mbamalu G, Harpal K, Shih TS, Jacks T, Pawson T. 1995. Vascular system defects and neuronal apoptosis in mice lacking ras GTPase-activating protein. *Nature* **377:** 695–701.

Hettich L, Marshall M. 1994. Structural analysis of the Ras GTPase activating protein catalytic domain by semirandom mutagenesis: Implications for a mechanism of interaction with Ras-GTP. *Cancer Res* **54:** 5438–5444.

Hiatt KK, Ingram DA, Zhang Y, Bollag G, Clapp DW. 2001. Neurofibromin GTPase-activating protein-related domains restore normal growth in $Nf1^{-/-}$ cells. *J Biol Chem* **276:** 7240–7245.

Hirata Y, Brems H, Suzuki M, Kanamori M, Okada M, Morita R, Llano-Rivas I, Ose T, Messiaen L, Legius E, et al. 2016. Interaction between a domain of the negative regulator of the Ras-ERK pathway, SPRED1 protein, and the GTPase-activating protein-related domain of neurofibromin is implicated in Legius syndrome and neurofibromatosis type 1. *J Biol Chem* **291:** 3124–3134.

Hollstein PE, Cichowski K. 2013. Identifying the ubiquitin ligase complex that regulates the NF1 tumor suppressor and Ras. *Cancer Discov* **3:** 880–893.

Honeybee Genome Sequencing C. 2006. Insights into social insects from the genome of the honeybee *Apis mellifera*. *Nature* **443:** 931–949.

Hota PK, Buck M. 2012. Plexin structures are coming: Opportunities for multilevel investigations of semaphorin guidance receptors, their cell signaling mechanisms, and functions. *Cell Mol Life Sci* **69:** 3765–3805.

Hsueh YP, Roberts AM, Volta M, Sheng M, Roberts RG. 2001. Bipartite interaction between neurofibromatosis type I protein (neurofibromin) and syndecan transmembrane heparan sulfate proteoglycans. *J Neurosci* **21:** 3764–3770.

Iwashita S, Song SY. 2008. RasGAPs: A crucial regulator of extracellular stimuli for homeostasis of cellular functions. *Mol Biosyst* **4:** 213–222.

Izawa I, Tamaki N, Saya H. 1996. Phosphorylation of neurofibromatosis type 1 gene product (neurofibromin) by cAMP-dependent protein kinase. *FEBS Lett* **382:** 53–59.

Jeong SY, Park SJ, Kim HJ. 2006. The spectrum of NF1 mutations in Korean patients with neurofibromatosis type 1. *J Korean Med Sci* **21:** 107–112.

Jeyabalan N, Clement JP. 2016. SYNGAP1: Mind the gap. *Front Cell Neurosci* **10:** 32.

Jin H, Wang X, Ying J, Wong AH, Cui Y, Srivastava G, Shen ZY, Li EM, Zhang Q, Jin J, et al. 2007. Epigenetic silencing of a Ca^{2+}-regulated Ras GTPase-activating protein RASAL defines a new mechanism of Ras activation in human cancers. *Proc Natl Acad Sci* **104:** 12353–12358.

Kaplan DR, Morrison DK, Wong G, McCormick F, Williams LT. 1990. PDGF β-receptor stimulates tyrosine phosphorylation of GAP and association of GAP with a signaling complex. *Cell* **61:** 125–133.

Kashishian A, Kazlauskas A, Cooper JA. 1992. Phosphorylation sites in the PDGF receptor with different specificities for binding GAP and PI3 kinase in vivo. *EMBO J* **11:** 1373–1382.

Kaufmann D. (ed.). 2008. *Neurofibromatoses.* Monographs in human genetics, Vol. 16. S. Karger, Basel, Switzerland.

Kehrer-Sawatzki H, Maier C, Moschgath E, Elgar G, Krone W. 1998. Genomic characterization of the neurofibromatosis type 1 gene of Fugu rubripes. *Gene* **222:** 145–153.

Kim JH, Liao D, Lau LF, Huganir RL. 1998. SynGAP: A synaptic RasGAP that associates with the PSD-95/SAP90 protein family. *Neuron* **20:** 683–691.

King PD, Lubeck BA, Lapinski PE. 2013. Nonredundant functions for Ras GTPase-activating proteins in tissue homeostasis. *Sci Signal* **6:** re1.

Klose A, Ahmadian MR, Schuelke M, Scheffzek K, Hoffmeyer S, Gewies A, Schmitz F, Kaufmann D, Peters H, Wittinghofer A, et al. 1998. Selective disactivation of neurofibromin GAP activity in neurofibromatosis type 1. *Hum Mol Genet* **7:** 1261–1268.

Kostenko EV, Mahon GM, Cheng L, Whitehead IP. 2004. The Sec14 homology domain regulates the cellular distribution and transforming activity of the Rho-specific guanine nucleotide exchange factor, Dbs. *J Biol Chem* **280:** 2807–2817.

Kotting C, Kallenbach A, Suveyzdis Y, Wittinghofer A, Gerwert K. 2008. The GAP arginine finger movement into the catalytic site of Ras increases the activation entropy. *Proc Natl Acad Sci* **105:** 6260–6265.

Krapivinsky G, Medina I, Krapivinsky L, Gapon S, Clapham DE. 2004. SynGAP-MUPP1-CaMKII synaptic complexes regulate P38 MAP kinase activity and NMDA receptor-dependent synaptic AMPA receptor potentiation. *Neuron* **43:** 563–574.

Kupzig S, Deaconescu D, Bouyoucef D, Walker SA, Liu Q, Polte CL, Daumke O, Ishizaki T, Lockyer PJ, Wittinghofer A, et al. 2006. GAP1 family members constitute bifunctional Ras and Rap GTPase-activating proteins. *J Biol Chem* **281:** 9891–9900.

Kupzig S, Bouyoucef-Cherchalli D, Yarwood S, Sessions R, Cullen PJ. 2009. The ability of GAP1IP4BP to function as a Rap1 GTPase-activating protein (GAP) requires its Ras GAP-related domain and an arginine finger rather than an asparagine thumb. *Mol Cell Biol* **29:** 3929–3940.

Kurella VB, Richard JM, Parke CL, Lecour LF Jr, Bellamy HD, Worthylake DK. 2009. Crystal structure of the GTPase-activating protein-related domain from IQGAP1. *J Biol Chem* **284:** 14857–14865.

Kweh F, Zheng M, Kurenova E, Wallace M, Golubovskaya V, Cance WG. 2009. Neurofibromin physically interacts with the N-terminal domain of focal adhesion kinase. *Mol Carcinog* **48:** 1005–1017.

LeCour L Jr, Boyapati VK, Liu J, Li Z, Sacks DB, Worthylake DK. 2016. The structural basis for Cdc42-induced dimerization of IQGAPs. *Structure* **24:** 1499–1508.

Lee MJ, Su YN, You HL, Chiou SC, Lin LC, Yang CC, Lee WC, Hwu WL, Hsieh FJ, Stephenson DA, et al. 2006. Identification of forty-five novel and twenty-three known NF1 mutations in Chinese patients with neurofibromatosis type 1. *Hum Mutat* **27:** 832.

Lemmon MA. 2004. Pleckstrin homology domains: Not just for phosphoinositides. *Biochem Soc Trans* **32:** 707–711.

Li Y, Bollag G, Clark R, Stevens J, Conroy L, Fults D, Ward K, Friedman E, Samowitz W, Robertson M, et al. 1992. Somatic mutations in the neurofibromatosis 1 gene in human tumors. *Cell* **69:** 275–281.

Ligeti E, Welti S, Scheffzek K. 2012. Inhibition and termination of physiological responses by GTPase activating proteins. *Physiol Rev* **92:** 237–272.

Lin YL, Hsueh YP. 2008. Neurofibromin interacts with CRMP-2 and CRMP-4 in rat brain. *Biochem Biophys Res Commun* **369:** 747–752.

Liu Q, Walker SA, Gao D, Taylor JA, Dai YF, Arkell RS, Bootman MD, Roderick HL, Cullen PJ, Lockyer PJ. 2005. CAPRI and RASAL impose different modes of information processing on Ras due to contrasting temporal filtering of Ca^{2+}. *J Cell Biol* **170:** 183–190.

Maertens O, Cichowski K. 2014. An expanding role for RAS GTPase activating proteins (RAS GAPs) in cancer. *Adv Biol Regul* **55:** 1–14.

Mangoura D, Sun Y, Li C, Singh D, Gutmann DH, Flores A, Ahmed M, Vallianatos G. 2006. Phosphorylation of neurofibromin by PKC is a possible molecular switch in EGF receptor signaling in neural cells. *Oncogene* **25:** 735–745.

Marshall MS, Hill WS, Ng AS, Vogel US, Schaber MD, Scolnick EM, Dixon RA, Sigal IS, Gibbs JB. 1989.

A C-terminal domain of GAP is sufficient to stimulate ras p21 GTPase activity. *EMBO J* **8:** 1105–1110.

Martin GA, Viskochil D, Bollag G, McCabe PC, Crosier WJ, Haubruck H, Conroy L, Clark R, O'Connell P, Cawthon RM, et al. 1990. The GAP-related domain of the neurofibromatosis type 1 gene product interacts with ras p21. *Cell* **63:** 843–849.

Martin-Garcia F, Mendieta-Moreno JI, Lopez-Vinas E, Gomez-Puertas P, Mendieta J. 2012. The role of Gln61 in HRas GTP hydrolysis: A quantum mechanics/molecular mechanics study. *Biophys J* **102:** 152–157.

Mattocks C, Baralle D, Tarpey P, ffrench-Constant C, Bobrow M, Whittaker J. 2004. Automated comparative sequence analysis identifies mutations in 89% of NF1 patients and confirms a mutation cluster in exons 11-17 distinct from the GAP related domain. *J Med Genet* **41:** e48.

McClatchey AI, Cichowski K. 2012. SPRED proteins provide a NF-ty link to Ras suppression. *Genes Dev* **26:** 1515–1519.

McLaughlin SK, Olsen SN, Dake B, De Raedt T, Lim E, Bronson RT, Beroukhim R, Polyak K, Brown M, Kuperwasser C, et al. 2013. The RasGAP gene, RASAL2, is a tumor and metastasis suppressor. *Cancer Cell* **24:** 365–378.

Mellert K, Lechner S, Ludeke M, Lamla M, Moller P, Kemkemer R, Scheffzek K, Kaufmann D. 2018. Restoring functional neurofibromin by protein transduction. *Sci Rep* **8:** 6171.

Merk A, Bartesaghi A, Banerjee S, Falconieri V, Rao P, Davis MI, Pragani R, Boxer MB, Earl LA, Milne JLS, et al. 2016. Breaking cryo-EM resolution barriers to facilitate drug discovery. *Cell* **165:** 1698–1707.

Miao W, Eichelberger L, Baker L, Marshall MS. 1996. p120 Ras GTPase-activating protein interacts with Ras-GTP through specific conserved residues. *J Biol Chem* **271:** 15322–15329.

Milburn MV, Tong L, deVos AM, Brunger A, Yamaizumi Z, Nishimura S, Kim SH. 1990. Molecular switch for signal transduction: Structural differences between active and inactive forms of protooncogenic ras proteins. *Science* **247:** 939–945.

Mishra AK, Lambright DG. 2016. Invited review: Small GTPases and their GAPs. *Biopolymers* **105:** 431–448.

Mittal R, Ahmadian MR, Goody RS, Wittinghofer A. 1996. Formation of a transition-state analog of the Ras GTPase reaction by Ras-GDP, tetrafluoroaluminate, and GTPase-activating proteins. *Science* **273:** 115–117.

Nassar N, Hoffman GR, Manor D, Clardy JC, Cerione RA. 1998. Structures of Cdc42 bound to the active and catalytically compromised forms of Cdc42GAP. *Nat Struct Biol* **5:** 1047–1052.

Neufeld G, Kessler O. 2008. The semaphorins: Versatile regulators of tumour progression and tumour angiogenesis. *Nat Rev Cancer* **8:** 632–645.

Oh JS, Manzerra P, Kennedy MB. 2004. Regulation of the neuron-specific Ras GTPase-activating protein, synGAP, by Ca^{2+}/calmodulin-dependent protein kinase II. *J Biol Chem* **279:** 17980–17988.

Oinuma I, Ishikawa Y, Katoh H, Negishi M. 2004. The Semaphorin 4D receptor Plexin-B1 is a GTPase activating protein for R-Ras. *Science* **305:** 862–865.

Oliveira AF, Yasuda R. 2014. Neurofibromin is the major Ras inactivator in dendritic spines. *J Neurosci* **34:** 776–783.

Padmanabhan A, Lee JS, Ismat FA, Lu MM, Lawson ND, Kanki JP, Look AT, Epstein JA. 2009. Cardiac and vascular functions of the zebrafish orthologues of the type I neurofibromatosis gene NFI. *Proc Natl Acad Sci* **106:** 22305–22310.

Pai EF, Kabsch W, Krengel U, Holmes KC, John J, Wittinghofer A. 1989. Structure of the guanine-nucleotide-binding domain of the Ha-ras oncogene product p21 in the triphosphate conformation. *Nature* **341:** 209–214.

Pamonsinlapatham P, Hadj-Slimane R, Lepelletier Y, Allain B, Toccafondi M, Garbay C, Raynaud F. 2009. p120-Ras GTPase activating protein (RasGAP): A multi-interacting protein in downstream signaling. *Biochimie* **91:** 320–328.

Pan X, Eathiraj S, Munson M, Lambright DG. 2006. TBC-domain GAPs for Rab GTPases accelerate GTP hydrolysis by a dual-finger mechanism. *Nature* **442:** 303–306.

Pascoe HG, Wang Y, Zhang X. 2015. Structural mechanisms of plexin signaling. *Prog Biophys Mol Biol* **118:** 161–168.

Patch AM, Christie EL, Etemadmoghadam D, Garsed DW, George J, Fereday S, Nones K, Cowin P, Alsop K, Bailey PJ, et al. 2015. Whole-genome characterization of chemoresistant ovarian cancer. *Nature* **521:** 489–494.

Patrakitkomjorn S, Kobayashi D, Morikawa T, Wilson MM, Tsubota N, Irie A, Ozawa T, Aoki M, Arimura N, Kaibuchi K, et al. 2008. Neurofibromatosis type 1 (NF1) tumor suppressor, neurofibromin, regulates the neuronal differentiation of PC12 cells via its associating protein, CRMP-2. *J Biol Chem* **283:** 9399–9413.

Pena V, Hothorn M, Eberth A, Kaschau N, Parret A, Gremer L, Bonneau F, Ahmadian MR, Scheffzek K. 2008. The C2 domain of SynGAP is essential for stimulation of the Rap GTPase reaction. *EMBO Rep* **9:** 350–355.

Phan VT, Ding VW, Li F, Chalkley RJ, Burlingame A, McCormick F. 2010. The RasGAP proteins Ira2 and neurofibromin are negatively regulated by Gpb1 in yeast and ETEA in humans. *Mol Cell Biol* **30:** 2264–2279.

Philpott C, Tovell H, Frayling IM, Cooper DN, Upadhyaya M. 2017. The NF1 somatic mutational landscape in sporadic human cancers. *Hum Genomics* **11:** 13.

Ratner N, Miller SJ. 2015. A RASopathy gene commonly mutated in cancer: The neurofibromatosis type 1 tumour suppressor. *Nat Rev Cancer* **15:** 290–301.

Rauen KA. 2013. The RASopathies. *Annu Rev Genomics Hum Genet* **14:** 355–369.

Revencu N, Boon LM, Mulliken JB, Enjolras O, Cordisco MR, Burrows PE, Clapuyt P, Hammer F, Dubois J, Baselga E, et al. 2008. Parkes Weber syndrome, vein of Galen aneurysmal malformation, and other fast-flow vascular anomalies are caused by RASA1 mutations. *Hum Mutat* **29:** 959–965.

Revencu N, Boon LM, Mendola A, Cordisco MR, Dubois J, Clapuyt P, Hammer F, Amor DJ, Irvine AD, Baselga E, et al. 2013. RASA1 mutations and associated phenotypes in 68 families with capillary malformation-arteriovenous malformation. *Hum Mutat* **34:** 1632–1641.

Riccardi VM. 1992. *Neurofibromatosis: Phenotype natural history, and pathogenesis*, 2nd ed. The Johns Hopkins University Press, Baltimore, MD.

Rittinger K, Walker PA, Eccleston JF, Nurmahomed K, Owen D, Laue E, Gamblin SJ, Smerdon SJ. 1997. Crystal structure of a small G protein in complex with the GTPase-activating protein ρGAP. *Nature* **388:** 693–697.

Rojas RJ, Yohe ME, Gershburg S, Kawano T, Kozasa T, Sondek J. 2007. Galphaq directly activates p63ρGEF and Trio via a conserved extension of the Dbl homology-associated pleckstrin homology domain. *J Biol Chem* **282:** 29201–29210.

Rojnueangnit K, Xie J, Gomes A, Sharp A, Callens T, Chen Y, Liu Y, Cochran M, Abbott MA, Atkin J, et al. 2015. High incidence of Noonan Syndrome features including short stature and pulmonic stenosis in patients carrying NF1 missense mutations affecting p.Arg1809: Genotype-phenotype correlation. *Hum Mutat* **36:** 1052–1063.

Rubinfeld B, Munemitsu S, Clark R, Conroy L, Watt K, Crosier WJ, McCormick F, Polakis P. 1991. Molecular cloning of a GTPase activating protein specific for the Krev-1 protein p21rap1. *Cell* **65:** 1033–1042.

Rudack T, Xia F, Schlitter J, Kotting C, Gerwert K. 2012. Ras and GTPase-activating protein (GAP) drive GTP into a precatalytic state as revealed by combining FTIR and biomolecular simulations. *Proc Natl Acad Sci* **109:** 15295–15300.

Saito K, Tautz L, Mustelin T. 2007. The lipid-binding SEC14 domain. *Biochim Biophys Acta* **1771:** 719–726.

Saito Y, Oinuma I, Fujimoto S, Negishi M. 2009. Plexin-B1 is a GTPase activating protein for M-Ras, remodelling dendrite morphology. *EMBO Rep* **10:** 614–621.

Sangha N, Wu R, Kuick R, Powers S, Mu D, Fiander D, Yuen K, Katabuchi H, Tashiro H, Fearon ER, et al. 2008. Neurofibromin 1 (NF1) defects are common in human ovarian serous carcinomas and co-occur with TP53 mutations. *Neoplasia* **10:** 1362–1372.

Santoro C, Maietta A, Giugliano T, Melis D, Perrotta S, Nigro V, Piluso G. 2015. Arg(1809) substitution in neurofibromin: Further evidence of a genotype-phenotype correlation in neurofibromatosis type 1. *Eur J Hum Genet* **23:** 1460–1461.

Scheffzek K, Ahmadian MR. 2005. GTPase activating proteins: Structural and functional insights 18 years after discovery. *Cell Mol Life Sci* **62:** 3014–3038.

Scheffzek K, Welti S. 2012. Pleckstrin homology (PH) like domains—Versatile modules in protein–protein interaction platforms. *FEBS Lett* **586:** 2662–2673.

Scheffzek K, Lautwein A, Kabsch W, Ahmadian MR, Wittinghofer A. 1996. Crystal structure of the GTPase-activating domain of human p120GAP and implications for the interaction with Ras. *Nature* **384:** 591–596.

Scheffzek K, Ahmadian MR, Kabsch W, Wiesmuller L, Lautwein A, Schmitz F, Wittinghofer A. 1997a. The Ras-RasGAP complex: Structural basis for GTPase activation and its loss in oncogenic Ras mutants. *Science* **277:** 333–338.

Scheffzek K, Lautwein A, Scherer A, Franken S, Wittinghofer A. 1997b. Crystallization and preliminary X-ray crystallographic study of the Ras- GTPase-activating domain of human p120GAP. *Proteins* **27:** 315–318.

Scheffzek K, Ahmadian MR, Wiesmuller L, Kabsch W, Stege P, Schmitz F, Wittinghofer A. 1998a. Structural analysis of the GAP-related domain from neurofibromin and its implications. *EMBO J* **17:** 4313–4327.

Scheffzek K, Ahmadian MR, Wittinghofer A. 1998b. GTPase-activating proteins: Helping hands to complement an active site. *Trends Biochem Sci* **23:** 257–262.

Scrima A, Thomas C, Deaconescu D, Wittinghofer A. 2008. The Rap–RapGAP complex: GTP hydrolysis without catalytic glutamine and arginine residues. *EMBO J* **27:** 1145–1153.

Seeburg PH, Colby WW, Capon DJ, Goeddel DV, Levinson AD. 1984. Biological properties of human c-Ha-*ras*1 genes mutated at codon 12. *Nature* **312:** 71–75.

Sermon BA, Eccleston JF, Skinner RH, Lowe PN. 1996. Mechanism of inhibition by arachidonic acid of the catalytic activity of Ras GTPase-activating proteins. *J Biol Chem* **271:** 1566–1572.

Sermon BA, Lowe PN, Strom M, Eccleston JF. 1998. The importance of two conserved arginine residues for catalysis by the Ras GTPase-activating protein, neurofibromin. *J Biol Chem* **273:** 9480–9485.

Serth J, Lautwein A, Frech M, Wittinghofer A, Pingoud A. 1991. The inhibition of the GTPase activating protein–Ha-*ras* interaction by acidic lipids is due to physical association of the C-terminal domain of the GTPase activating protein with micellar structures. *EMBO J* **10:** 1325–1330.

Shi Y. 2014. A glimpse of structural biology through X-ray crystallography. *Cell* **159:** 995–1014.

Simanshu DK, Nissley DV, McCormick F. 2017. RAS proteins and their regulators in human Disease. *Cell* **170:** 17–33.

Sirokmany G, Szidonya L, Kaldi K, Gaborik Z, Ligeti E, Geiszt M. 2005. Sec14 homology domain targets p50RhoGAP to endosomes and provides a link between Rab- and Rho GTPases. *J Biol Chem* **281:** 6096–6105.

Smith JM, Hedman AC, Sacks DB. 2015. IQGAPs choreograph cellular signaling from the membrane to the nucleus. *Trends Cell Biol* **25:** 171–184.

Sot B, Kotting C, Deaconescu D, Suveyzdis Y, Gerwert K, Wittinghofer A. 2010. Unravelling the mechanism of dual-specificity GAPs. *EMBO J* **29:** 1205–1214.

Sot B, Behrmann E, Raunser S, Wittinghofer A. 2013. Ras GTPase activating (RasGAP) activity of the dual specificity GAP protein Rasal requires colocalization and C2 domain binding to lipid membranes. *Proc Natl Acad Sci* **110:** 111–116.

Stebbins CE, Galan JE. 2000. Modulation of host signaling by a bacterial mimic: Structure of the *Salmonella* effector SptP bound to Rac1. *Mol Cell* **6:** 1449–1460.

Stowe IB, Mercado EL, Stowe TR, Bell EL, Oses-Prieto JA, Hernandez H, Burlingame AL, McCormick F. 2012. A shared molecular mechanism underlies the human rasopathies Legius syndrome and neurofibromatosis-1. *Genes Dev* **26:** 1421–1426.

Tan M, Zhao Y, Kim SJ, Liu M, Jia L, Saunders TL, Zhu Y, Sun Y. 2011. SAG/RBX2/ROC2 E3 ubiquitin ligase is essential for vascular and neural development by targeting NF1 for degradation. *Dev Cell* **21:** 1062–1076.

Tassabehji M, Strachan T, Sharland M, Colley A, Donnai D, Harris R, Thakker N. 1993. Tandem duplication within a neurofibromatosis type 1 (NF1) gene exon in a family with features of Watson syndrome and Noonan syndrome. *Am J Hum Genet* **53:** 90–95.

The I, Hannigan GE, Cowley GS, Reginald S, Zhong Y, Gusella JF, Hariharan IK, Bernards A. 1997. Rescue of a *Drosophila* NF1 mutant phenotype by protein kinase A. *Science* **276:** 791–794.

Thomas SL, Deadwyler GD, Tang J, Stubbs EB Jr, Muir D, Hiatt KK, Clapp DW, De Vries GH. 2006. Reconstitution of the NF1 GAP-related domain in NF1-deficient human Schwann cells. *Biochem Biophys Res Commun* **348:** 971–980.

Thomas L, Richards M, Mort M, Dunlop E, Cooper DN, Upadhyaya M. 2012. Assessment of the potential pathogenicity of missense mutations identified in the GTPase-activating protein (GAP)-related domain of the neurofibromatosis type-1 (NF1) gene. *Hum Mutat* **33:** 1687–1696.

Tokuo H, Yunoue S, Feng L, Kimoto M, Tsuji H, Ono T, Saya H, Araki N. 2001. Phosphorylation of neurofibromin by cAMP-dependent protein kinase is regulated via a cellular association of N^G,N^G-dimethylarginine dimethylaminohydrolase. *FEBS Lett* **494:** 48–53.

Tong J, Hannan F, Zhu Y, Bernards A, Zhong Y. 2002. Neurofibromin regulates G protein-stimulated adenylyl cyclase activity. *Nat Neurosci* **5:** 95–96.

Trahey M, McCormick F. 1987. A cytoplasmic protein stimulates normal N-ras p21 GTPase, but does not affect oncogenic mutants. *Science* **238:** 542–545.

Trahey M, Wong G, Halenbeck R, Rubinfeld B, Martin GA, Ladner M, Long CM, Crosier WJ, Watt K, Koths K, et al. 1988. Molecular cloning of two types of GAP complementary DNA from human placenta. *Science* **242:** 1697–1700.

Tsai MH, Yu CL, Wei FS, Stacey DW. 1989. The effect of GTPase activating protein upon ras is inhibited by mitogenically responsive lipids. *Science* **243:** 522–526.

Upadhyaya M, Cooper DN., ed. 2012. *Neurofibromatosis type 1—Molecular and cellular biology*. Springer, New York.

Upadhyaya M, Maynard J, Osborn M, Harper PS. 1997a. Six novel mutations in the neurofibromatosis type 1 (NF1) gene. *Hum Mutat* **10:** 248–250.

Upadhyaya M, Osborn MJ, Maynard J, Kim MR, Tamanoi F, Cooper DN. 1997b. Mutational and functional analysis of the neurofibromatosis type 1 (NF1) gene. *Hum Genet* **99:** 88–92.

Upadhyaya M, Huson SM, Davies M, Thomas N, Chuzhanova N, Giovannini S, Evans DG, Howard E, Kerr B, Griffiths S, et al. 2007. An absence of cutaneous neurofibromas associated with a 3-bp inframe deletion in exon 17 of the NF1 gene (c.2970-2972 delAAT): Evidence of a clinically significant NF1 genotype-phenotype correlation. *Am J Hum Genet* **80:** 140–151.

Vallee B, Doudeau M, Godin F, Gombault A, Tchalikian A, de Tauzia ML, Benedetti H. 2012. Nf1 RasGAP inhibition of LIMK2 mediates a new cross-talk between Ras and ρ pathways. *PLoS ONE* **7:** e47283.

van Minkelen R, van Bever Y, Kromosoeto JN, Withagen-Hermans CJ, Nieuwlaat A, Halley DJ, van den Ouweland AM. 2014. A clinical and genetic overview of 18 years neurofibromatosis type 1 molecular diagnostics in the Netherlands. *Clin Genet* **85:** 318–327.

Vetter IR, Wittinghofer A. 2001. The guanine nucleotide-binding switch in three dimensions. *Science* **294:** 1299–1304.

Viskochil D, Buchberg AM, Xu G, Cawthon RM, Stevens J, Wolff RK, Culver M, Carey JC, Copeland NG, Jenkins NA, et al. 1990. Deletions and a translocation interrupt a cloned gene at the neurofibromatosis type 1 locus. *Cell* **62:** 187–192.

Vogel US, Dixon RA, Schaber MD, Diehl RE, Marshall MS, Scolnick EM, Sigal IS, Gibbs JB. 1988. Cloning of bovine GAP and its interaction with oncogenic ras p21. *Nature* **335:** 90–93.

Volta M, Calza S, Roberts AM, Roberts RG. 2010. Characterisation of the interaction between syndecan-2, neurofibromin and CASK: Dependence of interaction on syndecan dimerization. *Biochem Biophys Res Commun* **391:** 1216–1221.

von Recklinghausen FD. 1882. *Über die multiplen Fibrome der Haut und ihre Beziehung zu den multiplen Neuromen*. Hirschwald, Berlin.

Wakioka T, Sasaki A, Kato R, Shouda T, Matsumoto A, Miyoshi K, Tsuneoka M, Komiya S, Baron R, Yoshimura A. 2001. Spred is a Sprouty-related suppressor of Ras signalling. *Nature* **412:** 647–651.

Walkup WGt, Washburn L, Sweredoski MJ, Carlisle HJ, Graham RL, Hess S, Kennedy MB. 2015. Phosphorylation of synaptic GTPase-activating protein (synGAP) by Ca^{2+}/calmodulin-dependent protein kinase II (CaMKII) and cyclin-dependent kinase 5 (CDK5) alters the ratio of its GAP activity toward Ras and Rap GTPases. *J Biol Chem* **290:** 4908–4927.

Wallace MR, Marchuk DA, Andersen LB, Letcher R, Odeh HM, Saulino AM, Fountain JW, Brereton A, Nicholson J, Mitchell AL, et al. 1990. Type 1 neurofibromatosis gene: Identification of a large transcript disrupted in three NF1 patients. *Science* **249:** 181–186.

Wang HF, Shih YT, Chen CY, Chao HW, Lee MJ, Hsueh YP. 2011. Valosin-containing protein and neurofibromin interact to regulate dendritic spine density. *J Clin Invest* **121:** 4820–4837.

Wang Y, He H, Srivastava N, Vikarunnessa S, Chen YB, Jiang J, Cowan CW, Zhang X. 2012. Plexins are GTPase-activating proteins for Rap and are activated by induced dimerization. *Sci Signal* **5:** ra6.

Welti S, Fraterman S, D'Angelo I, Wilm M, Scheffzek K. 2007. The sec14 homology module of neurofibromin binds cellular glycerophospholipids: Mass spectrometry and structure of a lipid complex. *J Mol Biol* **366:** 551–562.

Welti S, Kuhn S, D'Angelo I, Brugger B, Kaufmann D, Scheffzek K. 2011. Structural and biochemical consequences of NF1 associated nontruncating mutations in the Sec14-PH module of neurofibromin. *Hum Mutat* **32:** 191–197.

Wittinghofer A. 1997. Signaling mechanistics: Aluminum fluoride for molecule of the year. *Curr Biol* **7:** R682–R685.

Wittinghofer A, Vetter IR. 2011. Structure–function relationships of the G domain, a canonical switch motif. *Annu Rev Biochem* **80:** 943–971.

Wittinghofer A, Waldmann H. 2000. Ras—A molecular switch involved in tumor formation. *Angewandte Chemie* **39:** 4173–4390.

Wu R, Lopez-Correa C, Rutkowski JL, Baumbach LL, Glover TW, Legius E. 1999. Germline mutations in NF1 patients with malignancies. *Genes Chromosomes Cancer* **26:** 376–380.

Wurtele M, Renault L, Barbieri JT, Wittinghofer A, Wolf E. 2001. Structure of the ExoS GTPase activating domain. *FEBS Lett* **491:** 26–29.

Xie K, Colgan LA, Dao MT, Muntean BS, Sutton LP, Orlandi C, Boye SL, Boye SE, Shih CC, Li Y, et al. 2016. nf1 is a direct g protein effector essential for opioid signaling to Ras in the striatum. *Curr Biol* **26:** 2992–3003.

Xu H, Gutmann DH. 1997. Mutations in the GAP-related domain impair the ability of neurofibromin to associate with microtubules. *Brain Res* **759:** 149–152.

Xu GF, Lin B, Tanaka K, Dunn D, Wood D, Gesteland R, White R, Weiss R, Tamanoi F. 1990a. The catalytic domain of the neurofibromatosis type 1 gene product stimulates *ras* GTPase and complements *ira* mutants of *S. cerevisiae*. *Cell* **63:** 835–841.

Xu GF, O'Connell P, Viskochil D, Cawthon R, Robertson M, Culver M, Dunn D, Stevens J, Gesteland R, White R, et al. 1990b. The neurofibromatosis type 1 gene encodes a protein related to GAP. *Cell* **62:** 599–608.

Yarwood S, Bouyoucef-Cherchalli D, Cullen PJ, Kupzig S. 2006. The GAP1 family of GTPase-activating proteins: Spatial and temporal regulators of small GTPase signaling. *Biochem Soc Trans* **34:** 846–850.

Zhu Y, Parada LF. 2001. Neurofibromin, a tumor suppressor in the nervous system. *Exp Cell Res* **264:** 19–28.

Index